AF594934

Radiolabeled Compounds for Diagnosis and Treatment of Cancer

Radiolabeled Compounds for Diagnosis and Treatment of Cancer

Editor

Krishan Kumar

MDPI • Basel • Beijing • Wuhan • Barcelona • Belgrade • Manchester • Tokyo • Cluj • Tianjin

Editor
Krishan Kumar
Department of Radiology, The
Wright Center of Innovation in
Biomedical Imaging
The Ohio State University
Columbus
United States

Editorial Office
MDPI
St. Alban-Anlage 66
4052 Basel, Switzerland

This is a reprint of articles from the Special Issue published online in the open access journal *Molecules* (ISSN 1420-3049) (available at: www.mdpi.com/journal/molecules/special_issues/Radiolabeled_Compounds).

For citation purposes, cite each article independently as indicated on the article page online and as indicated below:

LastName, A.A.; LastName, B.B.; LastName, C.C. Article Title. *Journal Name* **Year**, *Volume Number*, Page Range.

ISBN 978-3-0365-2427-6 (Hbk)
ISBN 978-3-0365-2426-9 (PDF)

Contents

About the Editor

Krishan Kumar

Krishan Kumar is currently an Associate Professor of Radiology, a Director of Laboratory for Translational Research in Imaging Pharmaceuticals, and an Ohio Molecular Imaging Pharmaceutical Research Scholar at the Wright Center of Innovation in Biomedical Imaging, Ohio State University. Dr. Kumar obtained his Ph.D. degree from the Indian Institute of Technology, Kanpur (India), and worked as a post-doctoral researcher at Wayne State University and Purdue University (USA). He is a co-inventor of 12 patents or patent applications and has co-authored 80 publications. His current research interests include process development of cyclotron-produced radionuclides, discovery and development of radiopharmaceuticals for imaging and therapy of cancer, and the development of novel reagents and methods for radiolabeling of molecules and biomolecules. Before joining academia, he worked in industry, where he was involved in drug discovery and development, and the worldwide registrations of numerous pharmaceutical drug products. He is serving as a member of the Editorial Board for several journals, including the Bioorganic Chemistry Section published by MDPI, and is a reviewer for numerous journals.

Preface to "Radiolabeled Compounds for Diagnosis and Treatment of Cancer"

Radiopharmaceuticals are used in the diagnosis and treatment of various diseases, especially cancer. In general, radiopharmaceuticals are either salts of radionuclides or radionuclides bound to biologically active molecules, drugs, or cells. Tremendous progress has been made in discovering, developing, and commercializing numerous radiopharmaceuticals for the imaging and therapy of cancer. Significant research is ongoing in academia and the pharmaceutical industry to develop more novel radiolabeled compounds as potential radiopharmaceuticals for unmet needs. This Special Issue aims to focus on all aspects of the design, characterization, evaluation, and development of novel radiolabeled compounds for the diagnosis and treatment of cancer and the application of new radiochemistry and methodologies for the development of novel radiolabeled compounds. Outstanding contributions presented in this Special Issue will significantly add to the field of radiopharmaceuticals.

Krishan Kumar
Editor

Editorial

Radiolabeled Compounds for Diagnosis and Treatment of Cancer

Krishan Kumar

Laboratory for Translational Research in Imaging Pharmaceuticals, The Wright Center of Innovation in Biomedical Imaging, Department of Radiology, The Ohio State University, Columbus, OH 43212, USA; krishan.kumar@osumc.edu

Citation: Kumar, K. Radiolabeled Compounds for Diagnosis and Treatment of Cancer. *Molecules* **2021**, *26*, 6227. https://doi.org/10.3390/molecules26206227

Received: 29 September 2021
Accepted: 11 October 2021
Published: 15 October 2021

Publisher's Note: MDPI stays neutral with regard to jurisdictional claims in published maps and institutional affiliations.

Nuclear medicine was recognized as a potential medical field a long time ago when ^{131}I was used in thyroid cancer patients [1–3]. Diagnostics and Therapeutics are the two branches of Nuclear Medicine. Single-Photon Emission Computed Tomography (SPECT) and Positron Emission Tomography (PET) are in vivo molecular imaging modalities which are widely used in nuclear medicine for the diagnosis and follow-up of many major diseases after treatment [4,5]. Combining PET with Computed X-Ray Tomography (CT), PET-CT, enables better diagnosis than with a traditional gamma camera alone. It is a powerful tool that provides unique information on a wide variety of diseases.

These methods use radiolabeled target-specific molecules and biomolecules, including peptides, proteins, protein fragments, and monoclonal antibodies (mAbs) as probes or imaging pharmaceuticals or radiopharmaceuticals. Molecules and biomolecules are labeled with metallic or non-metallic radionuclides with the desired emission type and half-lives for the intended application. Imaging pharmaceuticals are being used routinely in cardiology, neurology, and oncology, etc. Their design and development is a rather interdisciplinary process covering many different areas of science: chemistry, radiochemistry, pharmaceutical, analytical, medicine, engineering, regulatory, etc.

The use of radionuclides for therapeutic applications was reported some time ago [1–3]. Several radionuclides have been used successfully for the treatment of many benign and malignant disorders [6]. For example, several new radionuclides and radiopharmaceuticals have been developed for the treatment of metastatic bone pain and neuroendocrine and other malignant or non-malignant tumors. Radioimmunotherapy is a targeting therapy for cancer that uses monoclonal antibodies (mAbs) labeled with a radionuclide directed against tumor-associated antigens. The ability for the antibody to specifically bind to a tumor-associated antigen increases the dose delivered to the cancer cells specifically, while decreasing the dose to normal tissues.

The concept of Theranostics has an integrated approach to diagnosis and therapy. A targeting vector is radiolabeled with a therapeutic radionuclide which also emits radiation for imaging. Alternatively, the targeting vector is labeled either with a diagnostic or a therapeutic radionuclide with similar chemical properties. One of the classic examples of theranostics is the use of ^{68}Ga-labeled tracers for diagnosis followed by therapy using a therapeutic radionuclide, i.e., ^{177}Lu, etc. In addition to their diagnostic and therapeutic applications in nuclear medicine, radiolabeled compounds are powerful tools for in vitro/in vivo evaluation during discovery and preclinical development and to evaluate the in vivo pharmacokinetics and pharmacodynamics of potential drug candidates.

Numerous radiopharmaceuticals based on,^{11}C, ^{64}Cu, ^{18}F, ^{67}Ga, ^{68}Ga, ^{111}In, ^{123}I, ^{125}I, ^{131}I, ^{177}Lu, ^{13}N, ^{223}Ra, ^{153}Sm, ^{99m}Tc, ^{201}Tl, ^{133}Xe, and ^{90}Y, radionuclides have been approved by the Food and Drug Administration (FDA) for various diagnostics and therapeutics applications [7]. Significant research is ongoing worldwide for use of novel radionuclides and radiolabeled molecules and biomolecules in oncology, neurology, and cardiology for imaging and therapy. A large number of human clinical trials using radionuclides and radiopharmaceuticals have been completed in the past and still are ongoing. Details of these clinical trials can be found in the Clinical Trials database (www.clinicaltrial.gov, accessed on 29 September 2021) published by the US National Library of Medicine of NIH.

The objective of the Special Issue entitled "Radiolabeled Compounds for Diagnosis and Treatment of Cancer" was to focus on all aspects of design, characterization, evaluation, and development of novel radiolabeled compounds for the diagnosis and treatment of cancer and the application of new radiochemistry and methodologies for the development of novel radiolabeled compounds. The Special Issue includes eleven outstanding papers, including seven research and four review articles. The following is an overview of these papers.

The main objective of the first paper by Kumar and Woolum was to develop and test a novel reagent, inorganic monochloramine (NH_2Cl) for radioiodine labeling of new chemical entities and biomolecules, which is cost-effective, easy to make and handle, and is selective to label amino acids, peptides, and proteins. The data presented in this report demonstrate that the yields of the non-radioactive iodine labeling reactions using monochloramine are >70% for an amino acid and a cyclic peptide. The reagent selectively iodinates the tyrosine residue in the biomolecules.

A new squaramide-containing AAZTA^5 (1,4-bis-(carboxymethyl)-6-[bis-(carboxy methyl)-amino-6-pentanoic-acid]-perhydro-1,4-diazepine) chelator for targeting FAP (Fibroblast Activation Protein) was evaluated by Rosch and coworkers. The ^{68}Ga-, ^{44}Sc-, and ^{177}Lu- AAZTA^5.SA.FAPi chelates were investigated for their in vitro properties and compared with those of DOTA.SA.FAPi. AAZTA^5.SA.FAPi and its derivatives showed sub-nanomolar IC_{50} values for FAP and sufficiently high stability in different media.

Vorobyeva et al. evaluated an ankyrin repeat protein (DARPin) Ec1, for imaging of EpCAM (Epithelial Cell Adhesion Molecule) in Triple-Negative Breast Cancer (TNBC). ^{125}I and ^{99m}Tc-labeled DARPin Ec1 imaging probes retained high binding specificity and affinity to EpCAM-expressing MDA-MB-468 TNBC cells. Biodistribution studies in Balb/c mice bearing MDA-MB-468 xenografts demonstrated specific uptake of both ^{125}I- and ^{99m}Tc-labeled Ec1 in TNBC tumors. ^{125}I-labeled Ec1 had appreciably lower uptake in normal organs compared with ^{99m}Tc-labeled Ec1, which resulted in significantly ($p < 0.05$) higher tumor-to-organ ratios. The biodistribution data were confirmed by micro-Single-Photon Emission Computed Tomography/Computed Tomography (microSPECT/CT) imaging.

A new minigastrin (MG) analog (DOTA-DGlu-Pro-Tyr-Gly-Trp-(*N*-Me)Nle-Asp-1Nal-NH_2 with site-specific amino acid substitutions and stabilized against enzymatic degradation) and possible metabolites were synthesized and investigated in preclinical studies by Hormann et al. A biodistribution study of the radiolabeled peptide in BALB/c mice showed low background activity, preferential renal excretion and prolonged uptake in CCK2R-expressing tissues. The in vivo stability study of the radiolabeled peptide was >56% intact radiopeptide in the blood of BALB/c mice 1 h post-injection. High CCK2R affinity and cell uptake were confirmed only for the intact peptide, whereas enzymatic cleavage within the receptor-specific C-terminal amino acid sequence resulted in a complete loss of affinity and cell uptake.

[^{68}Ga]Ga-DOTA-AmBz-MVK(Ac)-OH and its derivative, [^{68}Ga]Ga-DOTA-AmBz-MVK(HTK01166)-OH, coupled with the PSMA-targeting motif were synthesized and evaluated by Bendre et al. to determine if they could be recognized and cleaved by the renal brush border enzymes. [^{68}Ga]Ga-DOTA-AmBz-MVK(Ac)-OH was effectively cleaved specifically by neutral endopeptidase (NEP) of renal brush border enzymes at the Met-Val amide bond, and the radio-metabolite [^{68}Ga]Ga-DOTA-AmBz-Met-OH was rapidly excreted via the renal pathway with minimal kidney retention. [^{68}Ga]Ga-DOTA-AmBz-MVK(HTK01166)-OH retained its PSMA-targeting capability and was also cleaved by NEP. It appears that MVK can be a promising cleavable linker for use to reduce kidney uptake of radiolabeled DOTA-conjugated peptides and peptidomimetics.

Halik and coworkers developed and evaluated two novel ^{68}Ga and ^{177}Lu-labeled chelate conjugates for their lipophilicity and stability in human serum. Additionally, the fully stable conjugates were examined in molecular modeling with a human neurokinin 1 receptor structure and in a competitive radioligand binding assay using rat brain homogenates. This initial research is in the conceptual stage to give potential theranostic-like radiopharmaceutical pairs for the imaging and therapy of neurokinin 1 receptor-overexpressing cancers.

Lin et al. evaluated the therapeutic efficacy of ^{188}Re-liposome on Hypopharyngeal Cancer (HPC) tumors using bioluminescent imaging followed by next-generation sequencing (NGS) analysis to address the deregulated microRNAs and associated signaling pathways. Repeated doses of ^{188}Re-liposome were administered to tumor-bearing mice, and the tumor growth was suppressed after treatment. It was concluded that the ^{88}Re-liposome could suppress the HPC tumors in vivo, and the therapeutic efficacy was associated with the deregulation of microRNAs that could be considered as a prognostic factor.

Four review articles are included in this Special Issue. The first review by Kumar and Ghosh provides a comprehensive review of the physical properties of iodine and iodine radionuclides, production processes of ^{124}I, various ^{124}I-labeling methodologies for molecules and biomolecules, peptides, proteins, protein fragments, and mAbs, and the development of ^{124}I-labeled immunoPET imaging pharmaceuticals for various cancer targets in preclinical and clinical environments. The second review by Alluri et al. provides an overview of the development of positron emission tomography (PET) radiotracers for in vivo imaging of the adrenergic receptors (ARs) system in the brain. The third article by Eychenne et al. focuses on the development of radiolabeled somatostatin analogs (SSAs) to visualize the distribution of receptor overexpression in tumors and radiotherapy of many solid tumors, especially gastro-entero-pancreatic neuroendocrine tumors (GEP-NET). The fourth review by Gomes et al. focuses on the use of pyrazoles as suitable scaffolds for the development of ^{18}F-labeled radiotracers for PET imaging in the last 20 years.

In summary, radiolabeled compounds play an important role in the diagnosis and treatment of various cancers. Tremendous progress has been made in discovering, developing, and registering with the FDA numerous radiopharmaceuticals for imaging and therapy by targeting various receptors in cancer. The impact of radiolabeled compounds in academia and industry is profound, and this continuous research tends to develop more novel compounds for unmet needs. Contributions from this Special Issue in Molecules will add significantly to the field of Radiopharmaceuticals.

Finally, I would like to thank the authors for their contributions to this Special Issue, the reviewers for their critical review in evaluating the submitted articles, and the editorial staff of Molecules, especially the Assistant Editor of the journal, Emity Wang, for her kind assistance during the preparation and release of this Special Issue.

Funding: This work was supported by the Ohio Third Frontier TECH 13-060, TECH 09-028, and the Wright Center of Innovation Development Fund.

Acknowledgments: The author is grateful to Michael V. Knopp (Director and Principal Investigator of the Wright Center of Innovation in Biomedical Imaging) for his encouragement and support during this work.

Conflicts of Interest: The author declares no conflict of interest.

References

1. Hamilton, J.G.; Soley, M.H.; Eichorn, K.B. *Deposition of Radioactive Iodine in Human Thyroid Tissue*; University of California Publications in Pharmacology; University of California Press: Berkeley, CA, USA, 1940; Volume 1, pp. 339–368.
2. Hamilton, J.G. The Use of Radioactive Tracers in Biology and Medicine. *Radiology* **1942**, *39*, 541–572. [CrossRef]
3. Seidlin, S.M.; Marinelli, L.D.; Oshry, E. Radioactive Iodine Therapy: Effect on Functioning Metastases of Adenocarcinoma of the Thyroid. *J. Am. Med. Assoc.* **1946**, *132*, 838–847. [CrossRef] [PubMed]
4. Payolla, F.B.; Massabni, A.C.; Orvig, C. Radiopharmaceuticals for Diagnosis in Nuclear Medicine. *Eclet. Quim. J.* **2019**, *44*, 11–19. [CrossRef]
5. Gutfilen, B.; Valentini, G. Radiopharmaceuticals in Nuclear Medicine: Recent Developments for SPECT and PET Studies. *BioMed. Res. Int.* **2014**, *2014*, 426892. [CrossRef] [PubMed]

6. Yeong, C.-H.; Cheng, M.-h.; Ng, K.-H. Therapeutic Radionuclide in Nuclear Medicine: Current and Future Prospects. *J. Zhejiang Univ.-Sci. B* **2014**, *15*, 845–863. [CrossRef] [PubMed]
7. FDA-Approved Radiopharmaceuticals. Available online: https://www.cardinalhealth.com/content/dam/corp/web/documents/fact-sheet/cardinal-health-fda-approved-radiopharmaceuticals.pdf (accessed on 29 September 2021).

 MDPI

Communication

A Novel Reagent for Radioiodine Labeling of New Chemical Entities (NCEs) and Biomolecules

Krishan Kumar * and Karen Woolum 

Laboratory for Translational Research in Imaging Pharmaceuticals, The Wright Center of Innovation in Biomedical Imaging, Department of Radiology, The Ohio State University, Columbus, OH 43212, USA; Karen.Woolum@osumc.edu

* Correspondence: krishan.kumar@osumc.edu

Abstract: Radioiodine labeling of peptides and proteins is routinely performed by using various oxidizing agents such as Chloramine T, Iodobeads, and Iodogen reagent and radioactive iodide (I^-), although some other oxidizing agents were also investigated. The main objective of the present study was to develop and test a novel reagent, inorganic monochloramine (NH_2Cl), for radioiodine labeling of new chemical entities and biomolecules which is cost-effective, easy to make and handle, and is selective to label amino acids, peptides, and proteins. The data presented in this report demonstrate that the yields of the non-radioactive iodine labeling reactions using monochloramine are >70% for an amino acid (tyrosine) and a cyclic peptide (cyclo Arg-Gly-Asp-d-Tyr-Lys, cRGDyK). No evidence of the formation of *N*-chloro derivatives in cRGDyK was observed, suggesting that the reagent is selective in iodinating the tyrosine residue in the biomolecules. The method was successfully translated into radioiodine labeling of amino acid, a peptide, and a protein, Bovine Serum Albumin (BSA).

Keywords: radioiodine labeling; radioiodination; radiotracers; biomolecules; peptides; proteins; monoclonal antibodies; radiopharmaceuticals; imaging pharmaceuticals; ^{123,124,125,131}I-labeled molecules and biomolecules

Citation: Kumar, K.; Woolum, K. A Novel Reagent for Radioiodine Labeling of New Chemical Entities (NCEs) and Biomolecules. *Molecules* **2021**, *26*, 4344. https://doi.org/10.3390/molecules26144344

Academic Editor: Kazuma Ogawa

Received: 15 June 2021
Accepted: 13 July 2021
Published: 18 July 2021

Publisher's Note: MDPI stays neutral with regard to jurisdictional claims in published maps and institutional affiliations.

1. Introduction

Radioiodine labeling (Radioiodination) of molecules, using radioactive iodide, was first established a long time ago when ^{131}I isotope of iodide was used for the labeling of polyclonal anti-kidney serum [1]. Since then, the radioiodine labeling technique is being used in the evaluation of New Chemical Entities (NCEs) and small and large biomolecules for their biological and medical applications. Four isotopes of iodine (^{123}I, ^{124}I, ^{125}I, and ^{131}I) are used routinely for radioiodine labeling of NCEs and biomolecules, depending on the intended application. For example, ^{125}I radionuclide, with a long half-life of 59.9 days, is used for radioiodine labeling of a molecule or a biomolecule for its pharmacokinetics, metabolism, and biodistribution studies. On the other hand, ^{123}I, with a half-life of 13.1 h, and ^{124}I, with a half-life of 4.17 d, are used for the evaluation of a molecule or a biomolecule for its SPECT (Single Photon Emission Computed Tomography) and PET (Positron Emission Tomography) imaging applications, respectively. ^{131}I radionuclide, a beta particle emitter, is used for the therapeutic application of a radiolabeled molecule or biomolecule.

For direct radioiodine labeling of a molecule or a biomolecule, the presence of an aromatic moiety, tyrosine or histidine is required. The primary site of the iodine addition is tyrosine amino acid residue in NCEs or biomolecules; however, if the pH exceeds 8.5, the secondary site on the imidazole ring of histidine is preferred. The tyrosine moiety can be labeled twice, giving a mixture of mono and di-iodinated species. The formation of di-iodinated tyrosine is faster than the mono-iodinated tyrosyl moiety. Several reports related to the methods and reagents of radioiodine labeling of molecules and biomolecules have been published in the past [2,3]. Every effort must be made, regardless of the application of the radioiodine-labeled molecule or biomolecule, to maintain immunoreactivity and high molar activity of the biomolecule after labeling and purification.

Radioiodine labeling of a molecule or a biomolecule, peptides and proteins, involves an oxidizing agent and an iodide radionuclide which is usually available as Sodium Iodide (NaI) in neutral or basic aqueous solutions. A large number of inorganic and organic oxidizing reagents have been used, in the past, for radioiodine labeling of biomolecules, peptides and proteins. This includes I_2 [1], sodium hypochlorite [4], nitrous acid [5,6], ammonium persulfate [7], hydrogen peroxide [8], ferric sulfate [9], iodate [10], iodine monochloride [11,12], hypochlorite/hypochlorous acid [13], $IBPy_2BF_4$ [14], Penta-*O*-Acetyl-*N*-Chloro-*N*-Methylglucamine [15,16], *N*-chloro derivatives of secondary amines [17], *N*-chloromorpholine [18], Chloramine-T [19,20], iodobeads [21], Iodogen [22,23], and Iodogen reagent coated on the bottom of tubes (commonly known as the Iodination tubes) [24,25]. Some enzymes are known for catalyzing the mild oxidation of iodide for radioiodine labeling of tyrosine, and to some extent histidine also, in proteins [26,27]. If it is not possible to radioiodine label proteins by direct electrophilic addition to tyrosine and histidine residues, a most common alternative approach (indirect radioiodine labeling method) is using a prosthetic group for radioiodine labeling of NCEs and biomolecules [28,29].

Iodination tubes, the most popular and convenient, are used routinely in research laboratories. Iodogen in the iodination tube, like Chloramine T, converts I^- to I^+ (H_2OI^+) or ICl followed by an electrophilic substitution reaction on the aromatic moiety in peptides and proteins (Figure 1), forming mono- and di-iodinated tyrosyl residues [30].

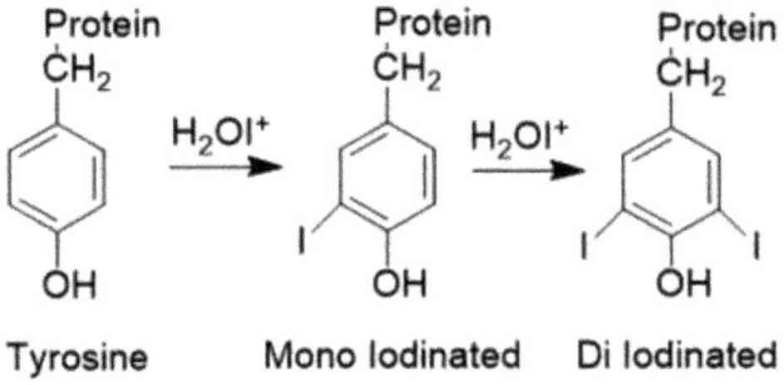

Figure 1. Radioiodine labeling scheme for tyrosine residue in peptides and proteins.

Iodogen, like Chloramine-T, has the potential to over radioiodine label or sometimes damage and form an *N*-chloro derivative of a lysine-residue in peptide and proteins [8]. For this reason, *N*-chloroderivatives of secondary amines with low oxidation potentials were tested as potential radio iodine labeling agents [17,18]. These *N*-chloroderivatives were prepared fresh and used immediately due to the instability of these materials.

The main objective of the present study was to evaluate an agent which is cost-effective, easy to make and handle, contains active chlorine for the conversion of I^- to I^+, and is selective for non-radioactive iodine/radioiodine labeling of tyrosine containing peptides and proteins, with iodine rather than forming *N*-chloro derivatives of amino acids in peptides and proteins containing primary amines. For this purpose, we selected an inorganic monochloramine (NH_2Cl) due to (1) easy in situ formation from the reaction of ammonium hydroxide and sodium hypochlorite, (2) the reagents, ammonium hydroxide and sodium hypochlorite, are inexpensive, (3) the redox potential of monochloramine is lower than Chloramine-T and, consequently, NH_2Cl may be less damaging than Chloramine-T to biomolecules [31], (4) oxidation of I^- to I^+ by monochloramine occurs via chlorine atom transfer and the rate of the oxidation reaction is very fast under neutral pH conditions [32], and (5) the routinely used oxidizing agents, chloramine T, iodobeads, and iodogen, probably, follow the same mechanism as NH_2Cl. In this present work, we have conducted a systematic nonradioactive iodine/radioiodine labeling ($^{127}I/^{125}I$ labeling) study of an amino acid, tyrosine, a cyclic peptide, cRGDyK (cyclo Arg-Gly-Asp-d-Tyr-Lys), and a protein, Bovine Serum Albumin (BSA), containing tyrosine and histidine residues, for the development of a novel reagent for non-radioactive iodine and radioiodine labeling.

A cyclic peptide, cRGDyK, selected in this study, is interesting in many ways: (1) monomeric, dimeric, and tetrameric cyclic RGD peptides have shown binding affinity to $\alpha_v\beta_3$ integrin, an angiogenic biomarker which is overexpressed in the endothelium of most solid tumors. (2) Several radiolabeled cyclic RGD peptides have been investigated as potential radiotracers for angiogenesis imaging [33,34]. (3) Presence of lysine amino group in cRGDyK provides additional possibilities for dual probes development, i.e., conjugating with dyes to produce optical probes. (4) Transfer of chlorine from NH_2Cl to nitrogen in amines, amino acids, and peptides is a thermodynamically favorable reaction [35]. NH_2Cl mediated labeling of cRGDyK, which contains lysine along with tyrosine residue, will also demonstrate the selectivity of the non-radioactive iodine/radioiodine labeling procedure.

The data presented in this report demonstrate that the yields of the non-radioactive iodine labeling reaction using monochloramine are high for an amino acid (tyrosine) and a cyclic peptide (cyclo Arg-Gly-Asp-d-Tyr-Lys, cRGDyK). No evidence of the formation of *N*-chloro derivative in cRGDyK was observed, suggesting that the reagent is selective in iodinating the tyrosine residue in the biomolecules. The method was successfully translated into radioiodine labeling of an amino acid, a peptide, and a protein, Bovine Serum Albumin (BSA).

2. Results and Discussion

2.1. Preparation and Characterization of Monochloramine

Monochloramine (NH_2Cl) was prepared fresh daily by mixing 10% to 20% excess ammonium hydroxide and sodium hypochlorite at pH 10 [30]. The concentrations of sodium hypochlorite stock solutions and the prepared monochloramine samples were determined from the measurement of their absorbances and molar extinction coefficients at the absorbance maxima [9]. The absorption maximum (nm) and the molar extinction coefficient ($M^{-1}cm^{-1}$) for sodium hypochlorite and monochloramine are 292, 350 and 243, 461, respectively [30]. The rate of formation of monochloramine from the reaction of hypochlorite and ammonia is fast and complete. The amount of hydrazine from the Raschig Synthesis and nitrogen is expected to be low due to the limited amount of excess ammonia. Raschig synthesis usually requires a large amount of excess ammonia [36]. Freshly prepared monochloramine under basic conditions was used immediately to avoid any formation of di or trichloramines from NH_2Cl by the disproportionation reactions.

2.2. Non-Radioactive Iodine Labeling of Tyrosine and cRGDyK

In several non-radioactive iodine labeling experiments with tyrosine, known amounts of tyrosine (0.8–2.2 µmole), sodium iodide (^{127}INa) solution (0.9–2.35 µmole), and monochloramine (1–3 µmole) were reacted in a small glass vial or Eppendorf tube containing 0.1 mL sodium phosphate buffer (0.1 M, pH 7.4). The reaction mixture was incubated at room temperature for 30 min. At the end of the incubation time, the reaction was quenched by adding the reducing agent, a freshly prepared sodium metabisulfite solution. The reaction mixture was analyzed using a gradient Reversed-Phase High-Performance Liquid Chromatography (RP-HPLC) method.

Four peaks (Retention Times, RT in min, given in parenthesis), unreacted Iodide (3.4) and tyrosine (6.4), mono-iodinated tyrosine (I-Tyr, 14.2,) and di-iodinated tyrosine (I_2-Tyr, 19.8), with a variable ratio dependent on the reaction conditions, were observed. The peak for unreacted iodide is in the form of oxidized iodide. The yield of the non-radioactive iodine labeling reaction was calculated based on the limiting reagent, tyrosine, as high as >85%. Figure 2 (top) shows a representative HPLC chromatogram for a reaction mixture in which 0.6 µmole of tyrosine, 0.52 µmole sodium iodide, and 0.55 µmole NH_2Cl were used. The percentage peak areas observed were 24.5, 20.8, 34.5, and 20.11 for unreacted sodium iodide, unreacted tyrosine, I-tyrosine, and I_2-tyrosine, respectively. The calculated yield for the formation of iodinated tyrosine is ~72%.

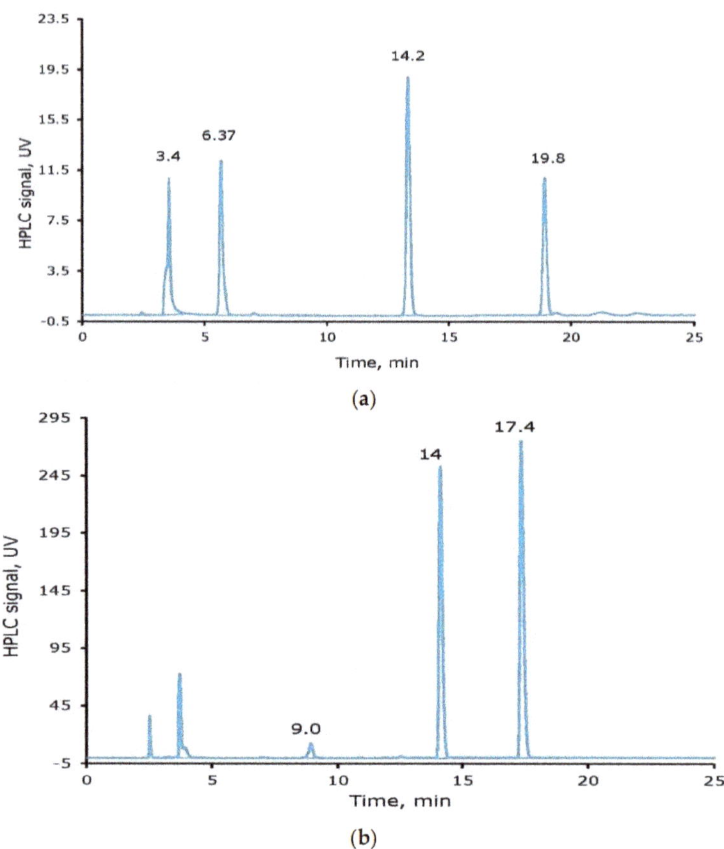

Figure 2. HPLC chromatograms of the reaction mixtures of non-radioactive iodine labeling of tyrosine (**a**) and cyclo Arg-Gly-Asp-d-Tyr-Lys (cRGDyK) (**b**).

Similarly, for non-radioactive iodine labeling of a cyclic peptide, 0.4 μmole of cRGDyK, 0.47 μmole sodium iodide, and 0.4 μmole NH_2Cl, were reacted in an Eppendorf tube for 30 min. At the end of the reaction, 0.58 μmole freshly prepared sodium metabisulfite was added. Four peaks (Retention Times, RT in min, and percentage peak areas given in parenthesis), unreacted Iodide (3.7, 1.85) and cRGDyK (9.0, 2.4), mono-iodinated-cRGDyK (I-cRGDyK, 14.0, 44.0) and di-iodinated –cRGDyK (I_2-cRGDyK, 17.4, 52. 0), were observed (Figure 2, bottom). There was also a small solvent front peak that was not integrated for further calculations. The yield of the non-radioactive iodine labeling reaction was calculated based on the limiting reagent, cRGDyK, as 97%. Additional, non-radioactive iodine labeling experiments with cRGDyK were performed in 1:1:1 (0.5 μmole scale cRGDyK) and 2:1:1 (1 μmole scale cRGDyK) mole ratios of cRGDyK:I:NH_2Cl. The yields of the formation of the iodinated cRGDyK (the sum of I-cRGDyK and I_2-cRGDyK) were 77% and 71% for the 1:1:1 and 2:1:1 reaction mixture, respectively. There is always a mixture of mono- and di-iodinated tyrosine and cRGDyK; however, the ratio of the two species is dependent on the concentrations of the reactants and reaction conditions.

The crude non-radioactive iodine labeling reaction mixtures of tyrosine and cRGDyK were purified by an RP Sep-Pak cartridge method, to remove unreacted iodide, and RP-HPLC purification method, to remove unreacted tyrosine or cRGDyK. The two RP-HPLC peaks, at 14.2 and 19.8 min retention times from the non-radioactive iodine labeling of tyrosine, were collected, concentrated, and confirmed by Electrospray Ionization (ESI) mass spectra as mono- and di-iodinated tyrosine with *m/e* peaks, for $(m + H)^+$, at 306.8 (calculated 307.09) and 433.8 (calculated 433.9), respectively. Similarly, the identity of I- cRGDyK and I_2- cRGDyK was confirmed by ESI mass spectra after collecting the 14 and 17.4 min peaks from the non-radioactive iodine labeling reaction mixture of cRGDyK. The *m/e* peaks, for $(m + H)^+$, were observed as 746.2 (calculated 746.68) and 872.1 (calculated (872.58) for I- cRGDyK and I_2- cRGDyK, respectively. As shown in Figure 3, the ESI mass spectrum did not show any evidence of the formation of the *N*-chloro derivative of cRGDyK. On the contrary, non-radioactive iodine labeling of cRGDyK using the Iodogen method showed evidence of *N*-chlorination of the lysine residue in the cyclic peptide [13].

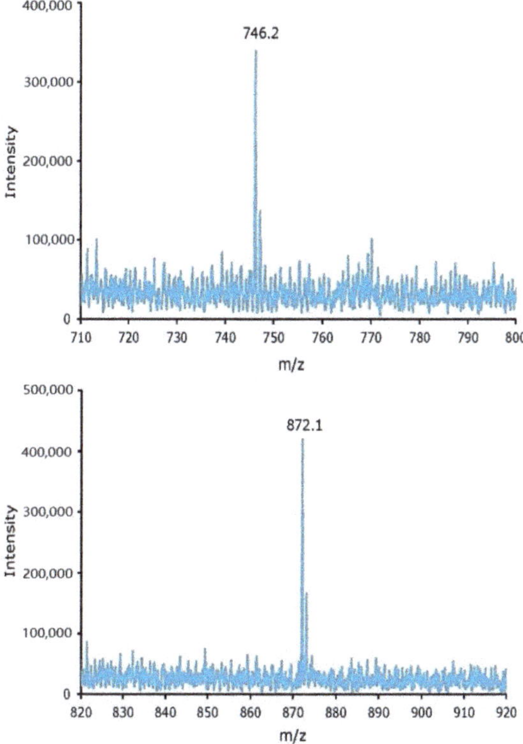

Figure 3. ESI mass spectra of two HPLC chromatogram peaks collected from the non-radioactive iodine labeling reaction mixture of cyclo Arg-Gly-Asp-d-Tyr-Lys (cRGDyK).

2.3. *Radioiodine Labeling of Tyrosine and Cyclo Arg-Gly-Asp-d-Tyr-Lys (cRGDyK)*

The non-radioactive iodine labeling reaction protocols were translated into radioiodine labeling of tyrosine and cRGDyK. For radioiodine labeling of tyrosine and cRGDyK, 0.64 µmole of tyrosine, 212 µCi ^{125}I, and 0.45 µmole NH_2Cl and 0.2 µmole of cRGDyK, 198 µCi ^{125}I, and 0.2 µmole NH_2Cl, respectively, were mixed in 0.1 mL sodium phosphate buffer (0.1 M, pH 7.0). The reaction mixtures were incubated at room temperature for 30 min. At the end of the incubation time, the radioiodine labeling reactions were quenched by the addition of a freshly prepared sodium metabisulfite solution. From the RP-HPLC (the method conditions given in the experimental section) analysis of the crude material, it was observed that the radioiodine incorporation into tyrosine and cRGDyK was 83.2% and ~99%, respectively.

Purification of the reaction mixture of a radioiodine-labeled tyrosine or cRGDyK was accomplished initially by using the Sep-Pak method. Several fractions containing approximately 10 drops were collected and counted for radioactivity during Sep-Pak purification. All major fractions were combined and concentrated to near dryness under a stream of nitrogen at room temperature. The final product was reconstituted in water or Phosphate Buffer Saline (PBS). Seventy to eighty percent of the radioiodine-labeled materials were recovered after Sep-Pak purification. Figure 4 shows HPLC chromatograms of radioiodine labeled tyrosine and cRGDyK after Sep-Pak purification. Due to the non-carrier-added nature of the radioiodine-labeling reactions, the formation of $^{125}I_2$-tyrosine and $^{125}I_2$-cRGDyK is low. For example, the percentages (given in the parenthesis) are: ^{125}I-tyrosine (96), $^{125}I_2$-tyrosine (4), ^{125}I-cRGdyK (93.4), and $^{125}I_2$-cRGDyK (6.6). Higher percentages of di-iodinated tyrosine and cRGDyK were seen in non-radioactive iodine labeling experiments.

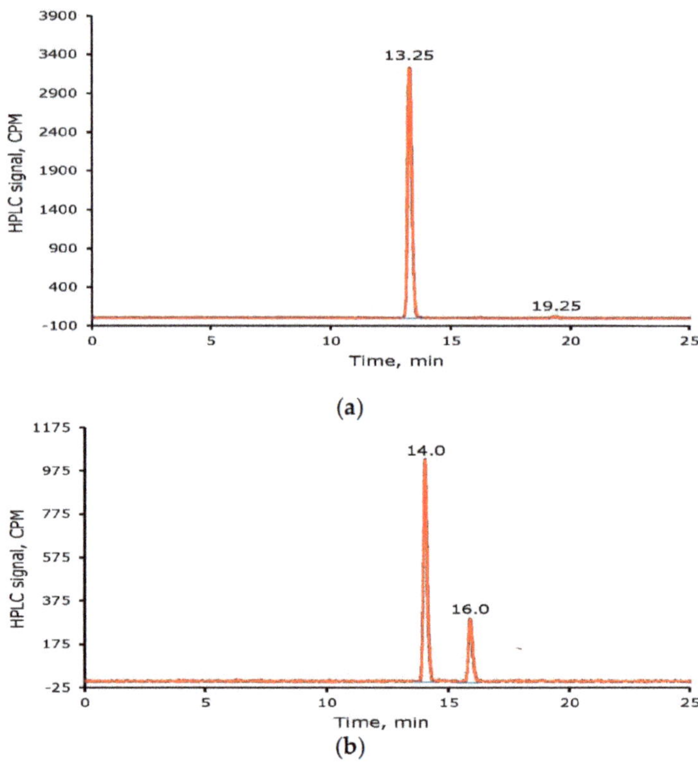

Figure 4. HPLC Chromatogram of radioiodine labeled tyrosine (**a**) and cyclo Arg-Gly-Asp-d-Tyr-Lys (cRGDyK) (**b**) after Sep-Pak purification.

Further HPLC purification was performed to separate mono and di- radioiodine-labeled tyrosine or cRGDyK. HPLC fractions were collected, concentrated, and analyzed. The fractions of the mixture of mono- and di- ^{125}I-labeled tyrosine and cRGDyK were collected. The calculated recovery (sum of the two) after purification was 65% to 75%.

To optimize the reaction time for radioiodine labeling of tyrosine and cRGDyK, a mixture of 0.16 µmole of cRGDyK, 356 µCi ^{125}I, and 0.22 µmole of NH_2Cl were reacted. An aliquot of the reaction mixture was analyzed at 10 min and 30 min after quenching the reaction. From the time-dependent radioiodine incorporation into cRGDyK, it was concluded that 10 min of incubation of the reaction mixture is sufficient for the completion of radioiodine labeling of the tyrosine residue in cRGDyK. Longer period incubation converted mono-iodinated cRGDyK to an increased percentage of di-iodinated species.

For example, the ratios of ^{125}I-cRGDyK:$^{125}I_2$-cRGDyK were observed as ~67:33 and ~42:58 after 10 and 30 min incubation, respectively (Figure 5). The amount of $^{125}I_2$-cRGDyk at 30 min in this study is higher than the study above, possibly due to the higher amount of ^{125}I (giving higher ^{125}I radioactivity/mass ratio) used in this study.

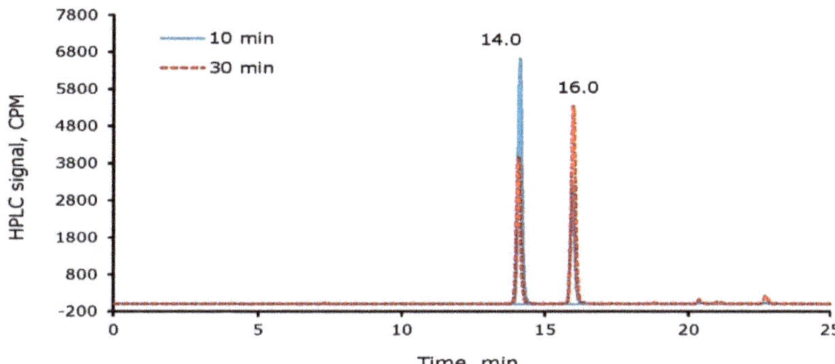

Figure 5. The percentage of ^{125}I-cRGDyK and $^{125}I_2$-cRGDyK after incubation of the reaction mixture for 10 and 30 min, respectively.

2.4. Radioiodine Labeling of Bovine Serum Albumin (BSA)

For radioiodine labeling of BSA, 50 µg (50 µL of 1 mg/mL) of the BSA solution was transferred into an Eppendorf tube containing 100 µL of sodium phosphate buffer (0.1 M pH 7.4). Carrier-free ^{125}INa (~93 µCi) followed by monochloramine (200 µL of 4.14 mM, 0.8 µmole) were added to the tube and incubated at room temperature for 30 min. The reaction was quenched by the addition of sodium metabisulfite. The crude reaction mixture was purified using a PD-10 column by loading radioiodine-labeled BSA onto the column and eluting with PBS. Small fractions were collected into pre-labeled microcentrifuge tubes. The fractions containing most of the activity were pooled and counted for radioactivity.

The incorporation of radioiodine into BSA (i.e., yield) was calculated as 82.4% from the ratio of radioactivity recovered from the elution of the PD-10 and the amount of radioactivity taken initially for radioiodine labeling of BSA. The PD-10 column purified radioiodine-labeled BSA was analyzed for radiochemical purity (RCP), and free radioiodide by using a Paper Chromatography and a Size-Exclusion High-Performance Liquid Chromatography (SEC-HPLC) methods. In the paper chromatography method, the radioiodine labeled BSA precipitates at the origin in the 85:15 methanol: water developing phase and radioiodide moves to the solvent front. The percent RCP and free radioiodide of radioiodine-labeled BSA were calculated from the measured CPM, by Capintec well counter, of the bottom half and top half of the paper strip, respectively (Equations (1) and (2)).

$$\%\text{ RCP of radioiodine-labeled BSA} = (\text{Counts on the bottom half/Total counts from top and the bottom halves}) \times 100 \quad (1)$$

$$\%\text{ Free radioiodide} = (\text{Counts on the top half/Total counts from top and the bottom halves}) \times 100 \quad (2)$$

For a representative radioiodine-labeled BSA sample, top and bottom halves had 4.58 and 212.2 kCPM, respectively. The RCP and free radioiodide of radioiodine labeled BSA were calculated from the Paper Chromatography method as 97.9% and 2.1%, respectively. An SEC-HPLC chromatogram for the same radioiodine labeled BSA sample is shown in Figure 6. Radioiodine labeled BSA and free radioiodide eluted at 6.7 and 12.9 min, respectively. Consistent with the Paper Chromatography results, the SEC-HPLC showed RCP of radioiodine labeled BSA as 98.7% with free radioiodide as 1.3%.

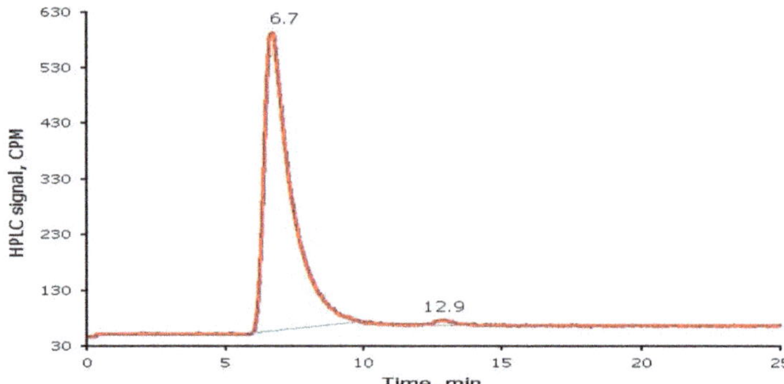

Figure 6. HPLC chromatogram of radioiodine labeled Bovine Serum Albumin (BSA).

2.5. *Comparison of Monochloramine with Other Oxidizing Agents*

Like other oxidizing agents, Chloramine-T, iodobeads, Iodogen, etc., monochloramine is a useful and effective oxidizing agent for radioiodine labeling of amino acids, peptides, and proteins. Monochloramine is easy to prepare and handle and is cost-effective due to the use of inexpensive reagents, ammonium hydroxide and sodium hypochlorite. The yields of radioiodine labeling and molar activities using monochloramine and other oxidizing agents are expected to be comparable. Using all oxidizing agents requires the purification steps to remove unreacted radioiodide and the amino acid or peptide. Radioiodine labeling using monochloramine is selective, i.e., no evidence of the formation of *N*-chloro derivative of cRGDyK while previous studies, using Iodogen, have shown the formation of *N*-chloroderivative of cRGDyK [13]. The rate of radioiodine labeling, using NH_2Cl, is faster, such as Chloramine-T, as these are solution–solution phase reactions. Chloramine-T and other oxidizing agents have shown damage to proteins under certain conditions [8]. Similarly, NH_2Cl also has the potential to oxidize –SH groups in proteins. However, the lower redox potential of NH_2Cl than Chloramine-T and faster rates of oxidation of I^- to I^+ than oxidation of –SH groups by NH_2Cl makes it less likely [31,37]. Like other oxidizing agents, radioiodine labeling conditions, i.e., amount of NH_2Cl and incubation time, must be optimized before routine radiolabeling of proteins using inorganic monochloramine.

3. Materials and Methods

3.1. *General*

All chemicals and reagents, tyrosine and BSA (Sigma-Aldrich, St. Louis, MO, USA), cRGDyK (Peptide International, Louisville, KY, USA), sodium iodide (Acros, Sommerville, NJ, USA), ammonium hydroxide and sodium hypochlorite (Fisher Scientific, Fair Lawn, NJ, USA), and sodium bisulfite (Sigma-Aldrich) were used as received. Sodium monobasic phosphate, sodium dibasic phosphate, sodium hydroxide, hydrochloric acid and sodium chloride (all from Fisher Scientific) were used for buffer and mobile phase preparations and pH and ionic strength control. Gibco 1X PBS (pH 7.4) buffer was supplied by Fisher Scientific. Crude reaction mixtures from the non-radioactive iodine/radioiodine labeling reactions of tyrosine and cRGDyK and BSA were purified to remove unreacted non-radioactive/radioactive sodium iodide by using a Reversed-Phase Sep-Pak C_{18} Light cartridge (Waters, Milford, MA, USA) and a PD-10 column (GE Healthcare, Chicago, IL, USA), respectively. For radioiodine labeling experiments, ^{125}INa was purchased from Perkin Elmer (Shelton, CT, USA).

3.2. *Chemistry*

For the preparation of the tyrosine stock solution, it was necessary to add diluted HCl to lower the pH ~5 initially for its solubilization followed by solution pH adjustment to 7 by the addition of a sodium phosphate buffer. Sodium iodide and cRGDyK solutions were prepared in water. Since the ^{125}I Na sample is supplied in a 0.1 N sodium hydroxide base solution, it was occasionally necessary to adjust the pH of the solution to ~7 by the addition of a small amount of hydrochloric acid. Monochloramine (NH_2Cl) was prepared fresh daily as described elsewhere by mixing 10% to 20% excess ammonium hydroxide and sodium hypochlorite at a pH ~10 [30]. The final pH of the monochloramine solution was adjusted to ~8 with HCl. The concentration of sodium hypochlorite stock solution and monochloramine sample was determined spectrophotometrically [9].

3.3. Analytics

An Agilent 8453 model spectrophotometer was used for all UV/Vis spectral and absorbance measurements. A Capintec dose calibrator Model CRC-R (Capintec, Ramsey, NJ, USA) was used for the determination of radioactivity amounts in the ^{125}INa source and the radioiodine-labeled materials. Agilent model 1100 HPLC systems (Agilent, Wilmington, DE, USA) were used for purification and analysis of non-radioactive iodine- and radioiodine-labeled tyrosine, cRGDyK, and BSA samples. These systems consisted of quaternary pumps, degasser, temperature-controlled column compartment, auto-injector, and multi-wavelength/diode array detectors and control by Agilent's Chem Station or Lab Logics' (Sheffield, UK) Laura software. For detection of radioiodine labeled materials, a Flow Scintillation Analyzer (FSA 150) from Perkin Elmer or a Flow Ram from Lab Logic Systems was used. A Capintec well counter model CRC-25W was used for the analysis of paper chromatography samples. ESI mass spectral analysis was used for the characterization of the nonradioactive-iodine labeled samples of tyrosine and cRGDyK. A Bruker amazon ETD mass spectrometer at Campus Chemical Instrument Center (CCIC) Mass Spectrometry and Proteomics Facility at The Ohio State University (OSU) was used.

3.4. Radiochemistry

In a typical non-radioactive iodine/radioiodine labeling experiment, known amounts of tyrosine or cRGDyK solution and ^{127}INa or ^{125}INa were mixed in a vial or Eppendorf tube containing 0.1–0.5 mL sodium phosphate buffer (0.1 M, pH 7.4). A known amount of monochloramine was added to the vial. The reaction mixture was agitated and mixed after the addition of each reagent by using a pipette, and incubated at room temperature for the desired time. At the end of the incubation period, the non-radioactive iodine or radioiodine labeling reaction was quenched by the addition of an excess of a freshly-prepared reducing agent, sodium metabisulfite. Purification of the reaction mixture of the non-radioactive iodine or the radioiodine labeled tyrosine or cRGDyK was accomplished in two steps. An RP-Sep-Pak C_{18} Light cartridge was used to remove any unreacted sodium iodide or ^{125}I Na, followed by an RP-HPLC purification method to remove any unlabeled or unreacted tyrosine or cRGDyK.

The Sep-Pak purification method involved conditioning the Sep-Pak C_{18} Light cartridge with 3 mL of ethanol, washing with 3 mL water, loading of the crude material, washing with 1.5 mL water, followed by elution with 100 µL portions of 0.5 mL 100% ethanol. All major fractions were combined and concentrated to near dryness under a stream of nitrogen at room temperature. The final product was reconstituted in water or Phosphate Buffer Saline (PBS). The semi-purified mixture was analyzed and further purified by an RP-HPLC method involving a Zorbax C18 5 µm, 4.6 × 250 mm column, a flow rate of 1 mL/min, a UV detection at $\lambda = 280$ nm, a radioisotope detector, and a gradient mobile phase. The following gradient of water containing 0.1% TFA (A) and acetonitrile containing 0.1% TFA (B) was used: 95% A and 5% B initially, ramping the concentration of B to 25% in 20 min and then keeping it at 25% for 4 min. The concentration of B was brought down to 5% next in one min and kept up to 30 min.

For radioiodine labeling of BSA, a known amount of BSA was transferred into an Eppendorf tube containing 100 µL of sodium phosphate buffer (0.1 M pH 7.4). Carrier-free ^{125}INa followed by monochloramine in sodium phosphate buffer were added to the tube. The reaction mixture was agitated and mixed with the pipette and incubated at room temperature for the desired time. The reaction was quenched by the addition of sodium metabisulfite in excess. The crude reaction mixture was purified using a PD-10 column. For purification of radioiodine-labeled BSA, the reaction mixture was loaded onto the conditioned column and eluted with PBS. Small fractions were collected into pre-labeled microcentrifuge tubes. The fractions containing most of the activity were pooled and counted for radioactivity.

The radioiodine-labeled BSA was analyzed by two methods, paper chromatography and SEC-HPLC. The Paper Chromatography method involved a 3 MM cellulose chromatography paper strip and an 85:15 methanol: water mixture as a developing solution. After the paper strip was developed and allowed to dry, the strip was then cut into two half pieces in the middle. Each piece of the strip was counted in a Capintec well counter. The Size-Exclusion HPLC method involved an Agilent SEC-3 100 Å column (4.6 × 300 mm), 150 mM sodium phosphate buffer pH 7.0, a flow rate of 0.35 mL/min, UV detection at $\lambda = 280$ nm, and a radioisotope detector.

4. Conclusions

The use of simple inorganic chloramine, NH_2Cl, for non-radioactive and radioiodine labeling of a tyrosine residue in NCEs and biomolecules has been demonstrated in this present work. The non-radioactive iodine labeling method is selective, i.e., no evidence of the formation of *N*-chloro derivative, and gives a high yield, >70%. The method was successfully translated for radioiodine labeling of biomolecules. As seen in the case of Chloramine-T and other oxidizing agents, NH_2Cl also has the potential to oxidize –SH groups in proteins [31,37]. However, the lower redox potential of NH_2Cl than Chloramine-T and faster rates of oxidation of I^- to I^+ than oxidation of –SH groups by NH_2Cl makes it less likely.

Author Contributions: K.K. conceived the idea, designed the study protocols, provided overall directions to K.W. involved in the studies, interpreted results, and completed the final draft of the manuscript. K.W. performed the execution of study protocols, data collection and analysis. Both authors have read and agreed to the published version of the manuscript.

Funding: This work was supported by the Ohio Third Frontier TECH 13-060, TECH 09-028, and the Wright Center of Innovation Development Fund.

Institutional Review Board Statement: Not applicable.

Informed Consent Statement: Not applicable.

Data Availability Statement: The data are included within the manuscript.

Acknowledgments: The authors are grateful to Michael V. Knopp (Director and Principal Investigator of the Wright Center of Innovation in Biomedical Imaging) for his encouragement and support during this work. The authors thank Arijit Ghosh for his assistance in the revision of the manuscript. The authors are also thankful to Nan Kleinholz and Campus Chemical Instrument Center (CCIC) Mass Spectrometry and Proteomics Facility at The Ohio State University for mass spectral measurements.

Conflicts of Interest: The authors declare no conflict of interest.

Sample Availability: Samples of the compounds are not available from the authors.

References

1. Pressman, D.; Keighley, G. The zone of activity of antibodies as determined by the use of radioactive tracers; the zone of activity of nephritoxic antikidney serum. *J. Immunol.* **1948**, *59*, 141–146.
2. Seevers, R.H.; Counsell, R.E. Radio-iodine labeling Techniques for Small Organic Molecules. *Chem. Rev.* **1982**, *82*, 575–590. [CrossRef]
3. Kumar, K.; Ghosh, A. Radiochemistry, Production Processes, Labeling Methods, and ImmunoPET Imaging Pharmaceuticals of Iodine-124. *Molecules* **2021**, *26*, 414. [CrossRef]
4. Redshaw, M.R.; Lynch, S.S. An improved method for the preparation of iodinated antigens for radioimmunoassay. *J. Endocrinol.* **1974**, *60*, 527–528. [CrossRef] [PubMed]

5. Eisen, H.N.; Keston, A.S. The Immunologic Reactivity of Bovine Serum Albumin Labelled with Trace-Amounts of Radioactive Iodine (I^{131}). *J. Immunol.* **1949**, *63*, 71–80. [PubMed]
6. Yalow, R.S.; Berson, S.A. Immunoassay of Endogenous Plasma Insulin in Man. *J. Clin. Investig.* **1960**, *39*, 1157–1175. [CrossRef] [PubMed]
7. Gilmore, R.C., Jr.; Robbins, M.C.; Reid, A.F. Labeling bovine and human albumin with I^{131}. *Nucleonics* **1954**, *12*, 65–68.
8. McFarlane, A.S. Labeling of plasma proteins with radioactive iodine. *Biochem. J.* **1956**, *62*, 135–143. [CrossRef]
9. Stadie, W.C.; Haugaard, N.; Vaughn, M. Studies of Insulin Binding with Isotopically Labeled Insulin. *J. Biol. Chem.* **1952**, *199*, 729–739. [CrossRef]
10. Francis, G.E.; Mulligan, W.; Wormall, A. Labeling of proteins with iodine-131, Sulphur-35 and phosphorus-32. *Nature* **1951**, *167*, 748–751. [CrossRef]
11. McFarlane, A.S. Efficient Trace-labeling of Proteins with Iodine. *Nature* **1958**, *182*, 53. [CrossRef]
12. Hung, L.T.; Fermandjian, S.; Morgat, J.L.; Fromageot, P. Peptide and protein labeling with iodine, iodine monochloride reaction with aqueous solution of L-tyrosine, L-histidine, L-histidine-peptides, and his effect on some simple disulfide bridges. *J. Label. Compd.* **1974**, *10*, 3–21. [CrossRef]
13. Doll, S.; Woolum, K.; Kumar, K. Radiolabeling of a Cyclic RGD (cyclo Arg-Gly-Asp-d-Tyr-Lys) Peptide Using Sodium Hypochlorite as an Oxidizing Agent. *J. Label. Comp. Radiopharm.* **2016**, *59*, 462–466. [CrossRef]
14. Barluenga, J.; Garcia-Martin, M.A.; Gonzalez, J.M.; Clapes, P.; Valencia, G. Iodination of aromatic residues in peptides by reaction with IPy_2BF_4. *Chem. Commun.* **1996**, *13*, 1505–1506. [CrossRef]
15. Tashtoush, B.M.; Traboulsi, A.A.; Dittert, L.; Hussain, A.A. Chloramine-T in radiolabeling techniques: IV. Pento-*O*-acetyl-*N*-chloro-*N*-methylglucamine as an oxidizing agent in radiolabeling techniques. *Anal. Biochem.* **2001**, *288*, 16–21. [CrossRef] [PubMed]
16. Hussain, A.A.; Bassam, T.; Dittert, L.W. Derivatives of N-chloro-N-Methyl Glucamine and N-Chloro-N-Methyl Glucamine Esters. U.S. Patent 5,985,239, 16 November 1999.
17. Kaminski, J.J.; Bodor, N.; Higuchi, T. *N*-Halo Derivatives IV: Synthesis of Low Chlorine Potential Soft *N*-Chloramine Systems. *J. Pharm. Sci.* **1976**, *65*, 1733–1737. [CrossRef] [PubMed]
18. Hussain, A.A.; Dittert, L.W. Non-Destructive Method for Radiolabeling Biomolecules by Halogenation. U.S. Patent 5,424,402, 13 June 1995.
19. Hunter, W.M.; Greenwood, F.C. Preparation of iodine-131 labelled human growth hormone of high specific activity. *Nature* **1962**, *194*, 495–496. [CrossRef]
20. Greenwood, F.C.; Hunter, W.M.; Glover, J.S. The preparation of ^{131}I-labelled human growth hormone of high specific radioactivity. *Biochem. J.* **1963**, *89*, 114–123. [CrossRef] [PubMed]
21. Markwell, M.A. A new solid-state reagent to iodinate proteins: Conditions for the efficient labeling of antiserum. *Anal. Biochem.* **1982**, *125*, 427–432. [CrossRef]
22. Fracker, P.J.; Speck, J.C., Jr. Protein and cell membrane iodinations with a sparingly soluble chloramide 1,3,4,6-tetrachloro-3a,6a-diphenylglycoluril. *Biochem. Biophys. Res. Commun.* **1978**, *80*, 849–857. [CrossRef]
23. Salacinski, P.; Hope, J.; McLean, C.; Clement-Jones, V.; Sykes, J.; Price, J.; Lowry, P.J. A new simple method which allows theoretical incorporation of radio-iodine into proteins and peptides without damage. *J. Endocrinol.* **1979**, *81*, 131.
24. Markwell, M.A.K.; Fox, C.F. Surface-specific iodination of membrane proteins of viruses and eucaryotic cells using 1,3,4,6-tetrachloro-3a,6a-diphenylglycouril. *Biochemistry* **1978**, *17*, 4807–4817. [CrossRef]
25. Boonkitticharoen, V.; Laohathai, K. Assessing performances of Iodogen-coated surfaces used for radio-iodine labeling of proteins. *Nucl. Med. Commun.* **1990**, *11*, 295–304. [CrossRef]
26. Holohan, K.N.; Murphy, R.F.; Flanagan, R.W.J.; Buchanan, K.D.; Elmore, D.T. Enzymic iodination of the histidyl residue of secretin: A radioimmunoassay of the hormone. *Biochim. Biophys. Acta* **1973**, *322*, 178–180. [CrossRef]
27. Holohan, K.N.; Murphy, R.F.; Elmore, D.T. The Site of Substitution in the Imidazole Nucleus after the Lactoperoxidase-Catalysed Iodination of Histidine Residues in Polypetides. *Biochem. Soc. Trans.* **1974**, *2*, 739–740. [CrossRef]
28. Navarro, L.; Berdal, M.; Cherel, M.; Pecorari, F.; Gestin, J.-F.; Guerard, F. Prosthetic groups for radio-iodine labeling and astatination of peptides and proteins: A comparative study of five potential bioorthogonal labeling strategies. *Bioorg. Med. Chem.* **2019**, *27*, 167–174. [CrossRef] [PubMed]
29. Santos, J.S.; Muramoto, E.; Colturato, M.T.; Siva, C.P.; Araujo, E.B. Radio-iodine labeling of proteins using prosthetic group: A convenient way to produce labelled proteins with in vivo stability. *Cell. Mol. Biol.* **2002**, *47*, 735–739.
30. Kumar, K. Radio-iodine Labeling Reagents and Methods for New Chemical Entities (NCEs) and Biomolecules. *Cancer Biother. Radiopharm.* **2021**. submitted.
31. Victorin, K.; Hellstrom, K.-G.; Rylander, R. Redox potential measurements for determining the disinfecting power of chlorinated water. *Epidemiol. Infect.* **1972**, *70*, 313–323. [CrossRef]
32. Kumar, K.; Day, R.A.; Margerum, D.W. Atom-Transfer redox kinetics: General-acid-assisted oxidation of iodide by chloramines and hypochlorite. *Inorg. Chem.* **1986**, *25*, 4344–4350. [CrossRef]
33. Haubner, R.; Wester, H.-J.; Reuning, U.; Senekowitsch-Schmidtke, R.; Diefenbach, B.; Kessler, H.; Stocklin, G.; Schwaiger, M. Radiolabeled $\alpha_v\beta_3$ Integrin Antagonists: A New Class of Tracers for Tumor Targeting. *J. Nucl. Med.* **1999**, *40*, 1061–1071.

34. Haubner, R. $\alpha_V\beta_3$-integrin imaging: A new approach to characterise angiogenesis? *Eur. J. Nucl. Med. Mol. Imaging* **2006**, *33*, 54–63. [CrossRef] [PubMed]
35. Snyder, M.P.; Margerum, D.W. Kinetics of chlorine transfer from chloramine to amines, amino acids, and peptides. *Inorg. Chem.* **1982**, *21*, 2545–2550. [CrossRef]
36. Cahn, J.W.; Powell, R.F. The Raschig Synthesis of Hydrazine. *J. Am. Chem. Soc.* **1954**, *76*, 2565–2567. [CrossRef]
37. Jacangelo, J.G.; Olivieri, V.P.; Kawata, K. Oxidation of Sulfydryl Group by Monochloramine. *Water Res.* **1987**, *21*, 1339–1344. [CrossRef]

Article

In Vitro Evaluation of the Squaramide-Conjugated Fibroblast Activation Protein Inhibitor-Based Agents AAZTA^5.SA.FAPi and DOTA.SA.FAPi

Euy Sung Moon [1], Yentl Van Rymenant [2], Sandeep Battan [1], Joni De Loose [2], An Bracke [2], Pieter Van der Veken [3], Ingrid De Meester [2] and Frank Rösch [1,*]

[1] Department of Chemistry–TRIGA, Johannes Gutenberg University Mainz, 55128 Mainz, Germany; emoon01@uni-mainz.de (E.S.M.); sbattan@students.uni-mainz.de (S.B.)
[2] Laboratory of Medical Biochemistry, Department of Pharmaceutical Sciences, University of Antwerp, 2610 Wilrijk, Belgium; Yentl.VanRymenant@uantwerpen.be (Y.V.R.); joni.deloose@uantwerpen.be (J.D.L.); an.bracke@uantwerpen.be (A.B.); ingrid.demeester@uantwerpen.be (I.D.M.)
[3] Laboratory of Medicinal Chemistry, Department of Pharmaceutical Sciences, University of Antwerp, 2610 Wilrijk, Belgium; pieter.vanderveken@uantwerpen.be
* Correspondence: frank.roesch@uni-mainz.de; Tel.: +49-61313925302

Abstract: Recently, the first squaramide-(SA) containing FAP inhibitor-derived radiotracers were introduced. DATA^{5m}.SA.FAPi and DOTA.SA.FAPi with their non-radioactive complexes showed high affinity and selectivity for FAP. After a successful preclinical study with [^{68}Ga]Ga-DOTA.SA.FAPi, the first patient studies were realized for both compounds. Here, we present a new squaramide-containing compound targeting FAP, based on the AAZTA^5 chelator 1,4-bis-(carboxylmethyl)-6-[bis-(carboxymethyl)-amino-6-pentanoic-acid]-perhydro-1,4-diazepine. For this molecule (AAZTA^5.SA.FAPi), complexation with radionuclides such as gallium-68, scandium-44, and lutetium-177 was investigated, and the in vitro properties of the complexes were characterized and compared with those of DOTA.SA.FAPi. AAZTA^5.SA.FAPi and its derivatives labelled with non-radioactive isotopes demonstrated similar excellent inhibitory potencies compared to the previously published SA.FAPi ligands, i.e., sub-nanomolar IC_{50} values for FAP and high selectivity indices over the serine proteases PREP and DPPs. Labeling with all three radiometals was easier and faster with AAZTA^5.SA.FAPi compared to the corresponding DOTA analogue at ambient temperature. Especially, scandium-44 labeling with the AAZTA derivative resulted in higher specific activities. Both DOTA.SA.FAPi and AAZTA^5.SA.FAPi showed sufficiently high stability in different media. Therefore, these FAP inhibitor agents could be promising for theranostic approaches targeting FAP.

Keywords: AAZTA; scandium-44; lutetium-177; FAP; SA; DPP; PREP

Citation: Moon, E.S.; Van Rymenant, Y.; Battan, S.; De Loose, J.; Bracke, A.; Van der Veken, P.; De Meester, I.; Rösch, F. In Vitro Evaluation of the Squaramide-Conjugated Fibroblast Activation Protein Inhibitor-Based Agents AAZTA^5.SA.FAPi and DOTA.SA.FAPi. *Molecules* **2021**, *26*, 3482. https://doi.org/10.3390/molecules26123482

Academic Editor: Krishan Kumar

Received: 31 March 2021
Accepted: 4 June 2021
Published: 8 June 2021

1. Introduction

Fibroblast activation protein (FAP) is a post-prolyl proteolytic enzyme that belongs to the S9 family of serine proteases [1]. In addition to FAP, this S9 family includes other proline-specific serine proteases, such as prolyl oligopeptidase (PREP) and the dipeptidyl peptidases 4, 8, and 9 (DPP4, DPP8, and DPP9). Targeting fibroblast activation protein (FAP), overexpressed selectively in cancer-associated fibroblasts (CAFs), has recently become an attractive goal for diagnostic imaging and first therapeutic trials. FAP is involved in the promotion and development of tumor growth and is typically overexpressed in activated fibroblasts in the tumor stroma, whereas it is absent in most normal healthy tissues. Furthermore, FAP is overexpressed in several pathological tissue sites that are characterized by active remodeling [2–5]. Expression of FAP is found in CAFs in approximately 90% of epithelial carcinomas such as breast, pancreatic, colon, and prostate tumors [6–8]. These properties make FAP a very interesting and universally applicable tumor target for a variety of tumor types.

PET tracers that operate as FAP-specific enzyme inhibitors (FAPi), have first been published by Lindner and Loktev et al. [9–11]. The FAP inhibitor used is a small molecule with an N-acylated glycyl(2-cyano-4,4-difluoropyrrolidine) that binds to FAP active site and blocks its enzymatic activity. This highly potent inhibitor, referred to as UAMC1110, shows high affinity for FAP but not for the DPPs and PREP [12]. Meanwhile, many clinical trials have been initiated with related PET tracers based on the same FAP inhibitor [13–21]. Lindner and Loktev et al. developed DOTA-based FAP inhibitor conjugates with heterocyclic units as spacer. The most prominent are FAPI-04 and FAPI-46, with piperazine between chelator and inhibitor. Other examples include a glycosylated fluorine-18 derivative and tracers for SPECT applications with technetium-99 and rhenium-188 introduced by tricarbonyl ligands with piperazine linker systems [22,23]. Recently, we developed FAP inhibitor agents using squaramide-combined bifunctional chelators [24]. The use of a squaramide linker facilitated the synthetic work and delivered compounds with good pharmacological properties. The latter results were illustrated by the excellent in vitro affinity of these products for FAP and by their in vivo behavior in preclinical and clinical applications. With respect to clinical studies, the DATA^{5m}.SA.FAPi derivative showed specific tracer uptake in focal nodular hyperplasia via ^{68}Ga-PET/CT [25]. The DOTA.SA.FAPi tracer displayed a high target-to-background ratio during ^{68}Ga-PET/CT studies in patients with various cancers [26].

The advantageous properties of gallium-68, such as its high positron energy with $\beta^+ = 89\%$ and $E_{\beta,avg} = 830$ keV and its good accessibility due to the availability of ^{68}Ge/^{68}Ga generators, make it a commonly used PET radionuclide [27]. However, the short physical half-life of the nuclide (1.1 h) may impede focusing on longer-lasting physiological processes in PET/CT measurements. Scandium-44, which is also characterized by a high branching ratio of $\beta^+ = 94\%$ and $E_{\beta,avg} = 632$ keV, could be a valuable alternative with a $t_{1/2}$ of 4.0 h. There are two ways to produce scandium-44: one uses a cyclotron via the ^{44}Ca(p,n)^{44}Sc reaction, and the other, that we chose, uses a ^{44}Ti/^{44}Sc generator [27–29]. An established post-processing elution protocol provides carrier-free scandium-44 from a 185 MBq generator, with ~90% elution efficiency and a titanium-44 breakthrough of only <7 MBq [30]. Due to its four-time longer half-life than gallium-68, the β^+-emitter scandium-44 could be better suited for pretherapeutic PET/CT measurements resulting in individual dosimetric calculations in endoradiotherapy with longer-lived therapy nuclides such as yttrium-90, lutetium-177, or scandium-47. Scandium-44 has already been used in both preclinical and clinical applications [31–35]. In particular, the first in human PET measurements in metastasized castrate-resistant prostate cancer with [^{44}Sc]Sc-PSMA-617 indicated its potential as PET nuclide and pre-therapeutic agent [36]. Furthermore, the β^--emitter lutetium-177, with a half-life of 6.7 days, is nowadays a very commonly used radionuclide in radioendotherapy. It is clinically used for neuroendocrine tumors in the somatostatin analogues [^{177}Lu]Lu-DOTATOC and [^{177}Lu]Lu-DOTATATE for peptide-mediated radioreceptor therapy and for treatment of prostate carcinomas by means of lutetium-177-PSMA therapy with PSMA derivatives such as PSMA-617 and PSMA-I&T [37–43].

In this work, we introduce a novel FAP inhibitor agent called AAZTA^5.SA.FAPi. AAZTA chelators allow fast and quantitative complexation under mild conditions and display high stability. This is in particular relevant for radionuclides with high needs of coordination capacity, such as scandium-44 and lutetium-177.

Together with the recently published DOTA.SA.FAPi, both precursors were radiochemically investigated in terms of labeling and stability with gallium-68, scandium-44, and lutetium-177 and tested for their in vitro properties.

2. Results and Discussion

2.1. Synthesis of Chelator Conjugates

For AAZTA^5.SA.FAPi, we first synthesized AAZTA^5(^{t}Bu)$_4$. The coupling of squaric acid to the terminal carboxyl group and the subsequent binding to the FAP inhibitor was performed in the same way as for the previously described DATA^{5m}.SA.FAPi [24]. Figure 1 shows the synthesis route of AAZTA^5.SA.FAPi, following the protocol of Sinnes et al. and Greifenstein et al. [44,45].

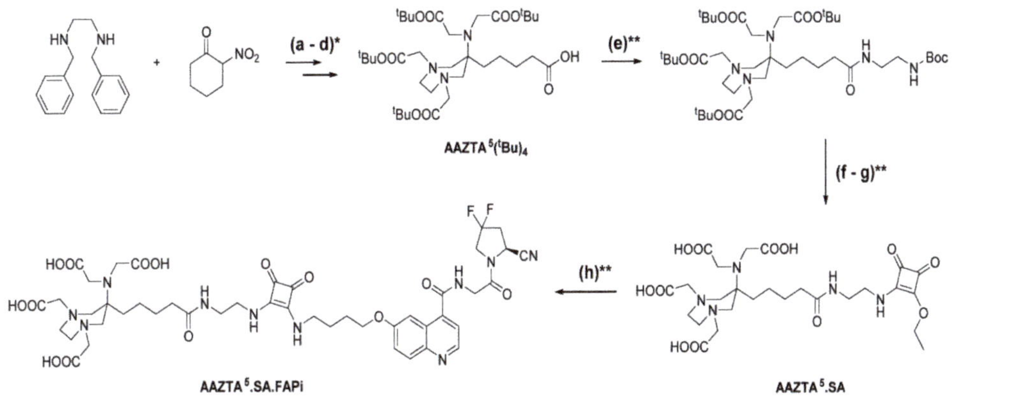

Figure 1. Synthesis of AAZTA^5.SA.FAPi via AAZTA^5(^{t}Bu)$_4$ and AAZTA^5.SA: (**a**) 2-nitrocyclohexanone, Amberlyst A21, paraformaldehyde, methanol, 80 °C, 16 h; (**b**) palladium hydroxide/C, acetic acid, hydrogen, ethanol, 25 °C, 16 h; (**c**) tert-butyl bromoacetate, potassium carbonate, potassium iodide, acetonitrile, 40 °C, 48-72 h; (**d**) 1 M lithium hydroxide, 1,4-dioxane/water (2:1), 25 °C, 16 h; (**e**) N-Boc-ethylenediamine, HATU, HOBt, DIPEA, acetonitrile, 25 °C, 16 h; (**f**) dichloromethane/TFA (80:20)%, 25 °C, 7 h; (**g**) 3,4-diethoxycyclobut-3-ene-1,2-dione, 0.5 M phosphate buffer pH = 7, 25 °C, 16 h; (**h**) NH_2-UAMC1110, 0.5 M phosphate buffer pH = 9, 25 °C, 16 h; (*) as reported [44,45]; (**) as reported [24]. DOTA.SA.FAPi was synthesized as previously described [24].

Figure 2 shows the structures of the FAP inhibitor probes DOTA.SA.FAPi and AAZTA^5.SA.FAPi.

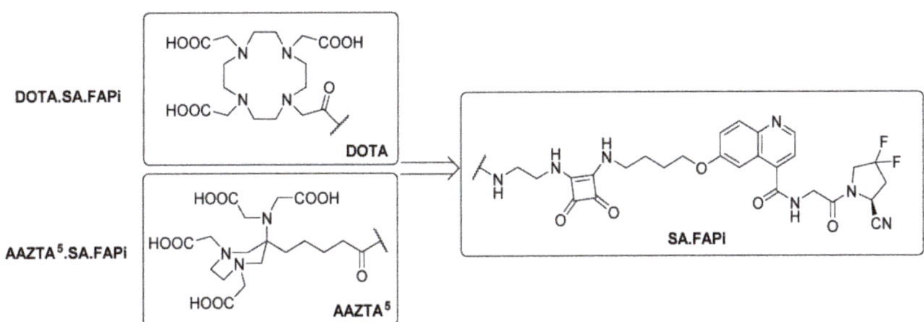

Figure 2. Structures of DOTA.SA.FAPi and AAZTA^5.SA.FAPi.

2.2. In Vitro Inhibition Measurements

The IC_{50} values for FAP, PREP, and the DPPs of the hybrid chelator conjugate AAZTA^5.SA.FAPi compared to those of DOTA.SA.FAPi are shown in Table 1. The IC_{50} values of AAZTA^5.SA.FAPi as well as those of its non-radioactive complexes [^{nat}Sc]Sc-AAZTA^5.SA.FAPi and [^{nat}Lu]Lu-AAZTA^5.SA.FAPi for FAP appeared to be in the low nanomolar range (0.55–0.57 nM), whereas the IC_{50} values for PREP resulted in the low micromolar range (2.4–3.6 µM). Screening against DPP4 and DPP9 for both SA.FAPi complexes revealed that the remaining activity was more than 50% at a final concentration of 1 µM. Hence, the IC_{50} values for the DPPs were reported as >1 µM. The absence of a basic amine in the FAP inhibitor is known to result in an enormous increase of selectivity for the target molecule FAP, whereas the affinity for the DPPs can be drastically reduced. [12,46]. The IC_{50} values for FAP and PREP were in the same order of magnitude of those for the previously reported SA.FAPi compounds, i.e., indicating high inhibition potency and excellent FAP-to-PREP selectivity indices. In addition, high selectivity towards DPP4 and DPP9 was achieved.

Table 1. IC_{50} values of AAZTA[5].SA.FAPi and DOTA.SA.FAPi derivatives for FAP and the related proteases DPPs and PREP. Data are presented as the mean with standard deviation (n = 3 for FAP and PREP and n = 2 for the DPPs).

Compound	DPPs IC_{50} (μM)	PREP IC_{50} (μM)	FAP IC_{50} (nM)	Selectivitiy Index (FAP/PREP)
AAZTA[5].SA.FAPi	>1	2.4 ± 0.4	0.56 ± 0.02	4286
[natSc]Sc-AAZTA[5].SA.FAPi	>1	3.6 ± 0.8	0.57 ± 0.04	6316
[natLu]Lu-AAZTA[5].SA.FAPi	>1	3.2 ± 0.6	0.55 ± 0.04	5818
DOTA.SA.FAPi	n.d.	5.4 ± 0.3 [a]	0.9 ± 0.1 [a]	6000
[natGa]Ga-DOTA.SA.FAPi	>1	8.7 ± 0.9 [a]	1.4 ± 0.2 [a]	6214
[natLu]Lu DOTA.SA.FAPi	>1	2.5 ± 0.4 [a]	0.8 ± 0.2 [a]	3125
DATA[5m].SA.FAPi	n.d.	1.7 ± 0.1 [a]	0.8 ± 0.2 [a]	2113
[natGa]Ga-DATA[5m].SA.FAPi	>1	4.7 ± 0.3 [a]	0.7 ± 0.1 [a]	6714
UAMC1110-FAP inhibitor	>10	1.8 ± 0.01 [b]	0.43 ± 0.07 [a]	4186

[a] data from Moon et al. [24]; [b] data from Jansen et al. [12]; n.d. not determined.

2.3. Radiolabeling and In Vitro Stability in Complexwith Gallium-68, Scandium-44, and Lutetium-177

Gallium-68: DOTA.SA.FAPi complexed with gallium-68 showed very high kinetics in quantitative radiochemical yields (RCYs) in our previous work [24]. Gallium labeling of AAZTA[5].SA.FAPi with diverse precursor amounts (10, 15 and 20 nmol) was performed at room temperature (Figure 3a). [^{68}Ga]Ga-AAZTA[5].SA.FAPi displayed quantitative complexation already after 3–5 min (Figure 3a, Figure S1, Supplementary Material). Compared to the DOTA derivative, complexation led to very high RCYs for tracer amounts ≥10 nmol, even at ambient temperature. In the case of the previously reported DOTA.SA.FAPi, high RCYs could only be achieved with an amount ≥15 nmol and at a high temperature of 95 °C. the stability of [^{68}Ga]Ga-AAZTA[5].SA.FAPi in human serum (HS), ethanol (EtOH), and saline (NaCl) was excellent, with >99.9% intact complexes over a measured time period of 2 h (Figure 3b, Figures S2–S4).

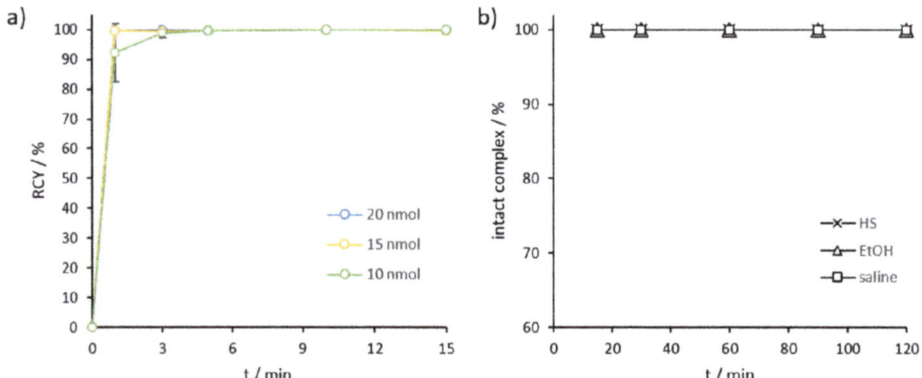

Figure 3. (**a**) Kinetics of [^{68}Ga]Ga-AAZTA[5].SA.FAPi at RT for tracer amounts ≥10 nmol (n = 3); (**b**) Stability of [^{68}Ga]Ga-AAZTA[5].SA.FAPi at 37 °C in HS, EtOH, and NaCl over a period of 120 min (n = 3).

Scandium-44: AAZTA[5].SA.FAPi demonstrated excellent complexation with scandium-44 even at RT. We tested 5–20 nmol of precursor, which resulted in quantitative labeling already after 5 min for all amounts (Figure 4a, Figure S5). Stability was tested in HS, phosphate-buffered saline (PBS), and NaCl at 37 °C, demonstrating in highly satisfactory values in all three media (Figure 4b, Figures S6–S11). After 1 h, [^{44}Sc]Sc-AAZTA^5.SA.FAPi conjugates were stable, with >99% intact conjugate in all three media. Even up to the end of the measurement (8 h), the intact conjugates were stable in PBS and saline (>99%) and in HS (>97%) (Figure 4b).

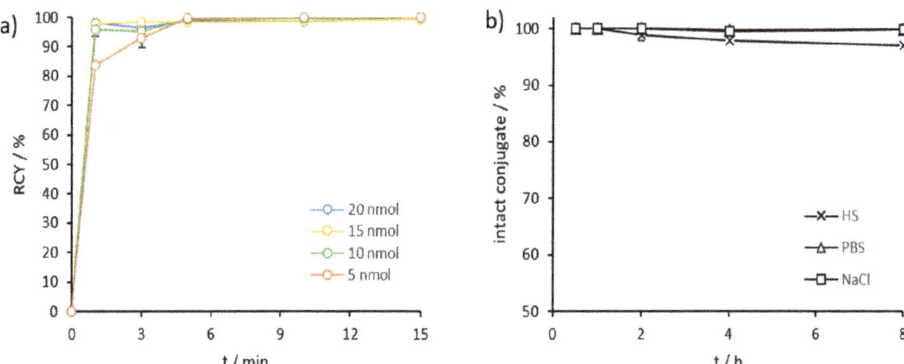

Figure 4. (**a**) Kinetics of [^{44}Sc]Sc-AAZTA^5.SA.FAPi at RT for tracer amounts ≥5 nmol (n = 3 for 10; n = 1 for 5, 15, and 20 nmol); (**b**) Stability of [^{44}Sc]Sc-AAZTA^5.SA.FAPi at 37 °C in HS, PBS, and NaCl over a period of 8 h (n = 3).

DOTA.SA.FAPi also showed good complexation with scandium-44. However, whereas [^{44}Sc]Sc-AAZTA^5.SA.FAPi already displayed quantitative RCYs with 5 nmol (GBq/0.17 µmol) of precursors, [^{44}Sc]Sc-DOTA.SA.FAPi showed very low complexation with 20 nmol. Only with a quantity of 30 nmol, DOTA.SA.FAPi high yields with scandium-44 were reached, with RCYs >83% and >95%, when, respectively, 30 and 40 nmol (GBq/1.33 µmol) were used (Figure 5a). Stability in HS, PBS, and NaCl were high over the measured period of 8 h, resulting in >97% intact complexes with ^{44}Sc in all three medias (Figure 5b).

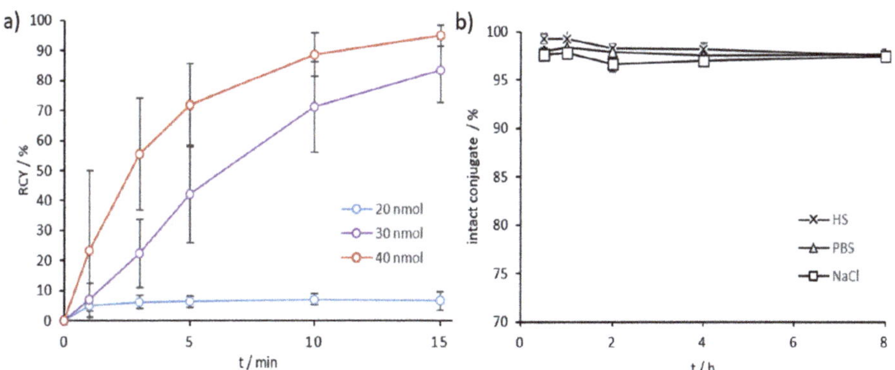

Figure 5. (**a**) Kinetics of [^{44}Sc]Sc-DOTA.SA.FAPi at 95 °C for tracer amounts ≥20 nmol (n = 5 for 20–40 nmol); (**b**) Stability of [^{44}Sc]Sc-DOTA.SA.FAPi at 37 °C in HS, PBS, and NaCl over a period of 8 h (n = 3).

Lutetium-177: For both DOTA.SA.FAPi and AAZTA^5.SA.FAPi, precursors at a concentration of 20 nmol were used for labeling with lutetium-177. Both derivatives presented quantitative complexations with the radiometal after 60 min (Figure 6, Figure S12). The ^{177}Lu-AAZTA derivative showed >99% RCY already after 1 min at RT, whereas the ^{177}Lu-DOTA derivative reached >99% complexation after 15 min at 95 °C (Figure 6).

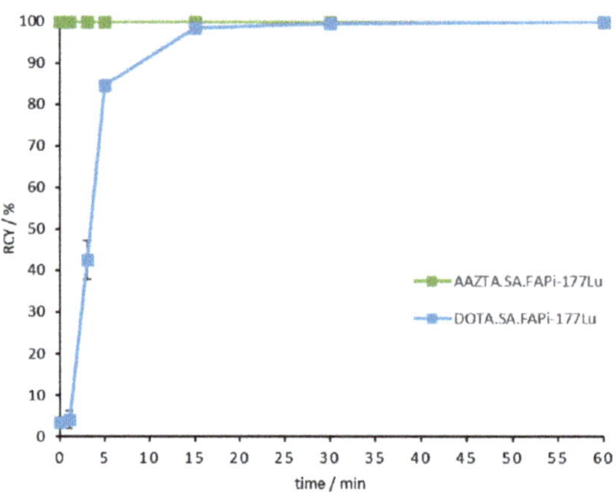

Figure 6. Kinetic measurements for [^{177}Lu]Lu-AAZTA^5.SA.FAPi up to 60 min (green); Kinetic measurements for [^{177}Lu]Lu-DOTA.SA.FAPi up to 60 min (blue); (n = 3, 20 nmol for both conjugates).

Stability studies of both conjugates were performed in HS, PBS, and saline over a period of 10 days at 37 °C. In PBS and NaCl, very high stability values could be achieved for [^{177}Lu]Lu-AAZTA^5.SA.FAPi, with >99% after 2 d, >98% after 3 d, and >95% intact conjugates after 10 days. In HS, the ^{177}Lu-AAZTA complex showed >99% of stability after 1 h, >98% after 3 h, and >96% after 6 h. However, the stability decreased significantly with time. After 1 d, the remaining stability of [^{177}Lu]Lu-AAZTA^5.SA.FAPi in HS was >83%, after 2 d it was >64%, and after 3 d it was >55% (Figure 7a). Nevertheless, the stability of [^{177}Lu]Lu-AAZTA^5.SA.FAPi in HS was satisfactory, with >95% intact conjugate after 6 h. If it is assumed that small molecules accumulate in the target tissue within the first few hours, and therefore their stability in HS over a long period is not relevant. [^{177}Lu]Lu-DOTA.SA.FAPi showed very high stability, with >99% of intact conjugate in HS within the measured time period of 10 days. In PBS and NaCl, the stability was high, i.e., >98% after 3 d and still >93% after 10 d (Figure 7b).

2.4. Lipophilicity Measurements

Lipophilicity (logD value) was determined via the "shake-flask" method. For both precursors AAZTA^5.SA.FAPi and DOTA.SA.FAPi logD (pH = 7.4), values were measured for the ^{68}Ga- and ^{44}Sc complexes. Table 2 shows the logD values for the respective radiotracers.

Table 2. LogD values (pH = 7.4) of [^{68}Ga]Ga-AAZTA^5.SA.FAPi, [^{44}Sc]Sc-AAZTA^5.SA.FAPi and [^{68}Ga]Ga-DOTA.SA.FAPi.

Compound	LogD$_{7.4}$
[^{68}Ga]Ga-AAZTA^5.SA.FAPi	−2.53 ± 0.13
[^{44}Sc]Sc-AAZTA^5.SA.FAPi	−2.50 ± 0.11
[^{68}Ga]Ga-DOTA.SA.FAPi	−2.68 ± 0.06

The lipophilicity of the radiolabeled compounds [^{68}Ga]Ga-AAZTA^5.SA.FAPi, [^{68}Ga]Ga-AAZTA^5.SA.FAPi and [^{44}Sc]Sc-AAZTA^5.SA.FAPi resulted located in hydrophilic ranges. Both gallium-68 derivatives [^{68}Ga]Ga-DOTA.SA.FAPi and [^{68}Ga]Ga-AAZTA^5.SA.FAPi showed almost identical logD values of −2.68 and −2.53, respectively. The carboxyl groups and the ionic bonds between chelator and radiometal favor the hydrophilic character of these radiotracers. The logD value of FAPI-04 is reported in the literature as −2.4 ± 0.28, confirming the hydrophilic character of ^{68}Ga-DOTA complexes [22]. [^{44}Sc]Sc-AAZTA^5.SA.FAPi display a similar logD value of −2.50 compared to gallium-68 derivatives. There seems to be no great influence of the DOTA and AAZTA chelators in the presence of gallium-68 and scandium-44 radiometals on the lipophilicity of the FAPi radiopharmaceuticals.

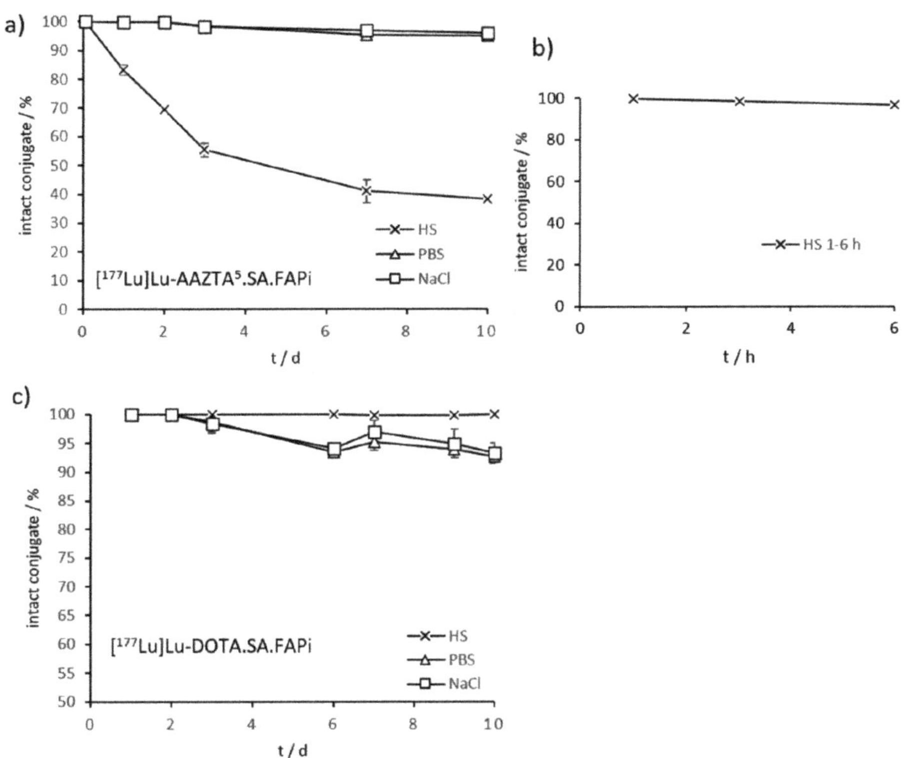

Figure 7. (**a**) Stability of $[^{177}Lu]Lu$-AAZTA^5.SA.FAPi at 37 °C in HS, PBS, and NaCl over a period of 10 d (n = 3); (**b**) Stability of $[^{177}Lu]Lu$-AAZTA^5.SA.FAPi at 37 °C in HS after 1, 3, and 6 h (n = 3); (**c**) Stability measurements of $[^{177}Lu]Lu$-DOTA.SA.FAPi at 37 °C in HS, PBS, and NaCl during 10 d (n = 3).

3. Materials and Methods

3.1. General

All basic chemicals were purchased from Merck KGaA (Darmstadt, Germany), TCI Deutschland GmbH (Eschborn, Germany), Fisher Scientific GmbH (Schwerte, Germany), Thermo Fisher GmbH (Kandel, Germany) and VWR International GmbH (Darmstadt, Germany) and used without further purification. (*S*)-6-(4-aminobutoxy)-*N*-(2-(2-cyano-4,4-difluoropyrrolidin-1-yl)-2-oxoethyl)-quinoline-4-carboxamide (called NH_2-UAMC1110) was purchased from KE Biochem Co. (Shanghai, China). Thin-layer chromatography was performed with silica gel 60 F254-coated aluminum plates that were acquired from Merck KGaA (Darmstadt, Germany). Detection was carried out by fluorescence extinction at λ = 254 nm and by staining with potassium permanganate. The LC/MS spectra were measured on an Agilent Technologies 1220 Infinity LC system coupled to an Agilent Technologies 6130B Single Quadrupole LC/MS system. NMR measurements were performed at 400 MHz (400 MHz FT NMR spectrometer AC 400, Bruker Analytik GmbH). For HPLC (high-performance liquid chromatography) a 7000 series Hitachi LaChrom with a Hitachi L7100 pump, an L7400 UV detector, and a Phenomenex Synergi C18 (250 × 10 mm, 4 µm) column (Aschaffenburg, Germany) were used.

3.2. Organic Synthesis

Synthesis of DOTA.SA.FAPi was reported recently [24]. Synthesis of $AAZTA^5$.SA.FAPi was realized by first generating $AAZTA^5(^tBu)_4$ according to the procedure by Sinnes et al. and Greifenstein et al. [44,45]. Subsequent coupling to the SA.FAPi conjugate was performed using the protocol published earlier for the analogous $DATA^{5m}$.SA.FAPi precursor [24]. After HPLC purification with a gradient of 10–20%, MeCN (+0.1% TFA)/90–80% water (+0.1% TFA) in 20 min, $AAZTA^5$.SA.FAPi was obtained as a yellowish solid (16.8 mg; 0.02 mmol; 41%). MS (ESI^+): 500.3 ($M+2H^{2+}$); 999.3 ($M+H^+$); calculated for $C_{45}H_{56}F_2N_{10}O_{14}$: 998.40.

3.3. Non-Radioactive Compounds and In Vitro Inhibition Studies

$^{nat}Sc/^{nat}Lu$-$AAZTA^5$.SA.FAPi were synthesized by reaction of 5.0 mg (5 µmol) $AAZTA^5$.SA.FAPi with 1.5 eq $^{nat}ScCl_3$ and $^{nat}LuCl_3$, respectively, in 500 µL 0.5 M NaAc buffer pH 4.5 for 2 h at room temperature. Complexation was confirmed by ESI–MS, and HPLC purification was performed with a flow rate 5 mL/min, H_2O (+0.1% TFA)/MeCN (+0.1% TFA), with a linear gradient condition of 5–95% MeCN in 10 min. The products (4.1 mg; 3.9 µmol; 79% for ^{nat}Sc-complex and 4.2 mg; 3.6 µmol; 72%) were obtained as a yellowish powder. MS (ESI^+) for ^{nat}Sc-complex: m/z (%): 521.3 $(M+2H)^{2+}$; 1041.4 $(M+H)^+$ calculated for $C_{45}H_{52}F_2ScN_{10}O_{14}$: 1039.9 and ^{nat}Lu-complex: m/z (%): 586.2 $(M+2H)^{2+}$; 1171.4 $(M+H)^+$ calculated for $C_{45}H_{52}F_2LuN_{10}O_{14}$: 1169.9.

3.4. Inhibition Assays

Enzymes: Recombinant human FAP and PREP were expressed and purified as published [24]. Recombinant human dipeptidyl 9 (DPP9) was purified as described by De Decker et al. [46]. Human dipeptidyl peptidase 4 was purified from seminal plasma as published [47].

IC_{50} measurements and counter-screening: IC_{50}-measurements of the probes for FAP and PREP were carried out as published, using, respectively, Z-Gly-Pro-AMC and Suc-Gly-Pro-AMC as the substrate [24]. IC_{50} experiments were repeated in triplicate, and the results are presented as mean ± standard deviation. Methods and data fitting were performed as published earlier [24]. Screening against DPP4 and DPP9 was performed at final probe concentrations of 10 µM and 1 µM using Ala-Pro-*para*nitroanilide (*p*NA) as the substrate at the respective final concentrations of 25 µM (DPP4) and 150 µM (DPP9) at pH 7.4 (0.05 M HEPES-NaOH buffer with 0.1% Tween-20, 0.1 mg/mL BSA, and 150 mM NaCl). Probes were pre-incubated with the respective enzyme for 15 min at 37 °C; afterwards, the substrate was added, and the velocities of pNA release were measured kinetically at 405 nm for at least 10 min at 37 °C. Measurements were executed using the Infinite 200 (Tecan Group Ltd., Mennendorf, Switzerland), and the Magellan software was used to process the data. If the remaining activity was more than 50% at 1 µM, the IC_{50} values for the DPPs were reported as >1 µM.

3.5. Radiolabeling and Stability Measurements

Gallium-68: $^{68}Ge/^{68}Ga$ generators (ITG Garching, Germany) were used with ethanol-based post-processing evaluated by Eppard et al. [48]. Elution of gallium-68 was performed with 0.05 M HCl trapped on a micro-chromatography CEX column AG 50W-X4. The column was washed with 80% EtOH/0.15 M HCl, and $^{68}Ga^{3+}$ was eluted with 90% EtOH/0.9 M HCl.

Scandium-44: Scandium-44 was obtained by a $^{44}Sc/^{44}Ti$ generator [29,30,36]. A solution of 0.005 M $H_2C_2O_4$/0.07 M HCl was eluted through the $^{44}Ti/^{44}Sc$ generator to adsorb $[^{44}Sc]Sc^{3+}$ onto the cation exchanger AG 50 W-X8. Elution of scandium-44 was executed with 0.25 M ammonium acetate buffer pH 4.

Lutetium-177: n.c.a. $[^{177}Lu]LuCl_3$ in 0.04 M HCl was provided by ITG Garching, Germany.

Radioactivity was measured using a PC-based dose calibrator (ISOMED 2010, Nuklear Medizintechnik Dresden GmbH, Dresden, Germany). Reaction controls were done using radio-TLC, with 0.1 M citrate buffer pH 4 and an analytical HPLC 7000 series Hitachi LaChrom with a Phenomenex Luna C18 column (250 × 4.6 mm, 5 μm), linear gradient of 5–95% MeCN (+0.1% TFA)/H_2O (+0.1% TFA), flow rate 1 mL/min in 10 min. TLCs were measured in a CR-35 Bio Test-Imager from Duerr-ndt (Bietigheim-Bissingen, Germany) with the analysis software AIDA Elysia-Raytest (Straubenhardt, Germany).

Labeling was carried out with 100–150 MBq gallium-68 in 300 μL of 1 M ammonium acetate (AmAc) buffer pH 5.5 and with 30–40 MBq scandium-44 in 1 mL of 0.25 M AmAc pH 4.0, and aliquots were taken at 1, 3, 5, 10, and 15 min. For lutetium-177, activity of 30–40 MBq in 300 μL of 1 M AmAc pH 5.5 was used, and aliquots were taken at 1, 3, 5, 15, 30, and 60 min. Stability was tested in 500 μL of human serum, phosphate-buffered saline, ethanol, and saline (0.9% isotonic NaCl solution) using ~5 MBq of tracer solution with >95% radiochemical purity. The measured time points were adjusted to the physical half-lifes, i.e., gallium-68 (15, 30, 60, 90, 120 min), scandium-44 (0.5, 1, 2, 4, 8 h), and lutetium-177 (1–6 h, 1, 2, 3, 7, 10 d). HS (human male AB plasma, USA origin) and PBS were purchased from Sigma Aldrich, and 0.9% saline from B. Braun Melsungen AG (Melsungen, Germany).

3.6. Lipophilicity Determination

Lipophilicity of [^{68}Ga]Ga-AAZTA^5.SA.FAPi, [^{44}Sc]Sc-AAZTA^5.SA.FAPi and [^{68}Ga]Ga-DOTA.SA.FAPi was determined using the "shake-flask" methodology. After reaction of the precursor with the respective radionuclide, the reaction solution was adjusted to pH 7.4 with NaOH. Aliquots of ~5 MBq for the ^{68}Ga complexes and of ~3 MBq for the ^{44}Sc-complexed were taken and adjusted to a total volume of 700 μL with PBS (n= 4). 700 μL 1-octanol was added, and the solution was shaken for 2 min (1500 rpm). Afterwards, each tube was centrifuged for 2 min. 400 μL of the octanol- and PBS phases were pipetted in new tubes, and aliquots of each phase (3 μL of the PBS phase and 6 μL of the octanol phase) were measured via radio-TLC. The PBS phases were adjusted to 700 μL, and 700 μL octanol was added to each tube. The procedure was repeated twice. LogD values were calculated as the logarithm of the octanol/PBS ratio.

4. Conclusions

In this work, a new squaramide FAPi conjugate to the AAZTA chelator is introduced. After successful preparative synthesis, the complex was tested for its in vitro binding characteristics and compared to the analogue DOTA.SA.FAPi derivative, published recently [24]. The inhibitory potency studies of AAZTA^5.SA.FAPi showed excellent sub-nanomolar affinities for FAP, in the same order of magnitude of those of the already published SA.FAPi monomeric structures DATA^{5m}.SA.FAPi and DOTA.SA.FAPi. Furthermore, high selectivity against PREP and the DPPs was achieved. AAZTA^5.SA.FAPi labeling with gallium-68, scandium-44, and lutetium-177, as well DOTA.SA.FAPi complexation with scandium-44 and lutetium-177, were successfully performed. Remarkably, for AAZTA^5.SA.FAPi, compared to the DOTA derivative, [^{44}Sc]Sc-AAZTA.SA.FAPi required significantly less precursor for quantitative labeling, resulting in higher specific activities, and performed complete complexation at ambient temperatures. The stability of the radiometal-complexed AAZTA^5.SA.FAPi in various media was excellent, as demonstrated by the presence of highly intact conjugates. Complexation with the β^+-emitting scandium-44 may offer a good alternative to gallium-68 usage in diagnosis due to the longer half-life of 4 h and the favorable traits of this nuclide. Interesting is also the remarkable labeling and stability with lutetium-177, allowing therapeutical application. A first theranostic approach of DOTA.SA.FAPi was reported by Ballal et al. [49]. Therefore, the new FAP inhibitor-based probes DOTA.SA.FAPi and AAZTA^5.SA.FAPi, complexed with gallium-68, scandium-44, and lutetium-177, are promising radiopharmaceuticals for use in a theranostic settings.

Supplementary Materials: Figure S1: radio-HPLC spectra of $[^{68}Ga]$Ga-AAZTA^5.SA.FAPi after reaction of 15 min with linear gradient condition of 10–95% MeCN (+0.1% TFA)/95–10% water (+0.1% TFA) in 10 min, 1 mL/min, t_R = 8.4 min; Figure S2: Stability test: radio-HPLC spectra of $[^{68}Ga]$Ga-AAZTA^5.SA.FAPi in ethanol after 1 h with linear gradient condition of 5–95% MeCN (+0.1% TFA)/95–5% water (+0.1% TFA) in 10 min, 1 mL/min, t_R = 9.1 min; Figure S3: Stability test: radio-HPLC spectra of $[^{68}Ga]$Ga-AAZTA^5.SA.FAPi in saline after 1 h with linear gradient condition of 5–95% MeCN (+0.1% TFA)/95–5% water (+0.1% TFA) in 10 min, 1 mL/min, t_R = 9.1 min; Figure S4: Stability test: radio-HPLC spectra of $[^{68}Ga]$Ga-AAZTA^5.SA.FAPi in human serum after 1 h with linear gradient condition of 5–95% MeCN (+0.1% TFA)/95–5% water (+0.1% TFA) in 10 min, 1 mL/min, t_R = 9.1 min; Figure S5: radio-HPLC spectra of $[^{44}Sc]$Sc-AAZTA^5.SA.FAPi after a 15 min reaction with linear gradient condition of 5–95% MeCN (+0.1% TFA)/95-5% water (+0.1% TFA) in 10 min, 1 mL/min, t_R = *8.9* min; Figure S6: Stability test: radio-HPLC spectra of $[^{44}Sc]$Sc-AAZTA^5.SA.FAPi in phosphate-buffered saline after 1 h with linear gradient condition of 5–95% MeCN (+0.1% TFA)/95–5% water (+0.1% TFA) in 10 min, 1 mL/min, t_R = 9.3 min; Figure S7: Stability test: radio-HPLC spectra of $[^{44}Sc]$Sc-AAZTA^5.SA.FAPi in saline after 1 h with linear gradient condition of 5–95% MeCN (+0.1% TFA)/95–5% water (+0.1% TFA) in 10 min, 1 mL/min, t_R = 9.1 min; Figure S8: Stability test: radio-HPLC spectra of $[^{44}Sc]$Sc-AAZTA^5.SA.FAPi in human serum after 1 h with linear gradient condition of 5–95% MeCN (+0.1% TFA)/95–5% water (+0.1% TFA) in 10 min, 1 mL/min, t_R = 9.1 min; Figure S9: Stability test: radio-HPLC spectra of $[^{44}Sc]$Sc-AAZTA^5.SA.FAPi in human serum after 4 h with linear gradient condition of 5–95% MeCN (+0.1% TFA)/95–5% water (+0.1% TFA) in 10 min, 1 mL/min, t_R = 9.5 min; Figure S10: Stability test: radio-HPLC spectra of $[^{44}Sc]$Sc-AAZTA^5.SA.FAPi in saline after 4 h with linear gradient condition of 5–95% MeCN (+0.1% TFA)/95–5% water (+0.1% TFA) in 10 min, 1 mL/min, t_R = 9.3 min; Figure S11: Stability test: radio-HPLC spectra of $[^{44}Sc]$Sc-AAZTA.SA.FAPi in phosphate-buffered saline after 4 h with linear gradient condition of 5–95% MeCN (+0.1% TFA)/95–5% water (+0.1% TFA) in 10 min, 1 mL/min, t_R = 9.1 min; Figure S12: radio-HPLC spectra of $[^{177}Lu]$Lu-AAZTA^5.SA.FAPi after a 60 min. reaction with linear gradient condition of 5–95% MeCN (+0.1% TFA)/95–5% water (+0.1% TFA) in 10 min, 1 mL/min, t_R = 9.1 min.

Author Contributions: Conceptualization, F.R., P.V.d.V., I.D.M.; methodology, E.S.M., Y.V.R., A.B.; validation, E.S.M., S.B., Y.V.R., J.D.L., A.B.; formal analysis, E.S.M., S.B., Y.V.R., J.D.L., A.B.; investigation, E.S.M., S.B., Y.V.R., J.D.L., A.B.; resources, F.R., P.V.d.V., I.D.M.; data curation, E.S.M., Y.V.R., F.R., P.V.d.V., I.D.M.; writing—original draft preparation, E.S.M.; writing—review and editing, E.S.M., Y.V.R., A.B., F.R., P.V.d.V., I.D.M.; supervision and project administration, F.R., P.V.d.V., I.D.M.; All authors have read and agreed to the published version of the manuscript.

Funding: This work was supported by the Fonds Wetenschappelijk Onderzoek Vlaanderen (FWO, Grant 1S64220N); Y. Van Rymenant is a SB PhD fellow at FWO. This project also received funding from the Agentschap Innoveren en Ondernemen (VLAIO HCB 2019. 2446).

Institutional Review Board Statement: Not applicable.

Informed Consent Statement: Not applicable.

Data Availability Statement: The study did not report any data.

Acknowledgments: Lutetium-177 (n.c.a. $[^{177}Lu]LuCl_3$ in 0.04 M HCl) was kindly provided by ITG Garching, Germany.

Conflicts of Interest: The authors declare no conflict of interest.

Sample Availability: Not applicable.

References

1. Rawlings, N.D.; Barrett, A.J.; Thomas, P.D.; Huang, X.; Bateman, A.; Finn, R.D. The MEROPS database of proteolytic enzymes, their substrates and inhibitors in 2017 and a comparison with peptidases in the PANTHER database. *Nucleic Acids Res.* **2018**, *46*, D624–D632. [CrossRef] [PubMed]
2. Huang, Y.; Simms, A.E.; Mazur, A.; Wang, S.; León, N.R.; Jones, B.; Aziz, N.; Kelly, T. Fibroblast activation protein-α promotes tumor growth and invasion of breast cancer cells through non-enzymatic functions. *Clin. Exp. Metastasis* **2011**, *28*, 567–579. [CrossRef]

3. Liu, R.; Li, H.; Liu, L.; Yu, J.; Ren, X. Fibroblast activation protein: A potential therapeutic target in cancer. *Cancer Biol. Ther.* **2012**, *13*, 123–129. [CrossRef] [PubMed]
4. Liu, T.; Zhou, L.; Li, D.; Andl, T.; Zhang, Y. Cancer-associated fibroblasts build and secure the tumor microenvironment. *Front. Cell Dev. Biol.* **2019**, *7*, 1–14. [CrossRef]
5. De Vlieghere, E.; Verset, L.; Demetter, P.; Bracke, M.; De Wever, O. Cancer-associated fibroblasts as target and tool in cancer therapeutics and diagnostics. *Virchows Arch.* **2015**, *467*, 367–382. [CrossRef] [PubMed]
6. Tao, L.; Huang, G.; Song, H.; Chen, Y.; Chen, L. Cancer associated fibroblasts: An essential role in the tumor microenvironment (review). *Oncol. Lett.* **2017**, *14*, 2611–2620. [CrossRef] [PubMed]
7. Lindner, T.; Loktev, A.; Giesel, F.; Kratochwil, C.; Altmann, A.; Haberkorn, U. Targeting of activated fibroblasts for imaging and therapy. *EJNMMI Radiopharm. Chem.* **2019**, *4*, 1–15. [CrossRef]
8. Busek, P.; Mateu, R.; Zubal, M.; Kotackova, L.; Sedo, A. Targeting Fibroblast activation protein in cancer—Prospects and caveats. *Front. Biosci. Landmark* **2018**, *23*, 1933–1968. [CrossRef]
9. Loktev, A.; Lindner, T.; Mier, W.; Debus, J.; Altmann, A.; Jaeger, D.; Giesel, F.; Kratochwil, C.; Barthe, P.; Roumestand, C.; et al. A Tumor-Imaging Method Targeting Cancer-Associated Fibroblasts. *J. Nucl. Med.* **2018**, *59*, 1423–1429. [CrossRef] [PubMed]
10. Lindner, T.; Loktev, A.; Altmann, A.; Giesel, F.; Kratochwil, C.; Debus, J.; Jäger, D.; Mier, W.; Haberkorn, U. Development of Quinoline-Based Theranostic Ligands for the Targeting of Fibroblast Activation Protein. *J. Nucl. Med.* **2018**, *59*, 1415–1422. [CrossRef]
11. Loktev, A.; Lindner, T.; Burger, E.M.; Altmann, A.; Giesel, F.; Kratochwil, C.; Debus, J.; Marmé, F.; Jäger, D.; Mier, W.; et al. Development of fibroblast activation protein-targeted radiotracers with improved tumor retention. *J. Nucl. Med.* **2019**, *60*, 1421–1429. [CrossRef]
12. Jansen, K.; Heirbaut, L.; Verkerk, R.; Cheng, J.D.; Joossens, J.; Cos, P.; Maes, L.; Lambeir, A.M.; De Meester, I.; Augustyns, K.; et al. Extended structure-activity relationship and pharmacokinetic investigation of (4-quinolinoyl)glycyl-2-cyanopyrrolidine inhibitors of fibroblast activation protein (FAP). *J. Med. Chem.* **2014**, *57*, 3053–3074. [CrossRef]
13. Kratochwil, C.; Flechsig, P.; Lindner, T.; Abderrahim, L.; Altmann, A.; Mier, W.; Adeberg, S.; Rathke, H.; Röhrich, M.; Winter, H.; et al. ^{68}Ga-FAPI PET/CT: Tracer uptake in 28 different kinds of cancer. *J. Nucl. Med.* **2019**, *60*, 801–805. [CrossRef]
14. Giesel, F.L.; Kratochwil, C.; Lindner, T.; Marschalek, M.M.; Loktev, A.; Lehnert, W.; Debus, J.; Jäger, D.; Flechsig, P.; Altmann, A.; et al. ^{68}Ga-FAPI PET/CT: Biodistribution and preliminary dosimetry estimate of 2 DOTA-containing FAP-targeting agents in patients with various cancers. *J. Nucl. Med.* **2019**, *60*, 386–392. [CrossRef]
15. Shi, X.; Xing, H.; Yang, X.; Li, F.; Yao, S.; Zhang, H.; Zhao, H.; Hacker, M.; Huo, L.; Li, X. Fibroblast imaging of hepatic carcinoma with ^{68}Ga-FAPI-04 PET/CT: A pilot study in patients with suspected hepatic nodules. *Eur. J. Nucl. Med. Mol. Imaging* **2021**, *48*, 196–203. [CrossRef] [PubMed]
16. Luo, Y.; Pan, Q.; Zhang, W.; Li, F. Intense FAPI Uptake in Inflammation May Mask the Tumor Activity of Pancreatic Cancer in ^{68}Ga-FAPI PET/CT. *Clin. Nucl. Med.* **2020**, *45*, 310–311. [CrossRef] [PubMed]
17. Khreish, F.; Rosar, F.; Kratochwil, C.; Giesel, F.L.; Haberkorn, U.; Ezziddin, S. Positive FAPI-PET/CT in a metastatic castration-resistant prostate cancer patient with PSMA-negative/FDG-positive disease. *Eur. J. Nucl. Med. Mol. Imaging* **2020**, *47*, 2040–2041. [CrossRef] [PubMed]
18. Chen, H.; Pang, Y.; Wu, J.; Zhao, L.; Hao, B.; Wu, J.; Wei, J.; Wu, S.; Zhao, L.; Luo, Z.; et al. Comparison of [^{68}Ga]Ga-DOTA-FAPI-04 and [^{18}F]FDG PET/CT for the diagnosis of primary and metastatic lesions in patients with various types of cancer. *Eur. J. Nucl. Med. Mol. Imaging* **2020**, *47*, 1820–1832. [CrossRef]
19. Chen, H.; Zhao, L.; Ruan, D.; Pang, Y.; Hao, B.; Dai, Y.; Wu, X.; Guo, W.; Fan, C.; Wu, J.; et al. Usefulness of [^{68}Ga]Ga-DOTA-FAPI-04 PET/CT in patients presenting with inconclusive [^{18}F]FDG PET/CT findings. *Eur. J. Nucl. Med. Mol. Imaging* **2021**, *48*, 73–86. [CrossRef] [PubMed]
20. Varasteh, Z.; Mohanta, S.; Robu, S.; Braeuer, M.; Li, Y.; Omidvari, N.; Topping, G.; Sun, T.; Nekolla, S.G.; Richter, A.; et al. Molecular imaging of fibroblast activity after myocardial infarction using a ^{68}Ga-labeled fibroblast activation protein inhibitor, FAPI-04. *J. Nucl. Med.* **2019**, *60*, 1743–1749. [CrossRef]
21. Koerber, S.A.; Staudinger, F.; Kratochwil, C.; Adeberg, S.; Haefner, M.F.; Ungerechts, G.; Rathke, H.; Winter, E.; Lindner, T.; Syed, M.; et al. The role of FAPI-PET/CT for patients with malignancies of the lower gastrointestinal tract—First clinical experience. *J. Nucl. Med.* **2020**, *61*, 1331–1336. [CrossRef]
22. Toms, J.; Kogler, J.; Maschauer, S.; Daniel, C.; Schmidkonz, C.; Kuwert, T.; Prante, O. Targeting Fibroblast Activation Protein: Radiosynthesis and Preclinical Evaluation of an ^{18}F-labeled FAP Inhibitor. *J. Nucl. Med.* **2020**, *61*, 1806–1813. [CrossRef] [PubMed]
23. Lindner, T.; Altmann, A.; Kraemer, S.; Kleist, C.; Loktev, A.; Kratochwil, C.; Giesel, F.; Mier, W.; Marme, F.; Debus, J.; et al. Design and development of ^{99}mTc labeled FAPI-tracers for SPECT-imaging and ^{188}Re therapy. *J. Nucl. Med.* **2020**, *61*, 1507–1513. [CrossRef] [PubMed]
24. Moon, E.S.; Elvas, F.; Gwendolyn, V.; De Lombaerde, S.; Vangestel, C.; De Bruycker, S.; Bracke, A.; Eppard, E.; Greifenstein, L.; Klasen, B.; et al. Targeting fibroblast activation protein (FAP): Next generation PET radiotracers using squaramide coupled bifunctional DOTA and DATA^{5m} chelators. *EJNMMI Radiopharm. Chem.* **2020**, *5*, 1–20. [CrossRef] [PubMed]

25. Kreppel, B.; Gärtner, F.; Marinova, M.; Attenberger, U.; Meisenheimer, M.; Toma, M.; Kristiansen, G.; Feldmann, G.; Moon, E.; Roesch, F.; et al. [^{68}Ga]Ga-DATA^{5m}.SA.FAPi PET/CT: Specific Tracer-uptake in Focal Nodular Hyperplasia and potential Role in Liver Tumor Imaging. *Nuklearmedizin* **2020**, *59*, 387–389. [CrossRef] [PubMed]
26. Ballal, S.; Yadav, M.P.; Moon, E.S.; Kramer, V.S.; Roesch, F.; Kumari, S.; Tripathi, M.; ArunRaj, S.T.; Sarswat, S.; Bal, C. Biodistribution, pharmacokinetics, dosimetry of [^{68}Ga]Ga-DOTA.SA.FAPi, and the head-to-head comparison with [^{18}F]F-FDG PET/CT in patients with various cancers. *Eur. J. Nucl. Med. Mol. Imaging* **2020**. [CrossRef]
27. Kostelnik, T.I.; Orvig, C. Radioactive Main Group and Rare Earth Metals for Imaging and Therapy. *Chem. Rev.* **2019**, *119*, 902–956. [CrossRef] [PubMed]
28. Roesch, F. Scandium-44: Benefits of a Long-Lived PET Radionuclide Available from the ^{44}Ti/^{44}Sc Generator System. *Curr. Radiopharm.* **2012**, *5*, 187–201. [CrossRef] [PubMed]
29. Filosofov, D.V.; Loktionova, N.S.; Rösch, F. A ^{44}Ti/^{44}Sc radionuclide generator for potential application of ^{44}Sc-based PET-radiopharmaceuticals. *Radiochim. Acta* **2010**, *98*, 149–156. [CrossRef]
30. Pruszyński, M.; Loktionova, N.S.; Filosofov, D.V.; Rösch, F. Post-elution processing of ^{44}Ti/^{44}Sc generator-derived ^{44}Sc for clinical application. *Appl. Radiat. Isot.* **2010**, *68*, 1636–1641. [CrossRef]
31. Hernandez, R.; Valdovinos, H.; Yang, Y.; Chakravarty, R.; Hong, H.; Barnhart, T.; Cai, W. Sc: An Attractive Isotope for Peptide-Based PET Imaging. *Mol. Pharm.* **2014**, *11*, 2954–2961. [CrossRef] [PubMed]
32. Koumarianou, E.; Loktionova, N.S.; Fellner, M.; Roesch, F.; Thews, O.; Pawlak, D.; Archimandritis, S.C.; Mikolajczak, R. ^{44}Sc-DOTA-BN[2–14]NH_2 in comparison to ^{68}Ga-DOTA-BN[2–14]NH_2 in pre-clinical investigation. Is ^{44}Sc a potential radionuclide for PET? *Appl. Radiat. Isot.* **2012**, *70*, 2669–2676. [CrossRef] [PubMed]
33. Domnanich, K.A.; Müller, C.; Farkas, R.; Schmid, R.M.; Ponsard, B.; Schibli, R.; Türler, A.; van der Meulen, N.P. ^{44}Sc for labeling of DOTA- and NODAGA-functionalized peptides: Preclinical in vitro and in vivo investigations. *EJNMMI Radiopharm. Chem.* **2017**, *1*, 1–19. [CrossRef] [PubMed]
34. Umbricht, C.A.; Benešová, M.; Schmid, R.M.; Türler, A.; Schibli, R.; van der Meulen, N.P.; Müller, C. ^{44}Sc-PSMA-617 for radiotheragnostics in tandem with ^{177}Lu-PSMA-617—preclinical investigations in comparison with ^{68}Ga-PSMA-11 and ^{68}Ga-PSMA-617. *EJNMMI Res.* **2017**, *7*, 1–10. [CrossRef] [PubMed]
35. Thorp-Greenwood, F.L.; Coogan, M.P. Multimodal radio- (PET/SPECT) and fluorescence imaging agents based on metallo-radioisotopes: Current applications and prospects for development of new agents. *J. Chem. Soc. Dalton Trans.* **2011**, *40*, 6129–6143. [CrossRef] [PubMed]
36. Eppard, E.; de la Fuente, A.; Benešová, M.; Khawar, A.; Bundschuh, R.A.; Gärtner, F.C.; Kreppel, B.; Kopka, K.; Essler, M.; Rösch, F. Clinical translation and first in-human use of [^{44}Sc]Sc-PSMA-617 for pet imaging of metastasized castrate-resistant prostate cancer. *Theranostics* **2017**, *7*, 4359–4369. [CrossRef]
37. Fröss-Baron, K.; Garske-Román, U.; Welin, S.; Granberg, D.; Eriksson, B.; Khan, T.; Sandström, M.; Sundin, A. ^{177}Lu-DOTATATE therapy of advanced pancreatic neuroendocrine tumors heavily pretreated with chemotherapy; analysis of outcome, safety and their determinants. *Neuroendocrinology* **2020**, *111*, 330–334. [CrossRef]
38. Forrer, F.; Uusijärvi, H.; Storch, D.; Maecke, H.R.; Mueller-Brand, J. Treatment with ^{177}Lu-DOTATOC of patients with relapse of neuroendocrine tumors after treatment with ^{90}Y-DOTATOC. *J. Nucl. Med.* **2005**, *46*, 1310–1316. [PubMed]
39. Baum, R.P.; Kluge, A.W.; Kulkarni, H.; Schorr-Neufing, U.; Niepsch, K.; Bitterlich, N.; van Echteld, C.J.A. [^{177}Lu-DOTA]0-D-Phe1-Tyr3-Octreotide (^{177}Lu-DOTATOC) for peptide receptor radiotherapy in patients with advanced neuroendocrine tumours: A Phase-II study. *Theranostics* **2016**, *6*, 501–510. [CrossRef]
40. Baum, R.P.; Kulkarni, H.R.; Schuchardt, C.; Singh, A.; Wirtz, M.; Wiessalla, S.; Schottelius, M.; Mueller, D.; Klette, I.; Wester, H.J. ^{177}Lu-labeled prostate-specific membrane antigen radioligand therapy of metastatic castration-resistant prostate cancer: Safety and efficacy. *J. Nucl. Med.* **2016**, *57*, 1006–1013. [CrossRef]
41. Iravani, A.; Violet, J.; Azad, A.; Hofman, M.S. Lutetium-177 prostate-specific membrane antigen (PSMA) theranostics: Practical nuances and intricacies. *Prostate Cancer Prostatic Dis.* **2020**, *23*, 38–52. [CrossRef] [PubMed]
42. Heck, M.M.; Retz, M.; Tauber, R.; Knorr, K.; Kratochwil, C.; Eiber, M. Radionuklidtherapie des Prostatakarzinoms mittels PSMA-Lutetium. *Urologe* **2017**, *56*, 32–39. [CrossRef] [PubMed]
43. Heck, M.M.; Tauber, R.; Schwaiger, S.; Retz, M.; D'Alessandria, C.; Maurer, T.; Gafita, A.; Wester, H.J.; Gschwend, J.E.; Weber, W.A.; et al. Treatment Outcome, Toxicity, and Predictive Factors for Radioligand Therapy with ^{177}Lu-PSMA-I&T in Metastatic Castration-resistant Prostate Cancer. *Eur. Urol.* **2019**, *75*, 920–926. [CrossRef]
44. Sinnes, J.; Nagel, J.; Rösch, F. AAZTA^5/AAZTA^5-TOC: Synthesis and radiochemical evaluation with ^{68}Ga, ^{44}Sc and ^{177}Lu. *EJNMMI Radiopharm. Chem.* **2019**, *4*, 1–10. [CrossRef]
45. Greifenstein, L.; Grus, T.; Nagel, J.; Sinnes, J.P.; Rösch, F. Synthesis and labeling of a squaric acid containing PSMA-inhibitor coupled to AAZTA^5 for versatile labeling with ^{44}Sc, ^{64}Cu, ^{68}Ga and ^{177}Lu. *Appl. Radiat. Isot.* **2020**, *156*, 108867. [CrossRef]
46. De Decker, A.; Vliegen, G.; Van Rompaey, D.; Peeraer, A.; Bracke, A.; Verckist, L.; Jansen, K.; Geiss-Friedlander, R.; Augustyns, K.; De Winter, H.; et al. Novel Small Molecule-Derived, Highly Selective Substrates for Fibroblast Activation Protein (FAP). *ACS Med. Chem. Lett.* **2019**, *10*, 1173–1179. [CrossRef] [PubMed]
47. De Meester, I.; Vanhoof, G.; Lambeir, A.M.; Scharpé, S. Use of immobilized adenosine deaminase (EC 3.5.4.4) for the rapid purification of native human CD26/dipeptidyl peptidase IV (EC 3.4.14.5). *J. Immunol. Methods* **1996**, *189*, 99–105. [CrossRef]

48. Eppard, E.; Wuttke, M.; Nicodemus, P.L.; Rösch, F. Ethanol-based post-processing of generator-derived ^{68}Ga Toward kit-type preparation of ^{68}Ga-radiopharmaceuticals. *J. Nucl. Med.* **2014**, *55*, 1023–1028. [CrossRef]
49. Ballal, S.; Yadav, M.P.; Kramer, V.; Moon, E.S.; Roesch, F.; Tripathi, M.; Mallick, S.; ArunRaj, S.T.; Bal, C. A theranostic approach of [^{68}Ga]Ga-DOTA.SA.FAPi PET/CT-guided [^{177}Lu]Lu-DOTA.SA.FAPi radionuclide therapy in an end-stage breast cancer patient: New frontier in targeted radionuclide therapy. *Eur. J. Nucl. Med. Mol. Imaging* **2021**, *48*, 942–944. [CrossRef]

 MDPI

Review

Radiochemistry, Production Processes, Labeling Methods, and ImmunoPET Imaging Pharmaceuticals of Iodine-124

Krishan Kumar * and Arijit Ghosh

Laboratory for Translational Research in Imaging Pharmaceuticals, The Wright Center of Innovation in Biomedical Imaging, Department of Radiology, The Ohio State University, Columbus, OH 43212, USA; Arijit.Ghosh@osumc.edu
* Correspondence: krishan.kumar@osumc.edu or kumar@wcibmi.org

Abstract: Target-specific biomolecules, monoclonal antibodies (mAb), proteins, and protein fragments are known to have high specificity and affinity for receptors associated with tumors and other pathological conditions. However, the large biomolecules have relatively intermediate to long circulation half-lives (>day) and tumor localization times. Combining superior target specificity of mAbs and high sensitivity and resolution of the PET (Positron Emission Tomography) imaging technique has created a paradigm-shifting imaging modality, ImmunoPET. In addition to metallic PET radionuclides, ^{124}I is an attractive radionuclide for radiolabeling of mAbs as potential immunoPET imaging pharmaceuticals due to its physical properties (decay characteristics and half-life), easy and routine production by cyclotrons, and well-established methodologies for radioiodination. The objective of this report is to provide a comprehensive review of the physical properties of iodine and iodine radionuclides, production processes of ^{124}I, various ^{124}I-labeling methodologies for large biomolecules, mAbs, and the development of ^{124}I-labeled immunoPET imaging pharmaceuticals for various cancer targets in preclinical and clinical environments. A summary of several production processes, including ^{123}Te(d,n)^{124}I, ^{124}Te(d,2n)^{124}I, ^{121}Sb(α,n)^{124}I, ^{123}Sb(α,3n)^{124}I, ^{123}Sb(^{3}He,2n)^{124}I, ^{nat}Sb(α, xn)^{124}I, ^{nat}Sb(^{3}He,n)^{124}I reactions, a detailed overview of the ^{124}Te(p,n)^{124}I reaction (including target selection, preparation, processing, and recovery of ^{124}I), and a fully automated process that can be scaled up for GMP (Good Manufacturing Practices) production of large quantities of ^{124}I is provided. Direct, using inorganic and organic oxidizing agents and enzyme catalysis, and indirect, using prosthetic groups, ^{124}I-labeling techniques have been discussed. Significant research has been conducted, in more than the last two decades, in the development of ^{124}I-labeled immunoPET imaging pharmaceuticals for target-specific cancer detection. Details of preclinical and clinical evaluations of the potential ^{124}I-labeled immunoPET imaging pharmaceuticals are described here.

Keywords: positron emission tomography; PET; target-specific biomolecules; immunoPET imaging pharmaceuticals; production processes; ^{124}I-labeled monoclonal antibodies; cancer; radiolabeling; radiotracers

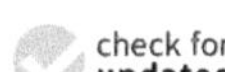 check for updates

Citation: Kumar, K.; Ghosh, A. Radiochemistry, Production Processes, Labeling Methods, and ImmunoPET Imaging Pharmaceuticals of Iodine-124. *Molecules* **2021**, *26*, 414. https://doi.org/10.3390/molecules26020414

Academic Editor: Kazuma Ogawa

Received: 8 December 2020
Accepted: 7 January 2021
Published: 14 January 2021

Publisher's Note: MDPI stays neutral with regard to jurisdictional claims in published maps and institutional affiliations.

1. Introduction

Several non-invasive imaging techniques are being used to identify, characterize, and quantify in vivo anatomical changes and biological processes that occur at the cellular and molecular levels. Radioisotope-based Positron Emission Tomography (PET), and Single-Photon Emission Computed Tomography (SPECT) are very sensitive imaging techniques. However, PET is considered to be superior to SPECT due to the availability of higher sensitivity scanners and better quantification of regional tissue concentrations of radiolabeled imaging pharmaceuticals [1]. Sufficient acquisition speed of the PET imaging technique allows the determination of pharmacokinetics and biodistribution of imaging pharmaceuticals and produces three-dimensional images of the functional processes in the body.

Various non-metallic (^{11}C, ^{13}N, ^{15}O, ^{18}F, and ^{124}I, etc.) and metallic (^{64}Cu, ^{68}Ga, and ^{89}Zr, etc.) radionuclides are used routinely for the preparation of PET imaging pharmaceuticals for preclinical and clinical environments [2]. Table 1 provides a summary of the physical characteristics and the production methods [2–5] for some PET radionuclides, produced by a generator and proton bombardment of solid, liquid, and gas targets, that are suitable for radiolabeling of small and large biomolecules and nanomaterials for the development of potential PET imaging pharmaceuticals.

Table 1. Physical properties and production methods for some cyclotron produced non- metallic and metallic positron (β^+) emitting radionuclides [2–5].

Radionuclide	Production Method	Half-Life	%Decay Mode	E_{max} (β^+), MeV	E_{mean} (β^+), MeV	Reference
^{15}O	$^{15}N(p,n)^{15}O$	2.0 min	β^+/99.9 EC/0.1	1.732	0.725	[2,3]
^{13}N	$^{13}C(p,n)^{13}N$ $^{16}O(p,\alpha)^{13}N$	10.0 min	β^+/99.8 EC/0.2	1.199	0.492	[2,3]
^{11}C	$^{14}N(p,\alpha)^{11}C$	20.4 min	β^+/99.8 EC/0.2	0.960	0.386	[2,3]
^{18}F	$^{18}O(p,n)^{18}F$	110 min	β^+/96.9 EC/3.1	0.634	0.250	[2,3]
^{124}I	$^{124}Te(p,n)^{124}I$	4.2 d	β^+/22.7 EC/77.3	2.138	0.975	[2,3]
^{64}Cu	$^{64}Ni(p,n)^{64}Cu$	12.7 h	β^+/17.5 EC/43.5	0.653	0.278	[2–4]
^{68}Ga	$^{68}Zn(p,n)^{68}Ga$ $^{68}Ge/^{68}Ga$ Generators	67.8 min	β^+/88.9 EC/11.1	1.899	0.836	[2,3]
^{44}Sc	$^{44}Ca(p,n)^{44}Sc$	3.97 h	β^+/94 EC/6	1.474	0.632	[5]
^{66}Ga	$^{66}Zn(p,n)^{66}Ga$	9.5 h	β^+/56.5 EC/43.5	4.15	1.75	[4]
^{86}Y	$^{86}Sr(p,n)^{86}Y$	14.7 h	β^+/31.9 EC/68.1	1.221	0.535	[3,4]
^{55}Co	$^{58}Ni(p,\alpha)^{55}Co$	17.5 h	β^+/76 EC/24	1.498	0.570	[4]
^{89}Zr	$^{89}Y(p,n)^{89}Zr$	78.4 h	β^+/22.7 EC/76.2	0.902	0.396	[2–4]
^{52}Mn	$^{52}Cr(p,n)^{52}Mn$ $^{nat}Cr(p,x)^{52}Mn$	5.59 d	β^+/29.4 EC/77	0.575	0.242	[4]

The clinical applications of PET imaging pharmaceuticals have increased tremendously over the past several years since the availability of the FDA (Food and Drug Administration) approved ^{11}C-, ^{18}F-, and ^{68}Ga-labeled imaging pharmaceuticals, ([^{11}C]Choline, [^{11}C]Acetate, [^{18}F]NaF, [^{18}F]FDG, [^{18}F]Florbetapir, [^{18}F]Fluemetamol, [^{18}F]Florbetaben, [^{18}F]Fluciclovine, [^{18}F]Flortaucipir [^{18}F]Fluoroestradiol, [^{68}Ga]DOTA-TATE (NETSPOT), [^{68}Ga]DOTA-TOC), and [^{68}Ga]PSMA-11, for various applications, including metabolism, neurology, and oncology, etc. Additional worldwide clinical trials with ^{68}Ga-labeled PSMA target-specific ligands, PSMA-11 and PSMA-617, are ongoing for prostate cancer imaging [6] The majority of clinical applications involve [^{18}F]FDG; however, its use for neurological, oncological, and cardiological applications has been limited [7]. Therefore, numerous radiolabeled biomolecules that can target receptors that are known to overexpress on certain tumors were discovered, developed, and tested in the past [8–10].

Target-specific biomolecules, known to have high specificity and affinity for receptors associated with tumors and other pathological conditions, include small biomolecules (e.g., folate), peptides, and larger biomolecules like monoclonal antibodies (mAb), proteins, antibody fragments, and RNA nanoparticles [11–13]. The large biomolecules (mAbs and proteins etc.), with higher tumor specificity and affinity, have relatively intermediate to long circulation half-lives (>day) and tumor localization times. Combining superior target specificity of mAbs and high sensitivity and resolution of the PET technique has created a paradigm-shifting imaging modality, ImmunoPET (Immuno Positron Emission Tomography) [14]. The concept of immunoPET was proposed more than two decades ago. Significant progress has been made, since then, in the development of immunoPET imaging pharmaceuticals as a result of FDA approval of several therapeutics mAbs in recent years [15–17]. Our understanding of tumor heterogeneity and clinical disease management has improved, in the recent past, due to the availability of immunoPET.

The critical factors that need to be considered for the selection of positron-emitting radionuclides for the development of immunoPET imaging pharmaceuticals are: (1) desirable decay characteristics of the radionuclide to yield high-quality images, (2) availability of methods to produce the isotope in sufficient and pure amounts, (3) availability of efficient radiolabeling methodologies, and most importantly, (4) physical half-life of the radionuclide that will allow sufficient time to monitor pharmacokintetics (tumor uptake and elimination) and biodistribution and for the transportation of the radiolabeled material to the preclinical or clinical site.

Short-lived and long-lived radioisotopes are considered suitable for the development of small-molecule- and large-biomolecule-based PET imaging pharmaceuticals, respectively, by using indirect and direct labeling techniques. Various strategies, including bifunctional chelating agents, prosthetic groups, click chemistry, enzyme-mediated, silicon- and boron- acceptor methodologies, the pre-targeting, the reporting gene methods, etc., are used for the design and development of immunoPET imaging pharmaceuticals [2,14]. The radionuclides with a short half-life (e.g., ^{68}Ga and ^{18}F) are unsuitable for the development of immunoPET imaging pharmaceuticals. Consequently, radionuclides with longer half-lives, with well-established radiolabeling methodologies, are used for the development of large biomolecules based imaging pharmaceuticals matching more closely with their longer circulation times. Some of the suitable metallic (e.g., ^{44}Sc, ^{52}Mn, ^{55}Co, ^{64}Cu, ^{66}Ga, ^{86}Y, ^{89}Zr, etc.) and non-metallic (e.g., ^{124}I) radionuclides with longer half-lives are listed in Table 1.

Thermodynamically stable and kinetically inert radiolabeled metal (e.g., ^{64}Cu, ^{89}Zr, ^{66}Ga, and ^{86}Y, etc.) chelate conjugates, using bifunctional chelating agents (BFC) to target-specific biomolecules, have been studied extensively for their potential applications as imaging pharmaceuticals. Two steps are involved in the development of metallic radionuclide-labeled large-biomolecules based imaging pharmaceuticals. The first step is the conjugation of a bifunctional chelating agent that forms a thermodynamically stable and kinetically inert metal chelate, with the target-specific large biomolecule. In the second step, the BFC-large biomolecule conjugate is labeled with a metallic radionuclide [18–20]. Linear (e.g., DTPA = Diethylenetriamine N, N, N′, N″, N‴ pentaacetic acid, HBED-CC = N,N′-bis[2-hydroxy-5-(carboxyethyl)benzyl] ethylenediamine-N,N′-diacetic acid, and DFO = Desferrioxamine B or Deferoxamine B, etc.) and macrocyclic polyaminocarboxylates (e.g., NOTA = 1,4,7-triazacyclononane-1,4,7-triacetic acid, DOTA = 1,4,7,10-tetraazacyclododecane-1,4,7,10-tetraacetic acid, etc.) and their analogs and derivatives are known to form thermodynamically stable and kinetically inert metal chelates. Alternatively, the radiolabeling of the BFC is accomplished in the first step followed by the conjugation to the target-specific biomolecule in the second step.

Based on the long half-life and physical properties of the positron-emitting isotope of iodine, ^{124}I may be used for both imaging (positron) and therapy (electron capture) as well as for ^{131}I dosimetry. The therapeutic effect of ^{124}I relies on the Auger electron emission responsible for the local action within nanometers. The relatively low percentage of high-energy positrons (22.7%) and a high percentage of cascade gamma photons in the background compared to the conventional PET isotopes makes imaging with ^{124}I technically challenging. However, optimizing image acquisition parameters and appropriate corrections within the image reconstruction process improve the image quality. ^{89}Zr and ^{52}Mn, with 3.27 and 5.59 d half-lives, respectively, are attractive choices for the development of immunoPET imaging pharmaceuticals. Labeling of large biomolecules with metallic radionuclides requires additional conjugation steps in the process and purification could be challenging. DFO has been the most popular bifunctional chelator for conjugation with mAbs and ^{89}Zr labeling. The production of high purity ^{52}Mn and in vivo stability of manganese chelates and their conjugates are still developing.

I-124 is an attractive radionuclide for the development of mAbs as potential immunoPET imaging pharmaceuticals due to its physical properties (decay characteristics and half-life), easy and routine production by cyclotrons [21], and well-established methodologies for radioiodination [22–24]. For example, ^{124}I has been used to label small molecules, peptides, mAbs, proteins, and antibody fragments for tumor imaging [25–28], in thyroid and parathyroid cancer imaging [29–31], to label single molecules like meta-iodobenzylguanidine (MIBG), amino acids, and fatty acids among others for investigation of several heart and brain diseases, as well as functional studies of neurotransmitter receptors [32–34]. It has also been used to label photosensitizers for photodynamic therapy [35]. Labeling of biomolecules with radioactive iodine was first established several decades ago when the ^{131}I isotope of iodide was used for labeling polyclonal antikidney serum [36]. The aromatic moieties present in the large biomolecule to be labeled are a tyrosine residue and to a lesser extent a histidyl group [37,38].

The objective of the present report is to provide a comprehensive review of the physical properties of iodine and ^{124}I radionuclide, production processes of ^{124}I radionuclide, various ^{124}I-labeling methodologies for large biomolecules, specifically mAbs, and application of ^{124}I-labeled mAb, as immunoPET imaging pharmaceuticals, for oncologic applications. A summary of several production processes, including $^{123}Te(d,n)^{124}I$, $^{124}Te(d,2n)^{124}I$, $^{nat}Sb(\alpha,xn)^{124}I$, $^{121}Sb(\alpha,n)$, natSb(^{3}He,n) reactions, a detailed overview of the $^{124}Te(p,n)^{124}I$ reaction (including target selection, preparation, processing, and recovery of ^{124}I), and a fully automated process that can be scaled up for GMPproduction of large quantities of ^{124}I is provided. Direct, using inorganic and organic oxidizing agents and enzyme catalysis, and indirect, using prosthetic groups, ^{124}I-labeling techniques, have been discussed. Significant research has been conducted, in more than the last two decades, in the development of ^{124}I-labeled immunoPET imaging pharmaceuticals for target-specific cancer detection. Details of preclinical and clinical evaluations of the potential ^{124}I-labeled immunoPET imaging pharmaceuticals are described here.

Overall, this is a first comprehensive review providing a thorough understanding of various areas that are essential for our understanding of the discovery and the development of novel ^{124}I-labeled immunoPET imaging pharmaceuticals.

2. Overview of Physical Properties of Iodine and Iodine Radionuclides

Iodine with symbol I, atomic number 53, and atomic weight 127, and with [Kr]$4d^{10}5s^{2}5p^{5}$ electronic configuration belongs to group 17 of the periodic table. Iodine exists as a diatomic molecule, I_2, in its elemental state and is known to exist in −1, +1, +3, 5, and 7 oxidation states. Atomic radii are 133 and 220 pm for Iodine (I_2) and iodide (I^-), respectively. Elemental iodine, with chemical formula I_2, where two iodine atoms share a pair of electrons to each achieve a stable octet for themselves. Similarly, the iodide anion, I^-, is the strongest reducing agent among the stable halogens, being the most easily oxidized back to diatomic I_2. In general, I_2 converts into I_3^- in the presence of excess iodide. The standard potential of the iodide/triiodide redox couple is 0.35 V (versus the normal hydrogen electrode, NHE) [39].

Iodine radioisotopes have long been used as theranostic agents in the field of thyroid cancer [40]. There are 37 known isotopes of iodine, ^{108}I to ^{144}I, that undergo radioactive decay, except ^{127}I which is a stable isotope. The longest-lived of the radioactive isotopes of iodine is ^{129}I with a 15.7 million years half-life, decaying via beta decay to stable ^{129}Xe [41]. However, the most well-known iodine radionuclides are ^{123}I, ^{124}I, ^{125}I, and ^{131}I, which are used in preclinical and clinical environments for medical applications. Background related to their physical characteristics and medical applications are given below.

The main gamma emission peak of ^{123}I, 159 keV, makes it suitable for SPECT imaging as it is close to the 140 keV peak of ^{99m}Tc peak. A short physical half-life (13.22 h) [41] of ^{123}I allows the study of compounds that have rapid radiolabeling methods, fast clearance, and short metabolic processes. Several ^{123}I-labeled imaging pharmaceuticals, including ^{123}I-Iobenguane, for detection of primary or metastatic pheochromocytoma or neuroblastoma as an adjunct to other diagnostic tests, ^{123}I-ioflupane for visualization of the striatal dopamine transporter, and [^{123}I]NaI capsules for evaluation of thyroid function and morphology, are approved by the Food and Drug Administration (FDA) for clinical use.

I-125 has mainly X-ray energy emission at 27 keV with low gamma emission at 35.5 keV. It has photon energy which is low for optimal imaging and its half-life is long (59.4 days) [41]. ^{125}I-labeled imaging pharmaceuticals, ^{125}I-HSA and ^{125}I-iothalamate, are approved by the FDA for clinical use for total blood/plasma determination and evaluation of glomerular filtration, respectively. This radionuclide is used routinely in discovery and preclinical environments. For example, ^{125}I has been used for NCEs (New Chemical Entities) labeling and their evaluation for in vitro cell binding assays, biodistribution, and pharmacokinetic properties in preclinical models.

I-131, a beta-emitting isotope (606 keV, 90%) and a half-life of 8.02 days [41], is often used for radiotherapy. The penetration range of the beta particle is 0.6 to 2.0 mm at the site of uptake. The ^{131}I is taken up into thyroid tissue. The beta particles emitted by the radioisotope destroy the associated thyroid tissue with little damage to surrounding tissues (more than 2.0 mm from the tissues absorbing the iodine). ^{131}I emits gamma photons that can be used for SPECT imaging. ^{131}Iodine meta-iodobenzylguanidine (^{131}I-MIBG) is a radiopharmaceutical used for both imaging and treating certain types of neuroendocrine tumors, including neuroblastomas, paragangliomas, and pheochromocytomas. FDA approved ^{131}I-labeled products are, iobenguane ^{131}I, a form of ^{131}I-MIBG, for the treatment of paragangliomas and pheochromocytomas, ^{131}I-labeled HSA for determination of total blood and plasma volumes, cardiac output, cardiac and pulmonary blood volumes and circulation times, protein turnover studies, heart and great vessel delineation, localization of the placenta, and localization of cerebral neoplasms, and [^{131}I]NaI for the diagnostics and the therapeutic applications.

Initially, ^{124}I was considered as an impurity in the production of ^{123}I, although it was recognized that this radionuclide has attractive properties for use in PET imaging. For example, the half-life of 4.18 d is long enough for clearance and localization of ^{124}I-labeled mAbs. Additionally, the 22.7% positron decay with maximum and mean positron energies of 2.138 and 0.975 MeV, respectively, allows PET imaging. In contrast, the most common PET radiotracer, ^{18}F, has a positron abundance of 97% with maximum and mean positron energies of 0.634 and 0.250 MeV, respectively. ^{124}I has potential as both diagnostics and therapeutic radionuclide and its use are becoming more widespread.

In addition to positron emissions, ^{124}I emits a rather large portion of gamma rays during its decay (Figure 1), with the majority (63%) of which is 603 keV energy (Table 2). Coincidences of this 603 keV photon and a 511 keV annihilation photon cannot be distinguished from the true coincidences involving two 511 keV annihilation photons. Multiple correction methods have been suggested to address this background activity but their effectiveness is limited in the setting of the low count rates observed in clinical scans.

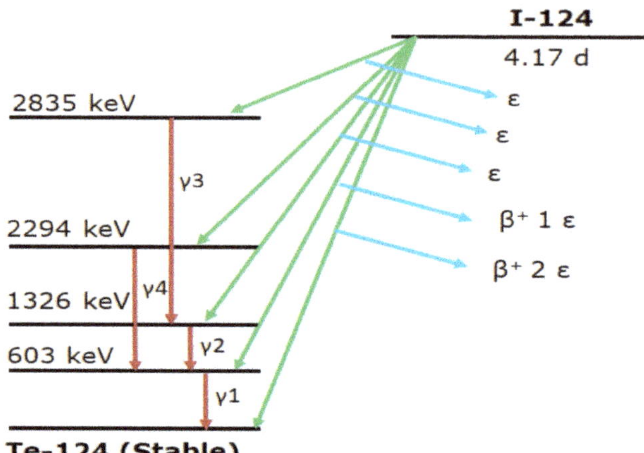

Figure 1. Simplified decay scheme of ^{124}I radionuclide (taken from reference [42]).

Table 2. A summary of the main emissions of ^{124}I (taken from reference [42]).

Decay Type	Energy, keV	Probability
β^+ 1	1535	12
β^+ 2	2138	11
γ 1	603	63
γ 2	723	10
γ 3	1510	3
γ 4	1691	11
X	27.2	17
X	27.5	31
ε	866	11
ε	2557	25
ε	3160	24

3. Overview of ^{124}I Production Processes

Routine availability of a long half-life radioisotope (^{124}I) for PET imaging that is economically, efficiently, and safely produced will enable the evaluation and development of numerous immunoPET imaging pharmaceuticals for research and clinical use. The planned strategies for the production of ^{124}I at a particular facility are decided by the availability of irradiating particles and their energy ranges. If multiple choices of beams are available at the production site, a reaction scheme is selected which produces ^{124}I with maximum yield and highest purity.

3.1. Production Reactions, Target Selection, and Preparation

Early investigations were focused on the production, including excitation functions determinations, of ^{124}I from the deuterium, α, ^{3}He irradiation of Te and Sb solid targets, including ^{123}Te(d,n)^{124}I, ^{124}Te(d,2n)^{124}I, ^{121}Sb(α,n)^{124}I, ^{123}Sb(α,3n)^{124}I, ^{123}Sb(^{3}He,2n)^{124}I, ^{nat}Sb(α, xn)^{124}I, ^{nat}Sb(^{3}He,n)^{124}I reactions [22,43–62]. Detailed background related to these production processes is reported in two excellent reviews [22,43]. As a result of less frequent availability of deuteron, alpha, and ^{3}He beams and high ^{125}I content in the produced materials by these reactions, these are not routinely used for ^{124}I production in the research and clinical environments. Significant interest grew in the ^{124}Te(p,n)^{124}I reaction, despite a slightly lower yield than the ^{124}Te(d,2n)^{124}I reaction, after a careful study of the process involving cross-section measurements and the production experiments [63–69]. The first ^{124}I production process was proposed based on the ^{123}I production method which involved 3–8 h irradiation of a ^{124}Te containing capsule, and irradiated by a ~26 MeV

8–18 μA proton beam. A similar capsule target was irradiated with a 12 MeV proton beam for the production of ^{124}I. The target was processed chemically to isolate $^{123/124}$I [70].

Numerous studies were conducted since the first report by Kondo et al. in 1977 [70]. The results from these studies suggested that ^{124}I can be successfully produced for research and clinical use by using low-energy cyclotrons that are available routinely for production of ^{11}C and ^{18}F-labeled imaging pharmaceuticals for standard care [71–86]. Consequently, the ^{124}Te(p,n)^{124}I production process is being used extensively in the research and clinical environments worldwide. The focus of this report will be to review the progress and the status of the ^{124}I production by using the ^{124}Te(p,n)^{124}I reaction. Other potential reactions that have been proposed and considered for the production of ^{124}I are ^{125}Te(p,2n)^{124}I and ^{126}Te(p,3n)^{124}I. [61,87–91] A summary of some of the recent studies is provided in Table 3.

The main goal of any production process is to ensure that it produces the final product with the highest purity and yield. Consequently, it is important to fully understand the background related to the enrichment and purity requirements of the tellurium target and the production process parameters, i.e., irradiation energy and current of the proton beam, and target processing and recovery. The percent of natural abundance of various isotopes (given in the parenthesis) of ^{nat}Te has been reported as ^{120}Te (0.09%), ^{122}Te (2.55%), ^{123}Te (0.89%), ^{124}Te (4.74%), ^{125}Te (7.07%), ^{126}Te (18.84%), ^{128}Te (31.74%), and ^{130}Te (34.08%) [69]. Irradiation of a ^{nat}Te target with proton beam will, consequently, produce a mixture of various unwanted iodine isotopes with long half-lives, making the production and purification process inefficient and challenging and the ^{124}I produced being unusable. Many reports exist on the proton-induced reactions on ^{nat}Te. These reports are valuable for testing nuclear model calculations, integral data validation and some other applications, but not for routine production of high purity ^{124}I for medical applications [91–96].

Therefore, a highly enriched ^{124}Te target (>99% or better) material must be used for the production of ^{124}I to minimize unwanted iodine isotopic impurities; although the cost of enriched tellurium increases significantly with the increased enrichment imposing the need for recycling the irradiated target material. In our laboratories, we have used the enriched target material with the following specifications: ^{124}Te (99.3%), ^{120}Te, ^{122}Te, ^{126}Te (<0.01%), ^{123}Te (<0.05), ^{128}Te (0.03%), ^{130}Te (0.02), and ^{125}Te (0.6%). Since the major contaminant in the enriched target is ^{125}Te, one should investigate the production of potential radionuclides from ^{125}Te(p,n)^{125}I, ^{125}Te(p,2n)^{124}I, and ^{125}Te(p,3n)^{123}I reactions also.

The tellurium target is available either as metallic tellurium or TeO_2. TeO_2 is used routinely for ^{124}I production due to better thermal characteristics than tellurium metal and to avoid evaporation of radioiodine. The melting points of TeO_2 and Te are 733 and 449.5 °C, respectively [74,82]. Additionally, tellurium tends to blow up upon heating. The tellurium target for proton irradiation has been prepared by the three different methods: (1) filling ^{124}Te in an aluminum capsule under He atmosphere [70,84], (2) introducing melted enriched tellurium onto a support plate [71–82,86], and (3) electroplating tellurium on a nickel-coated copper substrate [45,48,97,98].

Table 3. Summary of ^{124}I production reactions, yield, and impurity profile.

Nuclear Reaction	Effective Energy	Target Material	Enrichment, %	Yield MBq/μAh	Radionuclidic Impurities, %	Reference
$^{nat}Sb(\alpha,xn)^{124}I$	22→13	Sb	Nat	0.45 at 5 d EOB	^{123}I (4), ^{125}I(27), ^{126}I(27) at 5 d EOB	[54]
$^{121}Sb(\alpha,n)^{124}I$	22→13	Sb	99.45	0.92 at 5 d EOB	^{123}I (<4), ^{125}I(<0.2), ^{126}I(<0.2) at 5 d EOB	[54]
$^{nat}Sb(^{3}He,xn)^{124}I$	35→13	Sb	Nat	0.42 at 5 d EOB	^{123}I(14), ^{125}I(1.3), ^{126}I(1.6)	[55,60]
$^{123}Sb(\alpha,3n)^{124}I$	42→32	Sb	98.28	11.7 at EOB	^{123}I (<5), ^{125}I(<1.8), ^{126}I(<0.6) at 60h EOB	[59]
$^{123}Sb(^{3}He,2n)^{124}I$	45→32	Sb		15.5	^{123}I (14), ^{125}I(<1.19)	[62]
$^{123}Te(d,n)^{124}I$	11→6	Te	91.0, 85.4	2.8 *	^{123}I(3321) **	[53]
$^{124}Te(d,2n)^{124}I$	15→0	Te	95	0.55	^{126}I(0.5)	[44]
	15→8	Te	91.7	18.9	^{125}I(0.35) **, ^{126}I(0.39) **, ^{131}I(0.08) **	[45]
	16→6	Te	96.7	0.64 *	——	[46]
	16–6	TeO_2	89.6	-	Total Impurities < 5, At 40 h EOB	[49]
	14→0	TeO_2	89.6	15	^{125}I(1.41), ^{126}I(1.16), ^{130}I (7.87), ^{131}I(0.31)	[47]
	13.5→10	TeO_2	96	-	^{125}I(2)	[50]
	14→10	Te	99.8	17.5 *	^{125}I(1.7) *	[51]
$^{124}Te(p,n)^{124}I$						
	13→9	Te	99.51	20 *	^{123}I(41)	[64]
	12.2→0	TeO_2	99.8	13	^{123}I(10.039), ^{125}I(0.018), ^{126}I(0.041), ^{130}I(0.379)	[47]
	14.7	TeO_2/6.7%$A_{12}O_3$	96.7–99.9	20	^{123}I(30), ^{125}I(<0.1), ^{126}I(<0.1), ^{130}I(<0.1) ^{131}I(<0.1)	[74]
	12.3→9.8	TeO_2	98	12.5 at 2 d EOB	^{123}I(0.5), ^{125}I(<0.07) At 2 d EOB	[76]
	13.5→9	TeO_2/5%Al_2O_3	99.8	5.8	^{125}I(0.01), ^{126}I(<0.0001)	[77]
	12.5→5	TeO_2	99.8	9.0 ± 1.0	^{125}I(0.053)	[78]
	14→7	TeO_2/5%Al_2O_3	99.86	21.1	^{125}I(0.03), ^{126}I(0.007)	[80]
	11→2.5	TeO_2/6%Al_2O_3	99.5	6.40 ± 0.44	^{125}I(<0.02), ^{126}I(<0.001)	[81]
	11→2.5	Al_2Te_3	99.5	8.47	^{125}I(<0.001), ^{126}I(<0.001)	[82]
	11.6→0	TeO_2/5%Al_2O_3	99.8	6.88	^{123}I	[84]
	12.6	TeO_2	99.8	13.0	^{123}I (0.18), ^{125}I (0.037) ^{126}I (0.0099)	[85]
	16.5→12	TeO_2/5%Al_2O_3	99.93	4–5	^{123}I (<1.5), ^{125}I (<0.001)	[86]
$^{125}Te(p,2n)^{124}I$	20.1→10.5	TeO_2	93	43.3	^{123}I(8), ^{125}I(5)	[87]
	22→4	Te	98.3	111 *	^{125}I(0.89)	[89]
	21→15	Te	98.3	81 *	^{123}I(7.4), ^{125}I(0.9)	[65]
	22	TeO_2	98.5	104	^{123}I(<1)	[88]
$^{126}Te(p,3n)^{124}I$						
	38→28	Te	98.69–99.8	222 *	^{123}I(148), ^{125}I(1.0), ^{126}I(1.0)	[61]
	36.8→33.6	Te	nat	67 *	-	[91]

* Based on cross-section data, ** Percent calculated here from the ratio of the published saturation yield data.

In the second method of target preparation, the isotopically enriched tellurium is-melted onto a small platinum plate. The platinum surface should not be smooth rather be scratched with a scalpel or lancet before preparation of the target. An optimized amount of tellurium is critical for the quality of the target. Powdered Al_2O_3 (5–7%) is commonly mixed with TeO_2 [74,77,79–82,86] during target preparation by melting for (1) increasing the heat transfer characteristics, (2) enhancing the TeO_2 binding to the target plate [74,77,79], (3) giving the target material a glassy solid structure and eliminating the need for a cover foil [53,59], and (4) increasing the uniformity of the target material layer. Several binary tellurium compounds, to improve the thermal properties, were used, in the past, for the development of ^{124}I production processes, including Al_2Te_3 [82] and Cu_2Te [73,75] with 895 and 1132 °C melting points, respectively. Al_2Te_3 appeared to be a promising target material, providing a high tellurium mass fraction and a glassy melt material [82].

Higher beam currents can be used for the bombardment of the target when the tellurium is electroplated on a suitably large area of the target carrier and when a small beam/target angle irradiation is performed under the optimum cooling conditions. Large area electroplated tellurium targets are attractive for this application as long as the deposits are smooth, homogeneous, and free of other constituents. A new plating technology involving CCE (Constant Current Electrolysis) was developed to avoid the poor quality target layers during plating procedures [97]. In this method, tellurium targets were prepared by DC-CCE of the metal from alkaline plating solutions. 50 μm nickel-coated, needed for good adhesion of the target material, copper plates were used for target preparation. Details of this technique are presented elsewhere [97]. A mean weight of 90 $\pm$ 9 mg of enriched tellurium was deposited per target. The electroplating process is more expensive and requires more work and a precise set-up, but it may produce higher yields for the production of ^{124}I. On the other hand, the melting process is experimentally simpler and produces targets that can be reused several times.

In addition to the selection of the target material and method of target preparation, various support plates, in which the target material is deposited either by melting or electrodeposition, have been used. These include Aluminum [45], Platinum [71,74,77,78,80,99], tantalum and nickel electroplated tantalum [100], nickel electroplated copper [44,47,98], tungsten and silicon [73], platinum-coated tungsten [75], platinum/iridium [47,49,50,72], and rhodium electroplated stainless steel [101]. Nickel-coated, to ensure good adhesion of tellurium, copper is a good target material for electrodeposition of tellurium to provide a good cooling efficiency during irradiation. This is due to its high melting point (1084.62 °C) and the high thermal conductivity (401 W m^{-1} K^{-1}) of copper; although there are some disadvantages of using copper plating. Natural copper consists of ^{65}Cu(30.83%) and ^{63}Cu(69.17%), which have potential to produce different zinc isotopes from $^{65}Cu(p,n)^{65}Zn$, $^{65}Cu(p,2n)^{64}Zn$, $^{63}Cu(p,n)^{63}Zn$, and $^{63}Cu(p,2n)^{62}Zn$ with ^{65}Zn being long-lived (half-life = 244 d). The maximum cross-section for the $^{65}Cu(p, n)^{65}Zn$ reaction is around 11 MeV, which is in the same energy range as for the $^{124}Te(p,n)^{124}I$ reaction. Depending on the target thickness, the cross-section for the $^{65}Cu(p,n)^{65}Zn$ will be high enough to produce ^{65}Zn. Consequently, a careful target design is required while using the nickel electroplated copper backing for tellurium electroplating. Platinum is considered a better choice as a coating or backing material for target preparation due to the fact that (1) it is not dissolved during the chemical processing to recover ^{124}Te, (2) it is not necessary to make one target per irradiation, (3) the recovered ^{124}Te may have a higher chemical purity, (4) it has a high melting point (1768 °C), which makes it suitable for a dry distillation of iodine. But it also has some disadvantages: it is more expensive than Cu and it has a much lower thermal conductivity (71.6 W·m^{-1}·K^{-1}) than copper.

The target thickness optimization and its orientation are two critical parameters during irradiation of the target for ^{124}I production for high yield and purity. The optimized thickness is important to (1) ensure that the entire beam energy is not deposited within the target itself, (2) reduce the production of unwanted radioiodine impurities, and (3) reduce the cost of production. Additionally, the orientation of the target is also optimized to reduce the power density, which increases both the area over which the heat is deposited and the effective target thickness.

3.2. Proton Beam Energy and Current for Target Irradiation

Proton irradiation parameters, i.e., proton beam energy, current, and irradiation time, are important parameters in maximizing the yield and minimizing the number and amount of impurities even if the 100% enriched ^{124}Te target material is used. A recent study reported the calculation of the excitation functions for production of ^{123}I and ^{124}I from proton bombardment of ^{124}Te by using TALYS 1.6[67] and comparing with the experimental results reported previously [64,70]. The calculated ^{124}Te(p,n)^{124}I reaction cross sections were in good agreement with the experimental data with a peak at ~600 mb [64]. The production of ^{124}I is appropriate for small, medium-sized cyclotrons. Figure 2 shows a comparison of cross-section data for ^{124}Te(p,n)^{124}I and ^{124}Te(p,2n)^{123}I reactions [21]. The calculated cross-section data for ^{124}Te(p,2n)^{123}I reaction, shown in Figure 2, are in fairly good agreement with experimental data with a peak over 900 mb. However, there are some discrepancies in low and high energy regions (10–18 MeV, 25–30 MeV). Since there is an overlap between ^{124}Te(p,n)^{124}I and ^{124}Te(p,2n)^{123}I cross-section curves in the 12 to 16 MeV energy range; therefore, the proton bombardment of ^{124}Te always produces a mixture of ^{124}I and ^{123}I. Since the decay of ^{123}I is 7.6 times faster than ^{124}I, overnight storage of the mixture is required in the production process for removal of ^{123}I improving the purity of ^{124}I; although it decreases the overall yield of ^{124}I production.

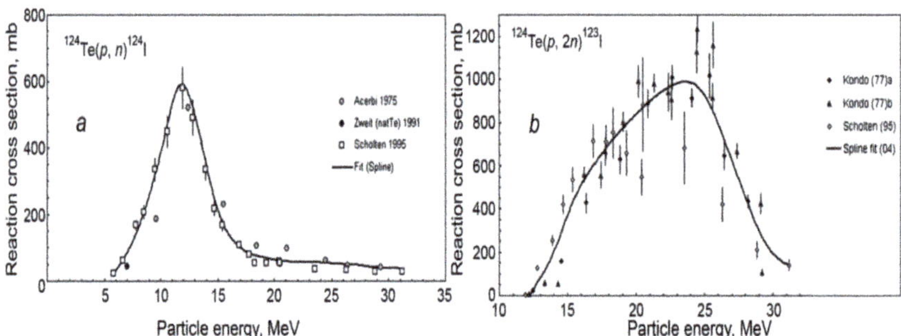

Figure 2. Comparison of reaction cross sections for the ^{124}Te(p,n)^{124}I and ^{124}Te(p,2n)^{123}I reactions (taken from reference [21]).

Excitation functions of the ^{125}Te(p,xn)^{123,124,125}I nuclear reactions were measured, using targets that were prepared by electrolytic deposition of 98.3% enriched ^{125}Te on a Ti-backing, in the threshold to 100 MeV energy range by using the stacked-foil techniques [65]. Additionally, the excitation functions were calculated by a modified hybrid model code ALICE-IPP. Figure 3 shows a plot of cross-section vs. incident proton energy for ^{125}Te(p,n)^{125}I, ^{125}Te(p,2n)^{124}I, and ^{125}Te(p,3n)^{123}I reactions. The experimental and calculated data agreed with each other. The data given in Figure 3 and integral yield data suggested that ^{124}I and ^{125}I are produced by low energy proton irradiation (<20 MeV). ^{123}I is produced at >20 MeV. The energy 21→15 MeV appears to be suitable for ^{124}I production from the ^{125}Te(p,2n)^{124}I reaction which is above the range of low energy cyclotrons. Below 15 MeV, the yield of ^{124}I from ^{125}Te(p,2n)^{124}I reaction is low, and ^{125}I from the ^{125}Te(p,n)^{125}I reaction is high. The ^{123}I impurity is not a problem for ^{125}Te(p,2n)^{124}I reaction as (1) ^{125}Te(p,3n)^{123}I reaction requires >20 MeV, and (2) it decays out rather fast. The formation of ^{125}I impurity, from ^{125}Te impurity, in the ^{124}Te(p,n)^{124}I nuclear reaction is more critical and must be controlled. It has been reported that the yield of ^{125}Te(p,2n)^{124}I reaction is four times higher than ^{124}Te(p,n)^{124}I reaction with some ^{125}I present making it an attractive route for ^{124}I production [65]. However, the proposed production energy range is too high for small cyclotrons requiring medium-sized commercial cyclotrons.

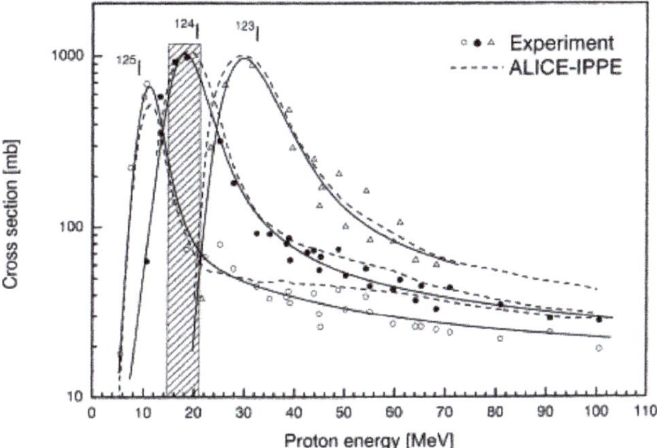

Figure 3. Excitation functions of ^{125}Te(p,xn)^{123,124,125}I reactions (taken from reference [65]). The broken lines show the results of nuclear model calculations using the code ALICE-IPPE. The shaded area gives a suitable energy range for the production of ^{124}I.

For the energy range window employed for proton irradiation of ^{124}Te enriched target using low energy cyclotrons, the primary reactions to consider are: ^{124}Te(p,n)^{124}I, ^{124}Te(p,2n)^{123}I, ^{125}Te(p,n)^{125}I, and ^{125}Te(p,2n)^{124}I. Given the difference in the half-lives of ^{123}I (13.2 h) and ^{125}I (59.4 days), a mixture of ^{123}I and ^{124}I will produce high purity ^{124}I upon storage of the crude product overnight. On the other hand, a mixture of ^{124}I and ^{125}I will give ^{124}I with lower purity with time, as the half-life ^{124}I decay is 15 times faster than ^{125}I decay. Consequently, it is critical to select an optimum proton beam energy to maximize the yield and purity of ^{124}I, i.e., the lowest amounts of ^{123}I and ^{125}I, as there is a possibility of competing reactions during proton irradiation.

The ^{123}I contaminant arising from the ^{124}Te(p,2n)^{123}I reaction may be minimized by reducing the incident proton energy. A decrease in energy from 13 MeV to 11 MeV results in a nearly three-fold decrease in the ^{124}I yield [81]. To minimize these impurities, the exit energy is controlled by varying the thickness of the target material or by degrading the incident proton beam energy using aluminum foils. For example, it is expected that the 16.5 MeV proton energy is degraded to 14.4, 13.1 MeV, and 12.0 MeV by using 320μm, 500μm, or 640μm aluminum foils, respectively [102]. It is critical to use an optimum thickness of the aluminum foil to ensure the highest yield and purity of ^{124}I produced. Lamparter and coworkers [86] demonstrated that the irradiation of the ^{124}Te solid target with a 10–15 μA proton beam degraded by a 320 μ foil resulted in an unfavorable ^{123}I/^{124}I ratio of 0.6–0.9. Introduction of a 640 μ thick foil produced ^{124}I with extremely high radionuclidic purity but with low yield. Using 500 μ foil and 10 and 12 μA beam current produced acceptable results, Under these conditions, up to 150 MBq (n = 12) of no-carrier added [^{124}I]NaI was produced after a 2 h irradiation time [86].

In general, the radionuclide produced from the proton bombardment of the target is dependent on the current intensity of the beam. However, there are certain limitations as to how much maximum current can be used for irradiation of the target which is dependent on the target material and the properties of radionuclide produced. For example, ^{124}I produced from the ^{124}Te(p,n)^{124}I reaction is directly proportional to the amount of current at which the target is irradiated. Various studies reported in the literature have used 8–29 μA beam current. Lamparter et al. [86]. reported a process for ^{124}I production using 10 and 12 uA proton beam irradiation for 2 h. However, the maximum current at which the ^{124}Te target can be irradiated is dictated by the thermal performance of the target material, i.e., in some cases melting of tellurium and the volatility of ^{124}I have been observed [82,103,104].

Due to the thermal stability of the target and volatility of ^{124}I, extensive and efficient cooling of the target material and the support plate is accomplished by using water for the back of the target and helium for the front of the target material [74,77,80]. Front water cooling has been also tried but was found unsuitable for a target system design where the target was perpendicular to the proton beam. Relatively high losses of ^{124}I, during extended irradiation period, to the cooling water directly in contact with the target were observed [77]. Computer simulation studies have been conducted to model heat transport during target irradiation [103].

3.3. Target Processing and Recovery of ^{124}I

A chemical separation technique was used for the recovery of $^{123/124}I$ from the irradiated ^{124}Te target initially [70]. However, separation of ^{124}I from irradiated solid ^{124}Te target, which is fabricated by melting method, is routinely accomplished by the dry distillation method. The method is straightforward and allows easy recycling of the target [50]. To ensure maximum recovery of the target material and extracting maximum ^{124}I, wide variation in setup parameters for distillation were reported, i.e., distillation time [50,74,78] and temperatures [74,87] being between 5 to 20 min and 670 to 820 °C, respectively. Similarly, a variation of carrier gases (Air [47,77,80], Argon [79], Helium [101], and Oxygen [74,78,93]) and their flow rates (5–80 mL/min) [50,74,79–81,87] were also reported in the literature for optimization of the method. Glaser et al. [78]. preferred an oxygen atmosphere for converting any tellurium, due to reduction, to TeO_2 for recovery of the target. Two types of traps have been used in the past that includes a 100–1000 μL solution of 0.001–0.1 N Sodium Hydroxide [77,80,87] or stainless steel [47,50], pyrex [74], or quartz [49,81] capillary tube coated with sodium hydroxide. To increase the surface area of the capillary tube, a platinum wire was loaded into it [55]. The adsorbed ^{124}I inside the capillary tube was washed with a weak buffer solution [74,81].

The IAEA (International Atomic Energy Agency) technical reports described two procedures of extraction of ^{124}I from the irradiated targets prepared by the two methods, melting technique and electroplating [97]. In the first procedure, the irradiated target was introduced into a quartz tube horizontally mounted in a cylindrical mini-furnace with carrier gas flow. The carrier gas flow and the power supply of both the furnace and the heating element around the narrow quartz tube were turned on. The iodine was vaporized at about 620 °C from the target and trapped downstream in a vial that contained 0.01 N NaOH. The distillation rate of ^{124}I from the ^{124}Te target was controlled by the diffusion of iodine from the target surface. Between 710 and 740 °C (MP TeO_2, 733 °C), an iodine vapor releases from the target. Therefore, 10 min after the start of the distillation, the furnace power supply was switched on and off so that the temperature oscillates between 700 and 740 °C. Periodic melting and solidification of the target resulted in a 98% recovery of the radioiodine and losses of TeO_2 were limited to less than 0.2% [97].

In the second method described in the IAEA report, [44,97] the irradiated electroplated target layer was dissolved in an oxidized alkaline medium containing NaOH, H_2O_2, and water followed by a reduction of an enriched target to metal by aluminum powder. After processing the mixture and removal of tellurium and aluminum hydroxide, the solution was filtered through a 0.45 μm glass filter and an in-line AG 50 WX8 cation-exchange (H^+ form, 100–200 mesh grade, 1 cm × 5 cm) column. When more than 5% of the iodine activity remained on the column, the latter was washed with 5 mL Milli-Q water. The eluate was collected into a pre-weighted serum vial. The overall yield of the chemical processing was higher than 95%.

3.4. A Fully-Automated Production Process for ^{124}I

Tremendous progress has been made in the recent past in the development of a fully automated process. A fully-automated process, developed by Lamparter et al. [86], involves three different steps: (1) the preparation of the target in a shuttle, (2) the irradiation of the target, and (3) the processing of the irradiated target shuttle using a Comecer ALCEO

Halogen system. The processing of the shuttle consists of two steps, (1) the extraction of ^{124}I out of the target, and (2) elution of the trapped ^{124}I into a product vessel. The Comecer ALCEO system consists of two different parts, the evaporation unit (EVP), which is used for the preparation of the target in the shuttle and processing of the irradiated shuttle, and the irradiation unit (PTS) with a supporting cooling unit. There is no intervention of an operator, during irradiation, target processing, and recovery of ^{124}I [86]. A schematic process diagram is given in Figure 4.

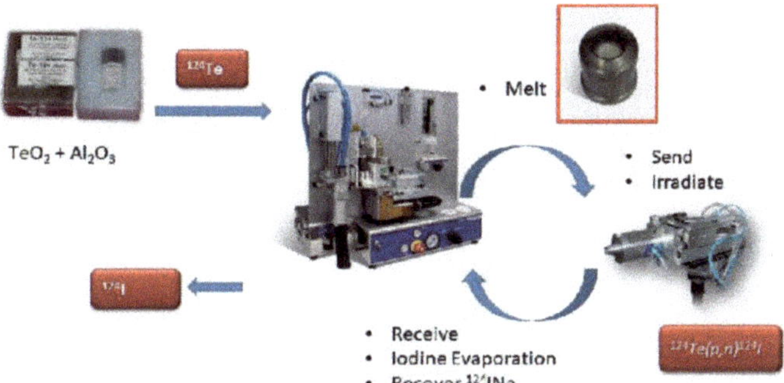

Figure 4. The schematic process diagram for the production of ^{124}I from $^{124}Te(p,n)^{124}I$ reaction using Comecer ALCEO halogen system (Courtesy Comecer S.p.A.).

In the automated process reported by Lamparter et al. [86], a solid target was prepared by mixing 300 mg enriched [^{124}Te]TeO_2 (99.93%) and 15 mg neutral alumina powder (Al_2O_3). The target material, $^{124}TeO_2/Al_2O_3$, was sintered into the shuttle as a 10 mm diameter circle with an estimated 4 mm^3 size melt. For the irradiation, the shuttle was transferred, fully automatically via a tube system, to the irradiation unit PTS, connected to a 16.5 MeV GE PETtrace cyclotron, while undercooling. The shuttle was irradiated using a 10 and 12 μA current for 2 h. A 500μ aluminum foil was used for an optimum thick target yield. The backside of the shuttle was cooled by water and the front was cooled by a constant Helium flow. After irradiation, the shuttle was transferred back to the ALCEO Halogen EVP module. ^{124}I was extracted by heating the shuttle to 740 °C for 10 min and trapping of the vaporized ^{124}I into a glass tube. The trapped ^{124}I was eluted in the form of [^{124}I]Iodide Sodium ([^{124}I]NaI) with 500 μL aqueous 0.05 N NaOH. The whole procedure, including evaporation and extraction of ^{124}I, was completed in 90 min. An extraction process of ^{124}I is shown in Figure 5.

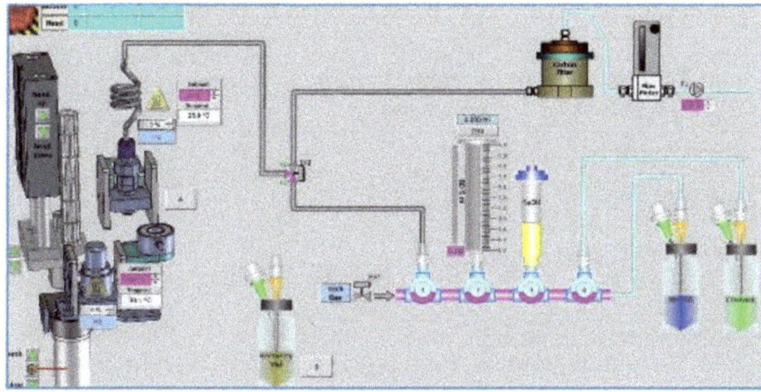

Figure 5. Recovery of ^{124}I from irradiated ^{124}Te target in a Comecer ALCEO halogen evaporation unit (EVP) module (courtesy of Comecer S.p. A.).

4. Overview of ^{124}I-Labeling Methods

Numerous methods for radioiodination of small and large biomolecules, i.e., mAbs, have been reported in the past [105,106]. Regardless of the application, a radioimmunoassay reagent for in vitro testing or in vivo use as a diagnostic or therapeutic agent, greater care and testing is required to maintain immunoreactivity of the biomolecule followed by radiolabeling and purification. Achieving high molar activity of the radioiodinated biomolecule remains very important due to the necessity to target very low concentrations of specific targets and to avoid non-specific binding.

4.1. Direct Labeling Methods

The basic radioiodination reactions are shown in Figure 6. The positive radioactive iodine species (I^+) generated in situ from the oxidation of radioiodide react with tyrosine and to some lesser extent to the histidine residues in the protein. Studies on the mechanism of the reaction of iodine with tyrosine and other phenols in stoichiometric amounts indicate that it is the phenolate anion which is radioiodinated. It is established that the primary site of the iodine addition is tyrosine amino acid residue in the large biomolecule; however, if the pH exceeds 8.5, the secondary site on the imidazole ring of histidine is favored. The oxidized I^+ electrophilic species hydrolyze rapidly in aqueous solution forming the hydrated iodonium ion, H_2OI^+, and/or hypoiodous acid, HOI. With tyrosine, the substitution of a hydrogen ion with the reactive iodonium ion occurs *ortho-* to the phenolic hydroxyl group. Mono and di-iodination of tyrosine residue are observed. With histidine, substitution occurs at the 2-position of the five-member imidazole ring. Following the desired reaction period, residual reactive I^+ species are reduced back to the I^- form and removed from the reaction solution by passage through either an anion exchange resin column or a gel filtration column. In this manner, high radiochemical purity can be achieved even if the labeling efficiency is low.

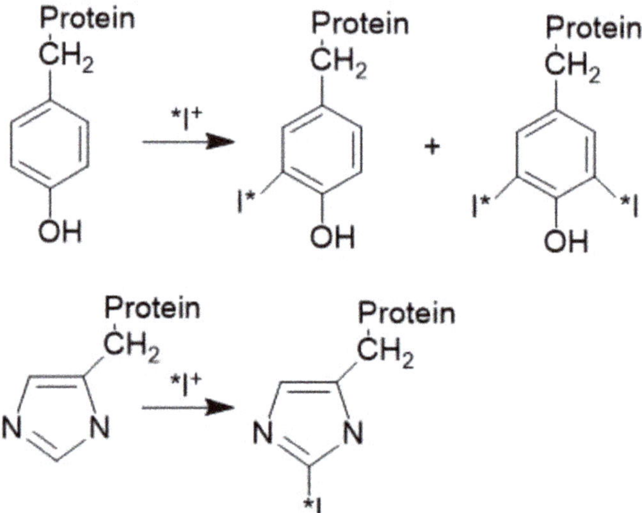

Figure 6. Radioiodination reactions of tyrosine and histidine residues in proteins.

4.1.1. Inorganic Oxidizing Agents Solution–Solution Phase Reactions

Numerous oxidizing reagents have been used for the direct radioiodination of proteins. Radioactive molecular Iodine was used as a labeling reagent in the early days of protein labeling. Since radioactive iodine is usually available as sodium iodide, Pressman and Keighly [36] used a mixture of ^{131}I and I_2 for the radioiodination of the protein. Later on, different oxidizing agents, e.g., sodium hypochlorite [107], nitrous acid [108,109] ammonium persulfate [110], hydrogen peroxide [111], ferric sulfate [112], and iodate [113] were used to generate radioactive molecular iodine before protein radiolabeling. Using molecular iodine as a radioiodination agent has some limitations, including: (1) the 50% maximum radiochemical yield and challenging purification. This is due to 50% conversion of iodide to iodine; (2) loss of radioactivity and increased exposure to the investigator due to volatility of molecular iodine; and (3) lower molar activity.

A technique using Iodine Monochloride (ICl), which eliminated the limitation of using molecular iodine, for protein radioiodination was developed [114–116]. The iodine-chlorine bond in ICl is polarized with a partial positive charge on the iodine, so the radiochemical yield is potentially 100%. The reagent was prepared by treating unlabeled ICl with radioactive sodium iodide. ICl, in the form of ICl_2^- is prepared from the reaction of sodium iodide and $NaIO_3$ in an acidic medium [117,118]. Studies on the iodination of phenol and substituted phenols with unlabeled ICl suggested a mechanism involving the electrophilic attack of iodide on the phenoxide ion followed by a slow loss of a proton. The electrophilic species has been suggested to be H_2OI^+ at low pH and ICl at higher pH or HOI [119,120].

More recently, a simple and rapid non-radioactive/radioactive iodide labeling method for peptides and proteins was developed [121,122]. In the method inorganic oxidizing agents, Hypochlorous acid/Hypochlorite and inorganic chloramines (NH_2Cl, $NHCl_2$, and NCl_3) were used to generate iodine monochloride in situ for radioiodination of a tyrosine residue in peptides and proteins. The radiolabeling yields were high with >99% radiochemical purity.

4.1.2. Organic Oxidizing Reagents for Solution–Solution Phase Reactions

The most widely used reagent for the radioiodination of peptides and proteins is chloramine-T, the sodium salt of N-monochloro-*p*-toluene-sulfonamide (Figure 7, Structure **1**), developed by Hunter and Greenwood [123,124]. In an aqueous solution, it forms HOCl, which is thought to be the actual oxidizing species [125]. This reacts with the radioactive iodide present to form an electrophilic iodine species, possibly H_2OI^+. On the other hand, the reaction of N-chloro derivatives with iodide was proposed via Cl^+ atom transfer or formation of an association complex to form ICl [126,127]. Only a few micrograms of Chloramine-T and a short reaction time are required to achieve nearly quantitative iodination of proteins as it is a very effective oxidizing agent. Longer reaction times may cause significant damage to the protein, including thiol oxidation, chlorination of aromatic rings and/or primary amines, and peptide bond cleavage [105]. In a typical radioiodination experiment, the protein solution is mixed with radioactive iodide in a slightly alkaline buffer (pH 7.5) and a freshly prepared solution of chloramine-T. The reaction mixture is incubated at room temperature for a specified time optimized for the reaction. At the end of the incubation period, a slight molar excess of reducing agent, sodium metabisulfite ($Na_2S_2O_5$), is immediately added to the mixture to reduce and inactivate the chloramine-T. It is important to note that the reducing agent, $Na_2S_2O_5$, used for reaction quenching can also cause cleavage of the disulfide bridges within the protein molecules and alter the tertiary structure of the protein.

N-chloro derivatives of secondary amines with lower oxidation potential, instead of chloramine-T, were used for radio iodination to reduce the oxidative damage to proteins [128]. For example, N-chloro-morpholine was found to produce higher radioiodination yields and less degradation than chloramine-T when reacting with L-tyrosine or leucine enkephalin (Leu-Gly-Gly-Phe-Leu) [129,130]. In situ preparation of the fresh reagent was required due to its instability. A water-soluble low oxidation potential reagent, Penta-*O*-acetyl-*N*-chloro-*N*-methylglucamine (NCMGE) (Figure 7, Structure **2**) was found to be stable, producing higher radiochemical yield, and less decomposition to model amino acids and small peptides than chloramine-T [131,132].

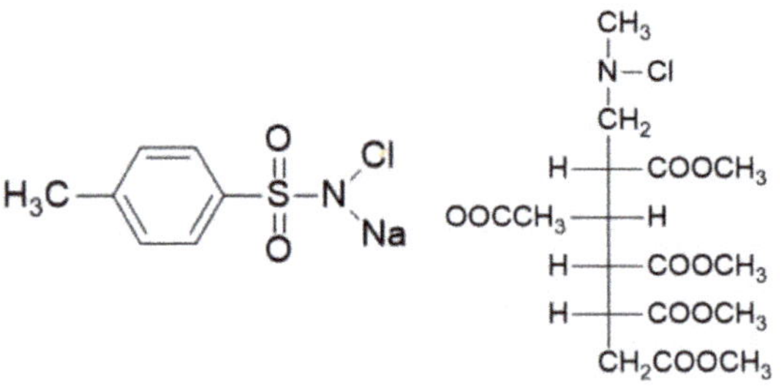

Figure 7. Structures of Chloramine-T (**1**) and Penta-*O*-acetyl-*N*-chloro-*N*-methylglucamine (NCMGE) (**2**).

4.1.3. Organic Oxidizing Reagents for Solid–Solution Phase Reactions

To minimize the oxidative damage to substrates caused by the chloramine-T/sodium metabisulfite method, a technique was developed by controlling the release of chloramine-T during the radio iodination reaction. This is accomplished by using the covalently attached chloramine-T to the surface of ~3 mm-diameter polystyrene beads with 0.55 ± 0.05 µmole/bead oxidizing capacity, known as IODO beads. (Figure 8, Structure **3**) These beads can be easily removed from the reaction mixture with a tweezer or by decanting the solution to stop the radio iodination reaction [133]. These beads are commercially available from Thermo Fisher Scientific. The technique has the following advantages: (1) the rates of radio iodination and oxidative damage of protein are slow as the reaction occurs on the solid surface rather than in solution; (2) the oxidative damage of the protein is low, but not eliminated; (3) reductive damage caused by metabisulfite is eliminated, as there is no need to stop the reaction chemically; and (4) the IODO beads are commercially available.

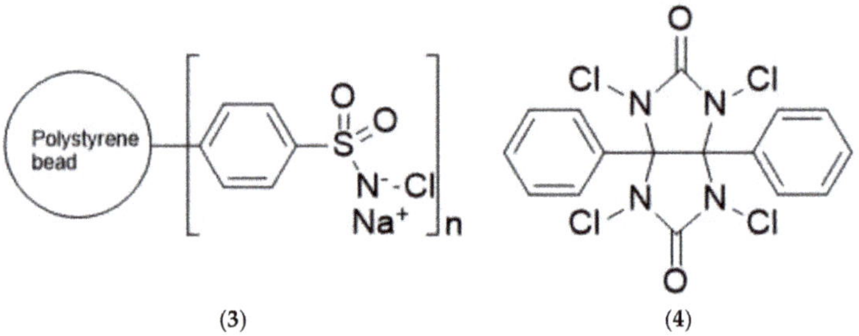

Figure 8. Structures of IODO beads (Chloramine-T attached to polystyrene bead, (**3**), and Iodogen (1,3,4,6-Tetrachloro-3α,6α-diphenyl-glycoluril, (**4**).

Radioiodination, using IODO beads, is accomplished by adding the buffered protein solution to a test tube containing IODO beads followed by the desired amount of sodium radioiodide solution. The rate of the reaction in the presence of IODO beads is slower than that with soluble chloramine-T and labeling efficiencies are somewhat lower for the same reasons. At the end of the incubation period, the IODO beads are removed and the reaction mixture is transferred to a gel column for purification.

A new reagent, 1,3,4,6-tetrachloro-3α, 6α-diphenyl glycoluril (Iodogen, Figure 8, Structure **4**), was introduced to iodinate proteins and to minimize their damage during radioiodination [134,135]. This reagent has several advantages: (1) Iodogen is virtually insoluble in aqueous media; the protein solution does not form a homogenous solution with the oxidizing agent; and (2) the radio iodination reaction could be stopped by simply removing the sample solution from the reaction tube, thus avoiding any use of reducing agent. Several proteins were radioiodinated by using the Iododen method [136] and the properties of the iodinated proteins were unaltered as confirmed by gel filtration, isoelectric focusing, and immunological reactivity. The stability of the labeled proteins during storage was good.

The Iodogen reagent is supplied by Thermo Fisher Scientific in the powder form. Iodogen-coated tubes can be prepared in advance by transferring aliquots of 20 μL (0.1 mg/mL concentration) Iodogen solution in methylene chloride into the suitable glass or polypropylene tubes. The tubes are allowed to dry under nitrogen at room temperature. Alternatively, the pre-coated Iodogen tubes, ready for single-use, can be purchased from Thermo Fisher. Each 12 × 75 mm tube is coated with ~50 μg of Iodogen at the bottom. The radioiodination procedure involves the same steps as for the IODO beads method. Variable rates of the radioiodination of proteins were observed depending on the solid surface on which Iodogen was coated [137]. For example, polypropylene test tubes resulted in the lowest rate of oxidation, followed by borosilicate glass, with polar soda-lime glass giving the highest rate of oxidation.

4.1.4. Enzyme Catalysts for Solution–Solution Phase Reactions

Some enzymes, e.g., peroxidases (lactoperoxidase, horseradish peroxidase, myeloperoxidase, and chloroperoxidase) are known for catalyzing the mild oxidation of iodide, in the presence of nanomolar concentrations of hydrogen peroxide (H_2O_2), for radioiodination of tyrosine, and to some extent histidine also, in proteins. The most extensively used enzyme is Lactoperoxidae for the radioiodination of proteins in the past. Hydrogen peroxide itself is capable of oxidizing radioiodide followed by radioiodination of proteins. Lactoperoxidase can also radioiodinate histidines in the proteins; however, the rate of iodination of histidines is much slower than the rate of iodination of tyrosines [138,139]. Lactoperoxidase is used as a catalyst, for peroxide oxidation of iodide, which permits extremely low H_2O_2 concentrations to be used [140]. In a typical radioiodination experiment, 2 to 10 μg protein are mixed with 1 to 10 mCi of radioiodide and 20 to 100 ng of lactoperoxidase. The reaction is initiated by the addition of 50 to 100 ng of H_2O_2 followed by the addition of 30 to 50 ng of H_2O_2 at 10 to 15 min intervals. After 30 to 60 min of incubation at room temperature, the reaction is quenched by the addition of cysteine or by dilution. Free iodide is removed by gel filtration or by other procedures. The rate of radioiodination is dependent on pH. Generally, a pH of 5.6 was found to be the optimal pH in most radiolabeling experiments. The immunological and biological properties of the original biomolecule are maintained as it is not exposed to strong oxidizing and reducing agents [141].

During radioiodination, the lactoperoxidase, containing 15 tyrosine and 14 histidine residues, is self-iodinated, leading to the loss of iodide and challenging separation. This problem was solved by attaching lactoperoxidase to Sephadex beads (Enzymobeads). In the labeling procedure, the Enzymobeads first were hydrated in distilled water for 2 to 4 h before use. 50 to 100 μg protein in 0.2 M phosphate buffer (50 μL, pH 7.2) was mixed with the Enzymobeads (25 μL suspension), 1 to 5 mCi of radioiodide, and 25 μL of 1% β-D-glucose. The radioiodination was allowed to proceed at room temperature for 15 to 25 min. The enzymobeads were removed by centrifugation or membrane filtration. The free iodide was removed by gel filtration or dialysis.

Bio-Rad Laboratories developed a novel commercial solid-state system, insoluble resin beads were covalently coated with a mixture of two enzymes: lactoperoxidase and glucose oxidase. When buffered solutions of protein and radioiodide were added to a suspension of the enzyme-coated beads in the presence of a small quantity of glucose, a chain of events was initiated: (1) the glucose oxidase enzyme used the glucose to generate a small amount of hydrogen peroxide at the surface of the beads; (2) the lactoperoxidase attached to the beads catalyzed the oxidation of iodide by the generated H_2O_2 in the solution; and (3) oxidized iodide radioiodinated the tyrosine residues in the protein. The reaction mixture was separated from the beads by decanting or centrifuging followed by loading onto a gel filtration column. Upon column elution, the desired labeled protein eluted first, the unreacted radioiodine was retained within the gel. BioRad, unfortunately, stopped supplying these beads in the early 1990s.

4.2. Indirect Labeling Methods

Sometimes it is not possible to radioiodinate proteins by direct electrophilic addition to tyrosine and histidine residues. This may be due to the fact that (1) a limited number of tyrosine and histidine residues may be present in the protein; (2) these may be buried within the tertiary structure of the protein and may not be readily available for radioiodination; and (3) these may be located at or near the active binding site of the molecule which cannot be disturbed. Consequently, several other labeling strategies have been developed to radioiodinate protein molecules at sites other than tyrosine and histidine.

A most common alternative approach is using a prosthetic group for radiolabeling of proteins. A prosthetic group for radioiodination contains an aromatic moiety, like tyrosine, which can be iodinated and covalently attached to the lysine moiety in the protein [142–144]. Two methods can be employed for the radioiodination of a protein. In the first method, the prosthetic group is radioiodinated, by using the methods given above, followed by coupling with the protein, thereby avoiding exposure of sensitive functionalities in the target molecule to oxidation. The coupling reaction must be efficient to avoid loss of radioiodide. In the second method, the prosthetic group is coupled with the protein followed by radioiodination using one of the methods given above. Overall radiolabeling efficiencies are lower with this approach for the simple reason that two separate labeling reactions are used. The main issue with this technique is to ensure the preservation of the immunological and biological properties of the protein. An early example of the use of prosthetic groups for radioiodination was the treatment of insulin with 4-[^{131}I]iodobenzenediazonium chloride [145].

The fact that some proteins are sensitive to oxidation and lack tyrosine for radioiodination prompted Bolton and Hunter to develop a reagent, *N*-hydroxysuccinimide Ester of 3-(4-Hydroxyphenyl) Propionic Acid [146] (Figure 9, structure **5**) which could be conjugated to a protein under milder conditions than those found in direct radioiodinations. Other reagents such as *p*-Hydroxybenzimidate (Wood's reagent) [147] (Figure 9, Structure **6**), *p*-Hydroxybenzaldehyde [148] (Figure 9, Structure **7**), and *p*-Hydroxybenzacetaldehyde [149] (Figure 9, Structure **8**) have also been studied as radioiodination reagents.

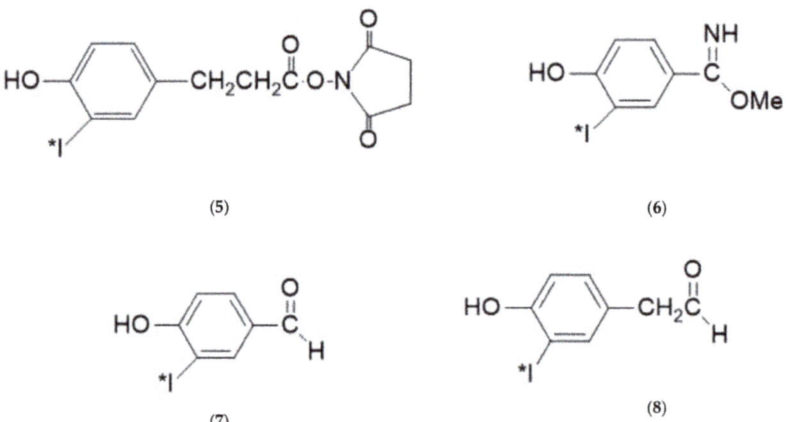

Figure 9. Structures of *N*-hydroxysuccinimide ester of 3-(4-Hydroxyphenyl) propionic acid (Bolton Hunter reagent, **5**), *p*-Hydroxybenzimidate (Wood's reagent, **6**), *p*-Hydroxy benzaldehyde (**7**), and *p*-Hydroxybenzaacetaldehyde (**8**).

5. Overview of ImmunoPET Imaging Pharmaceuticals for Cancer-Preclinical

PET imaging pharmaceuticals are routinely used for early detection of cancer and monitoring the progress of treatment following surgery, chemotherapy, and radiotherapy [150]. Numerous ^{124}I-labeled small molecules have been produced by nucleophilic and electrophilic substitution reactions and tested for various targets. Some of the ^{124}I-labeled PET imaging pharmaceuticals based on the small molecule (with the target given in the parenthesis) are ^{124}I-MIBG (adrenergic activity), ^{124}I-IAZA and ^{124}I-IAZG (hypoxia agent), ^{124}I-dRFIB, ^{124}I-IUdR and ^{124}I-CDK4/6 inhibitors (cell proliferation), ^{124}I-hypericin (protein-kinase C), ^{124}I-FIAU (herpes virus thymidine kinase), *m*-^{124}I-IPPM (opioid receptors), ^{124}I-IPQA ((EGFR kinase activity), ^{124}I-labeled-6-anilino-quinazoline (EGFR inhibitors), ^{124}I-purpurinimide (tumor imaging) [151].

As mentioned above, ImmunoPET is a paradigm-shifting molecular imaging modality that involves a combination of targeting specificity of mAbs and the high sensitivity of the PET imaging technique [14,152,153]. ImmunoPET imaging provides excellent specificity, sensitivity, and resolution in detecting primary tumors and is the method of choice for imaging specific tumor markers, immune cells, immune checkpoints, and inflammatory processes. Various ^{124}I-labeled antibodies, nanobodies, antibody fragments, and proteins have been used for molecular imaging of differentiated thyroid cancer, breast cancer, colorectal cancer, clear-cell renal cell carcinoma, ovarian cancer, and neuroblastoma, etc. Clinical feasibility of ^{124}I-labeled mAb (HMFGI) as an immunoPET imaging pharmaceutical, for quantitative measurement of distribution and blood flow in breast cancer patients by using ^{124}I and PET, was first demonstrated in 1991 [154]. Herein, we present an overview of the development strategies for target-specific ^{124}I-labeled ImmunoPET imaging pharmaceuticals and their preclinical and clinical applications over the past three decades.

5.1. *Receptor Tyrosine Kinase*

Receptor tyrosine kinases (RTKs) are high-affinity cell surface receptors which play an important role in a variety of cellular processes, including growth, motility, differentiation, and metabolism. RTKs are key regulators of normal cellular processes with a critical role in the development and progression of many types of cancers [155]. Approximately 20 different RTK classes have been identified, including the Epidermal Growth Factor Receptor (EGFR) family which includes HER1 (ErbB1), HER2 (Neu, ErB2), HER3(ErbB3), and HER4 (Erb4) and Vascular Endothelial Growth Factor (VEGFR). Two RTKs (EGFR and VEGFR) have been targeted the most for the development of immunoPET imaging pharmaceuticals.

5.1.1. Epidermal Growth Factor Receptor (EGFR)

The Epidermal Growth Factor Receptor, a transmembrane protein, is highly expressed in a variety of human cancers, including non-small-cell lung cancer (NSCLS). The overexpression of EGFR has been observed in both premalignant lesions and malignant tumors of the lung, in 40–80% patients with NSCLS, in 18–25% of all breast cancer carcinoma (specifically HER2 expression), and subsets of ovarian, lung, prostate and gastric cancers [156,157]. Breast cancers overexpressing HER2 have been associated with aggressive tumor growth, high relapse, poor prognosis, and being more resistant to endocrine therapy and chemotherapy. Consequently, substantial research has been conducted in the development of immunoPET imaging approaches and pharmaceuticals for the evaluation of the heterogeneous status of RTKs in cancers [158].

Trastuzumab (Herceptin), Cetuximab (Erbitux), Panitumumab (Vectibix), and Nimotuzumab (BioMAb) have been approved, recently, for the treatment of EGFR positive cancers by targeting the extracellular domain of EGFR. PET and SPECT techniques using radiolabeled antibodies, including trastuzumab, pertuzumab, and trastuzumab fragment, were able to detect HER2 expression; however, their large size resulted in slow tumor uptake and clearance from circulation [159,160].

Several ^{124}I-labeled mAbs have been investigated as potential immunoPET imaging pharmaceuticals for targeting a wide variety of tumors overexpressing the human EGFR. For example, ^{124}I-labeled ICR 12, a rat mAb recognizing the external domain of the human c-Erb B2 protooncogene, was evaluated as a potential PET imaging pharmaceutical for breast cancer patients. Biodistribution and imaging studies were performed in athymic mice bearing human breast carcinoma xenografts. Good tumor uptake (up to 12% ID/g at 120 h post-injection), with localization indices (3.4–6.2) was observed. Tumor xenografts of 6 mm diameter were successfully imaged with high resolution at 24, 48, and 120 h post-injection [161]. Two ^{124}I-labeled mAbs, MX35 and MH99, were also evaluated in nude rats bearing subcutaneous human SK-OV-7 and SK-OV-3 ovarian cancer xenografts. A melanoma cell line (SK-MEL-30) was used as a negative control tumor. Subcutaneous ovarian cancer nodules as small as 7 mm were identified with PET imaging. Tumor uptake was seen as high as six times to the normal tissue [162].

^{124}I-labeled C6.5 diabody, a small-engineered antibody fragment that is specific for the HER2 receptor tyrosine kinase, was investigated by using SCID mice bearing HER2-positive human ovarian carcinoma (SK-OV-3) xenografts. The diabody accumulated in SK-OV-3 tumors and blood at 48 h post-injection [163]. ^{124}I-labeled mAb, Trastuzumab, and a small 7-kDa scaffold protein, the affibody molecule, were evaluated and compared for the development of anti-HER2 targeting immunoPET imaging pharmaceuticals [164]. Both moieties were found to bind with HER2-expressing cells in vitro and xenografts in vivo. Total uptake of trastuzumab in tumors was higher than that of ^{124}I-labeled affibody. However, tumor-to-organ ratios were appreciably higher for ^{124}I-labeled affibody due to its more rapid clearance from blood and normal organs. A small-animal study was used to confirm ex vivo results. The study concluded that the use of the small scaffold targeting affibody provides better contrast in HER2 imaging than does the mAb.

A fragment of Trastuzumab (Fab) was modified via PASylation (PAS = Pro-Ala-Ser chain) for blood circulation optimization, radiolabeling with ^{89}Zr and ^{124}I, and their comparative performance assessment in CD1-*Foxn1*^{nu} mice bearing HER2-positive xenografts. The ^{89}Zr- and ^{124}I-labeled Fab-PAS$_{200}$ showed specific tumor uptakes of 11% ID/g and 2.3% ID/g 24 h post-injection with high contrast, respectively, with high tumor-to-blood (3.6 and 4.4) and tumor-to-muscle (20 and 43) ratios [165].

Several mouse mAbs have been screened in the past and it was found that mAb 806 specifically targets the overexpressed or activated forms of EGFR [166,167]. ch806, a chimeric form of mAb 806 which has been validated as an effective therapeutic antibody, showed specific accumulation of the antibody at multiple tumor sites and potential for molecular imaging. Biodistribution studies, in BALB/c nude mice bearing de2–7 EGFR-expressing xenografts, revealed that ^{125}I-labeled ch806 did not show significant tumor retention. However, specific and prolonged tumor localization of ^{111}In-labeled ch806 was demonstrated with the uptake of 31% ID/g and a tumor to blood ratio of 5:1 observed at 7 days post- injection [168].

The chimeric antibody, ch806, was conjugated with the residualizing ligand IMP-R4 for ^{124}I labeling, in vivo biodistribution, and small-animal PET imaging studies in BALB/c nude mice bearing U87MG.de2-7 glioma xenografts. The biodistribution data analysis showed 30.95 $\pm$ 6.01% (% ID/g) tumor uptake of ^{124}I-IMP-R4-ch806 injected dose at 48 h post-injection, with prolonged tumor retention (6.07 $\pm$ 0.80%I D/g at 216 h post-injection). The tumor-to-blood ratio increased from 0.44 at 4 h post-injection to a maximum of 4.70 at 168 h post-injection. PET images of ^{124}I-IMP-R4-ch806 were able to detect the U87MG.de2-7 tumors at 24 h post- injection and for at least 168 h post-injection [169]. Similarly, the mean uptake of ^{124}I-PEG4-tptddYddtpt-ch806 by U87MG.de2-7 glioma xenografts reached a maximum of 36.03% $\pm$5.08% ID/g at 72 h post injection. These studies suggest that the chimeric antibody, ch806, has potential for further studies [170].

5.1.2. Vascular Endothelial Growth Factor (VEGF)

Vascular endothelial growth factor (VEGF) is a signal protein produced by cells that stimulates the formation of blood vessels. They are important signaling proteins involved in both vasculogenesis (the de novo formation of the embryonic circulatory system) and angiogenesis (the growth of blood vessels from pre-existing vasculature). The VEGF and its receptor (VEGFR) have been shown to play major roles not only in physiological but also in most pathological angiogenesis, such as cancer [171]. Several therapeutic agents targeting VEGF (e.g., bevacizumab and ramucirumab) and VEGFR (e.g., sorafenib and sunitinib) have been approved for clinical use around the world [172]. Clinical immunoPET studies using ^{89}Zr-labeled bevacizumab were performed in a variety of tumors, including breast cancer [173], neuroendocrine tumors [174], renal cell carcinoma (RCC) [175], NSCLC [176], and glioma [177,178].

An IgG1 monoclonal antibody, VG76e, that binds to human VEGF, was labeled with ^{124}I (i.e., [^{124}I]-SHPPVG76e) and was investigated in the HT1080 human fibrosarcoma xenografts in immune-deficient mice for VEGF-specific localization. A single intravenous injection of [^{124}I]-SHPPVG76e into tumor-bearing mice showed a time-dependent and specific localization of the tracer to the tumor tissue. High tumor-to-background contrast and distribution of [^{124}I]-SHPPVG76e in the major organs were seen in the whole-body animal PET imaging studies. These studies support further development of [^{124}I]-SHPP-VG76e as an immunoPET imaging pharmaceutical for measuring tumor levels of VEGF in humans [179].

5.2. Clusters of Differentiation

The clusters of differentiation (CD) antigens are cell-surface receptors involved in cellular functions like activation, adhesion, and inhibition. These receptors express elevated levels of the CD on cells which can serve as key markers in several cancers and infectious diseases. CD markers are mostly useful for classifying white blood cells (WBC) and especially important for the diagnosis of lymphomas and leukemias. The CD nomenclature was proposed and established a long time ago. Since then, its use has expanded to many other cell types, and more than 320 CD unique clusters and sub clusters have been identified.

Several CD antigens have been investigated as diagnostics or therapeutics targets in the past. For example, CD20 and CD30 are common biomarkers for lymphoma imaging [180–182] and the Food and Drug Administration approved a CD20-specific chimeric mAb, Rituximab, for the treatment of non-Hodgkin's lymphoma (NHL) and rheumatoid arthritis (RA). Feasibility studies, related to ^{64}Cu-DOTA-rituximab and ^{89}Zr-labeled rituximab as immunoPET imaging pharmaceutical for CD20 expression in NHL-bearing humanized mouse models and translating later into the clinic, were reported followed by the translation of ^{89}Zr-labeled Rituximab into the clinic [183–185].

5.2.1. Cluster of Differentiation 20 (CD 20)

Tumor targeting of anti-CD20 diabodies (scFv dimers) for detection of low-grade B-cell lymphoma was investigated. The scFv-8 and Cys-Db were labeled with ^{124}I and ^{131}I for PET imaging and biodistribution, respectively, at 2, 4, 10, and 20 h. Mice bearing 38C13-huCD20 (positive) and wild-type 38C13 (negative) tumors were used. Both ^{124}I-labeled scFv-8 and Cys-Db exhibited similar tumor targeting at 8 h post-injection, with significantly higher uptakes than in control tumors. At 20 h, less than 1% ID/g of ^{131}I-labeled Cys-Db was present in tumors and tissues [186]. Two recombinant anti-CD20 rituximab fragments, a minibody, Mb (scFv-C_H3 dimer; 80 kDa) and a modified scFv-Fc fragment (105 kDa), designed to clear rapidly, were produced and labeled with ^{64}Cu and ^{124}I. Rapid and specific localization to CD20-positive tumors was observed with both radioiodinated fragments producing high-contrast images in vivo. The ^{124}I-labeled mini body showed higher uptake in CD-20 positive tumors than scFv-Fc [187].

In yet another similar report, cys-diabody (cDb) and cys-mini body (cMb) based on rituximab and obinutuzumab (GA101) were labeled with ^{124}I and used to target the CD20 antigen in transgenic mice and a CD20-expressing murine lymphoma model. Obinutuzumab-based imaging pharmaceuticals (^{124}I-GAcDb and ^{124}I-GAcMb) produced high-contrast immunoPET images of B-cell lymphoma and outperformed the respective rituximab-based tracers [188].

5.2.2. Cluster of Differentiation 274 (CD274)

The Cluster of differentiation 274, CD274, or Programmed Death-Ligand 1 (PD-L1), is a protein (40 kD type 1 transmembrane protein) that in humans is encoded by the CD274 gene. Upregulation of PD-L1 may allow cancers to evade the host immune system. An analysis of 196 tumor specimens from patients with renal cell carcinoma found that high tumor expression of PD-L1 was associated with increased tumor aggressiveness. Many PD-L1 inhibitors, durvalumab, pembrolizumab, atezolizumab, and avelumab, are in development as immuno-oncology therapies and are showing good results in clinical trials.

A novel heavy-chain antibody (HCAb) was constructed and labeled with ^{124}I to target the programmed cell death ligand-1 (hPD-L1) which is known to activate T cells associated with malignancies. Biodistribution studies in osteosarcoma OS-732 tumor-bearing mouse model showed a tumor uptake of $4.43 \pm 0.33\%$ ID/g at 24 h. Tumor lesions were detected on micro PET/CT 24 h post-injection [189]. In continuation for development of ^{124}I-labeled imaging pharmaceuticals, JS001 (Toripalimab, a humanized IgG mAb) was investigated for targeting human PD-L1 (hPD-L1) in a tumor mouse model [190].

5.3. *Carbohydrate Antigen*

Carbohydrate antigen 19-9 (CA19-9), also known as sialyl-Lewis A, is a tetrasaccharide that is usually attached to O-glycans on the surface of cells. It is known to play a vital role in cell-to-cell recognition processes [191] and as an established biomarker for several cancers, including, lung, breast, and PDAC (Pancreatic Ductal Adenocarcinoma). CA19-9 is the most highly expressed tumor antigen, present on cellular membrane proteins in more than 90% of pancreas cancer patients [192]. 5B1, a fully human IgG monoclonal antibody, is a known anti-CA 19-9 antibody that has been used as a theranostic agent [193–195]. For example, in a first-in-human clinical trial, ^{89}Zr-labeled 5B1 was used for immunoPET imaging of detected known PDACs, metastases [196].

On the contrary, Girgis and coworkers created several antibodies and diabodies for targeting CA 19-9 antigens expressed by pancreas cancer. An anti-CA19-9 monoclonal antibody and a cys-diabody, created by engineering C-terminal cysteine residue into the DNA single-chain Fv construct of CA19-9, were labeled with ^{124}I and injected into mice harboring CA19-9 antigen-positive and CA19-9 negative xenografts. MicroPET/CT imaging was performed at 72, 96, and 120 h post-injection. The average tumor to blood (% ID/g) ratio was 5.0 and 3.0 and the average positive tumor to negative tumor (% ID/g) ratio was 20.0 and 7.4 for mAb and cys-diabody, respectively [197,198]. Another diabody (~55 kDa) construct, which was created by isolation of variable region genes of the intact anti-CA 19-9 antibody, was created, ^{124}I-labeled, and tested in mice harboring an antigen-positive (BxPC3 or Capan-2) and a negative xenograft (MiaPaca-2). Pancreas xenograft imaging of BxPC3/MiaPaca-2 and Capan-2/MiaPaca-2 models with the anti-CA19-9 diabody demonstrated an average tumor: blood ratio of 5.0 and 2.0, respectively, and an average positive: negative tumor ratio of 11 and 6, respectively [199].

An Fc-mutated, anti-CA19-9 antibody fragment, scFv-Fc H310A, 105 kD dimer, was created for microPET imaging of pancreatic cancer xenografts. The ^{124}I-scFv-Fc H310A localized to the antigen-positive tumor xenografts and confirmed by microPET imaging. Higher % ID/g in the antigen-positive tumor compared to the blood, antigen-negative tumor, and liver was observed [200]. ^{124}I-JAA-F11 was investigated to target Thomsen–Friedenreich antigen (TF-Ag), a mucin-type disaccharide galactose-b1-3*N*-acetylgalactosamine conjugated to proteins by an alpha-*O*-serine or *O*-threonine linkage, which is found on human carcinomas of many types including those of the breast, colon, bladder, and prostate [201].

5.4. Carcinoembryonic Antigen (CEA)

Carcinoembryonic antigen (CEA), a protein, is normally present at very low levels in the adult blood but may be elevated with certain types of cancers. CEA serves as a vital tumor antigen and a serum tumor marker [202]. The normal CEA range in adult's blood is <2.5 ng/mL (non-smoker) and <5.0 ng/mL (smoker) and is elevated in cancer patients. The most common cancers that show elevated CEA levels are colon, rectum, and ovarian. Arcitumomab (CEAScan) is a ^{99m}Tc-labeled hapten peptide pre targeted imaging probe approved by the FDA and EMA (European Medicine Agency) for detecting colonic cancer metastases [203].

A series of antibody fragments were engineered from a murine mAb, T84.66, and have been radiolabeled with ^{64}Cu and ^{124}I [204,205]. The ^{124}I-labeled anti-CEA T84.66 mini body (single-chain Fv fragment [scFv]-C(H)$_3$ dimer, 80 kDa) and diabody (noncovalent dimer of scFv, 55 kDa) were evaluated in the athymic mouse/LS174T xenograft model PET images, 18 h post injection, using mini body and diabody, showed specific binding to the CEA-positive xenografts and relatively low activity in normal tissues. Target to background (T/B) ratios were 3.05, 3.95 and 11.03, and 10.93 at 4 and 8 h post injection for mini and diabody, respectively [205]. To improve the T/B ratio of the engineered antibody fragments, mutation of the residues in the Fc fragment was performed. A series of anti-CEA scFv-Fc fragments were evaluated for tumor localization and pharmacokinetics [206,207] in LS174T xenografted athymic mice by small-animal PET. The PET imaging with a ^{124}I-labeled scFv-Fc with double mutation (H310A/H435Q) quickly localized to the tumor site, rapidly cleared from animal circulation, and produced clear images [207].

A pretargeting technique for targeting CEA expressing tumors was developed [208,209]. The pretargeting technique uses a bispecific monoclonal antibody bs-mAb (a multivalent, recombinant anti-CEA, carcinoembryonic antigen/anti-HSG histamine-succinyl-glycine fusion protein) with the affinity for a tumor and a small hapten peptide. Typically, mice are implanted with CEA-expressing LS174T human colonic tumors, a bispecific monoclonal anti-CEA/anti-HSG/anti-hapten antibody is given to the mice, followed by an administration of a radiolabeled hapten peptide. A new peptide, IMP-325, In-DOTA-D-Tyr-D-Lys(HSG)-D-Glu-D-Lys(HSG)-NH$_2$, was labeled with ^{124}I and tested in nude mice bearing LS174T human colonic tumors that were given anti-CEA/anti-HSG bs-mAb. The ^{124}I-IMP-325 alone cleared quickly from the blood with no evidence of tumor targeting, but when pretargeted with the bs-mAb, tumor uptake increased 70-fold, with efficient and rapid clearance from normal tissues, allowing clear visualization of the tumor within 122 h [210].

5.5. Carbonic Anhydrase IX

Carbonic Anhydrase IX is a transmembrane protein that is overexpressed in clear cell renal cell carcinoma (ccRCC) and carcinomas of the uterine cervix, kidney, esophagus, lung, breast, colon, brain, and hypoxic solid tumors. Its overexpression in cancerous tissues compared to normal ones is due to hypoxic conditions in the tumor microenvironment. Consequently, it is a cellular biomarker of hypoxia [211].

A chimeric mAb, cG250, subclass IgG1, was reported in 1986 to recognize an antigen which preferentially expresses on cell membranes of renal cell carcinoma (RCC). Since that time, G250 has been shown to localize in primary (98%) and metastatic (88%) ccRCC lesions found on human histologic slides under light microscopy [212]. Oxygen tension measurements were used to investigate hypoxia and carbonic anhydrase IX expression, tumor uptake, and biodistribution, in a renal cell carcinoma SK-RC-52 xenograft model using ^{124}I-labeled cG250 and PET/CT. Oxygen tension was found to be significantly higher in normal tissues than in the xenograft tumor. Biodistribution studies of ^{124}I-cG250 demonstrated isotope uptake in the xenografts peaking at 23.45 $\pm$ 5.07% ID/g at 48 h post-injection [213]. ^{89}Zr-labeled, an alternative to ^{124}I, cG250 was evaluated in ccRCC xenograft models in mice. Greater uptake, retention, and superior PET images for ^{89}Zr-labeled cG250, due to trapping inside the tumor cell, compared to ^{124}I–labeled cG250, due to internalization and release of ^{124}I, were observed [214,215].

5.6. *Glycoproteins*

5.6.1. Glycoprotein A33

The glycoprotein A33, GPA 33, is a transmembrane glycoprotein with homology to the immunoglobulin superfamily. This antigen is expressed in >95% of colorectal cancer and a subset of gastric and pancreatic cancers [216]. It contains three distinct structural domains: a 213 amino acid extracellular region containing two immunoglobulin-like domains, a 23 amino acid hydrophobic transmembrane domain, and a highly polar 62 amino acid intracellular tail containing four consecutive cysteine residues. ^{125}I- and ^{131}I-labeled murine mAb have been investigated as SPECT and Radioimmunotherapy agents, respectively, in phase I/II clinical trials [217,218].

A recombinant humanized anti-colorectal cancer A33 antibody, huA33, was labeled with ^{124}I and used for biodistribution properties and PET imaging characteristics in SW1222 colorectal xenograft bearing BALB/c nude mice. Excellent tumor uptake, with a maximum of 50.0 $\pm$ 7.0% ID/g at 4 days post injection, was observed [219].

5.6.2. Glycoprotein CD44v6

When the CD44 gene is expressed, its pre-messenger RNA (mRNA) can be alternatively spliced into mature mRNAs that encode several CD44 isoforms. The mRNA assembles with ten standard exons, and the sixth variant exon encodes CD44v6, which engages in a variety of biological processes, including cell growth, apoptosis, migration, and angiogenesis Overexpression of the mature mRNA encoding CD44v6 can induce cancer progression. For example, CD44v6 assists in colorectal cancer stem cells in colonization, invasion, and metastasis [220]. CD44v6 is also expressed in thyroid carcinoma on the outer cell surface of squamous-cell carcinomas of head-and-cancer [221]. U36, an anti-CD44v6 chimeric (mouse/human) monoclonal antibody (cmAb), was found to target CD44v6 antigen. Biodistribution and scintigraphy studies in nude mice bearing tumors from the HNX-OE human head and neck tumor cell line were conducted. Co-injection of ^{124}I-cMAb U36 and ^{131}I-cMAb U36 provided similar tissue uptake values. Selective tumor uptake was confirmed with PET imaging at 24, 48, and 72 h post injection, which detected 15 out of 15 tumors [222]. A comparative biodistribution study of ^{89}Zr- and ^{124}I-labeled head and neck squamous cell carcinoma (HNSCC)-selective cMAb U36 versus ^{88}Y-, ^{131}I-, and ^{186}Re-labeled cMAb U36 conjugates was conducted in HNSCC xenograft bearing mice at 24, 48, and 72 h post injection. Tumor uptake was higher for the ^{89}Zr- and ^{88}Y-labeled cMAb U36 than the ^{124}I-, ^{131}I-, and ^{186}Re-labeled cMAb U36 [223]. Biodistribution and PET imaging studies of ^{124}I-cMAb U36 nude mice bearing KAT-4 tumors in the left flank and the right front leg were performed. ^{124}I-cmAb U36 uptake (%ID/g) in the flank tumors was 8.2 $\pm$ 3.6, 13.7 $\pm$ 0.7, 21.8 $\pm$ 2.8%, and 12.8 $\pm$ 5.2 at 4, 24, 48, and 72 h post injection, respectively. The tumors were visible in PET images at all-time points with the highest uptakes at 48 h post injection [224].

5.7. *Prostate-Specific Membrane Antigen (PSMA)*

Prostate cancer (PCa) is the most common cancer in men [225]; therefore, early detection of primary disease and its metastases is critical for clinical staging, prognosis, and therapy management. The prostate-specific membrane antigen (PSMA) is a transmembrane glycoprotein that is significantly over-expressed in most early-stage prostate cancer cells compared to benign prostatic tissues. Consequently, it has gained significant interest as a target for imaging and therapy in the past five years [226,227].

Capromab pendetide (ProstaScint®) is the murine mAb, 7E11-C5.3, conjugated to the DTPA chelator. The 7E11-C5.3 antibody is of the IgG1, kappa subclass (IgG1κ). This antibody is directed against Prostate-Specific Membrane Antigen (PSMA). ^{111}In-labeled

Capromab pendetide is approved by the FDA for prostate cancer imaging in newly-diagnosed patients by biopsy. ^{124}I-labeled Carpromab was proposed as a PET imaging pharmaceuticals to decrease the retention of radioactivity in healthy organs, due to the non-residualizing properties of the radiolabel. Carpromab was radioiodinated and its targeting properties were compared with the ^{111}In-labeled counterparts in LNCaP xenografts. PSMA-negative xenografts (PC3) were used as the negative control. Biodistribution of ^{125}I/^{111}In-capromab showed more rapid clearance of iodine radioactivity from liver, spleen, kidneys, bones, colon tissue, as well as tumors. Maximum tumor uptake (13 ± 8% ID/g for iodine and 29 ± 9% ID/g for indium) and tumor-to-non-tumor ratios for both agents were measured at 5 days post-injection. High tumor accumulation and low uptake of radioactivity in normal organs were confirmed using micro PET/CT at 5 days post-injection of ^{124}I-capromab. Although tumor uptake was relatively lower for the ^{124}I-Capromab than ^{111}In-Capromab in LNCaP xenografts (13 ± 8% vs. 29 ± 9% ID/g), it showed lower uptake in normal organs compared to its ^{111}In counterpart [228]. More recently, Frigerio et al. demonstrated the targeting specificity and sensitivity of ^{124}I-labeled anti-PSMA single-chain variable fragment (scFv) in a preclinical in vivo model. The uptake of ^{124}I-scFv was found to be very high and specific for PSMA-positive cells [229]. J591, a humanized mAb that binds to an extracellular domain of PSMA, has been investigated for both imaging and therapy [230–234]. It was demonstrated recently that ^{124}I- and ^{89}Zr-labeled J591 had comparable surface binding and internalization rates in preclinical prostate models [235]. These studies imply that PCa theranostics using ^{177}Lu- and ^{124}I- or ^{89}Zr- labeled J591 is feasible, safe, and may have superior targeting toward bone lesions relative to conventional imaging modalities.

5.8. Prostate Stem Cell Antigen (PSCA)

Prostate stem cell antigen (PSCA) is a protein that in humans is encoded by the PSCA gene. This gene encodes a glycosylphosphatidylinositol-anchored cell membrane glycoprotein. PSCA is expressed in 83%–100% of prostate cancers [236–240]. It is also highly expressed in most prostate cancer bone metastases (87–100%) and the local bladder, pancreatic carcinoma, bladder, placenta, colon, kidney, and stomach cancers [241–244].

The anti-PSCA murine mAb 1G8 showed anti-tumor activity [245]. An ^{124}I-labeled 2B3 anti-PSCA minibody, a hu1G8 minibody fragment dimer of scFvs-C_H3 with an 18 amino acids linker and ~80 kDa molecular weight, was evaluated in mice bearing LAPC-9 (PSCA-positive) and PC-3 (PSCA-negative) xenografts. Micro PET imaging of the PSCA positive tumors showed ^{124}I-2B3 minibody to target and image PSCA-expressing xenografts with high contrast at earlier time points than the ^{124}I-labeled intact hu1G8 anti-PSCA mAb. This was due to faster clearance of the minibody than the anti-PSCA mAb [246,247]. The parental 2B3 diabody (p2B3-Db) (molecular weight, 55 kDa) was back mutated with a linker of 8 amino acids to produce a high-affinity diabody (bm2B3-Db8). ^{124}I-p2B3-Db8 and bm2B3-Db8 were evaluated in bio-distribution and for tumor imaging studies in nude mice bearing xenografts of the LAPC-9 and PC-3 (PSCA-negative) tumor cell lines. The uptake of ^{124}I-p2B3-Db8 and ^{124}I-bm2B3-Db8 in PSCA-positive tumors was lower than that of ^{124}I-2B3 minibody in the same tumor model. PET imaging with ^{124}I-bm2B3-Db8 visualized the LAPC-9 tumor as early as 4 h post injection with a higher contrast at 12 h post injection [248]. Subsequent affinity maturation of the 2B3 minibody created the A11 anti-PSCA minibody, which showed improved immunoPET performance [249].

^{124}I- and ^{89}Zr-labeled anti-PSCA A11 minibodies (scFv-CH3 dimer, 80 kDa) were developed and evaluated for quantitative immunoPET imaging of prostate cancer in 22Rv1-PSCA or LAPC-9 xenograft bearing mice. The non-residualizing ^{124}I-labeled minibody had lower tumor uptake (3.62 ± 1.18% ID/g 22Rv1-PSCA, 3.63 ± 0.59% ID/g LAPC-9) than the residualizing ^{89}Zr-labeled minibody (7.87 ± 0.52% ID/g 22Rv1-PSCA, 9.33 ± 0.87% ID/g LAPC-9. However, the ^{124}I-labeled minibody achieved higher imaging contrast because of lower nonspecific uptake and better tumor-to-soft-tissue ratios [250]. In another study, ^{124}I-labeled A11 minibody immunoPET imaging was compared with ^{18}F-Fluoride bone scans for detecting prostate cancer bone tumors in osteoblastic, PSCA-expressing, and LAPC-9 intratibial xenografts. The ^{124}I-labeled A11 minibody demonstrated superior sensitivity and specificity over the ^{18}F-Fluoride bone scans in detecting the xenografts at all-time points [251].

A11 cMb was conjugated with the near-infrared fluorescence (NIRF) dye Cy5.5 and radiolabeled with ^{124}I or ^{89}Zr for evaluation as an immunoPET/fluorescence imaging agent to improve intraoperative prostate cancer margin visualization. ImmunoPET imaging using dual-labeled ^{124}I-A11 cMb-Cy5.5 showed specific targeting to both 22Rv1-PSCA and PC3-PSCA. xenografts in nude mice. Similarly, fluorescence imaging showed a strong signal from both 22Rv1-PSCA and PC3-PSCA tumors compared with non-PSCA expressing tumors [252]. Another dual probe, A2 cys-diabody (A2cDb)-IR800, targeting PSCA was labeled with ^{124}I (^{124}I-A2cDb-IR800) and evaluated in a prostate cancer xenograft model. Dual-modality imaging using the anti-PSCA cys-diabody resulted in high-contrast immuno-PET/NIRF images [253].

5.9. Other Biomarkers

5.9.1. Extra Domain-B (ED-B) of Fibronectin

The extracellular matrix protein fibronectin contains a domain, the extra domain B (ED-B) of fibronectin (~80 kDa molecular weight), that is rarely found in healthy adults and is almost exclusively expressed by newly formed blood vessels in tumors, i.e., angiogenesis and different types of lymphoma and leukemias.

The human mAb fragment L19-SIP ((Radretumab) is directed against extra domain B (ED-B) of fibronectin. ^{124}I-L19-SIP immunoPET was used to demonstrate its suitability for imaging of angiogenesis at early-stage tumor development and as a scouting procedure before clinical ^{131}I-L19-SIP radioimmunotherapy. Tumor uptake, in FaDu xenograft-bearing nude mice, was 7.3 ± 2.1, 10.8 ± 1.5, 7.8 ± 1.4, 5.3 ± 0.6, and $3.1 \pm 0.4\%$ ID/g at 3, 6, 24, 48, and 72 h post injection [254]. ImmunoPET imaging with ^{124}I-labeled L19SIP was used to predict doses delivered to tumor lesions and healthy organs by subsequent Radretumab RIT in patients with brain metastases from solid cancer. Although the fraction of injected activity in normal organs was similar in different patients, the antibody uptake in the neoplastic lesions varied by as much as a factor of 60 [255].

5.9.2. Phosphatidylserine

Phosphatidylserine (PS) is a marker normally absent that becomes exposed on tumor cells and tumor vasculature in response to oxidative stress in cancer cells (lung, breast, pancreatic, bladder, skin, brain metastasis, rectal adenocarcinoma, etc.) but not on the normal cells. ^{124}I-labeled PGN650, an $F(ab')_2$ antibody fragment, was evaluated as a biomarker of the tumor microenvironment. Pharmacokinetics, tumor uptake, and radiation dosimetry in cancer patients were assessed. Apart from the tumor, the liver was found to receive a high radiation dose [256]. Annexin-V, a calcium-dependent protein that binds with high specificity to phosphatidylserine exposed during apoptosis, was labeled with ^{124}I for use as a potential PET probe. The biological activity of radiolabeled Annexin-V was tested in control and camptothecin-treated (i.e., apoptotic) human leukemic HL60 cells. A significantly high binding (21%) was observed [257].

5.9.3. Placental Alkaline Phosphatase (PLAP)

Placental alkaline phosphatase (PLAP), also known as an allosteric enzyme that in humans is encoded by the ALPP gene. PLAP is a tumor marker, especially in seminoma and ovarian cancer (e.g., dysgerminoma). The ^{124}I-labeled murine mAb H17E2, detecting placental alkaline phosphatase (PLAP), was administered by intraperitoneal injection into nude mice bearing subcutaneous HEp2 human tumor xenografts (a PLAP expressing cell-line). Activity in tumor rose to 4.26% injected dose by 48 h post injection and remained at this level until day 7 post injection, giving a tumor: blood ratio of 0.78 at this time [258].

6. Overview of ImmunoPET Imaging Pharmaceuticals for Cancer–Clinical

6.1. Receptor Tyrosine Kinase

Trastuzumab

^{124}I-labeled trastuzumab was evaluated in animals and humans for its application as a potential PET tracer [259]. MicroPET imaging and biodistribution of ^{124}I-trastuzumab were performed to examine its specificity in HER2-positive and negative mouse models. Higher tumor uptake of ^{124}I-trastuzumab than ^{124}I-IgG1 in HER2-positive PDX mouse models at 24 h was seen. The low tumor uptake of ^{124}I-trastuzumab in HER2-negative PDX models further confirmed the specificity. ^{124}I-trastuzumab was evaluated for its distribution, internal dosimetry, and initial PET imaging of HER2-positive lesions in gastric cancer (GC) patients. PET/CT images of six gastric cancer patients with metastases were compared using ^{124}I-trastuzumab and [^{18}F]FDG PET/CT. 18 HER2-positive lesions and 11 HER2-negative lesions were evaluated in PET imaging analysis. The detection sensitivity of ^{124}I-trastuzumab was 100% (18/18) at 24 h post injection. The PET images showed a significant difference in tumor uptake between HER2-positive and HER2-negative lesions at 24 h post injection. Higher specificity of ^{124}I-trastuzumab than [^{18}F]FDG was observed.

6.2. Glycoproteins

Glycoproteins A33 (huA33)

In a clinical study, ^{124}I-labeled hu A33 was injected intravenously to 15 patients with colorectal cancer to examine the quantitative features of antibody–antigen interactions in tumors and normal tissues. PET/CT studies showed significant antibody targeting in tumors and normal bowel. There was a linear correlation between the amount of bound antibody and antigen concentration [260]. Targeting, biodistribution, and safety of ^{124}I-labeled huA33 in patients with colorectal cancer were evaluated using quantitative PET. Additionally, biodistribution was also determined when a large dose of human intravenous IgG (IVIG) was administered to manipulate the Fc receptor or when ^{124}I-labeled huA33 was given via hepatic arterial infusion (HAI). Ten of 12 primary tumors in 11 patients (0.016% ID/g in tumors vs. 0.004% ID/g in normal tissues) were visualized. The HAI route had no advantage over the intravenous route [261]. A novel, nonlinear compartmental model using PET-derived data from 11 patients was developed. The objective of the study was to determine the "best-fit" parameters and model-derived quantities for optimizing biodistribution of intravenously injected ^{124}I-labeled A33. Excellent agreement between fitted and measured parameters of tumor uptake was observed [262]. Red marrow activity concentration and the self-dose component of absorbed radiation dose to red marrow were estimated based on PET/CT results of ^{124}I-labeled cG250 and huA33. The red marrow-to-plasma activity concentration (RMPR) values were found to be patient-dependent and increase over a 7-day timescale for both the antibodies, indicating that individualized image-based dosimetry is required for optimal therapeutic delivery of radiolabeled antibodies [263].

6.3. Carbonic Anhydrase IX

cG250

^{124}I-labeled cG250 (Girentuximab) was investigated for PET assessment to predict clear-cell renal carcinoma in cancer patients. Twenty-six patients with renal masses who were scheduled to undergo surgical resection were given a single intravenous infusion of ^{124}I-cG250. 15 of 16 clear-cell carcinomas were identified accurately by antibody PET. The sensitivity of ^{124}I-cG250 PET for clear-cell kidney carcinoma in this trial was 94% [264,265]. Additional clinical studies involving 195 patients validated the safety and superior diagnostic value of ^{124}I-cG250 in ccRCC with an average sensitivity and specificity of 86.2% and 85.9%, respectively [266].

Multimodal imaging technique development study, using ^{124}I-cG250, concluded that it could realize precise intraoperative localization of ccRCC. This could be clinically very useful to urologic surgeons, urologic medical oncologists, nuclear medicine physicians, radiologists, and pathologists in further guiding and confirming complete evaluation and surgical resection of the diseases [267–269]. Furthermore, cG250 (Girentuximab) has been labeled with an assortment of radionuclides (^{124}I, ^{111}In, ^{89}Zr, ^{131}I, ^{90}Y, and ^{177}Lu) and is the most extensively investigated as CA-IX theranostics pharmaceuticals [270].

6.4. Other Biomarkers

Glypican 3

Glypican-3, a cell-surface glycoprotein in which heparan sulfate glycosaminoglycan chains are covalently linked to a protein core, is overexpressed in hepatocellular carcinoma (HCC) tissues but not in the healthy adult liver. Thus, Glypican-3 is becoming a promising candidate for liver cancer diagnosis and immunotherapy. In a clinical study, ^{124}I-codrituzumab (aka GC33), an antibody directed at Glypican 3, was evaluated in 14 patients with hepatocellular carcinoma (HCC). ^{124}I-codrituzumab detected tumor localization in most patients with HCC. Pharmacokinetics was similar to that of other intact iodinated humanized IgG [271].

7. Summary

In this report, a comprehensive review of the physical properties of iodine and iodine radionuclide, production processes (target selection, preparation, irradiation, and processing), various ^{124}I-labeling methodologies for radiolabeling of large biomolecules, (mAbs, proteins, and protein fragments), and the development of immunoPET imaging pharmaceuticals for various cancer targets in preclinical and clinical environments is provided. Several production processes, including ^{123}Te(d,n)^{124}I, ^{124}Te(d,2n)^{124}I, ^{121}Sb(α,n)^{124}I, ^{123}Sb(α,3n)^{124}I, ^{123}Sb(^{3}He,2n)^{124}I, ^{nat}Sb(α,xn)^{124}I, ^{nat}Sb(^{3}He,n)^{124}I reactions, have been used in the past. However, as a result of the less frequent availability of deuteron, alpha, and ^{3}He beams, ^{124}I is being produced, using ^{124}Te(p,n)^{124}I reaction, successfully for research and clinical use by low-energy cyclotrons. A fully-automated process for the production of ^{124}I which can be scaled up for GMP production of large quantities of ^{124}I was developed recently. Direct, using inorganic and organic oxidizing agents and enzyme catalysis, and indirect, using prosthetic groups, ^{124}I-labeling techniques have been developed and optimized in the past. The Iodogen method is used routinely in research and clinical environments. Significant research has been conducted over more than two decades in the development of immunoPET imaging pharmaceuticals for target-specific cancer detection. ^{124}I-labeled Trastuzumab, huA33, and cG250 have shown promise in human clinical trials. There is no FDA approved ^{124}I-labeled immunoPET imaging pharmaceutical available. It may be due to (1) availability of manual, difficult, and costly production and purification processes for I-124 in the past, (2) low resolution of PET images due to the high energy of available positrons from I-124, and (3) dehalogenation of ^{124}I-labeled mAbs. These bottlenecks have been resolved now by (1) development of a fully-automated process for I-124 production which can be scaled up for the cost-effective GMP production, (2) optimization of image acquisition parameters and appropriate corrections within the image reconstruction process to improve the image quality, and (3) using non internalizing mAbs for development target-specific immunoPET imaging pharmaceuticals. Further future studies in the improvement of safety and efficacy of immunoPET imaging pharmaceuticals and establishment of GMP-compliant I-124 production facilities may bring FDA-approved ^{124}I-labeled immnoPET imaging pharmaceuticals to the human clinic use in the future.

Funding: This work was supported by the Ohio Third Frontier TECH 13-060, TECH 09-028, and the Wright Center of Innovation Development Fund.

Institutional Review Board Statement: Not applicable.

Informed Consent Statement: Not applicable.

Data Availability Statement: No new data were created or analyzed in this study. Data sharing is not applicable to this article.

Acknowledgments: The authors are grateful to Michael V. Knopp (Director and Principal Investigator of the Wright Center of Innovation in Biomedical Imaging) for his encouragement and support during this work. We thank Mario Malinconico (Comecer S.p.A) for providing the schematic process diagrams for the automated-production of ^{124}I from ^{124}Te(p,n)^{124}I reaction using Comecer ALCEO halogen system and for helpful discussions.

Conflicts of Interest: The authors declare no conflict of interest.

References

1. Muehllehner, G.; Karp, J.S. Positron emission tomography. *Phys. Med. Biol.* **2006**, *51*, R117–R137. [CrossRef] [PubMed]
2. Kumar, K.; Ghosh, A. ^{18}F-AlF Labeled Peptide and Protein Conjugates as Positron Emission Tomography Imaging Pharmaceuticals. *Bioconj. Chem.* **2018**, *29*, 953–975. [CrossRef] [PubMed]
3. Conti, M.; Eriksson, L. Physics of pure and non-pure positron emitters for PET: A review and a discussion. *EJNMMI Phys.* **2016**, *3*, 8–25. [CrossRef] [PubMed]
4. Aluicio-Sarduy, E.; Ellison, P.A.; Barnhart, T.E.; Cai, W.; Nickels, R.J.; Engle, J.W. PET radiometals for antibody labeling. *J. Label. Comp. Radiopharm.* **2018**, *61*, 636–651. [CrossRef]
5. Hernandez, R.; Valdovinos, H.F.; Yang, Y.; Chakravarty, R.; Hong, H.; Barnhart, T.E.; Cai, W. ^{44}Sc: An Attractive Isotope for Peptide-Based PET Imaging. *Mol. Pharm.* **2014**, *11*, 2954–2961. [CrossRef]
6. Afshar-Oromieh, A.; Hetzheim, H.; Kratochwil, C.; Benesova, M.; Eder, M.; Neels, O.C.; Eisenhut, M.; Kubler, W.; Holland-Letz, T.; Giesel, F.L.; et al. The Theranostic PSMA Ligand PSMA-617 in the Diagnosis of Prostate Cancer by PET/CT: Biodistribution in Humans, Radiation Dosimetry, and First Evaluation of Tumor Lesions. *J. Nucl. Med.* **2015**, *56*, 1697–1705. [CrossRef]
7. Nanni, C.; Fantini, L.; Nicolini, S.; Fanti, S. Non FDG PET. *Clin. Radiol.* **2010**, *65*, 536–548. [CrossRef]
8. Lee, S.; Xie, J.; Chen, X. Peptide-based Probes for Targeted Molecular Imaging. *Biochemistry* **2010**, *49*, 1364–1376. [CrossRef]
9. Schottelius, M.; Wester, H.-J. Molecular imaging targeting peptide receptors. *Methods* **2009**, *48*, 161–177. [CrossRef]
10. Fani, M.; Maecke, H.R. Radiopharmaceutical development of radiolabeled peptides. *Eur. J. Nucl. Med. Mol. Imaging* **2012**, *39*, S11–S30. [CrossRef]
11. Tweedle, M.F. Peptide-Targeted Diagnostics and Radiotherapeutics. *Acc. Chem. Res.* **2009**, *42*, 958–968. [CrossRef] [PubMed]
12. Long, N.E.; Sullivan, B.J.; Ding, H.; Doll, S.; Ryan, M.A.; Hitchcock, C.L.; Martin, E.W., Jr.; Kumar, K.; Tweedle, M.F.; Magliery, T.J. Linker engineering in anti-TAG-72 antibody fragments optimizes biophysical properties, serum half-life, and high-specificity tumor imaging. *J. Biol. Chem.* **2018**, *293*, 9030–9040. [CrossRef] [PubMed]
13. Guo, S.; Xu, C.; Yin, H.; Hill, J.; Pi, F.; Guo, P. Tuning the size, shape and structure of RNA nanoparticles for favorable cancer targeting and immunostimulation. *Wiley Interdiscip. Rev. Nanomed. Nanobiotechnol.* **2020**, *12*, e1582–e1600. [CrossRef]
14. Wei, W.; Rosenkrans, Z.T.; Liu, J.; Huang, G.; Luo, Q.-Y.; Cai, W. ImmunoPET: Concept, Design, and Applications. *Chem. Rev.* **2020**, *120*, 3787–3851. [CrossRef] [PubMed]
15. Philpott, G.W.; Schwarz, S.W.; Anderson, C.J.; Dehdashti, F.; Connett, J.M.; Zinn, K.R.; Meares, C.F.; Cutler, P.D.; Welch, M.J.; Siegel, B.A. RadioimmunoPET: Detection of colorectal carcinoma with positron-emitting copper-64-labeled monoclonal antibody. *J. Nucl. Med.* **1995**, *36*, 1818–1824.
16. Goldenberg, D.M.; Nabi, H.A. Breast cancer imaging with radiolabeled antibodies. *Semin. Nucl. Med.* **1999**, *29*, 41–48. [CrossRef]
17. Knowles, S.M.; Wu, A.M. Advances in Immuno-Positron Emission Tomography: Antibodies for Molecular Imaging in Oncology. *J. Clin. Oncol.* **2012**, *30*, 3884–3892. [CrossRef]
18. Wadas, T.J.; Wong, E.H.; Weisman, G.R.; Anderson, C.J. Coordinating Radiometals of Copper, Gallium, Indium, Yttrium, and Zirconium for PET and SPECT Imaging of Disease. *Chem. Rev.* **2010**, *110*, 2858–2902. [CrossRef]
19. Deri, M.A.; Zeglis, B.M.; Francesconi, L.C.; Lewis, J.S. PET Imaging with ^{89}Zr: From Radiochemistry to the Clinic. *Nucl. Med. Biol.* **2013**, *40*, 3–14. [CrossRef]
20. Severin, G.W.; Engle, J.W.; Barnhart, T.E.; Nickels, R.J. ^{89}Zr Radiochemistry for Positron Emission Tomography. *Med. Chem.* **2011**, *7*, 389–394. [CrossRef]
21. Salodkin, S.S.; Golovkov, V.M. Cyclotron Production of Iodine-124. *Russ. Phys. J.* **2020**, *62*, 2347–2353. [CrossRef]
22. Koehler, L.; Gagnon, K.; McQuarrie, S.; Wuest, F. Iodine-124: A promising positron emitter for organic PET chemistry. *Molecules* **2010**, *15*, 2686–2718. [CrossRef] [PubMed]
23. Dubost, E.; McErlain, H.; Babin, V.; Sutherland, A.; Cailly, T. Recent Advances in Synthetic Methods for Radioiodination. *J. Org. Chem.* **2020**, *85*, 8300–8310. [CrossRef]
24. Ogawa, K.; Takeda, T.; Yokokawa, M.; Yu, J.; Makino, A.; Kiyono, Y.; Shiba, K.; Kinuya, S.; Odani, A. Comparison of Radioiodine- or Radiobromine-labeled RGD Peptides between Direct and Indirect Labeling Methods. *Chem. Pharm. Bull.* **2018**, *66*, 651–659. [CrossRef]
25. Rangger, C.; Haubner, R. Radiolabelled Peptides for Positron Emission Tomography and Endoradiotherapy in Oncology. *Pharmaceuticals* **2020**, *13*, 22. [CrossRef]
26. Wright, B.D.; Lapi, S.E. Designing the Magic Bullet? The Advancement of Immuno-PET into Clinical Use. *J. Nucl. Med.* **2013**, *54*, 1171–1174. [CrossRef] [PubMed]
27. Guenther, I.; Wyer, L.; Knust, E.J.; Finn, R.D.; Koziorowski, J.; Weinreich, R. Radiosynthesis and quality assurance of 5-[^{124}I]Iodo-2′-deoxyuridine for functional PET imaging of cell proliferation. *Nucl. Med. Biol.* **1998**, *25*, 359–365. [CrossRef]

28. Chacko, A.-M.; Divgi, C.R. Radiopharmaceutical Chemistry with Iodine-124: A Non-Standard Radiohalogen for Positron Emission Tomography. *Med. Chem.* **2011**, *7*, 395–412. [CrossRef] [PubMed]
29. Eschmann, S.M.; Reischl, G.; Bilger, K.; Kupferschlager, J.; Thelen, M.H.; Dohmen, B.M.; Besenfelder, H.; Bares, R. Evaluation of dosimetry of radioiodine therapy in benign and malignant thyroid disorders by means of iodine-124 and PET. *Eur. J. Nucl. Med. Mol. Imaging* **2002**, *29*, 760–767. [CrossRef]
30. Samnick, S.; Al-Momani, E.; Schmid, J.-S.; Mottok, A.; Buck, A.K.; Lapa, C. Initial Clinical Investigation of [^{18}F]Tetrafluoroborate PET/CT in Comparison to [^{124}I]iodine PET/CT for Imaging Thyroid Cancer. *Clin. Nucl. Med.* **2018**, *43*, 162–167. [CrossRef]
31. Phan, H.T.T.; Jager, P.L.; Paans, A.M.J.; Plukker, J.T.M.; Sturkenboom, M.G.G.; Sluiter, W.J.; Wolffenbuttel, B.H.R.; Dierckx, R.A.J.O.; Links, T.P. The diagnostic value of ^{124}I-PET in patients with differentiated thyroid cancer. *Eur. J. Nucl. Med. Mol. Imaging* **2008**, *35*, 958–965. [CrossRef]
32. Israel, I.; Brandau, W.; Farmakis, G.; Samnick, S. Improved synthesis of no-carrier-added *p*-[^{124}I]iodo-L-phenylalanine and *p*-[^{131}I]iodo-L-phenylalanine for nuclear medicine applications in malignant gliomas. *Appl. Radiat. Isot.* **2008**, *66*, 513–522. [CrossRef] [PubMed]
33. Kulkarni, P.V.; Corbett, J.R. Radioiodinated tracers for myocardial imaging. *Semin. Nucl. Med.* **1990**, *20*, 119–129. [CrossRef]
34. Seo, Y.; Gustafson, W.C.; Dannoon, S.F.; Nekritz, E.A.; Lee, C.-L.; Murphy, S.T.; VanBrocklin, H.F.; Hernandez-Pampaloni, M.; Haas-Kogan, D.A.; Weiss, W.A.; et al. Tumor dosimetry using [^{124}I]m-iodobenzylguanidine microPET/CT for [^{131}I]m-iodobenzylguanidine treatment of neuroblastoma in a murine xenograft model. *Mol. Imaging Biol.* **2012**, *14*, 735–742. [CrossRef]
35. Pandey, S.K.; Gryshuk, A.L.; Sajjad, M.; Zheng, X.; Chen, Y.; Abouzeid, M.M.; Morgan, J.; Charamisinau, I.; Nabi, H.A.; Oseroff, A.; et al. Multimodality Agents for Tumor Imaging (PET, Fluorescence) and Photodynamic Therapy. A Possible "See and Treat" Approach. *J. Med. Chem.* **2005**, *48*, 6286–6295. [CrossRef] [PubMed]
36. Pressman, D.; Keighley, G. The zone of activity of antibodies as determined by the use of radioactive tracers; the zone of activity of nephritoxic antikidney serum. *J. Immunol.* **1948**, *59*, 141–146.
37. Hughs, W.L. The Chemistry of Iodination. *Ann. N. Y. Acad. Sci.* **1957**, *70*, 3–18. [CrossRef]
38. Krohn, K.A.; Knight, L.C.; Harwig, J.F.; Welch, M.J. Differences in the sites of iodination of proteins following four methods of radioiodination. *Biochim. Biophys. Acta* **1977**, *490*, 497–505. [CrossRef]
39. Boschloo, G.; Hagfeldt, A. Characteristics of the Iodide/Triiodide Redox Mediator in Dye-Sensitized Solar Cells. *Acc. Chem. Res.* **2009**, *42*, 1819–1826. [CrossRef]
40. Silberstein, E.B. Radioiodine: The classic theranostic agent. *Semin. Nucl. Med.* **2012**, *42*, 164–170. [CrossRef]
41. Audi, G.; Bersillon, O.; Blachot, J.; Wapstra, A.H. The NUBASE evaluation of nuclear and decay properties. *Nucl. Phys. A* **2003**, *729*, 3–128. [CrossRef]
42. Kuker, R.; Sztejnberg, M.; Gulec, S. I-124 Imaging and Dosimetry. *Mol. Imaging Radionucl. Ther.* **2017**, *26*, 66–73. [CrossRef] [PubMed]
43. Braghirolli, A.M.S.; Waissmann, W.; da Silva, J.B.; dos Santos, G.R. Production of iodine-124 and its applications in nuclear medicine. *Appl. Radiat. Isot.* **2014**, *90*, 138–148. [CrossRef] [PubMed]
44. Lambrecht, R.M.; Sajjad, M.; Qureshi, M.A.; Al-Yanbawi, S.J. Production of iodine-124. *J. Radioanal. Nucl. Chem. Lett.* **1988**, *127*, 143–150. [CrossRef]
45. Sharma, H.L.; Zweit, J.; Downey, S.; Smith, A.M.; Smith, A.G. Production of ^{124}I for positron emission tomography. *J. Label. Compd. Rad.* **1988**, *26*, 165–167. [CrossRef]
46. Firouzbakht, M.L.; Schlyer, D.J.; Finn, R.D.; Laguzzi, G.; Wolf, A.P. Iodine-124 production: Excitation functions for the ^{124}Te(d,2n)^{124}I and ^{124}Te(d,3n)^{123}I reactions from 7 to 24 MeV. *Nucl. Instrum. Methods B* **1993**, *79*, 909–910. [CrossRef]
47. Knust, E.J.; Weinreich, R. Yields and impurities in several production reactions for ^{124}I. In Proceedings of the 7th Workshop on Targetry and Target Chemistry, Heidelberg, Germany, 8–11 June 1997; pp. 253–262.
48. Clem, R.G.; Lambrecht, R.M. Enriched Te-124 targets for production of I-123 and I-124. *Nucl. Instrum. Methods Phys. Res. Sect. A Accel. Spectrom. Detect. Assoc. Equip.* **1991**, *303*, 115–118. [CrossRef]
49. Weinreich, R.; Knust, E.J. Quality assurance of iodine-124 produced via the nuclear reaction ^{124}Te(d,2n)^{124}I. *J. Radioanal. Nucl. Chem. Lett.* **1996**, *213*, 253–261. [CrossRef]
50. Knust, E.J.; Dutschka, K.; Weinreich, R. Preparation of ^{124}I solutions after thermodistillation of irradiated $^{124}TeO_2$ targets. *Appl. Radiat. Isot.* **2000**, *52*, 181–184. [CrossRef]
51. Bastian, T.H.; Coenen, H.H.; Qaim, S.M. Excitation functions of ^{124}Te(d,xn)^{124,125}I reactions from threshold up to 14 MeV: Comparative evaluation of nuclear routes for the production of ^{124}I. *Appl. Radiat. Isot.* **2001**, *55*, 303–308. [CrossRef]
52. Zaidi, H.; Qaim, S.M.; Stocklin, G. Excitation functions of deuteron induced nuclear reactions on natural tellurium and enriched ^{122}Te: Production of ^{123}I via the ^{122}Te(d,n)^{123}I-process. *Int. J. Appl. Radiat. Isot.* **1983**, *34*, 1425–1430. [CrossRef]
53. Scholten, B.; Takács, S.; Kovács, Z.; Tárkányi, F.; Qaim, S.M. Excitation functions of deuteron induced reactions on ^{123}Te: Relevance to the production of ^{123}I and ^{124}I at low and medium sized cyclotrons. *Appl. Radiat. Isot.* **1997**, *48*, 267–271. [CrossRef]
54. Hassan, K.F.; Qaim, S.M.; Saleh, Z.A.; Coenen, H.H. Alpha-particle induced reactions on ^{nat}Sb and ^{121}Sb with particular reference to the production of the medically interesting radionuclide ^{124}I. *Appl. Radiat. Isot.* **2006**, *64*, 101–109. [CrossRef]
55. Hassan, K.F.; Qaim, S.M.; Saleh, Z.A.; Coenen, H.H. ^{3}He-particle-induced reactions on ^{nat}Sb for production of ^{124}I. *Appl. Radiat. Isot.* **2006**, *64*, 409–413. [CrossRef] [PubMed]

56. Watson, I.A.; Waters, S.L.; Silveste, D.J. Excitation-functions for reactions producing I-121, I-123 and I-124 from irradiation of natural antimony with He-3 and He-4 particles with energies upto 30-MeV. *J. Inorg. Nucl. Chem.* **1973**, *35*, 3047–3053. [CrossRef]
57. Ismail, M. Hybrid model analysis of the excitation-function for alpha-induced reaction on Sb-121 and Sb-123. *Pramana* **1989**, *32*, 605–618. [CrossRef]
58. Ismail, M. Measurement and analysis of the excitation-function for alpha-induced reactions on Ga and Sb isotopes. *Phys. Rev. C* **1990**, *41*, 87–108. [CrossRef]
59. Uddin, M.S.; Hermanne, A.; Sudar, S.; Aslam, M.N.; Scholten, B.; Coenen, H.H.; Qaim, S.M. Excitation functions of alpha-particle induced reactions on enriched ^{123}Sb and ^{nat}Sb for production of ^{124}I. *Appl. Radiat. Isot.* **2011**, *69*, 699–704. [CrossRef]
60. Tarkanyi, F.; Takacs, S.; Kiraly, B.; Szelecsenyi, F.; Ando, L.; Bergman, J.; Heselius, S.-J.; Solin, O.; Hermanne, A.; Shubin, Y.N.; et al. Excitation functions of He-3-and alpha-particle induced nuclear reactions on ^{nat}Sb for production of medically relevant ^{123}I and ^{124}I radioisotopes. *Appl. Radiat. Isot.* **2009**, *67*, 1001–1006. [CrossRef]
61. Scholten, B.; Hassan, K.F.; Saleh, Z.A.; Coenen, H.H.; Qaim, S.M. Comparative studies on the production of the medically important radionuclide ^{124}I via p-, d-, ^{3}He- and alpha-particle induced reactions. In Proceedings of the International Conference on Nuclear Data for Science and Technology, Nice, France, 22–27 April 2007; Bersillon, O., Gunsing, F., Bauge, E., Jacqmin, R., Leray, S., Eds.; EDP Sciences: Ulis, France, 2007; pp. 1359–1361.
62. Aslam, M.N.; Sudar, S.; Hussain, M.; Malik, A.A.; Qaim, S.M. Evaluation of excitation functions of ^{3}He- and alpha-particle induced reactions on antimony isotopes with special relevance to the production of iodine-124. *Appl. Radiat. Isot.* **2011**, *69*, 94–104. [CrossRef]
63. Qaim, S.M.; Rosch, F.; Scholten, B.; Stocklin, G.; Kovacs, Z.; Tarkanyi, F. Nuclear data relevant to the production of medically important $\beta^{\pm}$ emitting radioisotopes ^{75}Br, ^{86}Y, ^{94m}Tc and ^{124}I at a small cyclotron. In Proceedings of the International Conference on Nuclear Data for Science and Technology, Gatlinburg, TN, USA, 9–13 May 1994; Dickens, J.K., Ed.; American Nuclear Society, Inc.: La Grange Park, IL, USA, 1994; pp. 1035–1038.
64. Scholten, B.; Kovacs, Z.; Tarkanyi, F.; Qaim, S.M. Excitation functions of ^{124}Te(p,xn)^{124,123}I reactions from 6 to 31 MeV with special reference to the production of ^{124}I at a small cyclotron. *Appl. Radiat. Isot.* **1995**, *46*, 255–259. [CrossRef]
65. Hohn, A.; Nortier, F.M.; Scholten, B.; van der Walt, T.N.; Coenen, H.H.; Qaim, S.M. Excitation functions of ^{125}Te (p,xn) reactions from their respective thresholds up to 100 MeV with special reference to the production of ^{124}I. *Appl. Radiat. Isot.* **2001**, *55*, 149–156. [CrossRef]
66. Aslam, M.N.; Sudar, S.; Hussain, M.; Malik, A.A.; Shah, H.A.; Qaim, S.M. Evaluation of excitation functions of proton and deuteron induced reactions on enriched tellurium isotopes with special relevance to the production of iodine-124. *Appl. Radiat. Isot.* **2010**, *68*, 1760–1773. [CrossRef]
67. Unal, R.; Akcaalan, U. The Reaction Cross Sections for ^{124,125}Te(p,xn)^{123,124}I and ^{123,124}Te(d,xn)^{123,124}I. *Eur. J. Sci. Tech.* **2020**, *18*, 958–963. [CrossRef]
68. Hermanne, A.; Tarkanyi, F.T.; Ignatyuk, A.V.; Takacs, S.; Capote, S.R. *Upgrade of IAEA Recommended Data of Selected Nuclear Reactions for Production of PET and SPECT Isotopes*; IAEA Report; International Atomic Energy Agency: Vienna, Austria, 2018.
69. National Nuclear Decay Center, Brookhaven National Lab. 2009. Available online: http://www.nndc.bnl.gov/ (accessed on 1 December 2020).
70. Kondo, K.; Lambrecht, R.M.; Norton, E.F.; Wolf, A.P. Cyclotron isotopes and radiopharmaceuticals-XXII. Improved targetry and radiochemistry for production of ^{123}I and ^{124}I. *Int. J. Appl. Radiat. Isot.* **1977**, *28*, 765–771. [CrossRef]
71. Qaim, S.M.; Blessing, G.; Tarkanyi, F.; Lavi, N.; Brautigam, W.; Scholten, B.; Stocklin, G. Production of longer-lived positron emitters ^{73}Se, ^{82m}Rb and ^{124}I. In Proceedings of the 14th International Conference on Cyclotrons and their Applications, Cape Town, South Africa, 8–13 October 1995; Cornell, J.C., Ed.; World Scientific: Singapore, 1995; pp. 541–544.
72. Weinreich, R.; Wyer, L.; Crompton, N.; Nievergelt-Egido, M.C.; Guenther, I.; Roelcke, U.; Leender, K.L.; Knust, E.J. I-124 And Its Applicaions in Nuclear Medicine and Biology. In Proceedings of the International Symposium on Modern Trends in Radiopharmaceuticals for Diagnosis and Therapy, Lisbon, Portugal, 30 March–3 April 1998; International Atomic Energy Agency: Vienna, Austria, 1998; pp. 399–418.
73. McCarthy, T.J.; Laforest, R.; Downer, J.B.; Lo, A.-R.; Margenau, W.H.; Hughey, B.; Shefer, R.E.; Klinkowskein, R.E.; Welch, M.J. Investigation of I-124, Br-76, and Br-77 production using a small biomedical cyclotron–Can induction furnaces help in the preparation and separation of targets? In Proceedings of the 8th Workshop on Targetry and Target Chemistry, St. Louis, MO, USA, 23–26 June 1999; pp. 127–130.
74. Sheh, Y.; Koziorowski, J.; Balatoni, J.; Lom, C.; Dahl, J.R.; Finn, R.D. Low energy cyclotron production and chemical separation of no-carrier added iodine-124 from a usable, enriched tellurium-124 dioxide/aluminum oxide solid solution target. *Radiochim. Acta* **2000**, *88*, 169–173. [CrossRef]
75. Rowland, D.J.; Laforest, R.; McCarthy, T.J.; Hughey, B.J.; Welch, M.J. Conventional and induction furnace distillation procedures for the routine production of Br-76,77 and I-124 on disk and slanted targets. *J. Label. Compd. Radooharm.* **2001**, *44*, S1059–S1060. [CrossRef]
76. Glaser, M.; Brown, D.J.; Law, M.P.; Iozzo, P.; Waters, S.L.; Poole, K.; Knickmeier, M.; Camici, P.G.; Pike, V.W. Preparation of no-carrier-added [^{124}I]A$_{14}$-iodoinsulin as a radiotracer for positron emission tomography. *J. Label. Compd. Radiopharm.* **2001**, *44*, 465–480. [CrossRef]

77. Qaim, S.M.; Hohn, A.; Bastian, T.; El-Azoney, K.M.; Blessing, G.; Spellerberg, S.; Scholten, B.; Coenen, H.H. Some optimization studies relevant to the production of high-purity ^{124}I and ^{120g}I at a small sized cyclotron. *Appl. Radiat. Isot.* **2003**, *58*, 69–78. [CrossRef]
78. Glaser, M.; Mackay, D.B.; Ranicar, A.S.O.; Waters, S.L.; Brady, F.; Luthra, S.K. Improved targetry and production of iodine-124 for PET studies. *Radiochim. Acta.* **2004**, *92*, 951–956. [CrossRef]
79. Nye, J.A.; Dick, D.W.; Avila-Rodriguez, M.A.; Nickles, R.J. Radiohalogen targetry at the University of Wisconsin. *Nucl. Instrum. Methods B* **2005**, *241*, 693–696. [CrossRef]
80. Sajjad, M.; Bars, E.; Nabi, H.A. Optimization of ^{124}I production via ^{124}Te(p,n)^{124}I reaction. *Appl. Radiat. Isot.* **2006**, *64*, 965–970. [CrossRef] [PubMed]
81. Nye, J.A.; Avila-Rodriguez, M.A.; Nickles, R.J. Production of [^{124}I]-iodine on an 11 MeV cyclotron. *Radiochim. Acta* **2006**, *94*, 213–216. [CrossRef]
82. Nye, J.A.; Avila-Rodriguez, M.A.; Nickles, R.J. A new binary compound for the production of ^{124}I via the ^{124}Te(p,n)^{124}I reaction. *Appl. Radiat. Isot.* **2007**, *65*, 407–412. [CrossRef] [PubMed]
83. Rajec, P.; Reich, M.; Szöllős, O.; Baček, D.; Vlk, P.; Kováč, P.; Čomor, J.J. Production of ^{124}I on an 18/9 MeV cyclotron. In Proceedings of the 7th International Conference on Nuclear and Radiochemistry, Budapest, Hungary, 24–29 August 2008.
84. Nagatsu, K.; Fukada, M.; Minegishi, K.; Suzuki, H.; Fukumura, T.; Yamazaki, H.; Suzuki, K. Fully automated production of iodine-124 using a vertical beam. *Appl. Radiat. Isot.* **2011**, *69*, 146–157. [CrossRef]
85. Schmitz, J. The production of [^{124}I]iodine and [^{86}Y]yttrium. *Eur. J. Nucl. Med. Mol. Imaging* **2011**, *38*, S4–S9. [CrossRef]
86. Lamparter, D.; Hallmann, B.; Hänscheid, H.; Boschi, F.; Malinconico, M.; Samnick, S. Improved small scale production of iodine-124 for radiolabeling and clinical applications. *Appl. Radiat. Isot.* **2018**, *140*, 24–28. [CrossRef]
87. Vidyanathan, G.; Wieland, B.W.; Larsen, R.H.; Zweit, J.; Zalutsky, M.R. High yield production of iodine-124 using the ^{125}Te(p,2n)^{124}I reaction. In Proceedings of the Sixth International Workshop on Targetry and Target Chemistry, Vancouver, BC, Canada, 17–19 August 1995; pp. 87–88.
88. Kim, J.H.; Lee, J.S.; Lee, T.S.; Park, H.; Chum, K.S. Optimization studies on the production of high-purity ^{124}I using (p,2n) reaction. *J. Label. Comp. Radiopharm.* **2007**, *50*, 511–512. [CrossRef]
89. Qaim, S.M.; Hohn, A.; Nortier, F.M.; Blessing, G.; Schroeder, I.W.; Scholten, B.; van der Walt, T.N.; Coenen, H.H. Production of ^{124}I at small and medium sized cyclotrons. In Proceedings of the Eighth International Workshop on Targetry and Target Chemistry, St. Louis, MO, USA, 23–26 June 1999; pp. 131–133.
90. Azzam, A.; Hamada, M.S.; Said, S.A.; Mohamed, G.Y.; Al-abyad, M. Excitation functions for proton-induced reactions of Te and ^{nat}Te targets: Measurements and model calculations special relevant to the ^{128}Te(p,n)^{128}I reaction. *Nuclear Physics A* **2020**, *999*, 121790. [CrossRef]
91. Zweit, J.; Bakir, M.A.; Ott, R.J.; Sharma, H.L.; Cox, M.; Goodall, R. Excitaion functions of proton induced reactions in natural tellurium: Production of no-carrier added iodine-124 for PET applications. In Proceedings of the Fourth International Workshop on Targetry and Target Chemistry, Villigen, Switzerland, 9–12 September 1991; pp. 76–78.
92. Acerbi, E.; Birattari, C.; Casiglioni, M.; Resmini, F. Productionof ^{123}I for medical purposes at the Milan AVF cyclotron. *Int. J. Appl. Radiat. Isot.* **1975**, *26*, 741–747. [CrossRef]
93. Scholten, B.; Qaim, S.M.; Stocklin, G. Excitation functions of proton induced reactions on natural tellurium and enriched ^{123}Te: Production of ^{123}I via the ^{123}Te(p,n)^{123}I process at a low-energy cyclotron. *Appl. Radiat. Isot.* **1989**, *40*, 127–132. [CrossRef]
94. Kiraly, B.; Tarkanyi, F.; Takacs, S.; Kovacs, Z. Excitation functions of proton induced nuclear reactions on natural tellurium upto 18MeV for validation of isotopic cross sections. *J. Radioanal. Nucl. Chem.* **2006**, *270*, 369–378. [CrossRef]
95. Zarie, K.; Hammad, N.A.; Azzam, A. Excitation functions of (p,xn) reactions on natural tellurium at low energy cyclotron: Relevance to the production of medical radioisotope ^{123}I. *J. Nucl. Radiat. Phys.* **2006**, *1*, 93–105.
96. El-Azony, K.M.; Suzuki, K.; Fukumura, T.; Szelecsenyi, F.; Kovacs, Z. Proton induced reactions on natural tellurium up to 63 MeV: Data validation and investigation of possibility of ^{124}I production. *Radiochim. Acta* **2008**, *96*, 763–769. [CrossRef]
97. International Atomic Energy Agency. *Standardized High Current Solid Targets for Cyclotron Production of Diagnostic and Therapeutic Radionuclides*; IAEA Technical Report Series No. 432; IAEA: Vienna, Austria, 2004; pp. 27–28, 32–33.
98. Sadeghi, M.; Dastan, M.; Ensaf, M.R.; Tehrani, A.A.; Tenreiro, C.; Avila, M. Thick tellurium electrodeposition on nickel-coated copper substrate for ^{124}I production. *Appl. Radiat. Isot.* **2008**, *66*, 1281–1286. [CrossRef] [PubMed]
99. Van den Bosch, R.; De Goeij, J.J.M.; van der Heide, J.A.; Tertoolen, J.F.W.; Theelen, H.M.J.; Zegers, C.A. New approach to target chemistry for the iodine-123 production via the ^{124}Te(p,2n) reaction. *Int. J. Appl. Radiat. Isot.* **1977**, *28*, 255–261. [CrossRef]
100. Alekseev, I.E.; Darmograi, V.V.; Marchenkov, N.S. Development of diffusion-thermal methods for preparing ^{67}Cu and ^{124}I for radionuclide therapy and positron emission tomography. *Radiochemistry* **2005**, *47*, 460–466. [CrossRef]
101. Stevenson, N.R.; Buckley, K.; Gelbart, W.Z.; Hurtado, E.T.; Johnson, R.R.; Ruth, T.J.; Zeisler, S.K. On-line production of radioiodines with low energy accelerators. In Proceedings of the 6th Workshop on Targetry and Target Chemistry, Vancouver, BC, Canada, 17–19 August 1995; pp. 82–83.
102. Čomor, J.J.; Stevanović, Ž.; Rajčević, M.; Košutić, D. Modeling of thermal properties of a TeO_2 target for radioiodine production. *Nucl. Instrum. Methods A* **2004**, *521*, 161–170. [CrossRef]
103. Kudelin, B.K.; Gromova, E.A.; Gavrilina, L.V.; Solin, L.M. Purification of recovered tellurium dioxide for re-use in iodine radioisotope production. *Appl. Radiat. Isot.* **2001**, *54*, 383–386. [CrossRef]

104. Janni, J.F. Proton Range-Energy Table, 1 keV to 10 GeV. *Atomic Data Nuclear Data Tables* **1982**, *27*, 147–529. [CrossRef]
105. Seevers, R.H.; Counsell, R.E. Radioiodination Techniques for Small Organic Molecules. *Chem. Rev.* **1982**, *82*, 575–590. [CrossRef]
106. Mock, B.; Zheng, Q.-H. *Radiopharmaceutical Chemistry: Iodination Techniques in Nuclear Medicine*, 2nd ed.; Henkin, R.E., Ed.; Elsevier: New York, NY, USA, 2006; pp. 397–405.
107. Redshaw, M.R.; Lynch, S.S. An improved method for the preparation of iodinated antigens for radioimmunoassay. *J. Endocrinol.* **1974**, *60*, 527–528. [CrossRef] [PubMed]
108. Eisen, H.N.; Keston, A.S. The Immunologic Reactivity of Bovine Serum Albumin Labelled with Trace-Amounts of Radioactive Iodine (I^{131}). *J. Immunol.* **1949**, *63*, 71–80. [PubMed]
109. Yalow, R.S.; Berson, S.A. Immunoassay of Endogenous Plasma Insulin in Man. *J. Clin. Investig.* **1960**, *39*, 1157–1175. [CrossRef] [PubMed]
110. Gilmore, R.C., Jr.; Robbins, M.C.; Reid, A.F. Labeling bovine and human albumin with I^{131}. *Nucleonics* **1954**, *12*, 65–68.
111. McFarlane, A.S. Labelling of plasma proteins with radioactive iodine. *Biochem. J.* **1956**, *62*, 135–143. [CrossRef]
112. Stadie, W.C.; Haugaard, N.; Vaughn, M. Studies of Insulin Binding with Isotopically Labeled Insulin. *J. Biol. Chem.* **1952**, *199*, 729–739. [CrossRef]
113. Francis, G.E.; Mulligan, W.; Wormall, A. Labelling of proteins with iodine-131, Sulphur-35 and phosphorus-32. *Nature (London)* **1951**, *167*, 748–751. [CrossRef]
114. McFarlane, A.S. Efficient Trace-labelling of Proteins with Iodine. *Nature (London)* **1958**, *182*, 53. [CrossRef]
115. Hung, L.T.; Fermandjian, S.; Morgat, J.L.; Fromageot, P. Peptide and protein labelling with iodine, iodine monochloride reaction with aqueous solution of L-tyrosine, L-histidine, L-histidine-peptides, and his effect on some simple disulfide bridges. *J. Label. Compd. Radiopharm.* **1974**, *10*, 3–21. [CrossRef]
116. Doran, D.M.; Spar, I.L. Oxidative iodine monochloride iodination technique. *J. Immunol. Methods* **1980**, *39*, 155–163. [CrossRef]
117. Philbrick, F.A. Hydrochloric Acid Solutions of Iodine Monochloride. *J. Chem. Soc.* **1930**, 2254–2260. [CrossRef]
118. Margerum, D.W.; Dickson, P.N.; Nagy, J.C.; Kumar, K.; Bowers, C.P.; Fogelman, K.D. Kinetics of the Iodine Monochloride Reaction with Iodide Measured by Pulsed Accelerated-Flow Method. *Inorg. Chem.* **1986**, *25*, 4900–4904. [CrossRef]
119. Prasada Rao, M.D.; Padmanabha, J. Kinetics and Mechanism of Iodination of Phenol and Substituted Phenols by Iodine Monochloride in Aqueous Acetic Acid. *Indian J. Chem.* **1981**, *20A*, 133–135.
120. Helmkamp, R.W.; Contreras, M.A.; Bale, W.F. I^{131}-labeling of proteins by the iodine monochloride method. *Int. J. Appl. Radiat. Isot.* **1967**, *18*, 737–746. [CrossRef]
121. Doll, S.; Woolum, K.; Kumar, K. Radiolabeling of a cyclic RGD (cyclo Arg-Gly-Asp-d-Tyr-Lys) peptide using sodium hypochlorite as an oxidizing agent. *J. Label. Compd. Radiopharm.* **2016**, *59*, 462–466. [CrossRef]
122. Kumar, K. Methods for Iodination of Biomolecules. U.S. Patent Application 20190276490, 12 September 2019.
123. Hunter, W.M.; Greenwood, F.C. Preparation of iodine-131 labelled human growth hormone of high specific activity. *Nature* **1962**, *194*, 495–496. [CrossRef]
124. Greenwood, F.C.; Hunter, W.M.; Glover, J.S. The preparation of ^{131}I-labelled human growth hormone of high specific radioactivity. *Biochem. J.* **1963**, *89*, 114–123. [CrossRef]
125. Jennings, V.J. Analytical applications of Chloramine-T. *CRC Crit. Rev. Anal. Chem.* **1974**, *3*, 407–419. [CrossRef]
126. Kumar, K.; Day, R.A.; Margerum, D.W. Atom-Transfer Redox Kinetics: General-Acid-Assisted Oxidation of Iodide by Chloramines and Hypochlorite. *Inorg. Chem.* **1986**, *25*, 4344–4350. [CrossRef]
127. Nagy, J.C.; Kumar, K.; Margerum, D.W. Non-Metal Redox Kinetics: Oxidation of Iodide by Hypochlorous Acid and by Nitrogen Trichloride Measured by the Pulsed-Accelerated-Flow Method. *Inorg. Chem.* **1988**, *27*, 2773–2780. [CrossRef]
128. Kaminski, J.J.; Bodor, N.; Higuchi, T. N-Halo Derivatives. IV. Synthesis of Low Chlorine Potential Soft N-Chloramine Systems. *J. Pharm. Sci.* **1976**, *65*, 1733–1737. [CrossRef]
129. Hussain, A.A.; Jona, J.A.; Yamada, A.; Dittert, L.W. Chloramine-T in radiolabeling techniques. II. A non-destructive method for radiolabeling biomolecules by halogenation. *Anal. Biochem.* **1995**, *224*, 221–226. [CrossRef] [PubMed]
130. Hussain, A.A.; Dittert, L.W. Non-Destructive Method for Radiolabeling Biomolecules by Halogenation. U.S. Patent 5,424,402, 13 June 1995.
131. Tashtoush, B.M.; Traboulsi, A.A.; Dittert, L.; Hussain, A.A. Chloramine-T in radiolabeling techniques. IV. Pento-O-acetyl-N-chloro-N-methylglucamine as an oxidizing agent in radiolabeling techniques. *Anal. Biochem.* **2001**, *288*, 16–21. [CrossRef] [PubMed]
132. Hussain, A.A.; Bassam, T.; Dittert, L.W. Derivatives of N-chloro-N-Methyl Glucamine and N-Chloro-N-Methyl Glucamine Esters. U.S. Patent 5,985,239, 16 November 1999.
133. Markwell, M.A. A new solid-state reagent to iodinate proteins: Conditions for the efficient labeling of antiserum. *Anal. Biochem.* **1982**, *125*, 427–432. [CrossRef]
134. Fracker, P.J.; Speck, J.C. Protein and cell membrane iodinations with a sparingly soluble chloramide 1,3,4,6-tetrachloro-3a, 6a-diphenylglycoluril. *Biochem. Biophys. Res. Common.* **1978**, *80*, 849–857. [CrossRef]
135. Salacinski, P.; Hope, J.; McLean, C.; Clement-Jones, V.; Sykes, J.; Price, J.; Lowry, P.J. A new simple method which allows theoretical incorporation of radio-iodine into proteins and peptides without damage. *J. Endocrinol.* **1979**, *81*, 131.

136. Paus, E.; Bormer, O.; Nustad, K. Radioiodination of proteins with Iodogen method. In Proceedings of the International Symposium on Radioimmunoassay and Related Procedures in Medicine, Vienna, Austria, 21–25 June 1982; International Atomic Energy Agency: Vienna, Austria, 1982; pp. 161–170.
137. Boonkitticharoen, V.; Laohathai, K. Assessing performances of Iodogen-coated surfaces used for radioiodination of proteins. *Nucl. Med. Common.* **1990**, *11*, 295–304. [CrossRef]
138. Holohan, K.N.; Murphy, R.F.; Flanagan, R.W.J.; Buchanan, K.D.; Elmore, D.T. Enzymic iodination of the histidyl residue of secretin: A radioimmunoassay of the hormone. *Biochim. Biophys. Acta* **1973**, *322*, 178–180. [CrossRef]
139. Holohan, K.N.; Murphy, R.F.; Elmore, D.T. The Site of Substitution in the Imidazole Nucleus after the Lactoperoxidase-Catalysed Iodination of Histidine Residues in Polypetides. *Biochem. Soc. Trans.* **1974**, *2*, 739–740. [CrossRef]
140. Marchalonis, J.J. An enzymic method for the trace iodination of immunoglobulins and other proteins. *Biochem. J.* **1969**, *113*, 299–305. [CrossRef] [PubMed]
141. Dewanjee, M.K. *Radioiodination: Theory, Practice and Biomedical Applications*; Springer Science & Business Media: Berlin, Germany, 1992; pp. 151–152.
142. Sugiura, G.; Kuhn, H.; Sauter, M.; Haberkorn, U.; Mier, W. Radiolabeling Strategies for Tumor-Targeting Proteinaceous Drugs. *Molecules* **2014**, *19*, 2135–2165. [CrossRef] [PubMed]
143. Navarro, L.; Berdal, M.; Cherel, M.; Pecorari, F.; Gestin, J.-F.; Guerard, F. Prosthetic groups for radioiodination and astatination of peptides and proteins: A comparative study of five potential bioorthogonal labeling strategies. *Bioorg. Med. Chem.* **2019**, *27*, 167–174. [CrossRef] [PubMed]
144. Santos, J.S.; Muramoto, E.; Colturato, M.T.; Siva, C.P.; Araujo, E.B. Radioiodination of proteins using prosthetic group: A convenient way to produce labelled proteins with in vivo stability. *Cell. Mol. Biol.* **2002**, *47*, 735–739.
145. Reiner, L.; Keston, A.S.; Green, M. The Absorption and Distribution of Insulin Labelled with Radioactive Iodine. *Science* **1942**, *96*, 362–363. [CrossRef]
146. Bolton, A.E.; Hunter, W.M. The labeling of proteins to high specific radioactivities by conjugation to a I-125 containing acylating agent. *Biochem. J.* **1973**, *133*, 529–539. [CrossRef]
147. Wood, F.T.; Wu, M.M.; Gerhart, J.C. The radioactive labeling of proteins with an iodinated amidination reagent. *Anal. Biochem.* **1975**, *69*, 339–349. [CrossRef]
148. Su, S.N.; Jeng, I. Conversion of a primary amine to a labeled secondary amine by the addition of phenolic group and radioiodination. *Anal. Biochem.* **1983**, *128*, 405–411. [CrossRef]
149. Panuska, J.R.; Parker, C.W. Radioiodination of proteins by reductive alkylation. *Anal. Biochem.* **1987**, *160*, 182–201. [CrossRef]
150. Arruebo, M.; Vilaboa, N.; Saez-Gutierrez, B.; Lambea, J.; Tres, A.; Valladares, M.; Gonzalez-Fernandez, A. Assessment of the Evolution of Cancer Treatment Therapies. *Cancers* **2011**, *3*, 3279–3330. [CrossRef]
151. Cascini, G.L.; Asabella, A.N.; Notaristefano, A.; Restuccia, A.; Ferrari, C.; Rubini, D.; Altini, C.; Rubini, G. ^{124}Iodine: A Longer-Life Positron Emitter Isotope—New Opportunities in Molecular Imaging. *Biomed. Res. Int.* **2014**. [CrossRef] [PubMed]
152. Verel, I.; Visser, G.W.M.; van Dongen, G. The Promise of Immuno-PET in Radioimmunotherapy. *J. Nucl. Med.* **2005**, *46*, 164S–171S. [PubMed]
153. Marik, J.; Junutula, J.R. Emerging role of immunoPET in receptor targeted cancer therapy. *Curr. Drug Deliv.* **2011**, *8*, 70–78. [CrossRef] [PubMed]
154. Wilson, C.B.; Snook, D.E.; Dhokia, B.; Taylor, C.V.; Watson, I.A.; Lammertsma, A.A.; Lambrecht, R.; Waxman, J.; Jones, T.; Epenetos, A.A. Quantitative measurement of monoclonal antibody distribution and blood flow using positron emission tomography and ^{124}iodine in patients with breast cancer. *Int. J. Cancer* **1991**, *47*, 344–347. [CrossRef]
155. Du, Z.; Lovly, C.M. Mechanisms of receptor tyrosine kinase activation in cancer. *Mol. Cancer* **2018**, *17*, 58–71. [CrossRef]
156. Baselga, J.; Swain, S.M. Novel anticancer targets: Revisiting ERBB2 and discovering ERBB3. *Nat. Rev. Cancer* **2009**, *9*, 463–475. [CrossRef]
157. Meric-Bernstein, F.; Hung, M.C. Advances in targeting human epidermal growth factor receptor-2 signaling for cancer therapy. *Clin. Cancer Res.* **2006**, *12*, 6326–6330. [CrossRef]
158. Pool, M.; de Boer, H.R.; Hooge, M.N.L.; van Vugt, M.A.T.; de Vries, E.G.E. Harnessing Integrative Omics to Facilitate Molecular Imaging of the Human Epidermal Growth Factor Receptor Family for Precision Medicine. *Theranostics* **2017**, *7*, 2111–2133. [CrossRef]
159. Lub-de Hooge, M.N.; Kosternik, J.G.W.; Perik, P.J.; Nijnuis, H.; Tran, L.; Bart, J.; Suurmeijer, A.J.H.; de Jong, S.; Jager, P.L.; de Vries, E.G.E. Preclinical characterization of ^{111}InDTPA-trastuzumab. *Br. J. Pharmacol.* **2004**, *143*, 99–106. [CrossRef]
160. Paudyal, P.; Paudyal, B.; Hanaoka, H.; Oriuchi, N.; Lida, Y.; Yoshioka, H.; Tominaga, H.; Watanabe, S.; Watanabe, S.; Ishioka, N.S.; et al. Imaging and biodistribution of Her2/neu expression in non-small cell lung cancer xenografts with Cu-labeled trastuzumab PET. *Cancer Sci.* **2010**, *101*, 1045–1050. [CrossRef]
161. Bakir, M.A.; Eccles, S.A.; Babich, J.W.; Aftab, N.; Styles, J.M.; Dean, C.J.; Ott, R.J. c-erbB2 Protein Overexpression in Breast Cancer as a Target for PET Using Iodine-124-Labeled Monoclonal Antibodies. *J. Nucl. Med.* **1992**, *33*, 2154–2160. [PubMed]
162. Rubin, S.C.; Kairemo, K.J.A.; Brownell, A.-L.; Daghighjan, F.; Federici, M.G.; Pentlow, K.S.; Finn, R.D.; Lambrecht, R.M.; Hoskins, W.J.; Lewis, J.L., Jr.; et al. High-Resolution Positron Emission Tomography of Human Ovarian Cancer in Nude Rats Using ^{124}I-Labeled Monoclonal Antibodies. *Gynecol. Oncol.* **1993**, *48*, 61–67. [CrossRef]

163. Robinson, M.K.; Doss, M.; Shaller, C.; Narayanan, D.; Marks, J.D.; Adler, L.P.; Gonzalez Trotter, D.E.; Adams, G.P. Quantitative Immuno-Positron Emission Tomography Imaging of HER2-Positive Tumor Xenografts with an Iodine-124 Labeled Anti-HER2 Diabody. *Cancer Res.* **2005**, *65*, 1471–1478. [CrossRef] [PubMed]
164. Orlova, A.; Wallberg, H.; Stone-Elander, S.; Tolmachev, V. On the Selection of a Tracer for PET Imaging of HER2-Expressing Tumors: Direct Comparison of a ^{124}I-Labeled Affibody Molecule and Trastuzumab in a Murine Xenograft Model. *J. Nucl. Med.* **2009**, *50*, 417–425. [CrossRef] [PubMed]
165. Mendler, C.T.; Gehring, T.; Wester, H.-J.; Schwaiger, M.; Skerra, A. ^{89}Zr-Labeled vs. ^{124}I-Labeled αHER2 Fab with Optimized Plasma Half-Life for High-Contrast Tumor Imaging In Vivo. *J. Nucl. Med.* **2015**, *56*, 1112–1118. [CrossRef] [PubMed]
166. Johns, T.G.; Stockert, E.; Ritter, G.; Jungbluth, A.A.; Huang, H.J.; Cavenee, W.K.; Smyth, F.E.; Hall, C.M.; Watson, N.; Nice, E.C.; et al. Novel monoclonal antibody specific for the de2–7 epidermal growth factor receptor (EGFR) that also recognizes the EGFR expressed in cells containing amplification of the EGFR gene. *Int. J. Cancer* **2002**, *98*, 398–408. [CrossRef] [PubMed]
167. Jungbluth, A.A.; Stockert, E.; Huang, H.J.; Collins, V.P.; Coplan, K.; Iversen, K.; Kolb, D.; Johns, T.J.; Scott, A.M.; Gullick, W.J.; et al. A monoclonal antibody recognizing human cancers with amplification/overexpression of the human epidermal growth factor receptor. *Proc. Natl. Acad. Sci. USA* **2003**, *100*, 639–644. [CrossRef]
168. Panousis, C.; Rayzman, V.M.; Johns, T.G.; Renner, C.; Liu, Z.; Cartwright, G.; Lee, F.T.; Wang, D.; Gan, H.; Cao, D.; et al. Engineering and characterisation of chimeric monoclonal antibody 806 (ch806) for targeted immunotherapy of tumours expressing de2–7 EGFR or amplified EGFR. *Br. J. Cancer* **2005**, *92*, 1069–1077. [CrossRef]
169. Lee, F.T.; O'Keefe, G.J.; Gan, H.K.; Mountain, A.J.; Jones, G.R.; Saunder, T.H.; Sagona, J.; Rigopoulos, A.; Smyth, F.E.; Johns, T.G.; et al. Immuno-PET quantitation of de2–7 epidermal growth factor receptor expression in glioma using ^{124}I-IMP-R4-labeled antibody ch806. *J. Nucl. Med.* **2010**, *51*, 967–972. [CrossRef]
170. Lee, F.T.; Burvenich, I.J.; Guo, N.; Kocovski, P.; Tochon-Danguy, H.; Ackermann, U.; O'Keefe, G.J.; Gong, S.; Rigopoulos, A.; Liu, Z.; et al. L-tyrosine confers residualizing properties to a d-amino acid-rich residualizing peptide for radioiodination of internalizing antibodies. *Mol. Imaging* **2016**, *15*. [CrossRef]
171. Ferrara, N.; Adamis, A.P. Ten years of anti-vascular endothelial growth factor therapy. *Nat. Rev. Drug Discov.* **2016**, *15*, 385–403. [CrossRef] [PubMed]
172. Apte, R.S.; Chen, D.S.; Ferrara, N. VEGF in signaling and disease: Beyond discovery and development. *Cell* **2019**, *176*, 1248–1264. [CrossRef] [PubMed]
173. Gaykema, S.B.; Brouwers, A.H.; Lub-de Hooge, M.N.; Pleijhuis, R.G.; Timmer-Bosscha, H.; Pot, L.; van Dam, G.M.; van der Meulen, S.B.; de Jong, J.R.; Bart, J.; et al. ^{89}Zr-bevacizumab PET imaging in primary breast cancer. *J. Nucl. Med.* **2013**, *54*, 1014–1018. [CrossRef] [PubMed]
174. Van Asselt, S.J.; Oosting, S.F.; Brouwers, A.H.; Bongaerts, A.H.; de Jong, J.R.; Lub-de Hooge, M.N.; Oude Munnink, T.H.; Fiebrich, H.B.; Sluiter, W.J.; Links, T.P.; et al. Everolimus reduces ^{89}Zr-bevacizumab tumor uptake in patients with neuroendocrine tumors. *J. Nucl. Med.* **2014**, *55*, 1087–1092. [CrossRef]
175. Oosting, S.F.; Brouwers, A.H.; van Es, S.C.; Nagengast, W.B.; Oude Munnink, T.H.; Lub-de Hooge, M.N.; Hollema, H.; de Jong, J.R.; de Jong, I.J.; de Haas, S.; et al. ^{89}Zr-Bevacizumab PET Visualizes Heterogeneous Tracer Accumulation in Tumor Lesions of Renal Cell Carcinoma Patients and Differential Effects of Antiangiogenic Treatment. *J. Nucl. Med.* **2015**, *56*, 63–69. [CrossRef]
176. Bahce, I.; Huisman, M.C.; Verwer, E.E.; Ooijevaar, R.; Boutkourt, F.; Vugts, D.J.; van Dongen, G.A.; Boellaard, R.; Smit, E.F. Pilot study of ^{89}Zr-bevacizumab positron emission tomography in patients with advanced non-small cell lung cancer. *EJNMMI Res.* **2014**, *4*, 35–42. [CrossRef]
177. Jansen, M.H.; Veldhuijzen van Zanten, S.E.M.; van Vuurden, D.G.; Huisman, M.C.; Vugts, D.J.; Hoekstra, O.S.; van Dongen, G.A.; Kaspers, G.L. Molecular Drug Imaging: ^{89}Zr-bevacizumab PET in Children with Diffuse Intrinsic Pontine Glioma. *J. Nucl. Med.* **2017**, *58*, 711–716. [CrossRef]
178. Veldhuijzen van Zanten, S.E.M.; Sewing, A.C.P.; van Lingen, A.; Hoekstra, O.S.; Wesseling, P.; Meel, M.H.; van Vuurden, D.G.; Kaspers, G.J.L.; Hulleman, E.; Bugiani, M. Multiregional tumor drug-uptake imaging by PET and microvascular morphology in end-stage diffuse intrinsic pontine glioma. *J. Nucl. Med.* **2018**, *59*, 612–615. [CrossRef]
179. Collingridge, D.R.; Carroll, V.A.; Glaser, M.; Aboagye, E.O.; Osman, S.; Hutchinson, O.C.; Barthel, H.; Luthra, S.K.; Brady, F.; Bicknell, R.; et al. The development of [(^{124}I)]iodinated-VG76e: A novel tracer for imaging vascular endothelial growth factor in vivo using positron emission tomography. *Cancer Res.* **2002**, *62*, 5912–5919.
180. Rylova, S.N.; Del Pozzo, L.; Klingeberg, C.; Tonnesmann, R.; Illert, A.L.; Meyer, P.T.; Maecke, H.R.; Holland, J.P. Immuno-PET imaging of CD30-positive lymphoma using ^{89}Zr-desferrioxamine labeled CD30-specific AC-10 antibody. *J. Nucl. Med.* **2016**, *57*, 96–102. [CrossRef]
181. England, C.G.; Rui, L.; Cai, W. Lymphoma: Current status of clinical and preclinical imaging with radiolabeled antibodies. *Eur. J. Nucl. Med. Mol. Imaging* **2017**, *44*, 517–532. [CrossRef] [PubMed]
182. Kang, L.; Jiang, D.; Ehlerding, E.B.; Barnhart, T.E.; Ni, D.; Engle, J.W.; Wang, R.; Huang, P.; Xu, X.; Cai, W. Noninvasive trafficking of brentuximab vedotin and PET imaging of CD30 in lung cancer murine models. *Mol. Pharm.* **2018**, *15*, 1627–1634. [CrossRef] [PubMed]
183. Natarajan, A.; Gowrishankar, G.; Nielsen, C.H.; Wang, S.; Iagaru, A.; Goris, M.L.; Gambhir, S.S. Positron emission tomography of ^{64}Cu-DOTA-Rituximab in a transgenic mouse model expressing human CD20 for clinical translation to image NHL. *Mol. Imaging Biol.* **2012**, *14*, 608–616. [CrossRef]

184. Natarajan, A.; Gambhir, S.S. Radiation dosimetry study of [^{89}Zr]rituximab tracer for clinical translation of B cell NHL imaging using positron emission tomography. *Mol. Imaging Biol.* **2015**, *17*, 539–547. [CrossRef]
185. Muylle, K.; Flamen, P.; Vugts, D.J.; Guiot, T.; Ghanem, G.; Meuleman, N.; Bourgeois, P.; Vanderlinden, B.; van Dongen, G.A.; Everaert, H.; et al. Tumour targeting and radiation dose of radioimmunotherapy with (90)Y-rituximab in CD20+ B-cell lymphoma as predicted by (89)Zr-rituximab immuno-PET: Impact of preloading with unlabelled rituximab. *Eur. J. Nucl. Med. Mol. Imaging* **2015**, *42*, 1304–1314. [CrossRef] [PubMed]
186. Olafsen, T.; Sirk, S.J.; Betting, B.J.; Kenanova, V.E.; Bauer, K.B.; Ladno, W.; Raubitschek, A.A.; Timmerman, J.M.; Wu, A.M. ImmunoPET imaging of B-cell lymphoma using ^{124}I-anti-CD20 scFv dimers (diabodies). *Protein Eng. Des. Sel.* **2010**, *23*, 243–249. [CrossRef] [PubMed]
187. Olafsen, T.; Betting, D.; Kenanova, V.E.; Salazar, F.B.; Clarke, P.; Said, J.; Raubitschek, A.A.; Timmerman, J.M.; Wu, A.M. Recombinant Anti-CD20 Antibody Fragments for Small-Animal PET Imaging of B-Cell Lymphomas. *J. Nucl. Med.* **2009**, *50*, 1500–1508. [CrossRef]
188. Zettlitz, K.A.; Tavaré, R.; Knowles, S.M.; Steward, K.K.; Timmerman, J.M.; Wu, A.M. ImmunoPET of Malignant and Normal B Cells with ^{89}Zr- and ^{124}I-Labeled Obinutuzumab Antibody Fragments Reveals Differential CD20 Internalization In Vivo. *Clin. Cancer Res.* **2017**, *23*, 7242–7252. [CrossRef]
189. Huang, H.-F.; Zhu, H.; Li, G.-H.; Xie, Q.; Yang, X.-T.; Xu, X.-X.; Tian, X.-B.; Wan, Y.-K.; Yang, Z. Construction of Anti-hPD-L1 HCAb Nb6 and in Situ ^{124}I Labeling for Noninvasive Detection of PD-L1 Expression in Human Bone Sarcoma. *Bioconj. Chem.* **2019**, *30*, 2614–2623. [CrossRef]
190. Huang, H.; Zhu, H.; Xie, Q.; Tian, X.; Yang, X.; Feng, F.; Jiang, Q.; Sheng, X.; Yang, Z. Evaluation of ^{124}I-JS001 for hPD1 immuno-PET imaging using sarcoma cell homografts in humanized mice. *Acta Pharm. Sin. B* **2020**. [CrossRef]
191. Perkins, G.; Slater, E.; Sanders, G.; Prichard, J. Serum Tumor Markers. *Am. Fam. Physician* **2003**, *68*, 1075–1082. [PubMed]
192. Haglund, C.; Lindgren, J.; Roberts, P.J.; Nordling, S. Gastrointestinal cancer-associated antigen CA 19-9 in histological specimens of pancreatic tumours and pancreatitis. *Br. J. Cancer* **1986**, *53*, 189–195. [CrossRef] [PubMed]
193. Sawada, R.; Sun, S.M.; Wu, X.; Hong, F.; Ragupathi, G.; Livingston, P.O.; Scholz, W.W. Human monoclonal antibodies to sialyl-Lewis (CA19.9) with potent CDC, ADCC, and antitumor activity. *Clin. Cancer Res.* **2011**, *17*, 1024–1032. [CrossRef] [PubMed]
194. Escorcia, F.E.; Steckler, J.M.; Abdel-Atti, D.; Price, E.W.; Carlin, S.D.; Scholz, W.W.; Lewis, J.S.; Houghton, J.L. Tumor-Specific Zr-89 Immuno-PET Imaging in a Human Bladder Cancer Model. *Mol. Imaging Biol.* **2018**, *20*, 808–815. [CrossRef] [PubMed]
195. Houghton, J.L.; Abdel-Atti, D.; Scholz, W.W.; Lewis, J.S. Preloading with Unlabeled CA19.9 Targeted Human Monoclonal Antibody Leads to Improved PET Imaging with ^{89}Zr-5B1. *Mol. Pharmaceutics* **2017**, *14*, 908–915. [CrossRef]
196. Lohrmann, C.; O'Reilly, E.M.; O'Donoghue, J.A.; Pandit-Taskar, N.; Carrasquillo, J.A.; Lyashchenko, S.K.; Ruan, S.; Teng, R.; Scholz, W.; Maffuid, P.W.; et al. Retooling a Blood-Based Biomarker: Phase I Assessment of the High-Affinity CA19−9 Antibody HuMab-5B1 for Immuno-PET Imaging of Pancreatic Cancer. *Clin. Cancer Res.* **2019**, *25*, 7014–7023. [CrossRef]
197. Girgis, M.D.; Olafsen, T.; Kenanova, V.; McCabe, K.E.; Wu, A.M.; Tomlinson, J.S. CA19-9 as a Potential Target for Radiolabeled Antibody-Based Positron Emission Tomography of Pancreas Cancer. *Int. J. Mol. Imaging* **2011**. [CrossRef]
198. Girgis, M.D.; Federman, N.; Rochefort, M.M.; McCabe, K.E.; Wu, A.M.; Nagy, J.O.; Denny, C.; Tomlinson, J.S. An engineered anti-CA19-9 cys-diabody for positron emission tomography imaging of pancreatic cancer and targeting of polymerized liposomal nanoparticles. *J. Surg. Res.* **2013**, *185*, 45–55. [CrossRef]
199. Girgis, M.D.; Kenanova, V.; Olafsen, T.; McCabe, K.E.; Wu, A.M.; Tomlinson, J.S. Anti-Ca19−9 Diabody as a Pet Imaging Probe for Pancreas Cancer. *J. Surg. Res.* **2011**, *170*, 169–178. [CrossRef]
200. Rochefort, M.M.; Girgis, M.D.; Knowles, S.M.; Ankeny, J.S.; Salazar, F.; Wu, A.M.; Tomlinson, J.S. A Mutated Anti-CA19-9 scFv-Fc for Positron Emission Tomography of Human Pancreatic Cancer Xenografts. *Mol. Imaging Biol.* **2014**, *16*, 721–729. [CrossRef]
201. Chaturvedi, R.; Heimburg, J.; Yan, J.; Koury, S.; Sajjad, M.; Abdel-Nabi, H.H.; Rittenhouse-Olson, K. Tumor immunolocalization using ^{124}I-iodine-labeled JAA-F11 antibody to Thomsen–Friedenreich alpha-linked antigen. *Appl. Radiat. Isot.* **2008**, *66*, 278–287. [CrossRef] [PubMed]
202. Beauchemin, N.; Arabzadeh, A. Carcinoembryonic antigen-related cell adhesion molecules (CEACAMs) in cancer progression and metastasis. *Cancer Metastasis Rev.* **2013**, *32*, 643–671. [CrossRef] [PubMed]
203. Moffat, F.L., Jr.; Pinsky, C.M.; Hammershaimb, L.; Petrelli, N.J.; Patt, Y.Z.; Whaley, F.S.; Goldenberg, D.M. Clinical utility of external immunoscintigraphy with the IMMU-4 technetium-99m Fab′ antibody fragment in patients undergoing surgery for carcinoma of the colon and rectum: Results of a pivotal, phase III trial. The Immunomedics Study Group. *J. Clin. Oncol.* **1996**, *14*, 2295–2305. [CrossRef]
204. Wu, A.M.; Yazaki, P.J.; Tsai, S.; Nguyen, K.; Anderson, A.L.; McCarthy, D.W.; Welch, M.J.; Shively, J.E.; Williams, L.E.; Raubitschek, A.A.; et al. High-resolution microPET imaging of carcinoembryonic antigen-positive xenografts by using a copper-64-labeled engineered antibody fragment. *Proc. Natl. Acad. Sci. USA* **2000**, *97*, 8495–8500. [CrossRef]
205. Sundaresan, G.; Yazaki, P.J.; Shively, J.E.; Finn, R.D.; Larson, S.M.; Raubitschek, A.A.; Williams, L.E.; Chatziioannou, A.F.; Gamhir, S.S.; Wu, A.M. ^{124}I-Labeled Engineered Anti-CEA Minibodies and Diabodies Allow High-Contrast, Antigen- Specific Small-Animal PET Imaging of Xenografts in Athymic Mice. *J. Nucl. Med.* **2003**, *44*, 1962–1969. [PubMed]

206. Kenanova, V.; Olafsen, T.; Crow, D.M.; Sundaresan, G.; Subbarayan, M.; Carter, N.H.; Ikle, D.N.; Yazaki, P.J.; Chatziioannou, A.F.; Gambhir, S.S.; et al. Tailoring the Pharmacokinetics and Positron Emission Tomography Imaging Properties of Anti-Carcinoembryonic Antigen Single-Chain Fv-Fc Antibody Fragments. *Cancer Res.* **2005**, *65*, 622–631. [PubMed]
207. Girgis, M.D.; Olafsen, T.; Kenanova, V.; McCabe, K.E.; Wu, A.M.; Tomlinson, J.S. Targeting CEA in Pancreas Cancer Xenografts with a Mutated scFv-Fc Antibody fragment. *EJNMMI Res.* **2011**. [CrossRef] [PubMed]
208. Goldenberg, D.M.; Sharkey, R.M.; Paganelli, G.; Barbet, J.; Chatal, J.-M. Antibody Pretargeting Advances Cancer Radioimmunodetection and Radioimmunotherapy. *J. Clin. Oncol.* **2006**, *24*, 823–834. [CrossRef]
209. Sharkey, R.M.; Cardillo, T.M.; Rossi, E.A.; Chang, C.-H.; Karacay, H.; McBride, W.J.; Hansen, H.J.; Horak, I.D.; Goldenberg, D.M. Signal Amplification in Molecular Imaging by Pretargeting A Multivalent, Bispecific Antibody. *Nat. Med.* **2005**, *11*, 1250–1255. [CrossRef]
210. McBride, W.J.; Zanzonico, P.; Sharkey, R.M.; Noren, C.; Karacay, H.; Rossi, E.A.; Losman, M.J.; Brard, P.Y.; Chang, C.H.; Larson, S.M.; et al. Bispecific Antibody Pretargeting PET (ImmunoPET) with an ^{124}I-Labeled Hapten-Peptide. *J. Nucl. Med.* **2006**, *47*, 1678–1688.
211. Chiche, J.; Brahimi-Horn, M.C.; Pouysségur, J. Tumour hypoxia induces a metabolic shift causing acidosis: A common feature in cancer. *J. Cell. Mol. Med.* **2010**, *14*, 771–794. [CrossRef] [PubMed]
212. Oosterwdk, E.; Ruiter, D.J.; Hoedemaeker, P.J.; Pauwels, E.K.J.; Jonas, U.; Zwartendijk, I.; Warnaar, S.O. Monoclonal antibody G250 recognizes a determinant present in renal-cell carcinoma and absent from normal kidney. *Int. J. Cancer* **1986**, *38*, 489–494. [CrossRef] [PubMed]
213. Lawrentschuk, N.; Lee, F.T.; Jones, G.; Rigopoulos, A.; Mountain, A.; O'Keefe, G.; Papenfuss, A.T.; Bolton, D.M.; Davis, I.D.; Scott, A.M. Investigation of hypoxia and carbonic anhydrase IX expression in a renal cell carcinoma xenograft model with oxygen tension measurements and ^{124}I-cG250 PET/CT. *Urol. Oncol.* **2011**, *29*, 411–420. [CrossRef] [PubMed]
214. Stillebroer, A.B.; Franssen, G.M.; Mulders, P.F.A.; Oyen, W.J.G.; van Dongen, G.A.M.S.; Laverman, P.; Oosterwijk, E.; Boerman, O.C. ImmunoPET Imaging of Renal Cell Carcinoma with ^{124}I- and ^{89}Zr-Labeled Anti-CAIX Monoclonal Antibody cG250 in Mice. *Cancer Biother. Radiopharm.* **2013**, *28*, 510–515. [CrossRef]
215. Cheal, S.M.; Punzalan, B.; Doran, M.G.; Evans, M.J.; Osborne, J.R.; Lewis, J.S.; Zanzonico, P.; Larson, S.M. Pairwise comparison of ^{89}Zr- and ^{124}I-labeled cG250 based on positron emission tomography imaging and nonlinear immunokinetic modeling: In vivo carbonic anhydrase IX receptor binding and internalization in mouse xenografts of clear-cell renal cell carcinoma. *Eur. J. Nucl. Med. Mol. Imaging* **2014**, *41*, 985–994.
216. Sakamoto, J.; Kojima, H.; Kato, J.; Hamashima, H.; Suzuki, H. Organ-specific expression of the intestinal epithelium-related antigen A33, a cell surface target for antibody-based imaging and treatment in gastrointestinal cancer. *Cancer Chemother. Pharmacol.* **2000**, *46*, S27–S32. [CrossRef]
217. Welt, S.; Scott, A.M.; Divgi, C.R.; Kemeny, N.E.; Finn, R.D.; Daghighian, F.; Germain, J.S.; Richards, E.C.; Larson, S.M.; Old, L.J. Phase I/II study of iodine 125-labeled monoclonal antibody A33 in patients with advanced colon cancer. *J. Clin. Oncol.* **1996**, *14*, 1787–1797. [CrossRef]
218. Welt, S.; Divgi, C.R.; Kemeny, N.; Finn, R.D.; Scott, A.M.; Graham, M.; Germain, J.S.; Richards, E.C.; Larson, S.M.; Oettgen, H.F. Phase I/II study of iodine 131-labeled monoclonal antibody A33 in patients with advanced colon cancer. *J. Clin. Oncol.* **1994**, *12*, 1561–1571. [CrossRef]
219. Lee, F.T.; Hall, C.; Rigopoulos, A.; Zweit, J.; Pathmaraj, K.; O'Keefe, G.J.; Smyth, F.E.; Welt, S.; Old, S.J.; Scott, A.M. Immuno-PET of Human Colon Xenograft- Bearing BALB/c Nude Mice Using ^{124}I-CDR–Grafted Humanized A33 Monoclonal Antibody. *J. Nucl. Med.* **2001**, *42*, 764–769.
220. Ma, L.; Dong, L.; Chang, P. CD44v6 engages in colorectal cancer progression. *Cell Death Dis.* **2019**, *10*. [CrossRef]
221. Van Hal, N.L.W.; Van Dongen, G.A.M.S.; Rood-Knippels, E.M.C.; Van Der Valk, P.; Snow, G.B.; Brakenhoff, R.H. Monoclonal antibody U36, a suitable candidate for clinical immunotherapy of squamous-cell carcinoma, recognizes a CD44 isoform. *Int. J. Cancer* **1996**, *15*, 520–527. [CrossRef]
222. Verel, I.; Visser, G.W.M.; Vosjan, M.J.W.D.; Finn, R.; Boellaard, R.; van Dongen, G.A.M.S. High-quality ^{124}I-labelled monoclonal antibodies for use as PET scouting agents prior to ^{131}I-radioimmunotherapy. *Eur. J. Nucl. Med. Mol. Imaging* **2004**, *31*, 1645–1652. [CrossRef] [PubMed]
223. Verel, I.; Visser, G.W.M.; Boerman, O.C.; van Eerd, J.E.M.; Finn, R.; Boellaard, R.; Vosjan, M.J.W.D.; Stinger-van Walsum, M.; Snow, G.B.; van Dongen, G.A.M.S. Long-Lived Positron Emitters Zirconium-89 and Iodine-124 for Scouting of Therapeutic Radioimmunoconjugates with PET. *Cancer Biother. Radiopharm.* **2003**, *18*, 655–661. [CrossRef] [PubMed]
224. Fortin, M.-A.; Salnikov, A.V.; Nestor, M.; Heldin, N.-E.; Rubin, K.; Lundqvist, H. Immuno-PET of undifferentiated thyroid carcinoma with radioiodine-labelled antibody cMAb U36: Application to antibody tumour uptake studies. *Eur. J. Nucl. Med. Mol. Imaging* **2007**, *34*, 1376–1387. [CrossRef] [PubMed]
225. Torre, L.A.; Bray, F.; Siegel, R.L.; Ferlay, J.; Lortet-Tieulent, J.; Jemal, A. Global Cancer Statistics, 2012. *CA Cancer J. Clin.* **2015**, *65*, 87–108. [CrossRef]
226. Maurer, T.; Eiber, M.; Schwaiger, M.; Gschwend, J.E. Current use of PSMA-PET in prostate cancer management. *Nat. Rev. Urol.* **2016**, *13*, 226–235. [CrossRef]
227. Wustemann, T.; Haberkorn, U.; Babich, J.; Mier, W. Targeting prostate cancer: Prostate-specific membrane antigen based diagnosis and therapy. *Med. Res. Rev.* **2019**, *39*, 40–69. [CrossRef]

228. Tolmachev, V.; Malmberg, J.; Estrada, S.; Eriksson, O.; Orlova, A. Development of a ^{124}I-labeled version of the anti-PSMA monoclonal antibody capromab for immunoPET staging of prostate cancer: Aspects of labeling chemistry and biodistribution. *Int. J. Oncol.* **2014**, *44*, 1998–2008. [CrossRef]
229. Frigerio, B.; Morlino, S.; Luison, E.; Seregni, E.; Lorenzoni, A.; Satta, A.; Valdagni, R.; Bogni, A.; Chiesa, C. Anti-PSMA ^{124}I-scFvD2B as a new immuno-PET tool for prostate cancer: Preclinical proof of principle. *J. Exp. Clin. Cancer Res.* **2019**, *38*, 326–334. [CrossRef]
230. Milowsky, M.I.; Nanus, D.M.; Kostakoglu, L.; Vallabhajosula, S.; Goldsmith, S.J.; Bander, N.H. Phase I Trial of Yttrium-90-Labeled Anti-Prostate-Specific Membrane Antigen Monoclonal Antibody J591 for Androgen-Independent Prostate Cancer. *J. Clin. Oncol.* **2004**, *22*, 2522–2531. [CrossRef]
231. Tagawa, S.T.; Milowsky, M.I.; Morris, M.; Vallabhajosula, S.; Christos, P.; Akhtar, N.H.; Osborne, J.; Goldsmith, S.J.; Larson, S.; Taskar, N.P.; et al. Phase II Study of Lutetium-177-Labeled Anti-Prostate-Specific Membrane Antigen Monoclonal Antibody J591 for Metastatic Castration-Resistant Prostate Cancer. *Clin. Cancer Res.* **2013**, *19*, 5182–5191. [CrossRef] [PubMed]
232. Pandit-Taskar, N.; O'Donoghue, J.A.; Beylergil, V.; Lyashchenko, S.; Ruan, S.; Solomon, S.B.; Durack, J.C.; Carrasquillo, J.A.; Lefkowitz, R.A.; Gonen, M.; et al. ^{89}Zr-huJ591 Immuno-PET imaging in patients with advanced metastatic prostate cancer. *Eur. J. Nucl. Med. Mol. Imaging* **2014**, *41*, 2093–2105. [CrossRef] [PubMed]
233. Pandit-Taskar, N.; O'Donoghue, J.A.; Durack, J.C.; Lyashchenko, S.K.; Cheal, S.M.; Beylergil, V.; Lefkowitz, R.A.; Carrasquillo, J.A.; Martinez, D.F.; Fung, A.M.; et al. A Phase I/II Study for Analytic Validation of ^{89}Zr-J591 ImmunoPET as a Molecular Imaging Agent for Metastatic Prostate Cancer. *Clin. Cancer Res.* **2015**, *21*, 5277–5285. [CrossRef] [PubMed]
234. Pandit-Taskar, N.; O'Donoghue, J.A.; Divgi, C.R.; Wills, E.A.; Schwartz, L.; Gonen, M.; Smith-Jones, P.; Bander, N.H.; Scher, H.I.; Larson, S.M.; et al. Indium 111-labeled J591 anti-PSMA antibody for vascular targeted imaging in progressive solidtumors. *EJNMMI Res.* **2015**, *5*, 28. [CrossRef] [PubMed]
235. Fung, E.K.; Cheal, S.M.; Fareedy, S.B.; Punzalan, B.; Beylergil, V.; Amir, J.; Chalasani, S.; Weber, W.A.; Spratt, D.E.; Veach, D.R.; et al. Targeting of radiolabeled J591 antibody to PSMA-expressing tumors: Optimization of imaging and therapy based on non-linear compartmental modeling. *EJNMMI Res.* **2016**, *6*, 7. [CrossRef]
236. Reiter, R.E.; Gu, Z.; Watabe, T.; Thomas, G.; Szigeti, K.; Davis, E.; Wahl, M.; Nisotani, S.; Yamashiro, J.; Le Beau, M.M.; et al. Prostate stem cell antigen: A cell surface marker overexpressed in prostate cancer. *Proc. Natl. Acad. Sci. USA* **1998**, *95*, 1735–1740. [CrossRef]
237. Han, K.R.; Seligson, D.B.; Liu, X.; Hovarth, S.; Shintaku, P.I.; Thomas, G.V.; Said, J.W.; Reiter, R.E. Prostate Stem Cell Antigen Expression is Associated with Gleason Score, Seminal Vesicle Invasion and Capsular Invasion in Prostate Cancer. *J. Urol.* **2004**, *171*, 1117–1121. [CrossRef]
238. Barbisan, F.; Mazzucchelli, R.; Santinelli, A.; Scarpelli, M.; Lopez-Beltran, A.; Cheng, L.; Montironi, R. Expression of prostate stem cell antigen in high-grade prostatic intraepithelial neoplasia and prostate cancer. *Histopathology* **2010**, *57*, 572–579. [CrossRef]
239. Gu, Z.; Thomas, G.; Yamashiro, J.; Shintaku, I.P.; Dorey, F.; Raitano, A.; Wite, O.N.; Said, J.W.; Loda, M.; Reiter, R.E. Prostate stem cell antigen (PSCA) expression increases with high gleason score, advanced stage and bone metastasis in prostate cancer. *Oncogene* **2000**, *19*, 1288–1296. [CrossRef]
240. Zhigang, Z.; Wenlv, S. Prostate Stem Cell Antigen (PSCA) Expression in Human Prostate Cancer Tissues: Implications for Prostate Carcinogenesis and Progression of Prostate Cancer. *JPN J. Clin. Oncol.* **2004**, *34*, 414–419. [CrossRef]
241. Lam, J.S.; Yamashiro, J.; Shintaku, I.P.; Vessella, R.L.; Jenkins, R.B.; Horvath, S.; Said, J.W.; Reiter, R.E. Prostate Stem Cell Antigen is Overexpressed in Prostate Cancer Metastases. *Clin. Cancer Res.* **2005**, *11*, 2591–2596.
242. Amara, N.; Palapattu, G.S.; Schrage, M.; Gu, Z.; Thomas, G.V.; Dorey, F.; Said, J.; Reiter, R.E. Prostate Stem Cell Antigen is Overexpressed in Human Transitional Cell Carcinoma. *Cancer Res.* **2001**, *61*, 4660–4665. [PubMed]
243. Wente, M.N.; Jain, A.; Kono, E.; Berberat, P.O.; Giese, T.; Reber, H.A.; Friess, H.; Buchler, M.W.; Reiter, R.E.; Hines, O.J. Prostate Stem Cell Antigen Is a Putative Target for Immunotherapy in Pancreatic Cancer. *Pancreas* **2005**, *31*, 119–125. [CrossRef]
244. Ananias, H.J.K.; van den Heuvel, M.C.; Helfrich, W.; de Jong, I.J. Expression of the gastrin-releasing peptide receptor, the prostate stem cell antigen and the prostate specific membrane antigen in lymph node and bone metastases of prostate cancer. *Prostate* **2009**, *69*, 1101–1108. [CrossRef] [PubMed]
245. Gu, Z.; Yamashiro, J.; Kono, E.; Reiter, R.E. Anti-Prostate Stem Cell Antigen Monoclonal Antibody 1G8 Induces Cell Death In vitro and Inhibits Tumor Growth In vivo via a Fc-Independent Mechanism. *Cancer Res.* **2005**, *65*, 9495–9500. [CrossRef]
246. Leyton, J.V.; Olafsen, T.; Lepin, E.J.; Hahm, S.; Bauer, K.B.; Reiter, R.E.; Wu, A.M. A Humanized Radioiodinated Minibody for Imaging of Prostate Stem Cell Antigen-Expressing Tumors. *Clin. Cancer Res.* **2008**, *14*, 7488–7496. [CrossRef]
247. Olafsen, T.; Gu, Z.; Sherman, M.A.; Leyton, J.V. Targeting, Imaging, and Therapy Using a Humanized Antiprostate Stem Cell Antigen (PSCA) Antibody. *J. Immunother.* **2007**, *30*, 396–405. [CrossRef]
248. Leyton, J.V.; Olafsen, T.; Sherman, M.A.; Bauer, K.B.; Aghajanian, P.; Reiter, R.E.; Wu, A.M. Engineered humanized diabodies for microPET imaging of prostate stem cell antigen-expressing tumors. *Protein Eng. Des. Sel.* **2009**, *22*, 209–216. [CrossRef]
249. Lepin, E.J.; Leyton, J.V.; Zhou, Y.; Olafsen, T.; Salazar, F.B.; McCabe, K.E.; Hahm, S.; Marks, J.D.; Reiter, R.E.; Wu, A.M. An affinity matured minibody for PET imaging of prostate stem cell antigen (PSCA)-expressing tumors. *Eur. J. Nucl. Med. Mol. Imaging* **2010**, *37*, 1529–1538. [CrossRef]

250. Knowles, S.M.; Zettlitz, K.J.; Tavare, R.; Rochefortl, M.M.; Salazar, F.B.; Stout, D.B.; Yazaki, P.J.; Reiter, R.E.; Wu, A.M. Quantitative ImmunoPET of Prostate Cancer Xenografts with ^{89}Zr- and ^{124}I-Labeled Anti-PSCA A11 Minibody. *J. Nucl. Med.* **2014**, *55*, 452–459. [CrossRef]
251. Knowles, S.M.; Tavare, R.; Zettlitz, K.J.; Rochefortl, M.M.; Salazar, F.B.; Jiang, Z.K.; Reiter, R.E.; Wu, A.M. Applications of ImmunoPET: Using ^{124}I-Anti-PSCA A11 Minibody for Imaging Disease Progression and Response to Therapy in Mouse Xenograft Models of Prostate Cancer. *Clin. Cancer Res.* **2014**, *20*, 6367–6378. [CrossRef]
252. Tsai, W.K.; Zettlitz, K.A.; Tavare, R.; Kobayashi, N.; Reiter, R.E.; Wy, A.M. Dual-Modality ImmunoPET/Fluorescence Imaging of Prostate Cancer with an Anti-PSCA Cys-Minibody. *Theranostics* **2018**, *8*, 5903–5914. [CrossRef] [PubMed]
253. Zettlitz, K.A.; Wen-Ting, K.T.; Knowles, S.M.; Kobayashi, N.; Donahue, T.R.; Reiter, R.E.; Wu, A.M. Dual-Modality Immuno-PET and Near-Infrared Fluorescence Imaging of Pancreatic Cancer Using an Anti–Prostate Stem Cell Antigen Cys-Diabody. *J. Nucl. Med.* **2018**, *59*, 1398–1406. [CrossRef] [PubMed]
254. Tijink, B.M.; Perk, L.R.; Budde, M.; Stigter-van Walsum, M.S.; Visser, G.W.M.; Kloet, R.W.; Dinkelborg, L.M.; Leemans, C.R.; Neri, D.; van Dongen, G.A.M.S. ^{124}I-L19-SIP for immuno-PET imaging of tumour vasculature and guidance of ^{131}I-L19-SIP radioimmunotherapy. *Eur. J. Nucl. Med. Mol. Imaging* **2009**, *36*, 1235–1244. [CrossRef] [PubMed]
255. Poli, G.L.; Bianchi, C.; Virotta, G.; Bettini, A.; Moretti, R.; Trachsel, E.; Elia, G.; Giovannoni, L.; Neri, D.; Bruno, A. Radretumab Radioimmunotherapy in Patients with Brain Metastasis: A ^{124}I-L19SIP Dosimetric PET Study. *Cancer Immunol. Res.* **2013**, *1*, 134–143. [CrossRef] [PubMed]
256. Laforest, R.L.; Dehdashti, F.; Liu, Y.; Frye, J.; Frye, S.; Luehmann, H.; Sultan, D.; Shan, J.S.; Freimarl, B.D.; Siegel, B.A. First-in-Man Evaluation of ^{124}I-PGN650: A PET Tracer for Detecting Phosphatidylserine as a Biomarker of the Solid Tumor Microenvironment. *Mol. Imaging* **2017**, *16*. [CrossRef]
257. Glaser, M.; Collingridge, D.R.; Aboagye, E.O.; Bouchier-Hayes, L.; Hutchinson, O.C.; Martin, S.J.; Price, P.; Brady, F.; Luthra, S.K. Iodine-124 labelled Annexin-V as a potential radiotracer to study apoptosis using positron emission tomography. *Appl. Radiat. Isot.* **2003**, *58*, 55–62. [CrossRef]
258. Snook, D.E.; Rowlinson-Busza, G.; Sharma, H.L.; Epenetos, A.A. Preparation and in vivo study of ^{124}I-labelled monoclonal antibody H17E2 in a human tumour xenograft model. A prelude to positron emission tomography (PET). *Br. J. Cancer* **1990**, *62*, 89–91.
259. Guo, X.; Zhou, N.; Chen, Z.; Liu, T.; Xu, X.; Lei, X.; Shen, L.; Gao, J.; Yang, Z.; Zhu, H. Construction of ^{124}I-trastuzumab for noninvasive PET imaging of HER2 expression: From patient-derived xenograft models to gastric cancer patients. *Gastric Cancer* **2020**, *23*, 614–626. [CrossRef]
260. O'Donoghue, J.A.; Smith-Jones, P.M.; Humm, J.L.; Ruan, S.; Pryma, D.A.; Jungbluth, A.A.; Divgi, C.R.; Carrasquillo, J.A.; Pandit-Taskar, N.; Fong, Y.; et al. ^{124}I-huA33 Antibody Uptake Is Driven by A33 Antigen Concentration in Tissues from Colorectal Cancer Patients Imaged by Immuno-PET. *J. Nucl. Med.* **2011**, *52*, 1878–1885. [CrossRef]
261. Carrasquillo, J.A.; Pandit-Taskar, N.; O'Donoghue, J.A.; Humm, J.L.; Zanzonico, P.; Smith-Jones, P.M.; Divgi, C.R.; Pryma, D.A.; Ruan, S.; Kemeny, N.E.; et al. ^{124}I-huA33 Antibody PET of Colorectal Cancer. *J. Nucl. Med.* **2011**, *52*, 1173–1180. [CrossRef]
262. Zanzonico, P.; Carrasquillo, J.A.; Pandit-Taskar, N.; O'Donoghue, J.A.; Humm, J.L.; Smith-Jones, P.; Ruan, S.; Divgi, C.; Kemeny, N.E.; Fong, Y.; et al. PET-based compartmental modeling of ^{124}I-A33 antibody: Quantitative characterization of patient-specific tumor targeting in colorectal cancer. *Eur. J. Nucl. Med. Mol. Imaging* **2015**, *42*, 1700–1706. [CrossRef] [PubMed]
263. Schwartz, J.; Humm, J.L.; Divgi, C.R.; Larson, S.M.; O'Donoghue, J.A. Bone Marrow Dosimetry Using ^{124}I-PET. *J. Nucl. Med.* **2012**, *53*, 615–621. [CrossRef] [PubMed]
264. Divgi, C.R.; Pandit-Taskar, N.; Jungbluth, A.A.; Reuter, V.E.; Gonen, M.; Ruan, S.; Pierre, C.; Nagel, A.; Pryma, D.; Humm, J.; et al. Preoperative characterisation of clear-cell renal carcinoma using iodine-124-labelled antibody chimeric G250 (^{124}I-cG250) and PET in patients with renal masses: A phase I trial. *Lancet Oncol.* **2007**, *8*, 304–310. [CrossRef]
265. Pryma, D.A.; O'Donoghue, J.A.; Humm, J.L.; Jungbluth, A.A.; Old, L.J.; Larson, S.M.; Divgi, R.R. Correlation of In Vivo and In Vitro Measures of Carbonic Anhydrase IX Antigen Expression in Renal Masses Using Antibody ^{124}I-cG250. *J. Nucl. Med.* **2011**, *52*, 535–540. [CrossRef] [PubMed]
266. Divgi, C.R.; Uzzo, R.G.; Gatsonis, C.; Bartz, R.; Treutner, S.; Yu, J.Q.; Chen, D.; Carrasquillo, J.A.; Larson, S.M.; Bevan, P.; et al. Positron Emission Tomography/Computed Tomography Identification of Clear Cell Renal Cell Carcinoma: Results From the REDECT Trial. *J. Clin. Oncol.* **2013**, *31*, 187–194. [CrossRef] [PubMed]
267. Povoski, S.P.; Hall, N.C.; Murrey, D.A., Jr.; Sharp, D.S.; Hitchcock, C.L.; Mojzisik, C.M.; Bahnson, E.E.; Knopp, M.V.; Martin, E.W., Jr.; Bahnson, R.R. Multimodal Imaging and Detection Strategy With ^{124}I-Labeled Chimeric Monoclonal Antibody cG250 for Accurate Localization and Confirmation of Extent of Disease During Laparoscopic and Open Surgical Resection of Clear Cell Renal Cell Carcinoma. *Surg. Innov.* **2013**, *20*, 59–69. [CrossRef]
268. Bahnson, E.B.; Murrey, D.A.; Mojzisik, C.M.; Hall, N.C.; Martinez-Suarez, H.J.; Knopp, M.V.; Martin, E.W.; Povoski, S.P.; Bahnson, R.R. PET/CT imaging of clear cell renal cell carcinoma with ^{124}I labeled chimeric antibody. *Ther. Adv. Urol.* **2009**, *1*, 67–70. [CrossRef]
269. Smaldone, M.C.; Chen, D.Y.T.; Yu, J.Q.; Pilmack, E.R. Potential role of ^{124}I-girentuximab in the presurgical diagnosis of clear-cell renal cell cancer. *Biol. Target Ther.* **2012**, *6*, 395–407.
270. Lau, J.; Lin, K.S.; Benard, F. Past, Present, and Future: Development of Theranostic Agents Targeting Carbonic Anhydrase IX. *Theranostics* **2017**, *7*, 4322–4339. [CrossRef]

271. Carrasquillo, J.A.; O'Donoghue, J.A.; Beylergi, V.; Ruan, S.; Pandit-Taskar, N.; Larson, S.M.; Smith-Jnoes, P.M.; Lyashchenko, S.K.; Ohishi, N.; Ohtomo, T.; et al. I-124 codrituzumab imaging and biodistribution in patients with hepatocellular carcinoma. *EJNMMI Res.* **2018**, *8*, 20. [CrossRef] [PubMed]

Article

Radionuclide Molecular Imaging of EpCAM Expression in Triple-Negative Breast Cancer Using the Scaffold Protein DARPin Ec1

Anzhelika Vorobyeva [1,2,*], Ekaterina Bezverkhniaia [2,3], Elena Konovalova [4], Alexey Schulga [2,4], Javad Garousi [1], Olga Vorontsova [1], Ayman Abouzayed [5], Anna Orlova [2,5,6], Sergey Deyev [2,4,7,8] and Vladimir Tolmachev [1,2]

1 Department of Immunology, Genetics and Pathology, Uppsala University, 751 85 Uppsala, Sweden; javad.garousi@igp.uu.se (J.G.); olga.vorontsova@igp.uu.se (O.V.); vladimir.tolmachev@igp.uu.se (V.T.)
2 Research Centrum for Oncotheranostics, Research School of Chemistry and Applied Biomedical Sciences, National Research Tomsk Polytechnic University, 634 050 Tomsk, Russia; yekaterinabezv@mail.ru (E.B.); schulga@gmail.com (A.S.); anna.orlova@ilk.uu.se (A.O.); biomem@mail.ru (S.D.)
3 Department of Pharmaceutical Analysis, Siberian State Medical University, 634050 Tomsk, Russia
4 Molecular Immunology Laboratory, Shemyakin & Ovchinnikov Institute of Bioorganic Chemistry, Russian Academy of Sciences, 117997 Moscow, Russia; elena.ko.mail@gmail.com
5 Department of Medicinal Chemistry, Uppsala University, 751 23 Uppsala, Sweden; ayman.abouzayed@ilk.uu.se
6 Science for Life Laboratory, Uppsala University, 751 23 Uppsala, Sweden
7 Bio-Nanophotonic Lab, Institute of Engineering Physics for Biomedicine (PhysBio), National Research Nuclear University 'MEPhI', 115409 Moscow, Russia
8 Center of Biomedical Engineering, Sechenov University, 119991 Moscow, Russia
* Correspondence: anzhelika.vorobyeva@igp.uu.se

Academic Editor: Krishan Kumar
Received: 29 July 2020; Accepted: 13 October 2020; Published: 14 October 2020

Abstract: Efficient treatment of disseminated triple-negative breast cancer (TNBC) remains an unmet clinical need. The epithelial cell adhesion molecule (EpCAM) is often overexpressed on the surface of TNBC cells, which makes EpCAM a potential therapeutic target. Radionuclide molecular imaging of EpCAM expression might permit selection of patients for EpCAM-targeting therapies. In this study, we evaluated a scaffold protein, designed ankyrin repeat protein (DARPin) Ec1, for imaging of EpCAM in TNBC. DARPin Ec1 was labeled with a non-residualizing $[^{125}I]I$-*para*-iodobenzoate (PIB) label and a residualizing $[^{99m}Tc]Tc(CO)_3$ label. Both imaging probes retained high binding specificity and affinity to EpCAM-expressing MDA-MB-468 TNBC cells after labeling. Internalization studies showed that Ec1 was retained on the surface of MDA-MB-468 cells to a high degree up to 24 h. Biodistribution in Balb/c nu/nu mice bearing MDA-MB-468 xenografts demonstrated specific uptake of both $[^{125}I]I$-PIB-Ec1 and $[^{99m}Tc]Tc(CO)_3$-Ec1 in TNBC tumors. $[^{125}I]I$-PIB-Ec1 had appreciably lower uptake in normal organs compared with $[^{99m}Tc]Tc(CO)_3$-Ec1, which resulted in significantly ($p < 0.05$) higher tumor-to-organ ratios. The biodistribution data were confirmed by micro-Single-Photon Emission Computed Tomography/Computed Tomography (microSPECT/CT) imaging. In conclusion, an indirectly radioiodinated Ec1 is the preferable probe for imaging of EpCAM in TNBC.

Keywords: EpCAM; radionuclide; molecular imaging; SPECT; iodine; PIB; breast; cancer

1. Introduction

Breast cancer is one of the most common types of cancer among women worldwide. It is a heterogeneous disease, which is categorized into four subtypes: Luminal A, luminal B, human epidermal growth factor receptor 2-positive (HER2-positive), and "basal-like" or "triple-negative" [1]. Triple-negative breast cancer (TNBC) does not express estrogen, progesterone, or human epidermal growth factor (HER2) receptors and is characterized by having an aggressive course, early metastatic spread, and poor prognosis [2]. Although initial response to chemotherapy in TNBC might be better compared to other breast cancer subtypes, early relapse is commonly observed [3–5]. Furthermore, the high rate of relapse among TNBC patients after surgery because of incomplete eradication of the tumor highlights the need for more effective therapies [2].

Targeted systemic treatment is well proven for estrogen/progesterone- or HER2-expressing breast cancer but is still not established to the same extent for TNBC [1]. Therefore, disseminated TNBC is particularly challenging to treat. The standard of care for TNBC patients is sequential chemotherapy. However, its indiscriminate toxicity to healthy tissues results in a narrow therapeutic window and limited efficacy. Targeted delivery of a cytotoxic payload (drug, toxin, or radionuclide) specifically to cancer cells would reduce the off-target toxicity and increase the therapeutic window and the efficiency of the treatment. In April 2020 the US Food and Drug Administration (FDA) approved the antibody-drug conjugate sacituzumab govitecan-hziy, which targets the tumor-associated calcium signal transducer 2 (TROP-2) antigen, for treatment of disseminated TNBC refractory to previous chemotherapies [6]. It showed a 33% response rate in patients with metastatic TNBC pretreated with chemotherapy [7]. Still, the development of therapeutics specific to other molecular targets could further increase the success rate in treatment of TNBC.

The epithelial cell adhesion molecule (EpCAM) is overexpressed in a large number of TNBC cases. The fraction of EpCAM overexpression in TNBC is 36–88%, depending on the scoring system [8–10], which makes it an attractive target for this malignancy. In patients with TNBC, EpCAM overexpression is associated with unfavorable prognosis [9] and correlates with poor survival, lymph node metastasis, and distant metastasis [11]. Several therapeutic strategies targeting EpCAM in TNBC are under preclinical and clinical development [12–14]. The anti-EpCAM monoclonal antibody adecatumumab has been evaluated in a phase II clinical study in patients with metastatic breast cancer [15]. Three of the 18 patients with high EpCAM expression and adecatumumab treatment developed new metastases up to week 6, compared with 14 of 29 patients with low EpCAM expression, indicating that response was related to EpCAM expression.

Due to heterogeneity of EpCAM expression in patients with TNBC, it is necessary to select the patients with high expression for EpCAM-targeted therapy. Radionuclide molecular imaging allows for non-invasive, whole-body evaluation of targeted protein expression. In the past years, a number of biomolecules against various targets have been investigated for radionuclide molecular imaging of breast cancer, such as somatostatin (SST) analogues targeting SST receptors, vasoactive intestinal peptide targeting affibody molecules against HER2, arginine-glycine-aspartate (RGD) peptides targeting integrin receptors, bombesin analogues targeting gastrin-releasing peptide receptors (GRPRs), and peptide analogues of alpha-melanocyte stimulating hormone-targeting melanocortin receptors [16].

Previously developed probes for radionuclide molecular imaging of EpCAM were mainly based on a monoclonal antibody (mAb) scaffold [17–19]. In comparison to mAbs, engineered scaffold proteins (ESP) have more favorable properties for imaging because of their small size, which enables rapid localization in tumors and fast decrease of blood-associated background activity due to renal clearance [20,21]. Additionally, ESPs can be genetically engineered to incorporate peptide-based chelators, e.g., histidine tags for labeling with technetium-99m tricarbonyl, and can generally tolerate harsh radiolabeling conditions, such as high temperature or changes in pH. A class of ESP, the designed ankyrin repeat proteins (DARPins), demonstrated excellent results for radionuclide molecular imaging of HER2 in preclinical studies, providing high tumor-to-nontumor tissue ratio shortly after injection [22–24]. The DARPin scaffold consists of four to six helix-turn-helix units and has molecular weight from 14 to 18 kDa, depending on the number of units. DARPins are currently the only class of ESPs with binders selected against EpCAM [25]. According to surface plasmon resonance, DARPin Ec1 has affinity of 68 pM to EpCAM [25], which meets requirements for a high-affinity imaging probe.

High affinity of an imaging probe is an important precondition for successful imaging, but it is not sufficient. Selection of a radionuclide and chemistry for its conjugation to a targeting probe is essential. Experience with other targeting proteins (affibody molecules [26] and DARPins targeting HER2) [22–24,27] demonstrated that the selection of an optimal labeling approach can increase tumor-to-organ ratios by an order of magnitude. Modification of the protein surface by a radionuclide in combination with a chelator or linker for coupling results in the alteration of off-target interactions with blood vessels and normal tissues. This has an essential impact on unspecific uptake in normal organs and tissues. Furthermore, accumulation of activity in both tumors and normal tissues depends on physicochemical properties of radioactive metabolites, which are formed after internalization and intracellular proteolysis of a labeled protein. Labels having charged or bulky polar radiometabolites are trapped inside the cells after proteolysis. Such labels are called *residualizing* [28]. The residualizing labels provide a long retention of activity both in tumors and in normal tissues if the imaging probe is internalized. In the case when the radiometabolites are lipophilic, they are capable of diffusing through cellular membranes and leaving the cell. These so-called *non-residualizing* labels are associated with low cellular retention of activity after internalization. When the internalization of an imaging probe by malignant cells is rapid, the use of residualizing labels is the only option for sufficient accumulation of activity in a tumor. However, the use of non-residualizing labels might be appropriate when internalization of a targeting probe by the cancer cells is slow, and an unspecific uptake, first and foremost, in liver and kidneys results in rapid internalization in normal tissues. In this case, tumor retention of activity depends mainly on the high affinity of a probe to its molecular target. A successful use of non-residualizing radiohalogen labels was demonstrated earlier for ESPs such as HER2-binding Albumin-binding domain (ABD)-Derived Affinity ProTein 6 (ADAPT6) [29] and HER2-binding DARPins 9_29 and G3 [22,23], as well as for Ec1 for imaging of EpCAM expression in pancreatic and ovarian cancer models [30,31]. Importantly, good retention in tumors was accompanied by a rapid clearance from normal tissues, which increased tumor-to-organ ratios. Apparently, slow internalization of Ec1 after binding to TNBC cells is critical for the use of a non-residualizing label for imaging in this cancer. Such a slow internalization of radiolabeled Ec1 was observed in the case of binding to pancreatic and ovarian cell lines [30,31], which was an indication that the cellular processing pattern would be similar for TNBC as well. However, our previous observations with ESP suggest that an internalization rate might depend on origin of cancer cells. For example, a rate of HER2-binding affibody [^{111}In]In-DOTA-ZHER2:342-pep2 internalization by ovarian cancer cells was twice higher compared with a rate of internalization by breast cancer cells [32]. HER3-binding affibody-ABD fusion proteins were internalized by pancreatic cancer cells appreciably rapider than by prostate cancer cells [33]. Thus, evaluation of internalization of radiolabeled Ec1 by TNBC cells was necessary before in vivo experiments.

The goal of this study was to investigate whether our previous findings could be translated to triple-negative breast cancer and to evaluate the potential of DARPin Ec1 for imaging of EpCAM in a TNBC model in vivo. To evaluate the influence of residualizing properties of the radiolabel, technetium-99m tricarbonyl [^{99m}Tc]Tc(CO)$_3$ with residualizing properties was used as a comparator to the non-residualizing [^{125}I]I-*para*-iodobenzoate ([^{125}I]I-PIB) label. The ^{99m}Tc ($T_{1/2}$ = 6.01 h) is the most commonly used radionuclide in clinical single-photon emission computed tomography (SPECT). The ^{125}I ($T_{1/2}$ = 59.4 d) is a chemical analogue and a convenient preclinical surrogate for ^{123}I ($T_{1/2}$ = 13.27 h), which is used for SPECT, or ^{124}I ($T_{1/2}$ = 4.18 d), which is used for positron emission tomography (PET).

2. Materials and Methods

2.1. General Procedures

Sodium iodide [^{125}I]NaI was purchased from PerkinElmer Sverige AB (Upplands Väsby, Sweden). Technetium-99m was obtained as pertechnetate by elution of Ultra-TechneKow generator (Mallinckrodt, Petten, The Netherlands) with sterile 0.9% sodium chloride (Mallinckrodt, Petten, The Netherlands). The CRS (Center for Radiopharmaceutical Sciences) kits for production of tricarbonyl technetium were purchased from the Center for Radiopharmaceutical Sciences (PSI, Villigen, Switzerland; contact e-mail: crs-kit@psi.ch). Instant thin-layer chromatography (iTLC) analysis was performed using iTLC silica gel strips (Varian, Lake Forest, CA, USA). The radioactivity distribution along iTLC strips was measured using a Cyclone storage phosphor system (Packard) and analyzed by OptiQuant image analysis software. Purification of radiolabeled proteins was performed using NAP-5 size-exclusion columns (GE Healthcare, Buckinghamshire, UK). Radioactivity was measured using an automated gamma-spectrometer with a NaI (TI) detector (1480 Wizard, Wallac, Finland). MDA-MB-468 breast cancer cells and Ramos lymphoma cells were purchased from the American Type Culture Collection (ATCC) and were cultured in Roswell Park Memorial Institute (RPMI) medium supplemented with 10% fetal bovine serum (FBS), 2 mM L-glutamine, 100 IU/mL penicillin, and 100 μg/mL streptomycin, in a humidified incubator with 5% CO_2 at 37 °C, unless mentioned otherwise. Binding specificity and cellular processing experiments were performed using 35-mm Petri dishes (Nunclon Delta Surface, ThermoFisher Scientific, Roskilde, Denmark). Ligand Tracer experiments were performed using 89-mm Petri dishes (Nunclon Delta Surface, ThermoFisher Scientific, Roskilde, Denmark).

2.2. Protein Production and Radiolabeling

The EpCAM-targeting DARPin Ec1-H_6 (containing a hexahistidine tag at C-terminus) was produced based on sequences published previously [25]. Production and purification of DARPin Ec1 was performed as described previously [30].

Indirect radioiodination of Ec1 using N-succinimidyl-para-(trimethylstannyl)benzoate was performed as described earlier [28,31]. Acetic acid in water (0.1%, 10 μL) was added to radioiodine (5–15 μL, 17–42 MBq). Then, N-succinimidyl-p-(trimethylstannyl)benzoate (13 nmoles, 5 μg, 5 μL of 1 mg/mL in 5% acetic acid in methanol) and chloramine-T (40 μg, 10 μL, 4 mg/mL in water) were added. The reaction was stopped by addition of sodium metabisulfite (60 μg, 10 μL, 6 mg/mL in water) after 5 min of incubation at room temperature. Then, DARPin Ec1 (7.6 nmoles, 140 μg, 39 μL of 3.6 mg/mL in phosphate-buffered saline (PBS)) in 140 μL of 0.07 M borate buffer (pH 9.3) was added and incubated at room temperature for 30 min. The radiolabeled conjugate was purified on a NAP-5 column, pre-equilibrated with 1% bovine serum albumin (BSA) in PBS, and eluted with PBS. The labeling yield and purity were determined using radio-iTLC analysis in 4:1 acetone:water system.

Site-specific radiolabeling of DARPin Ec1-H_6 bearing a C-terminal His_6-tag with tricarbonyl technetium-99m was performed as described earlier [31]. The solution eluted from the technetium generator (500 μL) containing 3–4 GBq of [^{99m}Tc]Tc pertechnetate was added to the CRS kit, and the mixture was incubated at 100 °C for 30 min. The obtained solution of [^{99m}Tc]Tc$(CO)_3$ (12 μL, 83–108 MBq) was mixed with a solution of DARPin Ec1 (40 μg, 2.18 nmol) in 33 μL of PBS and incubated at 60 °C for 60 min. The radiochemical yield and purity were determined using iTLC strips eluted with PBS. The radiolabeled DARPin Ec1 was purified using NAP-5 columns pre-equilibrated and eluted with PBS.

Radio high-performance liquid chromatography (HPLC) analysis was performed using a Hitachi Chromaster HPLC system with a radioactivity detector and Phenomenex Luna® C18 column (100 Å; 150 × 4.6 mm; 5 μm) at room temperature (20 °C). Solvent A was 0.1% trifluoroacetic acid (TFA) in H_2O, solvent B was 0.1% TFA in acetonitrile, and the flow rate was 1 mL/min. For identity and purity analysis, the 20-min method with a gradient from 5 to 95% solvent B over 18 min and from 95% to 5% solvent B from 18 to 20 min was used.

The label stability under challenge conditions (excess of histidine for [^{99m}Tc]Tc(CO)$_3$-Ec1, NaI, or in 30% ethanol for [^{125}I]I-PIB-Ec1) was assessed by analysis with iTLC silica gel (SG) strips eluted by PBS or by 4:1 acetone:water. The stability in complete cell culture medium containing 10% fetal bovine serum after 24 h of incubation at 37 °C was assessed by passing the media through a NAP-5 size-exclusion column and collecting the high-molecular-weight fraction (containing molecules over 5 kDa) and the low-molecular-weight fraction (containing molecules below 5 kDa). The activity in the column, the high- and low-molecular-weight fractions, was measured using a gamma-spectrometer.

2.3. *Binding Specificity and Cellular Processing Assays*

In vitro studies were performed using EpCAM-expressing breast cancer cell line MDA-MB-468. One day before the experiment, cells were seeded in 3-cm Petri dishes (ca. 1×10^6 cells per dish) and three dishes per group were used.

Binding specificity to EpCAM was evaluated as described previously [30,31]. To saturate EpCAM receptors, 100-fold excess of nonlabeled Ec1 DARPin (200 nM) in cell culture medium was added to one group of cells and an equal volume of media only was added to the second group. After 30 min of incubation at room temperature, radiolabeled DARPins [^{125}I]I-PIB-Ec1 or [^{99m}Tc]Tc(CO)$_3$-Ec1 were added at 2 nM final concentration. After 6 h of incubation at room temperature, the medium was collected, cells were washed, and trypsin was added to detach the cells. The cell suspension was collected, and the radioactivity of cells and medium was measured to calculate the percent of cell-bound radioactivity. The data were analyzed using unpaired two-tailed *t*-test.

Cellular retention and processing were studied during continuous incubation using an acid-wash method [32]. To study cellular processing during continuous incubation, radiolabeled [^{99m}Tc]Tc(CO)$_3$-Ec1 or [^{125}I]I-PIB-Ec1 (1 nM) were added to cells, which were incubated at 37 °C in a humidified incubator for 1, 2, 4, 6, and 24 h. At these time points, the media were collected from one group and cells were washed once with serum-free media. To collect the membrane-bound fraction, the cells were treated with 0.2 M glycine buffer containing 4 M urea (pH 2.0) on ice for 5 min causing dissociation of membrane-bound protein. The buffer was collected, and the cells were washed once with the same buffer. Then the cells were treated with 1 M NaOH for 30 min to lyse the cells containing internalized fraction, and the solution was collected. The activity in every fraction was measured. The maximum value of cell-associated activity for each dataset was taken as 100% and the other dataset values were normalized to it. To study cellular retention and processing after interrupted incubation, the cells were incubated with [^{99m}Tc]Tc(CO)$_3$-Ec1 or [^{125}I]I-PIB-Ec1 (10 nM) for 1 h at 4 °C. Then the media were removed, the cells were washed, fresh medium was added, and the cells were placed in a humidified incubator at 37 °C. At 1, 4, and 24 h, the medium was collected and cells were washed and treated as described above to evaluate the membrane-bound and internalized fractions.

For analysis of radiocatabolites in the supernatant after interrupted incubation, a part of it (500 μL) was separated using a NAP-5 size-exclusion column, pre-equilibrated with 1% BSA in PBS. Fractions containing activity associated with the high-molecular-weight compounds (first 900 μL) and low-molecular-weight compounds (3.6 mL) were collected. The activity in each fraction and the column were measured using a gamma-spectrometer. The residual activity left on the columns after separation was below 4% from the total activity. As a control for stability of the label, [^{99m}Tc]Tc(CO)$_3$-Ec1 and [^{125}I]I-PIB-Ec1 were incubated in complete media at 37 °C in a humidified incubator for 24 h and analyzed as described above.

2.4. *Affinity Measurements Using LigandTracer*

The kinetics of [^{125}I]I-PIB-Ec1 and [^{99m}Tc]Tc(CO)$_3$-Ec1 binding to living MDA-MB-468 and Ramos cells were measured using LigandTracer and evaluated using the TraceDrawer Software (both from Ridgeview Instruments, Vänge, Sweden) as described earlier [34]. Briefly, 2×10^6 MDA-MB-468 cells were seeded to a local area of an 89-mm Petri dish one day before the experiment. Ramos cells growing in suspension were attached to a Petri dish following a method developed by Bondza et al. [35].

Biomolecular anchor molecule (BAM) (SUNBRIGHT® OE-040CS, NOF Corporation) was dissolved in water to a concentration of 2 mg/mL. An area of a 98-mm Petri dish (Nunc, Cat No 263991) about 1.5 cm in diameter and 5 mm from the rim of the dish was covered with 400 μL of BAM solution (0.8 mg) and was incubated under sterile conditions at room temperature for 1 h. The BAM solution was aspirated and 400 μL of Ramos cell suspension (5×10^6 cells/mL, 2×10^6) were added dropwise to the BAM-coated area. Cells were allowed to attach to the dish for 40 min. Then the dish was tilted to remove the remaining cell suspension. The dish was covered with complete cell culture medium (10 mL) and placed into the incubator overnight. Cell attachment was confirmed the next day by observing the cells under a microscope.

To measure the binding during association phase, three concentrations of [^{125}I]I-PIB-Ec1 (1.8, 5.4, and 14.5 nM) or [^{99m}Tc]Tc(CO)$_3$-Ec1 (0.2, 0.6, and 1.8 nM) were added to cells, followed by exchange of media and measurement of retention in the dissociation phase. Binding kinetics were recorded at room temperature and dissociation constants were calculated based on association and dissociation rates.

2.5. Ethical Statement

The described procedures were reviewed and approved by the Animal Research Committee at Uppsala University (ethical permission 4/16 from 26 February 2016) and were performed in accordance with the Swedish national legislation on protection of laboratory animals.

2.6. Animal Studies

To select an optimal label, a dual-label biodistribution study was performed. To establish MDA-MB-468 xenografts, 10^7 cells were implanted subcutaneously in 8-week-old Balb/c nu/nu mice. For specificity control, 10^7 EpCAM-negative Ramos cells were implanted. At the time of experimentation (two to three weeks after implantation), the weights of the animals were 17 ± 2 g in the MDA-MB-468 group and 17 ± 0 g in the Ramos group. Average tumor weights were 0.05 ± 0.04 g for MDA-MB-468 and 0.05 ± 0.04 g for Ramos. Groups of four animals per data point were used.

A well-established, dual-label approach [36,37] was selected for animal studies. In this methodology, a mixture of compounds labeled with different nuclides is co-injected into animals, and the distribution of each labeled compound is determined by gamma-spectrometry of tissue samples. A precondition for this approach is that the gamma-spectra of nuclides can be resolved. This is the case for ^{125}I and ^{99m}Tc (Figure S7). An advantage of the dual-label methodology is that the factors related to a host animal (e.g., individual features of metabolic rate and blood circulation) and xenografts (e.g., vascularization or presence of necrotic areas) act in the same way on both tracers. This method enables the use of a paired t-test, which provides high statistic power with a small number of animals.

Mice were injected with a mixture of both [^{125}I]I-PIB-Ec1 (non-residualizing label) and [^{99m}Tc]Tc(CO)$_3$-Ec1 (residualizing label) and the biodistribution was measured 6 h and 24 h post injection (pi). The injected activity was 40 kBq/mouse for technetium-99m and 20 kBq/mouse for iodine-125. The injected protein dose was adjusted to 4 μg/mouse using unlabeled protein. The labeled proteins were injected into the tail vein. Before dissection, mice were anesthetized by an intraperitoneal (i.p.) injection of ketamine and xylazine solution and sacrificed by heart puncture. The dose of ketamine was 250 mg/kg and the dose of xylazine was 25 mg/kg. The organs and tissues were collected and weighed and the activity was measured using an automated gamma-spectrometer. Whole submandibular salivary gland, lung, liver, spleen, pancreas, stomach, and kidneys were sampled for measurements. A small section of small intestines was emptied of content to measure the uptake in intestinal walls. Activity in the rest of the intestinal tract was measured to estimate hepatobiliary excretion. The rest of the body was also collected and its activity was measured. The energy ranges for measurements of ^{125}I and ^{99m}Tc were 18–85 keV and 110–160 keV, respectively. Correction for counts' spillover was performed automatically by the software of the gamma-spectrometer. Activity in a sample was considered as nonmeasurable (NM) if a count rate for a sample plus background was less than two-fold higher than for background (approximately 0.005% of the injected activity). The percentage of injected dose per gram of sample (%ID/g) was calculated.

In addition, an in vivo saturation experiment was performed. EpCAM receptors in MDA-MB-468 xenografts were saturated by co-injection of unlabeled protein at 0.5 mg/mouse (125-fold molar excess to 4 μg Ec1 dose used for biodistribution) together with the injection of [^{125}I]I-PIB-Ec1 and [^{99m}Tc]Tc(CO)$_3$-Ec1, and the biodistribution measurement was performed 6 h pi. To confirm specificity, the uptake of [^{125}I]I-PIB-Ec1 and [^{99m}Tc]Tc(CO)$_3$-Ec1 was measured in EpCAM-negative Ramos lymphoma xenografts 6 h pi.

Whole body SPECT/CT scans of mice bearing MDA-MB-468 xenografts were performed using nanoScan SPECT/CT (Mediso Medical Imaging Systems, Budapest, Hungary). Mice were injected with [^{125}I]I-PIB-Ec1 (4 μg, 0.3 MBq) and [^{99m}Tc]Tc(CO)$_3$-Ec1 (4 μg, 9.4 MBq). Imaging at 6 h pi was performed after mice were sacrificed by CO_2. The acquisition time was 20 min. CT scans were acquired using X-ray energy peak of 50 keV, 670 μA, 480 projections, and 5.26-min acquisition time. SPECT raw data were reconstructed using Tera-Tomo™ 3D SPECT reconstruction technology (version 3.00.020.000; Mediso Medical Imaging Systems Ltd.): High dynamic range, 30 iterations, one subset. CT data were reconstructed using Filter Back Projection and fused with SPECT files using Nucline 2.03 Software (Mediso Medical Imaging Systems Ltd.). Images are presented as maximum-intensity projections in the red, green, and blue (RGB) color scale.

To analyze the radioactive species in urine after injection of [^{125}I]I-PIB-Ec1, it was injected intravenously (i.v.) into two healthy Naval Medical Research Institute (NMRI) mice (Scanbur AS, Karlslunde, Denmark) (10 μg, 1.9 MBq per mouse). Mice were kept in separate cages covered with absorbent paper, which was later checked for the activity indicating any release of urine. One hour after the injection, mice were anesthetized by i.p. injection of ketamine and xylazine solution and sacrificed by cervical dislocation. Urinary bladders containing urine were excised, cut through, the urine was collected in Eppendorf tubes, and the activity was measured (0.58 MBq in 50 μL for mouse 1; 0.68 MBq in 100 μL for mouse 2). An equal volume of ice-cold acetonitrile was added to each tube and the tubes were centrifuged at 15,000 rpm at 4 °C for 15 min. The solution was filtered through a 0.45-μm filter and analyzed by radio-HPLC. The same procedure with addition of acetonitrile and centrifugation was performed for the intact [^{125}I]I-PIB-Ec1 as a control to check that it would be detectable during HPLC analysis.

Radio-HPLC analysis was performed using a Hitachi Chromaster HPLC system with a radioactivity detector and Phenomenex Luna® C18 column (100 Å; 150 × 4.6 mm; 5 μm) at room temperature (20 °C). Solvent A was 0.1% trifluoroacetic acid (TFA) in H2O, solvent B was 0.1% TFA in acetonitrile, and the flow rate was 1 mL/min. The 30-min method with a gradient from 5 to 95% solvent B over 28 min and from 95% to 5% solvent B from 28 to 30 min was used.

2.7. Statistical Analysis

The in vitro specificity and cellular processing data are presented as the mean ± standard deviation (SD) of three samples. Statistical analysis was performed using GraphPad Prism (version 7.02; GraphPad Software, Inc., La Jolla, CA, USA). The $p < 0.05$ was considered a statistically significant difference. The data were analyzed using an unpaired two-tailed *t*-test. The biodistribution data for dual-label experiments at 6- or 24-h time points were analyzed using a paired two-tailed *t*-test.

3. Results

3.1. Radiolabeling

DARPin Ec1 was labeled site-specifically with ^{99m}Tc(CO)$_3$ using a hexahistidine tag at C-terminus to provide a residualizing label. Labeling of DARPin Ec1 with [^{125}I]I-*para*-iodobenzoate was performed by attaching the N-hydroxysuccinimide ester derivative of [^{125}I]I-PIB to amino groups of lysines. Data concerning radiolabeling of DARPin Ec1 with [^{125}I]I-PIB and technetium-99m tricarbonyl are presented in Table 1. Size-exclusion chromatography provided radiochemical purities over 99%. Both labeling methods provided stable labels (Tables 2 and 3).

Table 1. Labeling and characterization of radiolabeled Ec1 variants.

DARPins	Radiochemical Yield of Non-Isolated Compound (%)	Radiochemical Yield of Isolated Compound (%)	Radiochemical Purity (%)	Binding Affinity to MDA-MB-468 Cells (K_D, pM)
[^{125}I]I-PIB-Ec1	23 ± 2 (*n* = 3)	19 ± 1 (*n* = 3)	99 ± 0 (*n* = 3)	121 ± 21 (*n* = 2)
[^{99m}Tc]Tc(CO)$_3$-Ec1	92 ± 1 (*n* = 3)	69 ± 7 (*n* = 3)	99 ± 0 (*n* = 3)	58 ± 5 (*n* = 2)

Table 2. In vitro stability of [^{99m}Tc]Tc(CO)$_3$-Ec1.

	Protein-Associated Activity, %	
	1000× Histidine	PBS
1 h	99 ± 0	99 ± 0
4 h	99 ± 0	99 ± 1
24 h	98 ± 0	99 ± 1

Samples were incubated in PBS or with 1000-fold molar excess of histidine at 37 °C. Analysis was performed in duplicates.

Table 3. In vitro stability of [^{125}I]I-PIB-Ec1.

	Protein-Associated Activity, %		
	1000× NaI	30% EtOH	PBS
1 h	98 ± 1	99 ± 0	99 ± 0
4 h	99 ± 0	99 ± 0	99 ± 0
24 h	99 ± 0	98 ± 0	99 ± 0

Samples were incubated in PBS, 30% ethanol or with 1000-fold molar excess of NaI at 37 °C. Analysis was performed in duplicates.

3.2. *Characterization of Radiolabeled DARPins In Vitro*

In vitro evaluation was performed using EpCAM-expressing MDA-MB-468 breast cancer cells. To demonstrate binding specificity of [^{125}I]I-PIB-Ec1 and [^{99m}Tc]Tc(CO)$_3$-Ec1 to EpCAM, the EpCAM receptors were saturated with 100-fold molar excess of nonlabeled Ec1 before addition of the radiolabeled compound. Blocking the EpCAM receptors resulted in a significant ($p < 0.001$) decrease of both [^{125}I]I-PIB-Ec1 and [^{99m}Tc]Tc(CO)$_3$-Ec1 uptake (Figure 1). This demonstrated a saturable character of radiolabeled Ec1 binding to MDA-MB-468 cells.

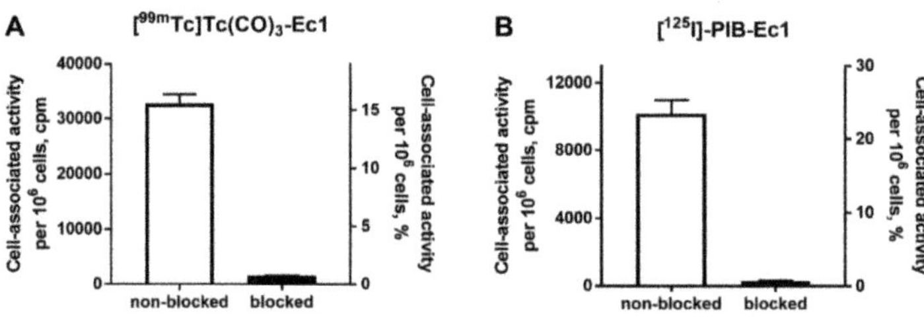

Figure 1. In vitro specificity of epithelial cell adhesion molecule (EpCAM) targeting using [^{99m}Tc]Tc(CO)$_3$-Ec1 (**A**) and [^{125}I]I-PIB-Ec1 (**B**) in EpCAM-expressing MDA-MB-468 cells. Uptake by cells was significantly ($p < 0.001$) reduced when 100-fold molar excess of nonlabeled Ec1 designed ankyrin repeat protein (DARPin) was added to the blocked groups. Final concentration of radiolabeled compound was 2 nM. Data are presented as mean from three samples ± SD.

The binding kinetics of [^{99m}Tc]Tc(CO)$_3$-Ec1 and [^{125}I]I-PIB-Ec1 to MDA-MB-468 cells were measured using LigandTracer (Figure 2, Figures S1 and S2). A rapid binding and slow dissociation were observed. The equilibrium dissociation constant (K_D) values for both probes were in the picomolar range (Table 1). As a control for nonspecific interactions with cells, binding of [^{99m}Tc]Tc(CO)$_3$-Ec1 and [^{125}I]I-PIB-Ec1 to EpCAM-negative Ramos cells was also measured using LigandTracer (Figures S3 and S4). The signal detected from Ramos cells was comparable to the background and it was much lower than the signal detected from MDA-MB-468 cells.

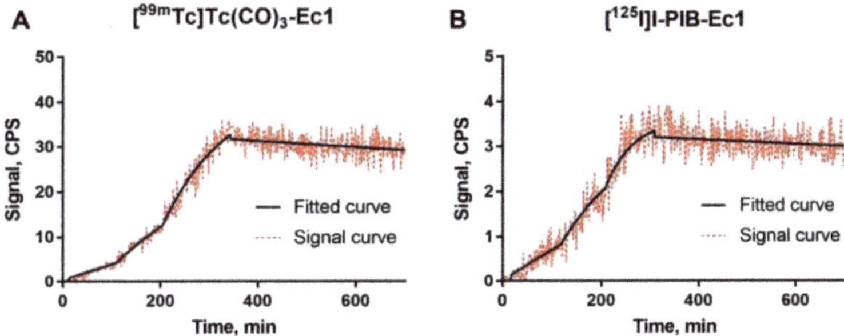

Figure 2. LigandTracer sensorgrams of [^{99m}Tc]Tc(CO)$_3$-Ec1 (**A**) and [^{125}I]I-PIB-Ec1 (**B**) binding to MDA-MB-468 cells. The association was measured at 0.2, 0.6, and 1.8 nM concentrations for [^{99m}Tc]Tc(CO)$_3$-Ec1 and at 1.8, 5.4, and 14.5 nM concentrations [^{125}I]I-PIB-Ec1.

The processing of [^{99m}Tc]Tc(CO)$_3$-Ec1 and [^{125}I]I-PIB-Ec1 by MDA-MB-468 breast cancer cells during continuous incubation is shown in Figure 3. For [^{99m}Tc]Tc(CO)$_3$-Ec1, the total cell-associated activity increased continuously over 24 h incubation and the internalized fraction had also a tendency to a slow increase. For [^{125}I]I-PIB-Ec1, the maximum of total cell-associated activity was reached at 6 h and it slowly decreased by 24 h. This could be explained by the non-residualizing properties of the [^{125}I]I-PIB label and the diffusion of iodine catabolites from the cells after internalization. A characteristic feature of radiolabeled Ec1 was the quite low internalization. The internalized fraction for [^{99m}Tc]Tc(CO)$_3$-Ec1 was approximately 15% of the total cell-associated activity at 24 h.

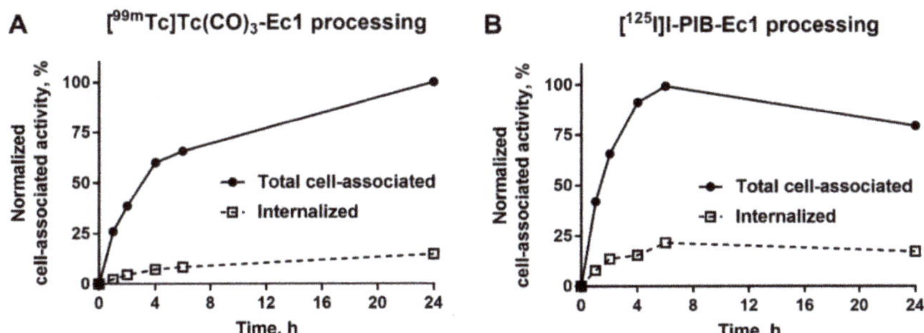

Figure 3. Cellular processing of [^{99m}Tc]Tc(CO)$_3$-Ec1 (**A**) and [^{125}I]I-PIB-Ec1 (**B**) by MDA-MB-468 cells during continuous incubation. Cells were incubated with the DARPins (1 nM) at 37 °C. Data are presented as the mean of three samples ± standard deviation (SD). Error bars might not be seen when they are smaller than data point symbols.

Cellular processing and retention of [^{99m}Tc]Tc(CO)$_3$-Ec1 and [^{125}I]I-PIB-Ec1 by MDA-MB-468 cells after interrupted incubation is shown in Figure 4. After initial release of both probes from the membranes due to the shift in equilibrium, the cell-associated activity for [^{99m}Tc]Tc(CO)$_3$-Ec1 remained constant until 24 h (Figure 4A). On the opposite, the cell-associated activity continued to decrease in the case of [^{125}I]I-PIB-Ec1 (Figure 4B). This decrease was associated with the gradual build-up of low-molecular-weight (<5 KDa) radioactive compounds in the media (25 ± 1% at 4 h and 70 ± 1% at 24 h) (Figure 4D). For [^{99m}Tc]Tc(CO)$_3$-Ec1, the low-molecular-weight activity fraction was only 18 ± 3% at 24 h. Most likely, these low-molecular-weight compounds are the products of intracellular degradation released from cells, since only less than 5% of activity was in the low-molecular-weight fraction after 24-h incubation of both compounds in cell-free complete media. The difference in the amount of low-molecular-weight-associated activity reflects, most likely, the difference in intracellular retention of radiometabolites of iodine and technetium labels. Identity and purity of the radiolabeled [^{99m}Tc]Tc(CO)$_3$-Ec1 and [^{125}I]I-PIB-Ec1 was confirmed by radio-HPLC analysis (Figure S5).

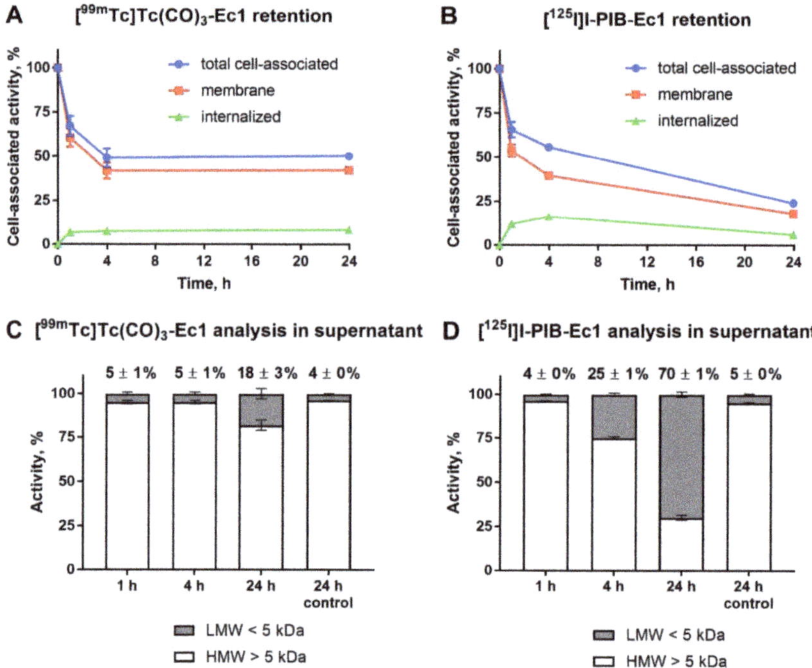

Figure 4. Cellular retention of activity after interrupted incubation of [^{99m}Tc]Tc(CO)$_3$-Ec1 (**A**) and [^{125}I]I-PIB-Ec1 (**B**) with MDA-MB-468 cells. Cells were first incubated with the DARPin variants (10 nM) at 4 °C for 1 h and then the media were exchanged and the cells were incubated at 37 °C for 1, 4, or 24 h. A fraction of the supernatant at every time point was analyzed using NAP-5 size-exclusion columns and compared to the control when the radiolabeled DARPins were incubated in complete media for 24 h (**C**,**D**). Numbers in panels (**C**,**D**) show a percentage of activity associated with low-molecular-weight compounds at each time point. Data for retention are presented as the mean of three samples ± SD and data for NAP-5 analysis are presented as the mean of two samples ± SD. LMW = low molecular weight, HMW = high molecular weight. Error bars might not be seen when they are smaller than data point symbols.

3.3. *In Vivo Studies*

The results of the specificity test (Figure 5) demonstrated that the uptake of both [^{99m}Tc]Tc(CO)$_3$-Ec1 and [^{125}I]I-PIB-Ec1 in EpCAM-negative Ramos xenografts was much lower ($p < 0.001$, unpaired t-test) than in EpCAM-positive MDA-MB-468 xenografts. In addition, saturation of EpCAM by co-injecting a large excess of unlabeled Ec1 resulted in a significant ($p < 0.01$, unpaired t-test) reduction in tumor uptake of both variants.

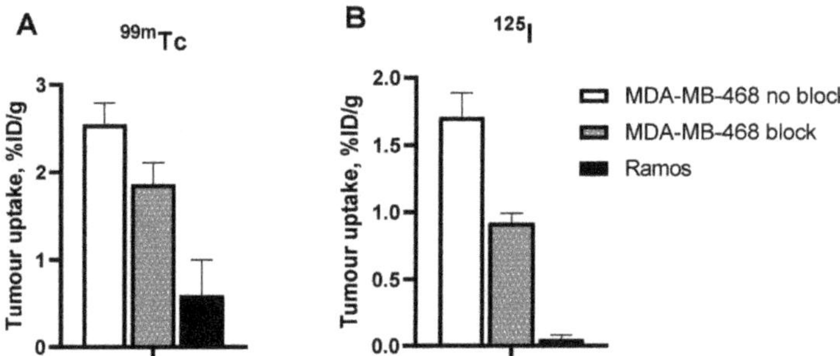

Figure 5. In vivo specificity of EpCAM targeting using [^{99m}Tc]Tc(CO)$_3$-Ec1 (**A**) and [^{125}I]I-PIB-Ec1 (**B**). Uptake of both DARPin variants was significantly ($p < 0.01$, unpaired t-test) higher in EpCAM-positive MDA-MB-468 xenografts than in EpCAM-negative Ramos xenografts 6 h post injection (pi). EpCAM blocking in MDA-MB-468 xenografts by co-injecting a large excess of unlabeled Ec1 also resulted in a significant decrease of tracer uptake. Data are presented as mean ± SD for four mice.

The tumor uptake and retention of [^{99m}Tc]Tc(CO)$_3$-Ec1 was significantly ($p < 0.05$, paired t-test) higher than tumor uptake and retention of [^{125}I]I-PIB-Ec1 (2.6 ± 0.2 vs. 1.7 ± 0.2%ID/g 6 h pi). Distribution of activity in normal organs and tissues was quite different for ^{99m}Tc- and ^{125}I-labeled Ec1 (Table 4). The renal uptake of [^{99m}Tc]Tc(CO)$_3$-Ec1 was more than 100-fold higher compared to [^{125}I]I-PIB-Ec1 even 24 h pi. A combination of a high kidney retention of activity and low activity accumulation in gastrointestinal tract indicated a renal excretion pathway of DARPin Ec1. The renal and hepatic uptake of [^{99m}Tc]Tc(CO)$_3$-Ec1 was higher than the tumor uptake. [^{125}I]I-PIB-Ec1 had appreciably lower hepatic and renal uptake. Overall, [^{125}I]I-PIB-Ec1 had the lowest uptake in normal tissues. Already by 6 h after injection of [^{125}I]I-PIB-Ec1 only ca. 4% of ID was left in mice and over 95% of ID was excreted in comparison with [^{99m}Tc]Tc(CO)$_3$-Ec1, when about 30% of ID was excreted by 6 h (Table S1). Accordingly, [^{125}I]I-PIB-Ec1 provided significantly higher tumor-to-organ ratios compared with [^{99m}Tc]Tc(CO)$_3$-Ec1 (Table 5). At 6 h pi, [^{125}I]I-PIB-Ec1 had two-fold higher tumor-to-blood, ca. 100-fold higher tumor-to-liver, 20-fold higher tumor-to-spleen and tumor-to-pancreas, and eight-fold higher tumor-to-muscle ratios than [^{99m}Tc]Tc(CO)$_3$-Ec1. By 24 h pi, the tumor-to-blood and tumor-to-kidney ratios increased further for the [^{125}I]I-PIB label, while no improvement was observed for the technetium-99m label.

Table 4. Biodistribution of [^{99m}Tc]Tc(CO)$_3$-Ec1 and [^{125}I]I-PIB-Ec1 in Balb/c nu/nu mice bearing MDA-MB-468 xenografts 6 and 24 h pi.

Tissue	[^{99m}Tc]Tc(CO)$_3$-Ec1	[^{125}I]I-PIB-Ec1	[^{99m}Tc]Tc(CO)$_3$-Ec1	[^{125}I]I-PIB-Ec1
	6 h		24 h	
Blood	0.24 ± 0.03 ^{a,b}	0.09 ± 0.01 ^{c}	0.11 ± 0.01 ^{a}	0.009 ± 0.001
Salivary glands	1.7 ± 0.2 ^{a}	0.11 ± 0.04	1.3 ± 0.3	NM
Lungs	0.9 ± 0.2 ^{a,b}	0.23 ± 0.03 ^{c}	0.6 ± 0.2 ^{a}	0.04 ± 0.01
Liver	18 ± 2 ^{a,b}	0.11 ± 0.02 ^{c}	9 ± 2 ^{a}	0.026 ± 0.002
Spleen	3.1 ± 0.3 ^{a,b}	0.10 ± 0.02 ^{c}	2.2 ± 0.3	0.044 ± 0.003
Pancreas	1.2 ± 0.2 ^{a,b}	0.04 ± 0.01	0.8 ± 0.2	NM
Small intestine	1.5 ± 0.3 ^{a,b}	0.11 ± 0.04	0.9 ± 0.3	NM
Stomach	1.8 ± 0.4 ^{a,b}	0.15 ± 0.03	0.9 ± 0.2	NM
Kidney	192 ± 15 ^{a,b}	2.7 ± 1.0 ^{c}	114 ± 13 ^{a}	0.08 ± 0.01
Tumor	2.6 ± 0.2 ^{a}	1.7 ± 0.2 ^{c}	1.5 ± 0.5 ^{a}	0.27 ± 0.05
Muscle	0.5 ± 0.1 ^{a,b}	0.04 ± 0.01	0.3 ± 0.1	NM
Bone	1.9 ± 0.3 ^{a,b}	0.9 ± 0.2 ^{c}	1.2 ± 0.3 ^{a}	0.5 ± 0.2
Intestines with content	1.8 ± 0.3 ^{b}	0.4 ± 0.1 ^{c}	1.1 ± 0.1 ^{a}	0.07 ± 0.01
Rest of the body	11.6 ± 1.2 ^{a,b}	1.7 ± 0.2 ^{c}	8.5 ± 1.5 ^{a}	0.9 ± 0.2

Data are presented as mean percent of injected dose (%ID)/g ± SD for four mice. Data for the rest of the intestines with contents and rest of the body are presented as %ID per whole sample. ^{a} Significant difference between [^{99m}Tc]Tc and [^{125}I]I at the same time point (paired t-test). ^{b} Significant difference between values for [^{99m}Tc]Tc at 6- and 24-h time point (unpaired t-test). ^{c} Significant difference between values for [^{125}I]I at 6- and 24-h time point (unpaired *t*-test). NM = nonmeasurable.

Table 5. Tumor-to-organ ratios of [^{99m}Tc]Tc(CO)$_3$-Ec1 and [^{125}I]I-PIB-Ec1 in Balb/C nu/nu mice bearing MDA-MB-468 xenografts at 6 and 24 h pi.

Tissue	[^{99m}Tc]Tc(CO)$_3$-Ec1	[^{125}I]I-PIB-Ec1	[^{99m}Tc]Tc(CO)$_3$-Ec1	[^{125}I]I-PIB-Ec1
	6 h		24 h	
Blood	11 ± 1 ^{a}	19 ± 3 ^{c}	13 ± 4 ^{a}	31 ± 6
Salivary glands	1.5 ± 0.1 ^{a}	17 ± 6	1.2 ± 0.3	NM
Lungs	3 ± 1 ^{a}	8 ± 2	3 ± 1	NM
Liver	0.14 ± 0.01 ^{a}	15 ± 2 ^{c}	0.2 ± 0.1 ^{a}	10 ± 1
Spleen	0.8 ± 0.1 ^{a}	18 ± 5	0.7 ± 0.2	NM
Pancreas	2.1 ± 0.2 ^{a}	43 ± 8	2 ± 1	NM
Small intestine	1.7 ± 0.4 ^{a}	17 ± 6	2 ± 1	NM
Stomach	1.4 ± 0.3 ^{a}	12 ± 4	2 ± 1	NM
Kidney	0.013 ± 0.002 ^{a}	0.7 ± 0.3 ^{c}	0.014 ± 0.005 ^{a}	3.5 ± 0.5
Muscle	5 ± 1 ^{a}	42 ± 10	5 ± 1	NM
Bone	1.3 ± 0.2	2 ± 1 ^{c}	1.3 ± 0.4	0.5 ± 0.1

Data are presented as mean ± SD for four mice. ^{a} Significant difference between [^{125}I]I and [^{99m}Tc]Tc at the same time point (paired t-test). ^{b} Significant difference between values for [^{99m}Tc]Tc at 6- and 24-h (unpaired t-test). ^{c} Significant difference between values for [^{125}I]I at 6- and 24-h (unpaired *t*-test). NM= nonmeasurable.

To study if the lower activity retention in kidneys in case of [^{125}I]I-PIB-Ec1 is caused by the excretion of radiometabolites, we performed radio-HPLC analysis of urine collected 1 h after [^{125}I]I-PIB-Ec1 injection in healthy NMRI mice (Figure S6). It was observed that the majority of activity excreted with urine was in the form of radiocatabolites already 1 h after the injection of $[^{125}I]$I-PIB-Ec1.

MicroSPECT/CT imaging using $[^{99m}Tc]Tc(CO)_3$-Ec1 and $[^{125}I]$I-PIB-Ec in Balb/c nu/nu mice bearing MDA-MB-468 xenografts at 6 h pi confirmed the results of the biodistribution studies (Figure 6). The tumor was visualized using both radiolabeled variants of DARPin Ec1. The use of non-residualizing label $[^{125}I]$I-PIB-Ec1 provided lower retention of activity in normal organs and kidneys and higher imaging contrast compared to $[^{99m}Tc]Tc(CO)_3$-Ec1.

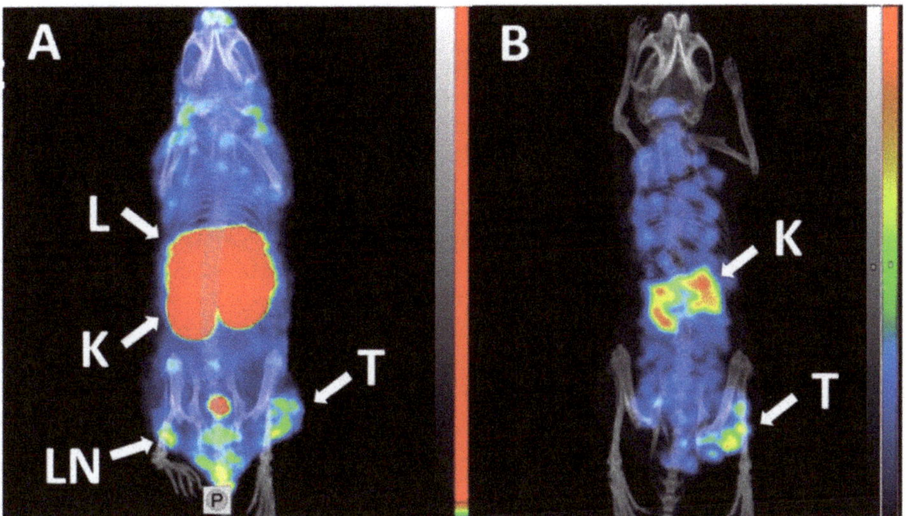

Figure 6. Micro-Single-Photon Emission Computed Tomography/Computed Tomography (microSPECT/CT) imaging of EpCAM expression in BALB/C nu/nu mice bearing EpCAM-positive MDA-MB-468 xenografts at 6 h pi using $[^{99m}Tc]Tc(CO)_3$-Ec1 (**A**) and $[^{125}I]$I-PIB-Ec (**B**). Arrows indicate: T—tumor, K—kidneys, L—liver, LN—lymph node. The scale in panel A was adjusted to the first red pixel in the tumor.

4. Discussion

Efficacy of targeted therapies is critically dependent on expression level of the molecular target in tumors. In the case of absence or too low expression level of the target, there will be no therapeutic effect. Unfortunately, an unspecific toxicity to normal organs and tissues will be preserved. This is a particularly apparent risk for a targeted delivery of cytotoxic payloads, such as drugs, toxins, or alpha or beta particle-emitting radionuclides. Thus, personalizing treatment by stratification of patients according to expression of a molecular target in tumors is an essential precondition for successful targeted therapy. Although the percentage of EpCAM-overexpressing TNBC tumors is high, identification of patients eligible for therapy is necessary to avoid overtreatment of patients having tumors with low expression level. Radionuclide molecular imaging is a promising approach, as it enables visualization of multiple metastases addressing heterogeneity of expression.

Imaging data can be used to select patients for targeted therapy in different ways. The most common method is the determination of tumor-to-reference organ ratio. A typical example is the so-called Krenning score, which is applied for selection of patients with neuroendocrine tumors for somatostatin receptor-targeted radionuclide therapy using $[^{177}Lu\text{-}DOTA^0,Tyr^3]$octreotate [38,39]. The scale grades the tumor uptake of $[^{111}$ In-DTPA$^0]$octreotide in planar gamma camera images: Grade 1, no tumor uptake; grade 2, tumor uptake is equal to normal liver tissue; grade 3, tumor uptake is greater than normal liver tissue; and grade 4, tumor uptake is higher than normal spleen or kidney uptake. It has been demonstrated that the response strongly correlates with the grade [38]. It has been shown that tumor-to-spleen uptake ratio for ^{111}In and ^{68}Ga-labeled affibody molecules [40] and tumor-to-contralateral breast ratio for ^{99m}Tc-labeled ADAPT6 [41] correlates strongly with the level of HER2 expression in breast cancer. Alternatively, the uptake of an imaging probe in tumor might be quantified using PET and correlated with expression level [42].

An essential factor for successful radionuclide imaging is high imaging contrast because it determines the diagnostic sensitivity. Thus, imaging agents providing high tumor-to-organ ratios (ratios of activity concentration in tumors to the concentration in normal organs), especially to organs that are frequent metastatic sites, are required. TNBC has a predominant metastasis to visceral organs, first and foremost, liver and lungs [5,43]. Thus, sufficiently high tumor-to-liver and tumor-to-lung ratios are the preconditions for successful translation to clinics. In addition, a high tumor-to-blood ratio is an essential parameter for evaluation of an imaging agent as a blood-borne activity and might contribute to the background signal.

Selection of a fitting label is important for obtaining of a high contrast. One strategy to achieve a high imaging contrast is based on using non-residualizing labels. After binding of radiolabeled proteins to cell-surface receptors on cancer cells or to scavenger receptors on cells in excretory organs (e.g., kidneys and liver), a protein-receptor complex is internalized and the radiolabeled protein is degraded in lysosomes with formation of radiocatabolites. Radiocatabolites of residualizing labels (typically, radiometals) are retained inside the cells as they are not able to diffuse through the lipophilic membranes. However, the radiocatabolites of non-residualizing labels rapidly diffuse from the cells, return to blood circulation, and are excreted with urine. When the internalization of radiolabeled proteins is slow in tumors and rapid in normal organs and tissues, the use of non-residualizing labels might enable fast clearance of activity from normal organs and tissues and provide higher imaging contrast than the use of residualizing labels.

Three different variants of non-residualizing radioiodine labels were evaluated for DARPins: A product of so-called direct electrophilic radioiodination, when the radioiodine is incorporated into the phenolic ring of tyrosine [22,23,30], a conjugate with [^{125}I]-PIB [30,31], and a site-specific conjugate of [^{125}I]I-iodo-[(4-hydroxyphenyl)ethyl]-maleimide (HPEM) to a unique cysteine engineered to a C-terminus of DARPin G3 [24]. It turned out that the use of the HPEM label results in a high level of hepatobiliary excretion and accumulation of activity in the content of gastrointestinal tract. This is undesirable, as it creates an unacceptable high background for imaging of visceral TNBC metastases. The use of direct iodination resulted in accumulation of radiocatabolites in Na/I-symporter-expressing organs (first and foremost, in salivary gland and stomach) but also in in visceral organs such as pancreas and intestines [30]. Overall, previous studies suggest that indirect radioiodination using PIB would be optimal for labeling of Ec1.

A potential disadvantage of the use of [^{125}I]I-PIB is a random attachment of N-succinimidyl-*para*-iodobenzoate to amino groups of lysines in DARPin Ec1. As DARPin Ec1 contains eight lysines, this labeling might result in a mixture of labeled proteins with a different number and positions of [^{125}I]I-PIB label. However, previous studies with proteins of similar size showed that the specificity and affinity is usually preserved with this type of labeling [27,44–46].

This study showed that the binding of both [^{125}I]I-PIB-Ec1 and [^{99m}Tc]Tc(CO)$_3$-Ec1 to TNBC cells in vitro (Figure 1) can be blocked with an excess of unlabeled Ec1, which confirmed its specificity.

Previous studies have demonstrated that kinetics and affinity of binding of radiolabeled proteins to their molecular targets depend on cell line origin [47,48]. This might be associated with the molecular context of cellular membranes, including such factors as co-expression of other receptors or cell-surface proteins, glycosylation patterns, and homo- and heterodimerization. The results of LigandTracer measurements (Figure 2) demonstrated that binding of both tracers to TNBC cells was characterized by rapid association and very slow dissociation rates. The equilibrium dissociation constants were 121 ± 21 pM and 58 ± 5 pM for [^{125}I]I-PIB-Ec1 and [^{99m}Tc]Tc(CO)$_3$-Ec1, respectively. Thus, affinity of [^{125}I]I-PIB-Ec1 binding to TNBC cells was somewhat lower compared to binding to ovarian cancer OvCAR-3 cells (35 ± 1 pM) [31] or pancreatic cancer BxPC-3 cells (58 ± 13 pM) [30]. According to stoichiometric calculations, the average number of pendant groups conjugated per Ec1 molecule should be 0.3, which should have a minimal impact on binding properties. However, it might happen that one of the pendant groups is conjugated to a lysine close to the binding site and might create a hindrance. This might explain the differences in affinities between [^{99m}Tc]Tc(CO)$_3$-Ec1, which was labeled site specifically, and [^{125}I]I-PIB-Ec1. Still, the picomolar affinity is a good precondition for a strong retention of an imaging probe by malignant cells in vivo.

Assessment of the internalization rate of Ec1 during continuous incubation of cells with the tracer was initially performed using [^{99m}Tc]Tc(CO)$_3$-Ec1 (Figure 3). A residualizing label is the most suitable for such kind of test because the data obtained by a non-residualizing label might be deceptive due to the "leakage" of radiometabolites from cells leading to underestimation of internalized radioactivity. The cellular processing study showed a slow internalization (approximately 15% of cell-bound activity at 24 h after incubation start) of Ec1 by TNBC cells. The difference between cellular processing of [^{99m}Tc]Tc(CO)$_3$-Ec1 and [^{125}I]I-PIB-Ec1 was further elucidated by studying cellular retention after interrupted incubation (Figure 4). The use of radioiodine label was associated with lower retention of activity by cells. In addition, a decrease of cell-associated radioiodine activity was accompanied with an increase of low-molecular-weight radioiodinated compounds in the incubation media. Most likely, this was caused by diffusion of radiometabolites of a non-residualizing label from cells. We considered the combination of the slow internalization and high affinity as a good rationale to proceed with in vivo studies.

Animal studies demonstrated clearly specific accumulation of both [^{125}I]I-PIB-Ec1 and [^{99m}Tc]Tc(CO)$_3$-Ec1 in MDA-MB-468 xenografts (Figure 5). Saturation of EpCAM by co-injection of a large excess of unlabeled Ec1 resulted in significantly lower uptake of the tracers in TNBC xenografts. In addition, the activity uptake in EpCAM-negative Ramos xenografts was much lower than in EpCAM-positive MDA-MB-468 xenografts.

Low accumulation in normal tissue is an important precondition for a high-contrast imaging. The results of the biodistribution experiments demonstrated advantages in the use of non-residualizing labels for slowly internalizing, high-affinity imaging probes (Table 4). The difference in the renal uptake of [^{125}I]I-PIB-Ec1 and [^{99m}Tc]Tc(CO)$_3$-Ec1 was spectacular already at 6 h after injection. A high re-absorption in the proximal tubuli of the kidneys is a common feature of imaging probes based on short peptides [49] and ESPs, such as affibody molecules, ADAPTs, or DARPins [20,50,51]. In the case of short peptides, the re-absorption could be suppressed by blocking scavenger receptors with cationic amino acids or succinylated bovine gelatin (Gelofusine) [49]. However, the use of these and a number of other substances potentially blocking or reducing renal reabsorption was unsuccessful in the case of DARPins [51]. Thus, there is no conventional way to reduce the renal uptake of DARPins labeled with a residualizing label. However, the use of [^{125}I]I-PIB-Ec1 resulted in more than a 70-fold lower renal uptake compared with [^{99m}Tc]Tc(CO)$_3$-Ec1. Radio-HPLC analysis of the urine 1 h after injection of [^{125}I]I-PIB-Ec1 demonstrated that the activity was excreted predominantly as radiometabolites (S6). This suggests that non-residualizing properties of the radioiodine label were critical for the reduction of renal retention. Even more impressive (and more relevant to imaging of EpCAM expression in TNBC metastases) was the reduction of hepatic uptake, which was 160-fold lower for [^{125}I]I-PIB-Ec1. While the uptake of [^{99m}Tc]Tc(CO)$_3$-Ec1 in tumor was nearly 7-fold lower than in liver, the tumor uptake of [^{125}I]I-PIB-Ec1was 15-fold higher than the hepatic uptake.

It has to be noted that the tumor uptake of [^{99m}Tc]Tc(CO)$_3$-Ec1 was significantly ($p < 0.05$, paired t-test) higher than tumor uptake and retention of [^{125}I]I-PIB-Ec1 (2.6 ± 0.2 vs. 1.7 ± 0.2%ID/g 6 h pi). This can be explained by the lower affinity of [^{125}I]I-PIB-Ec1 compared with its ^{99m}Tc-labeled counterpart. In addition, we can expect that some internalization would take place during six hours and leakage of catabolites would affect the retention of activity in tumors. However, the magnitude of the uptake reduction was appreciably smaller compared with the uptake in normal tissues. For this reason, tumor-to-organ ratios were several folds higher for [^{125}I]I-PIB-Ec1 than for [^{99m}Tc]Tc(CO)$_3$-Ec1 (Table 3). Particularly, tumor-to-blood, tumor-to-lung, tumor-to-liver, and tumor-to-muscle ratios were 19 ± 3, 8 ± 2, 15 ± 2, and 42 ± 10 at 6 h after injection, respectively. Accordingly, an experimental microSPECT imaging using [^{125}I]I-PIB-Ec1 permitted clear visualization of EpCAM expression in TNBC xenografts (Figure 5). It has to be noted that the tumor-associated activity decreased with time after injection of [^{125}I]I-PIB-Ec1. At 24 h pi, the tumor uptake of [^{125}I]I-PIB-Ec1was only 0.27 ± 0.05%ID/g. This was more than five-fold lower compared with the tumor uptake of [^{99m}Tc]Tc(CO)$_3$-Ec1. Such low uptake makes this tracer unsuitable for imaging the next day after injection. Thus, non-residualizing labels are suitable for imaging only a few hours after injection even in the cases of slow internalization by malignant cells.

It should also be noted that the observed differences in the biodistribution between [^{99m}Tc]Tc(CO)$_3$-Ec1 and [^{125}I]I-PIB-Ec1 might not only be due to residualizing properties of labels but also be influenced by other label properties (polarity, site specificity of labeling), as well as by differences in pharmacokinetics.

In this study, ^{125}I was used as a label because of its convenient half-life (60 days). However, its low-energy electromagnetic radiation (max. 35.5 keV) permits its use only in small rodents and is unsuitable for clinical translation. Two iodine radioisotopes, ^{123}I ($T_{1/2}$ = 13.3 h, $E\gamma$ = 159 keV) and positron-emitting ^{124}I ($T_{1/2}$ = 100 h, $\beta+$ 23%), are suitable for radionuclide imaging using single-photon computed tomography (SPECT) and positron emission tomography (PET), respectively. Since the chemical properties of isotopes are identical, only a minor re-optimization is required to change the labeling chemistry from ^{125}I to these nuclides [24,52].

Summarizing, this study demonstrated that [^{125}I]I-PIB-Ec1 has high (subnanomolar) affinity to EpCAM. It binds in a specific manner to EpCAM-expressing TNBC cell line in vitro and accumulates specifically in EpCAM-expressing TNBC xenografts in mice. The tumor-to-organ ratios, which are a measure of imaging contrast and sensitivity, are appreciably higher for [^{125}I]I-PIB-Ec1 than for [^{99m}Tc]Tc(CO)$_3$-Ec1. The [^{125}I]I-PIB-Ec1 is capable of visualization of EpCAM-expressing TNBC xenograft in a mouse. This creates a rationale for further clinical studies concerning imaging of EpCAM in TNBC and correlation of imaging and biopsy data.

5. Conclusions

Both [^{125}I]I-PIB-Ec1 and [^{99m}Tc]Tc(CO)$_3$-Ec1 demonstrated specific uptake in EpCAM-positive TNBC xenografts. Radioiodine provided better tumor-to-organ ratios compared to [^{99m}Tc]Tc(CO)$_3$ label. Radioiodinated DARPin Ec1 is a promising agent for same-day imaging of EpCAM expression in triple-negative breast cancer using SPECT.

Supplementary Materials: The following are available online, Figure S1: representative curves of the LigandTracer measurement of [^{99m}Tc]Tc(CO)$_3$-Ec1 binding to MDA-MB-468 cells, Figure S2: representative curves of the LigandTracer measurement of [^{125}I]I-PIB-Ec1 binding to MDA-MB-468 cells, Figure S3: representative curves of the LigandTracer measurement of [^{99m}Tc]Tc(CO)$_3$-Ec1 binding to Ramos cells, Figure S4: representative curves of the LigandTracer measurement of [^{125}I]I-PIB-Ec1 binding to Ramos cells; Figure S5: radio-HPLC analysis of [^{125}I]I-PIB-Ec1 and [^{99m}Tc]Tc(CO)$_3$-Ec1 in comparison to the non-labeled Ec1, Figure S6: radio-HPLC analysis of mouse urine 1 h after injection of [^{125}I]I-PIB-Ec1 in comparison to the intact [^{125}I]I-PIB-Ec1, Figure S7: resolution of gamma-spectra of iodine-125 and technetium-99m, Table S1: biodistribution of [^{99m}Tc]Tc(CO)$_3$-Ec1 and [^{125}I]I-PIB-Ec1 in Balb/c nu/nu mice bearing MDA-MB-468 xenografts at 6 and 24 h.

Author Contributions: A.V., S.D., and V.T. contributed to the concept and study design. E.K. and A.S. performed the production and purification of proteins. S.D. supervised the production and purification of proteins. J.G., O.V., A.A., A.O., and V.T. participated in planning and performing the in vivo experiments, data treatment, and interpretation. A.V. performed the labeling chemistry, in vitro and in vivo studies, data treatment, and interpretation. E.B. participated in data treatment and interpretation. A.V. and E.B. wrote the first draft of the manuscript. All authors revised the manuscript critically, read, and approved the final manuscript. All authors have read and agreed to the published version of the manuscript.

Funding: This research was funded by the grant of Ministry of Science and Higher Education of the Russian Federation (075-15-2019-1925). The funders had no role in the study design, data collection and analysis, decision to publish, or preparation of the manuscript.

Acknowledgments: The authors thank Sara S. Rinne for proofreading the paper.

Conflicts of Interest: The authors declare that they have no conflict of interest.

References

1. Goldhirsch, A.; Wood, W.C.; Coates, A.S.; Gelber, R.D.; Thürlimann, B.; Senn, H.J. Panel members. Strategies for subtypes–dealing with the diversity of breast cancer: Highlights of the St. Gallen International Expert Consensus on the Primary Therapy of Early Breast Cancer 2011. *Ann. Oncol.* **2011**, *22*, 1736–1747. [CrossRef] [PubMed]
2. Anders, C.K.; Carey, L.A. Biology, metastatic patterns, and treatment of patients with triple-negative breast cancer. *Clin. Breast Cancer* **2009**, *9*, S73–S81. [CrossRef] [PubMed]
3. Rouzier, R.; Perou, C.M.; Symmans, W.F.; Ibrahim, N.; Cristofanilli, M.; Anderson, K.; Hess, K.R.; Stec, J.; Ayers, M.; Wahner, P.; et al. Breast cancer molecular subtypes respond differently to preoperative chemotherapy. *Clin. Cancer Res.* **2005**, *11*, 5678–5685. [CrossRef] [PubMed]
4. Carey, L.A.; Dees, E.C.; Sawyer, L.; Gatti, L.; Moore, D.T.; Collichio, F.; Ollila, D.W.; Sartor, C.I.; Graham, M.L.; Perou, C.M. The triple negative paradox: Primary tumor chemosensitivity of breast cancer subtypes. *Clin. Cancer Res.* **2007**, *13*, 2329–2334. [CrossRef]
5. Liedtke, C.; Mazouni, C.; Hess, K.R.; André, F.; Tordai, A.; Mejia, J.A.; Symmans, W.F.; Gonzalez-Angulo, A.M.; Hennessy, B.; Green, M.; et al. Response to neoadjuvant therapy and long-term survival in patients with triple-negative breast cancer. *J. Clin. Oncol.* **2008**, *26*, 1275–1281. [CrossRef] [PubMed]
6. Syed, Y.Y. Sacituzumab Govitecan: First Approval. *Drugs* **2020**, *80*, 1019–1025. [CrossRef] [PubMed]
7. Bardia, A.; Mayer, I.A.; Vahdat, L.T.; Tolaney, S.M.; Isakoff, S.J.; Diamond, J.R.; O'Shaughnessy, J.; Moroose, R.L.; Santin, A.D.; Abramson, V.G.; et al. Sacituzumab Govitecan-hziy in Refractory Metastatic Triple-Negative Breast Cancer. *N. Engl. J. Med.* **2019**, *380*, 741–751. [CrossRef]
8. Schmidt, M.; Hasenclever, D.; Schaeffer, M.; Boehm, D.; Cotarelo, C.; Steiner, E.; Lebrecht, A.; Siggelkow, W.; Weikel, W.; Schiffer-Petry, I.; et al. Prognostic effect of epithelial cell adhesion molecule overexpression in untreated node-negative breast cancer. *Clin. Cancer Res.* **2008**, *14*, 5849–5855. [CrossRef]
9. Soysal, S.D.; Muenst, S.; Barbie, T.; Fleming, T.; Gao, F.; Spizzo, G. EpCAM expression varies significantly and is differentially associated with prognosis in the luminal B HER2$^+$, basal-like, and HER2 intrinsic subtypes of breast cancer. *Br. J. Cancer* **2013**, *8*, 1480–1487. [CrossRef]
10. Abd El-Maqsoud, N.M.; Abd El-Rehim, D.M. Clinicopathologic implications of EpCAM and Sox2 expression in breast cancer. *Clin. Breast Cancer* **2014**, *14*, e1–e9. [CrossRef]
11. Wu, Q.; Wang, J.; Liu, Y.; Gong, X. Epithelial cell adhesion molecule and epithelial-mesenchymal transition are associated with vasculogenic mimicry, poor prognosis, and metastasis of triple-negative breast cancer. *Int. J. Clin. Exp. Pathol.* **2019**, *12*, 1678–1689. [PubMed]
12. Martin-Killias, P.; Stefan, N.; Rothschild, S.; Plückthun, A.; Zangemeister-Wittke, U. A novel fusion toxin derived from an EpCAM-specific designed ankyrin repeat protein has potent antitumor activity. *Clin. Cancer Res.* **2011**, *17*, 100–110. [CrossRef] [PubMed]
13. Amoury, M.; Kolberg, K.; Pham, A.T.; Hristodorov, D.; Mladenov, R.; Di Fiore, S.; Helfrich, W.; Kiessling, F.; Fischer, R.; Pardo, A.; et al. Granzyme B-based cytolytic fusion protein targeting EpCAM specifically kills triple negative breast cancer cells in vitro and inhibits tumor growth in a subcutaneous mouse tumor model. *Cancer Lett.* **2016**, *372*, 201–209. [CrossRef] [PubMed]
14. Jenkins, S.V.; Nima, Z.A.; Vang, K.B.; Kannarpady, G.; Nedosekin, D.A.; Zharov, V.P.; Griffin, R.J.; Biris, A.S.; Dings, R.P.M. Triple-negative breast cancer targeting and killing by EpCAM-directed, plasmonically active nanodrug systems. *NPJ Precis. Oncol.* **2017**, *1*, 1–9. [CrossRef] [PubMed]
15. Schmidt, M.; Scheulen, M.E.; Dittrich, C.; Obrist, P.; Marschner, N.; Dirix, L.; Schmidt, M.; Rüttinger, D.; Schuler, M.; Reinhardt, C.; et al. An open-label, randomized phase II study of adecatumumab, a fully human anti-EpCAM antibody, as monotherapy in patients with metastatic breast cancer. *Ann. Oncol.* **2010**, *21*, 275–282. [CrossRef] [PubMed]
16. Ahmadpour, S.; Hosseinimehr, S.J. Recent developments in peptide-based SPECT radiopharmaceuticals for breast tumor targeting. *Life Sci.* **2019**, *239*, 116870. [CrossRef] [PubMed]

17. De Bree, R.; Roos, J.C.; Quak, J.J.; Den Hollander, W.; Snow, G.B.; Van Dongen, G.A. Clinical screening of monoclonal antibodies 323/A3, cSF-25 and K928 for suitability of targetting tumours in the upper aerodigestive and respiratory tract. *Nucl. Med. Commun.* **1994**, *15*, 613–627. [CrossRef]
18. Kosterink, J.G.; de Jonge, M.W.; Smit, E.F.; Piers, D.A.; Kengen, R.A.; Postmus, P.E.; Shochat, D.; Groen, H.J.; The, H.T.; de Leij, L. Pharmacokinetics and scintigraphy of indium-111-DTPA-MOC-31 in small-cell lung carcinoma. *J. Nucl. Med.* **1995**, *36*, 2356–2362. [CrossRef]
19. Breitz, H.B.; Tyler, A.; Bjorn, M.J.; Lesley, T.; Weiden, P.L. Clinical experience with Tc-99m nofetumomab merpentan (Verluma) radioimmunoscintigraphy. *Clin. Nucl. Med.* **1997**, *22*, 615–620. [CrossRef]
20. Garousi, J.; Orlova, A.; Frejd, F.Y.; Tolmachev, V. Imaging using radiolabelled targeted proteins: Radioimmunodetection and beyond. *EJNMMI Radiopharm. Chem.* **2020**, *5*, 1–26. [CrossRef]
21. Krasniqi, A.; D'Huyvetter, M.; Devoogdt, N.; Frejd, F.Y.; Sörensen, J.; Orlova, A.; Keyaerts, M.; Tolmachev, V. Same-Day Imaging Using Small Proteins: Clinical Experience and Translational Prospects in Oncology. *J. Nucl. Med.* **2018**, *59*, 885–891. [CrossRef] [PubMed]
22. Vorobyeva, A.; Bragina, O.; Altai, M.; Mitran, B.; Orlova, A.; Shulga, A.; Proshkina, G.; Chernov, V.; Tolmachev, V.; Deyev, S. Comparative Evaluation of Radioiodine and Technetium-Labeled DARPin 9_29 for Radionuclide Molecular Imaging of HER2 Expression in Malignant Tumors. *Contrast Media Mol. Imaging* **2018**, *2018*, 6930425. [CrossRef] [PubMed]
23. Deyev, S.; Vorobyeva, A.; Schulga, A.; Proshkina, G.; Güler, R.; Löfblom, J.; Mitran, B.; Garousi, J.; Altai, M.; Buijs, J.; et al. Comparative Evaluation of Two DARPin Variants: Effect of Affinity, Size, and Label on Tumor Targeting Properties. *Mol. Pharm.* **2019**, *3*, 995–1008. [CrossRef] [PubMed]
24. Vorobyeva, A.; Schulga, A.; Konovalova, E.; Güler, R.; Mitran, B.; Garousi, J.; Rinne, S.; Löfblom, J.; Orlova, A.; Deyev, S.; et al. Comparison of tumor-targeting properties of directly and indirectly radioiodinated designed ankyrin repeat protein (DARPin) G3 variants for molecular imaging of HER2. *Int. J. Oncol.* **2019**, *54*, 1209–1220. [CrossRef] [PubMed]
25. Stefan, N.; Martin-Killias, P.; Wyss-Stoeckle, S.; Honegger, A.; Zangemeister-Wittke, U.; Plückthun, A. DARPins recognizing the tumor-associated antigen EpCAM selected by phage and ribosome display and engineered for multivalency. *J. Mol. Biol.* **2011**, *413*, 826–843. [CrossRef]
26. Tolmachev, V.; Orlova, A. Affibody Molecules as Targeting Vectors for PET Imaging. *Cancers* **2020**, *12*, 651. [CrossRef]
27. Vorobyeva, A.; Schulga, A.; Rinne, S.S.; Günther, T.; Orlova, A.; Deyev, S.; Tolmachev, V. Indirect radioiodination of DARPin G3 using N-succinimidyl-para-iodobenzoate improves the contrast of HER2 molecular imaging. *Int. J. Mol. Sci.* **2019**, *20*, 3047. [CrossRef]
28. Tolmachev, V.; Orlova, A.; Lundqvist, H. Approaches to improve cellular retention of radiohalogen labels delivered by internalising tumour-targeting proteins and peptides. *Curr. Med. Chem.* **2003**, *10*, 2447–2460. [CrossRef]
29. Lindbo, S.; Garousi, J.; Mitran, B.; Altai, M.; Buijs, J.; Orlova, A.; Hober, S.; Tolmachev, V. Radionuclide Tumor Targeting Using ADAPT Scaffold Proteins: Aspects of Label Positioning and Residualizing Properties of the Label. *J. Nucl. Med.* **2018**, *59*, 93–99. [CrossRef]
30. Deyev, S.M.; Vorobyeva, A.; Schulga, A.; Abouzayed, A.; Günther, T.; Garousi, J.; Konovalova, E.; Dinf, H.; Gråslund, T.; Orlova, A.; et al. Effect of a radiolabel biochemical nature on tumor-targeting properties of EpCAM-binding engineered scaffold protein DARPin Ec1. *Int. J. Biol. Macromol.* **2020**, *145*, 216–225. [CrossRef]
31. Vorobyeva, A.; Konovalova, E.; Xu, T.; Schulga, A.; Altai, M.; Garousi, J.; Rinne, S.S.; Orlova, A.; Tolmachev, V.; Deyev, S. Feasibility of Imaging EpCAM Expression in Ovarian Cancer Using Radiolabeled DARPin Ec1. *Int. J. Mol. Sci.* **2020**, *21*, 3310. [CrossRef]
32. Wållberg, H.; Orlova, A. Slow internalization of anti-HER2 synthetic affibody monomer 111In-DOTA-ZHER2:342-pep2: Implications for development of labeled tracers. *Cancer Biother. Radio.* **2008**, *23*, 435–442. [CrossRef]
33. Altai, M.; Leitao, C.D.; Rinne, S.S.; Vorobyeva, A.; Atterby, C.; Ståhl, S.; Tolmachev, V.; Löfblom, J.; Orlova, A. Influence of Molecular Design on the Targeting Properties of ABD-Fused Mono- and Bi-Valent Anti-HER3 Affibody Therapeutic Constructs. *Cells* **2018**, *7*, 164. [CrossRef] [PubMed]
34. Tolmachev, V.; Orlova, A.; Andersson, K. Methods for Radiolabelling of Monoclonal Antibodies. *Methods Mol. Biol.* **2014**, *1060*, 309–330. [CrossRef]

35. Bondza, S.; Foy, E.; Brooks, J.; Andersson, K.; Robinson, J.; Richalet, P.; Buijs, J. Real-time Characterization of Antibody Binding to Receptors on Living Immune Cells. *Front. Immunol.* **2017**, *8*, 455. [CrossRef] [PubMed]
36. Wilbur, D.S.; Hadley, S.W.; Grant, L.M.; Hylarides, M.D. Radioiodinated iodobenzoyl conjugates of a monoclonal antibody Fab fragment. In Vivo comparisons with chloramine-T-labeled Fab. *Bioconjugate Chem.* **1991**, *2*, 111–116. [CrossRef]
37. Reist, C.J.; Archer, G.E.; Wikstrand, C.J.; Bigner, D.D.; Zalutsky, M.R. Improved targeting of an anti-epidermal growth factor receptor variant III monoclonal antibody in tumor xenografts after labeling using N-succinimidyl 5-iodo-3-pyridinecarboxylate. *Cancer Res.* **1997**, *57*, 1510–1515.
38. Kwekkeboom, D.J.; de Herder, W.W.; Kam, B.L.; van Eijck, C.H.; van Essen, M.; Kooij, P.P.; Feelders, R.A.; van Aken, M.O.; Krenning, E.P. Treatment with the radiolabeled somatostatin analog [177 Lu-DOTA 0,Tyr3]octreotate: Toxicity, efficacy, and survival. *J. Clin. Oncol.* **2008**, *26*, 2124–2130. [CrossRef] [PubMed]
39. Teunissen, J.J.M.; Kwekkeboom, D.J.; Valkema, R.; Krenning, E.P. Nuclear medicine techniques for the imaging and treatment of neuroendocrine tumours. *Endocr. Relat. Cancer* **2011**, *18*, S27–S51. [CrossRef]
40. Sandberg, D.; Tolmachev, V.; Velikyan, I.; Olofsson, H.; Wennborg, A.; Feldwisch, J.; Carlsson, J.; Lindman, H.; Sörensen, J. Intra-image referencing for simplified assessment of HER2-expression in breast cancer metastases using the Affibody molecule ABY-025 with PET and SPECT. *Eur. J. Nucl. Med. Mol. Imaging* **2017**, *44*, 1337–1346. [CrossRef]
41. Bragina, O.; von Witting, E.; Garousi, J.; Zelchan, R.; Sandström, M.; Medvedeva, A.; Orlova, A.; Doroshenko, A.; Vorobyeva, A.; Lindbo, S.; et al. Phase I study of 99mTc-ADAPT6, a scaffold protein-based probe for visualization of HER2 expression in breast cancer. *J. Nucl. Med.* **2020**. [CrossRef]
42. Sörensen, J.; Velikyan, I.; Sandberg, D.; Wennborg, A.; Feldwisch, J.; Tolmachev, V.; Orlova, A.; Sandström, M.; Lubberink, M.; Olofsson, H.; et al. Measuring HER2-Receptor Expression In Metastatic Breast Cancer Using [68Ga]ABY-025 Affibody PET/CT. *Theranostics* **2016**, *6*, 262–271.
43. Lin, N.U.; Vanderplas, A.; Hughes, M.E.; Theriault, R.L.; Edge, S.B.; Wong, Y.N.; Blayney, D.W.; Niland, J.C.; Winer, E.P.; Weeks, J.C. Clinicopathologic features, patterns of recurrence, and survival among women with triple-negative breast cancer in the National Comprehensive Cancer Network. *Cancer* **2012**, *118*, 5463–5472. [CrossRef] [PubMed]
44. Pruszynski, M.; Koumarianou, E.; Vaidyanathan, G.; Revets, H.; Devoogdt, N.; Lahoutte, T.; Lyerly, H.K.; Zalutsky, M.R. Improved tumor targeting of anti-HER2 nanobody through N-succinimidyl 4-guanidinomethyl-3-iodobenzoate radiolabeling. *J. Nucl. Med.* **2014**, *55*, 650–656. [CrossRef] [PubMed]
45. D'Huyvetter, M.; De Vos, J.; Xavier, C.; Pruszynski, M.; Sterckx, Y.G.J.; Massa, S.; Raes, G.; Caveliers, V.; Zalutsky, M.R.; Lahoutte, T.; et al. 131I-labeled Anti-HER2 Camelid sdAb as a Theranostic Tool in Cancer Treatment. *Clin. Cancer Res.* **2017**, *23*, 6616–6628. [CrossRef]
46. Tolmachev, V.; Mume, E.; Sjöberg, S.; Frejd, F.Y.; Orlova, A. Influence of valency and labelling chemistry on in vivo targeting using radioiodinated HER2-binding Affibody molecules. *Eur. J. Nucl. Med. Mol. Imaging* **2009**, *36*, 692–701. [CrossRef]
47. Björkelund, H.; Gedda, L.; Andersson, K. Comparing the epidermal growth factor interaction with four different cell lines: Intriguing effects imply strong dependency of cellular context. *PLoS ONE* **2011**, *6*, e16536. [CrossRef]
48. Barta, P.; Malmberg, J.; Melicharova, L.; Strandgård, J.; Orlova, A.; Tolmachev, V.; Laznicek, M.; Andersson, K. Protein interactions with HER-family receptors can have different characteristics depending on the hosting cell line. *Int. J. Oncol.* **2012**, *40*, 1677–1682. [CrossRef]
49. Vegt, E.; de Jong, M.; Wetzels, J.F.; Masereeuw, R.; Melis, M.; Oyen, W.J.; Gotthardt, M.; Boerman, O.C. Renal toxicity of radiolabeled peptides and antibody fragments: Mechanisms, impact on radionuclide therapy, and strategies for prevention. *J. Nucl. Med.* **2010**, *51*, 1049–1058. [CrossRef]
50. von Witting, E.; Garousi, J.; Lindbo, S.; Vorobyeva, A.; Altai, M.; Oroujeni, M.; Mitran, B.; Orlova, A.; Hober, S.; Tolmachev, V. Selection of the optimal macrocyclic chelators for labeling with 111In and 68Ga improves contrast of HER2 imaging using engineered scaffold protein ADAPT6. *Eur. J. Pharm. Biopharm.* **2019**, *140*, 109–120. [CrossRef] [PubMed]
51. Altai, M.; Garousi, J.; Rinne, S.S.; Schulga, A.; Deyev, S.; Vorobyeva, A. On the prevention of kidney uptake of radiolabeled DARPins. *EJNMMI Res.* **2020**, *10*, 7. [CrossRef] [PubMed]

52. Orlova, A.; Wållberg, H.; Stone-Elander, S.; Tolmachev, V. On the selection of a tracer for PET imaging of HER2-expressing tumors: Direct comparison of a 124I-labeled affibody molecule and trastuzumab in a murine xenograft model. *J. Nucl. Med.* **2009**, *50*, 417–425. [CrossRef] [PubMed]

Sample Availability: Samples of the compound DARPin Ec1 are available from the authors.

Article

Initial In Vitro and In Vivo Evaluation of a Novel CCK2R Targeting Peptide Analog Labeled with Lutetium-177

Anton Amadeus Hörmann [1], Maximilian Klingler [1], Maliheh Rezaeianpour [1,2], Nikolas Hörmann [3], Ronald Gust [3], Soraya Shahhosseini [2] and Elisabeth von Guggenberg [1,*]

1 Department of Nuclear Medicine, Medical University of Innsbruck, 6020 Innsbruck, Austria; anton.hoermann@i-med.ac.at (A.A.H.); Maximilian.Klingler@i-med.ac.at (M.K.); m.rezaeianpour@yahoo.com (M.R.)

2 Pharmaceutical Chemistry and Radiopharmacy Department, School of Pharmacy, Shahid Beheshti University of Medical Sciences, 1991953381 Tehran, Iran; soraya.shahhosseini@gmail.com

3 Department of Pharmaceutical Chemistry, University of Innsbruck, 6020 Innsbruck, Austria; nikolas.hoermann@uibk.ac.at (N.H.); ronald.gust@uibk.ac.at (R.G.)

* Correspondence: elisabeth.von-guggenberg@i-med.ac.at; Tel.: +43-512-504-80960

Academic Editor: Krishan Kumar

Received: 27 August 2020; Accepted: 1 October 2020; Published: 8 October 2020

Abstract: Targeting of cholecystokinin-2 receptor (CCK2R) expressing tumors using radiolabeled minigastrin (MG) analogs is hampered by rapid digestion of the linear peptide in vivo. In this study, a new MG analog stabilized against enzymatic degradation was investigated in preclinical studies to characterize the metabolites formed in vivo. The new MG analog DOTA-DGlu-Pro-Tyr-Gly-Trp-(*N*-Me)Nle-Asp-1Nal-NH_2 comprising site-specific amino acid substitutions in position 2, 6 and 8 and different possible metabolites thereof were synthesized. The receptor interaction of the peptide and selected metabolites was evaluated in a CCK2R-expressing cell line. The enzymatic stability of the ^{177}Lu-labeled peptide analog was evaluated in vitro in different media as well as in BALB/c mice up to 1 h after injection and the metabolites were identified based on radio-HPLC analysis. The new radiopeptide showed a highly increased stability in vivo with >56% intact radiopeptide in the blood of BALB/c mice 1 h after injection. High CCK2R affinity and cell uptake was confirmed only for the intact peptide, whereas enzymatic cleavage within the receptor specific C-terminal amino acid sequence resulted in complete loss of affinity and cell uptake. A favorable biodistribution profile was observed in BALB/c mice with low background activity, preferential renal excretion and prolonged uptake in CCK2R-expressing tissues. The novel stabilized MG analog shows high potential for diagnostic and therapeutic use. The radiometabolites characterized give new insights into the enzymatic degradation in vivo.

Keywords: cholecystokinin-2 receptor; minigastrin; molecular imaging; targeted radiotherapy; lutetium-177

1. Introduction

Radiolabeled peptide analogs for theranostic use in the diagnosis and treatment of cancer need to fulfill important prerequisites, such as high receptor affinity, appropriate metabolic stability, high and persistent tumor uptake, as well as low uptake in non-target tissue and fast blood clearance [1]. Up to date, targeting G-protein coupled receptors overexpressed on the surface of tumor cells for nuclear medicine applications is mainly limited to radiolabeled somatostatin analogs. The cyclic somatostatin analog octreotide with high affinity to the somatostatin receptor subtype 2 is used for symptomatic and biochemical control in the treatment of neuroendocrine tumors [2].

Radiolabeled octreotide derivatives have been successfully introduced in routine nuclear medicine applications for diagnosis and treatment of neuroendocrine tumors [3]. A major milestone in this respect is the recent approval of Lutathera® for peptide receptor radionuclide therapy by the European Medicines Agency (EMA) and by the Food and Drug Administration (FDA) [4,5]. So far, this success could not be translated to radiolabeled peptide analogs targeting other receptors. The reason often lies in the rapid metabolism in vivo of the linear peptide sequences derived from natural peptide hormones leading to cleavage of amino acids by enzymatic degradation and subsequent loss of affinity to the target receptor [1]. Radiolabeled peptide analogs targeting the cholecystokinin-2 receptor (CCK2R), overexpressed in different tumors, such as small cell lung cancer, stromal ovarian cancers, gastrointestinal stromal tumors, astrocytoma and especially medullary thyroid carcinoma (MTC), have shown to be very promising for application in diagnosis and therapy [6,7]. The reported clinical use has mainly focused on the diagnosis and treatment of patients with advanced MTC [8]. In the last decades, several attempts have been made to develop a radiolabeled CCK2R targeting peptide analog with suitable pharmacological properties for theranostic applications. First prove of principle studies with a radioiodinated gastrin analog confirmed the feasibility of CCK2R targeting [9]. Different CCK2R-targeting peptide analogs conjugated to the bifunctional chelators diethylenetriaminepentaacetic acid (DPTA) and 1,4,7,10-tetraazacyclododecane-1,4,7,10-tetraacetic acid (DOTA) radiolabeled with trivalent radiometals have been developed and evaluated in clinical studies [8]. The MG analog DTPA-DGlu-Glu-Glu-Glu-Glu-Glu-Ala-Tyr-Gly-Trp-Met-Asp-Phe-NH_2 (DTPA-MG0), derived from human MG, displayed a high renal uptake hindering the therapeutic use [10]. With the removal of the penta-Glu sequence in the truncated MG analog DOTA-DGlu-Ala-Tyr-Gly-Trp-Met-Asp-Phe-NH_2 (DOTA-MG11), the renal uptake was efficiently reduced, however, so was the stability in vivo [11,12]. Thus, the clinical applicability of the peptide analogs developed so far is limited.

CCK2R-targeting peptide analogs are potential substrates of various enzymes such as the angiotensin converting enzyme (ACE), neutral endopeptidase (NEP), aminopeptidase A (APA) and cathepsins [13–17]. Extensive preclinical research was undertaken to improve the stability in vivo as well as the targeting properties. Different modifications were introduced in the linear peptide sequence of different CCK2R targeting peptide analogs, such as the incorporation of unnatural amino acids, inversion of the configuration of amino acids, cyclisation of the linear peptide and dimerization [18–21]. Most of these developments have not led to the required improvements necessary for successful clinical application. Two MG analogs, DOTA-$(DGlu)_6$-Ala-Tyr-Gly-Trp-Met-Asp-Phe-NH_2 (PP-F11) labeled with indium-111 as well as DOTA-$(DGlu)_6$-Ala-Tyr-Gly-Trp-Nle-Asp-Phe-NH_2 (PP-F11N) labeled with lutetium-177, which are derived from MG0 by inversion of the configuration of the penta-Glu motif, are currently examined in clinical studies (ClinicalTrials.gov Identifier: NCT03246659 and NCT02088645) [22–24]. Besides chemical modification of the peptide, in situ stabilization by co-injection of enzyme inhibitors was investigated. For ^{111}In-labeled DOTA-MG0 and DOTA-MG11, the use of phosphoramidone improved the tumor uptake, whereas for DOTA-MG0, a concomitant increase of renal retention occurred [25]. Sauter et al. studied the effect of two NEP inhibitors, phosphoramidone and thiorphan, on the targeting properties of ^{177}Lu-labeled DOTA-MG11, PP-F11 and PP-F11N. Only for DOTA-MG11 an improved tumor uptake could be achieved, whereas no improvement was found for PP-F11 and PP-F11N [14]. The results suggest that in situ stabilization is highly dependent on the individual radiopeptide and cannot be generalized. Besides that, the long-term use of protease inhibitors, especially NEP inhibitors, can potentially cause side effects that are not yet well understood [26].

In our recent studies we could introduce new modifications within the C-terminal sequence Trp-Met-Asp-Phe-NH_2, known to be essential for CCK2R binding [27–29]. Most favorable properties were found for the new MG analog with the sequence DOTA-DGlu-Ala-Tyr-Gly-Trp-(*N*-Me)Nle-Asp-1Nal-NH_2 (DOTA-MGS5), in which methionine is replaced by *N*-methylated norleucine ((*N*-Me)Nle) and phenylalanine by 1-naphtylalanine (1Nal) [28]. Besides leading to highly improved stability in vivo, the introduced modifications also led to an enhanced receptor-specific cell uptake, as well as to a highly increased tumor uptake, when radiolabeled with different radiometals. In nude BALB/c mice bearing CCK2R-expressing tumor xenografts, a tumor uptake of more than 20% of the injected activity per gram (IA/g) was observed. This corresponds to a three-fold improvement compared to PP-F11 and PP-F11N (~6.7% and 6.9% IA/g) [14,28]. Nevertheless, some degradation products were still detected by radio-HPLC analysis of blood obtained from mice injected with ^{177}Lu-labeled DOTA-MGS5 [28].

With the aim of further improving the in vivo stability of DOTA-MGS5, we have explored additional modification of the peptide sequence. The amino acid proline (Pro) is a promising candidate for amino acid exchange and forms a tertiary amide bond which similarly to *N*-methylated peptide bonds and triazoles may improve the stability in vivo [27,30]. In this study, a preliminary preclinical characterization of the new MG analog DOTA-DGlu-Pro-Tyr-Gly-Trp-(*N*-Me)Nle-Asp-1Nal-NH_2 (**1**) was carried out focusing on the enzymatic stability of the ^{177}Lu-labeled radiopeptide in vivo. For this purpose, besides characterizing the stability in vitro in different media, metabolic stability studies were carried out in BALB/c mice giving first insights into the enzymatic degradation and biodistribution profile of this new radiolabeled MG analog. To characterize the ^{177}Lu-labeled metabolites formed in vivo, different possible metabolites of **1** were synthesized. Furthermore, the receptor affinity of the new MG analog, as well as selected metabolites, was studied in A431 human epidermoid carcinoma cells stably transfected with human CCK2R (A431-CCK2R). The same cell line was used to investigate the cell uptake of the ^{177}Lu-labeled peptide analog and selected radiometabolites. Mock-transfected A431 cells (A431-mock) were used as negative control.

2. Results

2.1. Peptide Synthesis and Radiolabeling

The amino acid sequence and chemical structure of **1** is displayed in Figure 1. **M1–M6** were synthesized by standard solid phase peptide synthesis starting from 100 mg of resin following the synthesis protocol described below. For **M7** and **M8**, which were synthesized by different strategies, conjugation with DOTA (DOTA-tris(tert-butyl) ester or DOTA mono-*N*-hydroxysuccinimide ester) was carried out in solution. After purification by reversed phase HPLC (RP-HPLC), characterization by mass spectrometry and lyophilization, the metabolites **M1–M8** were obtained with a purity >95% (with the exception of **M1** and **M4**, for which a purity of 93–94% was achieved). The amino acid sequences and analytical data for **1** and the different metabolites **M1–M8** are shown in Table 1.

DOTA-DGlu-Pro-Tyr-Gly-Trp-(N-Me)Nle-Asp-1Nal-NH_2 (**1**)

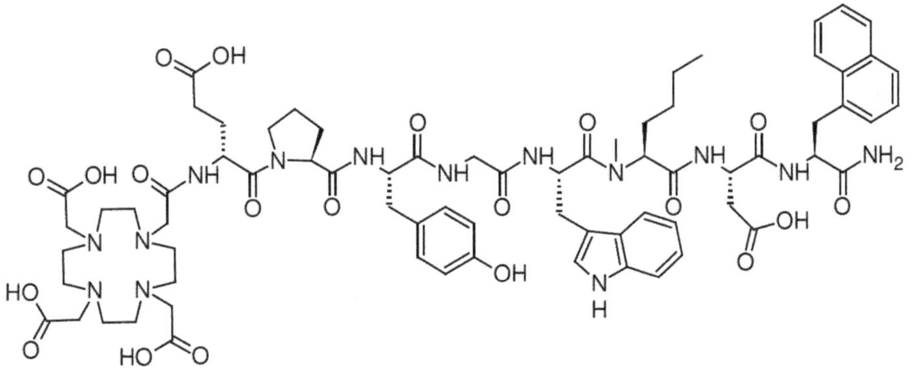

Figure 1. Amino acid sequence and chemical structure of **1**.

Table 1. Summary of the analytical data of **1** and the metabolites **M1–M8** (radio t_R obtained after labeling with lutetium-177).

Peptide	Peptide Sequence	Purity(%)	Radio t_R [min]	UV t_R [min]	MW calc *m/z* [M + H]+	MW found *m/z* [M + H]+
1	DOTA-DGlu-Pro-Tyr-Gly-Trp-(*N*-Me)Nle-Asp-1Nal-NH_2	>95	32.6	32.4	1475.6	1475.8
M1	DOTA-DGlu-Pro-Tyr-Gly-Trp-(*N*-Me)Nle-Asp-1Nal-COOH	94	33.1	32.8	1476.6	1475.9
M2	DOTA-DGlu-Pro-Tyr-Gly-Trp-(*N*-Me)Nle-Asp	99	24.9	24.9	1278.4	1279.2
M3	DOTA-DGlu-Pro-Tyr-Gly-Trp-(*N*-Me)Nle	98	25.8	26.0	1163.3	1164.2
M4	DOTA-DGlu-Pro-Tyr-Gly-Trp	93	23.5	23.4	1036.1	1037.4
M5	DOTA-DGlu-Pro-Tyr-Gly	99	13.7	15.2	849.9	851.4
M6	DOTA-DGlu-Pro-Tyr	96	14.4	16.7	792.8	794.5
M7	DOTA-DGlu-Pro	99	9.8	12.6	630.6	631.5
M8	DOTA-DGlu	99	6.3	6.4	533.5	534.5

For experiments in vitro, labeling with lutetium-177 was carried out at a low molar activity of 10–20 MBq/nmol, yielding nearly quantitative labeling and allowing the use of the radiolabeled conjugates without further purification. The radiolabeled conjugates used in animal studies were radiolabeled at a higher molar activity of ~40 MBq/nmol. Hydrophilic impurities were removed by solid phase extraction (SPE) to obtain the radiolabeled peptides with a radiochemical purity of >99%. The radio-HPLC chromatograms of the radiolabeled compounds are shown in Figure 2.

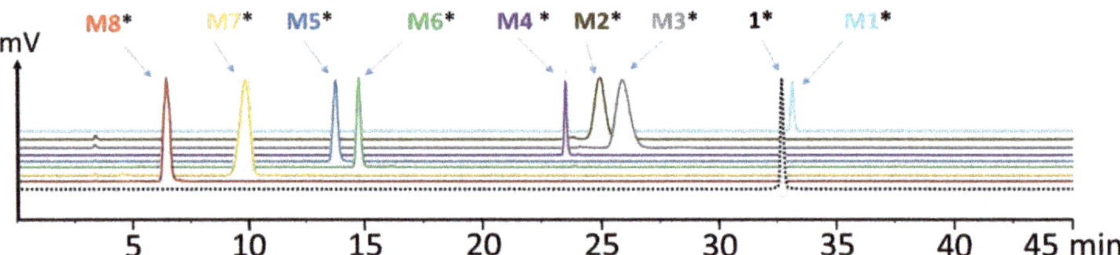

Figure 2. Radio-HPLC chromatograms of [^{177}Lu]Lu-**1** and its ^{177}Lu-labeled metabolites **M1–M8** (* indicating radiolabeled with lutetium-177).

2.2. Characterization In Vitro

The stability of ^{177}Lu-labeled **1** and **M1–4** in fresh human serum, as well as the stability of [^{177}Lu]Lu-**1** in liver and kidney homogenates, was analyzed for up to 24 h after incubation. The percentage of intact radiopeptide found over time is presented in Figure 3. [^{177}Lu]Lu-**1** showed a very high stability in human serum with values of 96.2 ± 1.3% intact peptide after 24 h incubation. A higher degree of degradation was found in liver homogenate with >90% intact radiopeptide up to 20 min after incubation and decreasing to 83.6 ± 0.2, 52.9 ± 2.0 and 49.3 ± 1.7%, at 30 min, 2 h and 24 h, respectively. In kidney homogenate, a faster metabolic breakdown was observed. However, still 65.6 ± 2.8 and 43.9 ± 1.0% intact radiopeptide were present 20 min and 30 min after incubation. At later time points, the radiolabeled conjugate was completely degraded (<5 and <1% at 2 and 24 h after incubation, respectively). In human serum, a high stability was also found for [^{177}Lu]Lu-**M1** (89.1 ± 1.3%), [^{177}Lu]Lu-**M2** (98.1 ± 0.3%) and [^{177}Lu]Lu-**M4** (98.4 ± 0.01%) after 24 h incubation in human serum, whereas [^{177}Lu]Lu-**M3** showed a different behavior. A fast enzymatic degradation occurred in serum with only 29.7 ± 0.8% of intact [^{177}Lu]Lu-**M3** 24 h after incubation.

The logD values calculated from the octanol/PBS distribution of the different ^{177}Lu-labeled peptides resulted in a hydrophilicity profile in the order of [^{177}Lu]Lu-**M4** (−4.19 ± 0.30) > [^{177}Lu]Lu-**M2** (−4.18 ± 0.17) > [^{177}Lu]Lu-**M3** (−4.13 ± 0.11) > [^{177}Lu]Lu-**M1** (−4.07 ± 0.35) > [^{177}Lu]Lu-**1** (−1.96 ± 0.07).

Protein binding in human serum as analyzed by size exclusion chromatography was found to be lowest for [^{177}Lu]Lu-**M3** (23.4 ± 1.9%), followed by [^{177}Lu]Lu-**1** (37.8 ± 2.9%), [^{177}Lu]Lu-**M4** (41.7 ± 0.2%), [^{177}Lu]Lu-**M1** (45.4 ± 1.8%) and [^{177}Lu]Lu-**M2** (48.1 ± 2.3%) for the time point of 24 h after incubation.

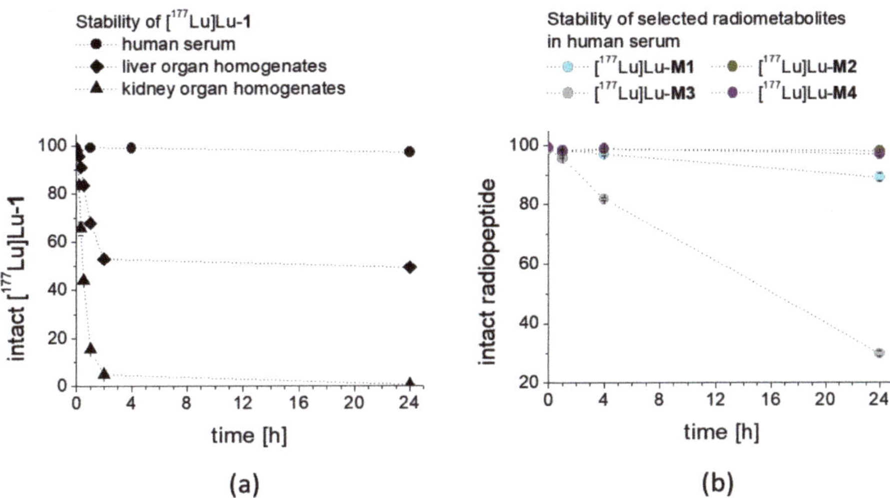

Figure 3. Stability of (**a**) [^{177}Lu]Lu-**1** after incubation in human serum as well as in rat liver and rat kidney homogenates, and of (**b**) selected radiolabeled metabolites **M1–M4** in human serum, as analyzed up to 24 h after incubation.

2.3. Cell Internalization and Receptor Binding Studies

For [^{177}Lu]Lu-**1** incubated with A431-CCK2R cells, a high internalization with uptake values of 44.4 ± 2.7% after 1 h incubation (Figure 4) was observed. As expected, the ^{177}Lu-labeled metabolites showed no internalization into A431-CCK2R cells ([^{177}Lu]Lu-**M1**: 0.18 ± 0.03%, [^{177}Lu]Lu-**M2**: 0.06 ± 0.02%, [^{177}Lu]Lu-**M3**: 0.05 ± 0.03% and [^{177}Lu]Lu-**M4**: 0.05 ± 0.02%). Additionally, after 2 h incubation a very high receptor-specific uptake could be confirmed for [^{177}Lu]Lu-**1**, with values further increasing to 66.6 ± 0.3%, whereas no internalization occurred for the ^{177}Lu-labeled metabolites. Specificity of the cell uptake was proven by contemporaneous incubation in A431-mock cells finding a cell uptake of <0.2% and <0.4% for [^{177}Lu]Lu-**1** at 1 and 2 h, respectively.

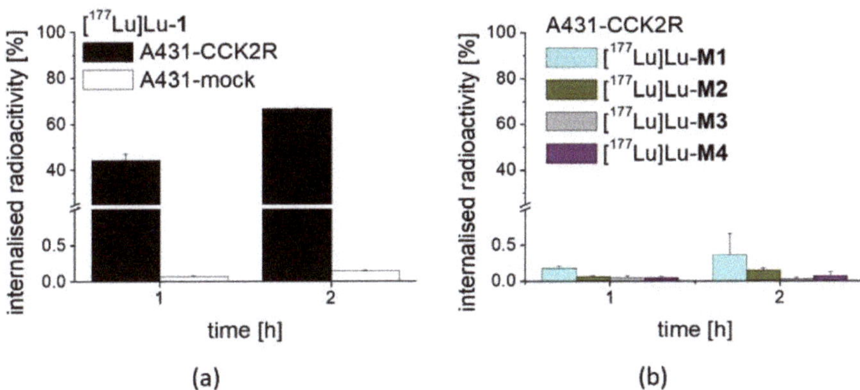

Figure 4. Cell uptake of (**a**) [^{177}Lu]Lu-**1** in A431-CCK2R and A431-mock cells and (**b**) ^{177}Lu-labeled **M1–M4** in A431-CCK2R cells, after 1 h and 2 h incubation.

In competition assays against [Leu^{15}]gastrin-I substituted with iodine-125, a high binding affinity to CCK2R was confirmed for **1** (IC_{50}: 0.69 ± 0.09 nM) on A431-CCK2R cells, comparable to pentagastrin used as a reference (IC_{50}: 0.76 ± 0.11 nM) (Figure 5). For **M1–M4** no binding affinity could be observed, confirming the complete loss of receptor binding after removal of the C-terminal amide function and the C-terminal amino acids (*N*-Me)Nle, Asp and 1Nal. Based on these findings, the cell uptake and receptor binding of the remaining metabolites were not tested.

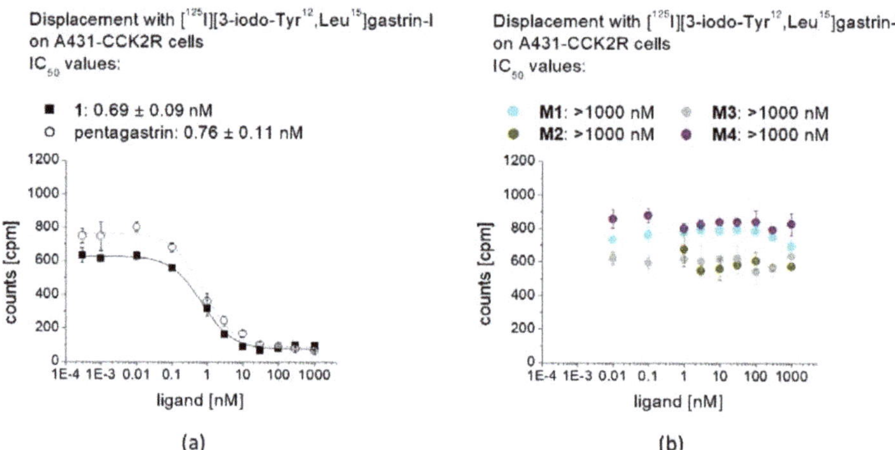

Figure 5. Competitive binding curves against [^{125}I][3-iodo-Tyr^{12},Leu^{15}]gastrin-I for (**a**) non-labeled **1** in comparison with pentagastrin, as well as (**b**) non-labeled **M1**–**M4.**

2.4. Stability In Vivo and Biodistribution Studies

The metabolic stability in vivo, as shown in Figure 6, was monitored after intravenous injection of [^{177}Lu]Lu-**1** in BALB/c mice. A high resistance against enzymatic degradation was found at 10 min post injection (p.i.) with 80.5% intact radiopeptide present in blood. A high percentage of intact radiopeptide in blood was also observable after 30 and 60 min p.i. with values of 64.1% and 56.9%, respectively. Analysis of the urine of the mice showed that only 29.1%, 19.4% and 18.0% intact [^{177}Lu]Lu-**1** was excreted at the different time points studied. In the soluble phase extracted from liver homogenate, 71.9%, 53.9% and 47.7% intact peptide were found after 10, 30 and 60 min p.i., respectively, reflecting the stability of the radiolabeled conjugate during circulation. Metabolism during excretion was confirmed by the analysis of the soluble phase extracted from kidney homogenate with values of 29.7%, 18.4% and 7.0% intact radiopeptide after 10, 30 and 60 min p.i., respectively.

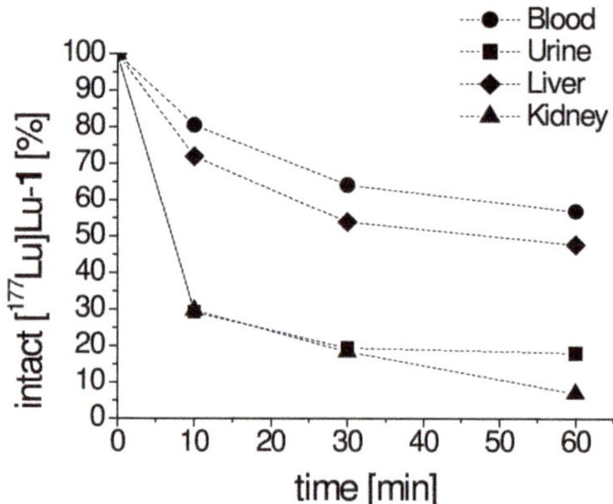

Figure 6. Intact [^{177}Lu]Lu-**1** detectable in blood and urine, as well as liver and kidney homogenates obtained from BALB/c mice, as analyzed up to 60 min p.i.

The preliminary evaluation of the biodistribution profile of [^{177}Lu]Lu-**1** showed a fast clearance from the blood with low non-specific uptake in most tissues. The observed uptake values are summarized in Table 2. The whole body activity ranged from 61.78% IA at 10 min p.i. to 12.27% IA at 1 h after injection. As shown in Figure 7a, the blood pool activity rapidly declined from 17.79% to 2.05% IA/g at 10 and 60 min, respectively. Consequently, a very low background activity was found in muscle with values ranging from 2.56% to 0.36% IA/g at 10 min and 60 min, respectively. Low non-specific uptake and rapid washout was observed also for lung (17.45% to 1.95% IA/g), heart (7.13% to 0.99% IA/g), femur (5.97% to 0.52% IA/g) and spleen (3.70% to 0.79% IA/g). Excretion occurred mainly through the kidneys with activity values decreasing from 14.49% to 7.01% IA/g from 10 to 60 min p.i. (Figure 7b). The activity values observed in liver (5.37% and 1.26% IA/g) and intestine (2.63% and 1.02% IA/g) for the same time points were much lower and comparable to the non-specific uptake in other organs. When looking at the washout of radioactivity from different tissues, a higher retention of radioactivity was observed for CCK2R-expressing stomach and pancreas, which was more prominent for stomach. The percentage decrease of radioactivity at 1 h versus 10 min p.i. was lower for stomach (59%) and pancreas (75%) in comparison to the washout of 79–91% from non-excretory organs (blood, lung, heart, femur, spleen and muscle), indicating a receptor-specific uptake in stomach and pancreas (Figure 7c, Table 2). No further blocking studies were performed to investigate the specificity of the uptake in more detail.

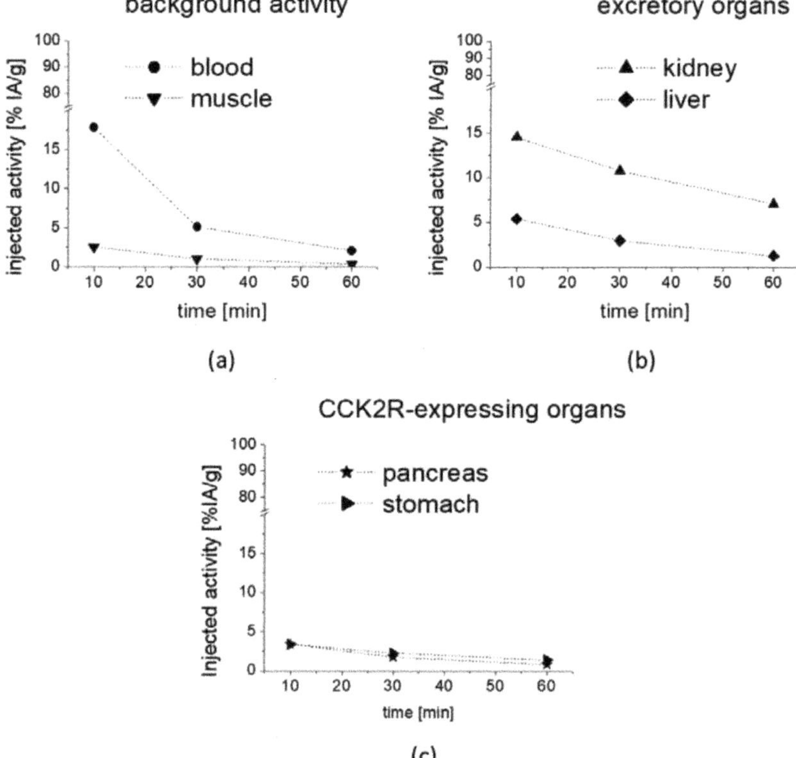

Figure 7. Uptake values in selected tissues as obtained from metabolic biodistribution studies with [^{177}Lu]Lu-**1** in BALB/c mice at 10, 30 and 60 min p.i.: (**a**) background activity in blood and muscle, (**b**) excretory organs kidney and liver as well as (**c**) CCK2R-expressing pancreas and stomach. Values are expressed as % IA/g.

Table 2. Biodistribution profile of [^{177}Lu]Lu-**1** in BALB/c mice after 10 min, 30 min and 1 h p.i. (30–40 MBq, 0.8 nmol). Values are expressed as % IA/g.

Organ	% IA/g		
	10 min p.i.	**30 min p.i.**	**1 h p.i.**
blood	17.79	5.08	2.05
lung	17.45	5.03	1.95
heart	7.13	2.72	0.99
femur	5.97	1.28	0.52
spleen	3.70	2.00	0.79
muscle	2.56	1.03	0.36
intestine	2.63	1.61	1.02
liver	5.37	2.95	1.26
kidneys	14.49	10.76	7.01
pancreas	3.39	1.76	0.84
stomach	3.36	2.29	1.39

2.5. Identification of the Radiometabolites Formed In Vivo

For the evaluation of the radiometabolites formed in vivo, blood and urine collected from BALB/c mice injected with [^{177}Lu]Lu-**1** were analyzed by radio-HPLC. In addition, the soluble phase extracted from kidney and liver homogenates was analyzed. [^{177}Lu]Lu-**1** showed a high stability in blood against enzymatic degradation in vivo with a total amount of radiometabolites of 19.5%, 35.9% and 43.1% at 10, 30 and 60 min p.i., respectively. The radiometabolites observed could be matched to [^{177}Lu]Lu-**M1**, [^{177}Lu]Lu-**M2**, [^{177}Lu]Lu-**M3**, [^{177}Lu]Lu-**M5**, with hydrolysis of the C-terminal amide, as well as cleavage of the peptide bonds of Asp-1Nal, (*N*-Me)Nle-Asp and Gly-Trp. The same metabolites were also observed in urine, whereas the presence of [^{177}Lu]Lu-**M1** was negligible. However, much higher levels of the radiometabolites were observed in urine, which was to be expected, given the renal pathway as main route of excretion. A similar pattern was observed also for liver and kidneys, where only minor additional radiometabolites could be detected. [^{177}Lu]Lu-**M6** with cleavage of the peptide bond of Tyr-Gly was confirmed for both, liver and kidneys, while [^{177}Lu]Lu-**M4** cleaved between Trp and (*N*-Me)Nle was detectable only in liver and [^{177}Lu]Lu-**M7** cleaved between Pro and Tyr only in kidneys. No cleavage between DGlu and Pro was observed, confirming a stabilizing effect for the tertiary amide bond introduced into the peptide backbone. Radiochromatograms of the different samples analyzed are displayed in Figure 8a–d. The percentage of the intact radiopeptide and of the different radiometabolites found for the different time points p.i. are summarized in Table 3.

Table 3. Quantification of the percentage of intact radiopeptide and of the radiometabolites as analyzed by radio-HPC of blood and urine, as well as liver and kidneys of BALB/c mice injected with [^{177}Lu]Lu-**1** for different time points p.i.

Peptide	Blood		Liver		Kidney		Urine	
	30 min	**60 min**	**30 min**	**60 min**	**30 min**	**60 min**	**30 min**	**60 min**
[^{177}Lu]Lu-**1**	64.1%	56.9%	53.9%	47.7%	18.0%	7.0%	19.4%	18.0%
[^{177}Lu]Lu-**M1**	9.9%	5.8%	4.8%	4.6%	2.5%	-	1.4%	-
[^{177}Lu]Lu-**M2**	3.8%	7.3%	7.2%	6.9%	20.7%	12.3%	40.5%	40.3%
[^{177}Lu]Lu-**M3**	17.0%	20.4%	8.5%	5.6%	7.9%	-	9.5%	9.8%
[^{177}Lu]Lu-**M4**	-	-	2.9%	2.6%	-	-	-	-
[^{177}Lu]Lu-**M5**	5.3%	9.6%	14.4%	12.9%	27.3%	23.7%	28.7%	31.3%
[^{177}Lu]Lu-**M6**	-	-	8.3%	19.7%	23.6%	53.2%	-	-
[^{177}Lu]Lu-**M7**	-	-	-	-	-	3.8%	-	-
[^{177}Lu]Lu-**M8**	-	-	-	-	-	-	-	-

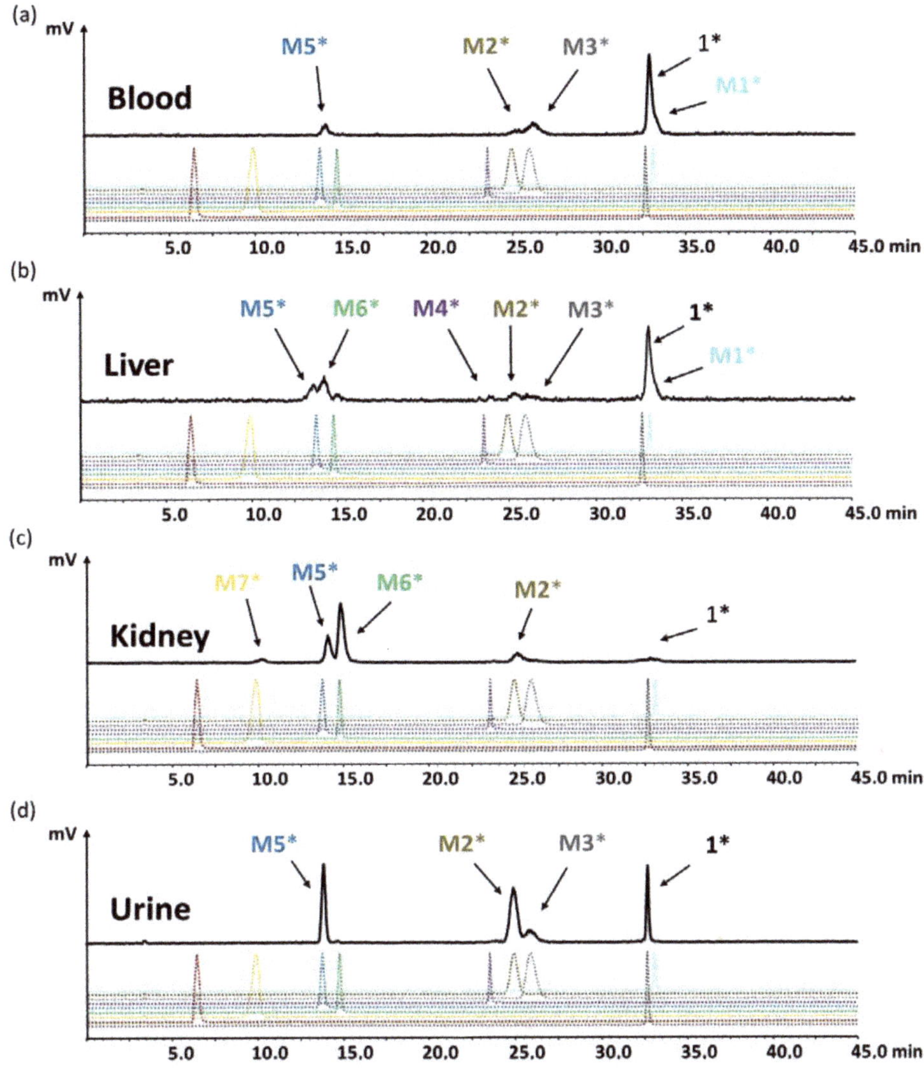

Figure 8. Radiochromatograms obtained from (**a**) blood, (**c**) liver, (**d**) kidneys and (**b**) urine of BALB/c mice injected with [^{177}Lu]Lu-**1** for the time point of 60 min p.i.; radiochromatograms of ^{177}Lu-labeled peptides **1** and **M1–M8** are shown for comparison (* indicating radiolabeled with lutetium-177).

3. Discussion

The strategy of CCK2R targeting with radiolabeled gastrin analogs for diagnostic and therapeutic application in patients with advanced tumors, in particular MTC, has been pursued for more than two decades. The successful clinical use of radiolabeled somatostatin analogs for targeting somatostatin receptors in patients with neuroendocrine tumors could not yet be translated to radiolabeled minigastrin analogs targeting CCK2R. In our recent studies, we could develop a new stabilization strategy leading to increased stability of the linear peptide sequence against enzymatic degradation in vivo and improving the targeting properties [28,29]. In this study, we have further explored our stabilization strategy by introducing an additional modification in the *N*-terminal region of the peptide backbone in our lead compound DOTA-MGS5. Alanine in position 2 was replaced by proline leading to the new peptide analog **1**. Furthermore, different possible metabolites were synthesized and analyzed in comparative studies with the aim to further investigate the enzymatic degradation in vivo. Pro was selected as a promising candidate for substitution, as the cyclic structure of the side chain with its conformational rigidity may protect the peptide against enzymatic degradation. The insertion of Pro into proteins influences the formation of α-helices and ß-sheets in dependence of the molecular environment conferring specific features to protein structure and folding [31]. It has been shown that the single change from the L- to D-configuration of the Glu residues in MG analogs alters the secondary structure of the peptide leading to improved serum stability [32].

In this study, the preclinical properties of [^{177}Lu]Lu-**1** were evaluated and the metabolites thereof formed during degradation in vivo were characterized to explore additional possible stabilization strategies in the development of CCK2R targeting radiopeptides. To enable the characterization of the radiometabolites, eight different metabolites of **1** were synthesized in moderate yields. For in vitro studies, **1** and the metabolites thereof were radiolabeled with [^{177}Lu]LuCl$_3$ at low molar activity allowing quantitative radiolabeling at high radiochemical purity >95%. Receptor affinity assays with the non-labeled peptides as well as cell uptake studies with the peptide derivatives radiolabeled with lutetium-177 were performed using A431-CCK2R cells. The results confirmed loss of receptor affinity as well as cell uptake for different metabolites studied, while the intact peptide showed a retained high CCK2R affinity and improved receptor-mediated cell uptake of more than 60% of the total activity added. The specificity of the cell uptake was verified via lack of uptake in A431-mock cells used as a control cell line. Hydrolysis of the C-terminal amide was sufficient to cause complete loss of receptor affinity, proving that the interaction of the amidated C-terminus with the binding pocket is essential for maintaining receptor affinity [33]. When incubated in fresh human serum, used as a model in vitro to measure the resistance against enzymatic degradation, radiolabeled **1** as well as **M2** and **M4** showed the highest stability after 24 h incubation (>98%). Radiolabeled **M1** showed a slightly lower stability of 89%, whereas for **M3**, less than 30% intact radiopeptide was detectable after 24 h incubation, suggesting a high enzymatic susceptibility. This could be explained by the lower protein binding of [^{177}Lu]Lu-**M3** leading to higher levels of free radiopeptide susceptible to proteases when compared to [^{177}Lu]Lu-**1** and radiolabeled **M1**, **M2** and **M4**. Thus, increased protein binding may play an additive role in protecting [^{177}Lu]Lu-**1** against enzymatic degradation. Highly improved stability of [^{177}Lu]Lu-**1** was observed also in rat tissue homogenates in vitro. It has been shown that unsubstituted DOTA-MG11 labeled with indium-111 is highly susceptible to proteolytic digestion and rapidly and completely degraded within 30 min in liver homogenate and within 10 min in kidney homogenate [34]. For [^{177}Lu]Lu-**1** still 84% and 44% intact radiopeptide could be observed in liver and kidney homogenate, respectively, after 30 min incubation. Only at the later time point of 2 h after incubation, an almost complete breakdown occurred in kidney homogenate, whereas in liver homogenate >50% intact radiopeptide was still present. The resistance against enzymatic degradation in rat liver homogenate was also improved when compared to DOTA-MGS5 labeled with different radiometals for which less than 30% intact radiopeptide was found for the same time point, suggesting a potential additive stabilizing effect of the insertion of Pro in position 2 [28].

Incubation in organ homogenates is connected with a higher breakdown of the radiopeptide due to the exposure to extracellular and intracellular proteases released after tissue homogenization, possibly leading to an underestimation of the stability in vivo [16]. Therefore, additional metabolic biodistribution studies in female BALB/c mice were performed to monitor the metabolites formed in vivo. After intravenous injection of [^{177}Lu]Lu-**1** in BALB/c mice, a highly improved stability with more than 80% intact radiopeptide in blood was observed at 10 min after injection. The in vivo stability during circulation was tested for up to 1 h p.i. still finding 56.9% intact [^{177}Lu]Lu-**1**. Much higher amounts of metabolites were present in the urine of the mice at the different studied time points, with the intact radiopeptide decreasing from 29% at 10 min to 18% at 1 h after injection. The metabolites found in liver homogenate resembled the metabolites observed in blood. The same correspondence was found for the metabolites in urine and kidney homogenate. A more rapid metabolism of the radiopeptide was observed in kidneys and urine (<30%), whereas in blood and liver, a much higher stability was observed. With the optimized radio-HPLC gradient allowing for a better separation of the different radiometabolites, possibly additional radiometabolites could be monitored, which were not detectable in previous studies with ^{177}Lu-labeled DOTA-MGS5 [28]. The additional substitution with Pro did, however, not show a considerable effect on in vivo stability. Still, the in vivo stability of [^{177}Lu]Lu-**1** is highly improved when compared to other MG analogs which are currently investigated in clinical trials [22,24]. For PP-F11 labeled with indium-111, the metabolic stability in the blood of mice was tested for the time point of 5 min p.i. finding >70% intact radiopeptide [28,35]. However, when analyzing the in vivo stability for a later time point, we found that ^{177}Lu-labeled PP-F11 is almost completely degraded at 30 min p.i. [28].

To identify the observed degradation products, different metabolites were synthesized and radiolabeled with lutetium-177 for comparative radio-HPLC analysis. The retention times of the different radiometabolites were matched with the metabolites found in vivo. In the blood of mice at different time points of up to 1 h p.i., only minor amounts of radiolabeled **M1** (<10%), **M2** (<8%), **M3** (<21%) and **M5** (<10%) resulting from hydrolysis of the C-terminal amide, as well as cleavage of the peptide bonds of Asp-1Nal, (*N*-Me)Nle-Asp and Gly-Trp were found. The same metabolites, except **M1**, were detected in urine, with [^{177}Lu]Lu-**M2** (~40%) and [^{177}Lu]Lu-**M5** (~30%) being the most prominent. In liver, besides similar amounts of the metabolites found in blood, additionally, radiolabeled **M4** with cleavage of the peptide bond of Trp-(*N*-Me)Nle was detectable at very low concentration (<3%), as well as radiolabeled **M6** with cleavage between Tyr and Gly (~20%). This could reflect the higher enzymatic turnover in the liver. In kidneys, only 7% of intact radiopeptide was found at 1 h p.i. even though still 18% were present in the urine at the same time point, whereas much higher amounts of [^{177}Lu]Lu-**M2** (12%), [^{177}Lu]Lu-**M5** (24%) and [^{177}Lu]Lu-**M6** (53%) were detected. Furthermore, minor amounts of [^{177}Lu]Lu-**M7** (~4%) with cleavage of the peptide bond of Pro-Tyr were observed. Despite the different radiometabolites found, the new minigastrin analog, with 56.9% intact radiopeptide still present in the blood at 1 h after injection, shows a highly improved stability in vivo.

It is well known from the literature that different enzymes are involved in the metabolism of members of the gastrin/CCK family such as the angiotensin converting enzyme (ACE), endopeptidase (NEP), aminopeptidase A (APA) and cathepsins. ACE is a zinc- and chloride-dependent peptidyl dipeptidase, widely distributed throughout the body including the lungs, gastrointestinal tract, vascular endothelium and blood, with broad specificity and besides inactivating vasoactive peptides also acts on other bioactive peptides [16,36]. In the degradation process of CCK and gastrin analogs with eight or less amino acids, ACE initially cleaves the amidated C-terminal dipeptide Asp-Phe-NH_2 and releases a further C-terminal di- or tripeptide in a secondary step [13]. Thirteen amino acid long MG analogs containing the penta-Glu motif seem to be ACE-resistant [14]. It has however been reported that radiolabeled MG analogs derived from MG11 and MG0 are not cleaved by ACE and ACE inhibitors cannot prevent the degradation in vivo [14,15]. NEP is a zinc-dependent cell-surface enzyme with wide distribution in the body, including the presence on granulocytes and endothelial cells of the vasculature compartment, as well as in major organs such as liver, kidneys and gastrointestinal tract, and is involved in the degradation of many bioactive peptides [16,37]. In the degradation process of CCK and gastrin analogs, NEP cleaves the peptide at Asp-Phe, Trp-Met, Gly-Trp, as well as Ala-Tyr in the case of gastrin [38,39]. The fact that co-injection of ^{111}In-labeled DOTA-MG11 together with a NEP inhibitor clearly increased the amount of intact radiopeptide in the blood of mice and led to a significant improvement of tumor uptake, points out the importance of NEP in the metabolism of radiolabeled MG analogs [15]. Incubation in vitro with neprylisin-1 and neprylisin-2 confirmed cleavage at Asp-Phe and Gly-Trp for DOTA-MG11, whereas for DOTA-PP-F11, only cleavage at Asp-Phe was observed and DOTA-PP-F11N with Met replaced by Nle seemed to be resistant against neprylisins. Interestingly, NEP inhibition did not result in improved tumor uptake of DOTA-PP-F11 and DOTA-PP-F11N [14]. APA is a membrane-bound type II zinc metalloprotease with broad tissue distribution [16,40]. It cleaves *N*-terminal glutamyl and aspartyl residues and is known to be involved in the degradation of CCK-8. *N*-terminal modification of CCK8 analogs led to resistance against APA, suggesting that also radiolabeled peptide analogs with *N*-terminal conjugation of a bifunctional chelator are APA-resistant [41]. During intracellular trafficking also other proteases may be involved in the degradation process. Cathepsins are a group of enzymes whose primarily function is to act as intralysosomal enzymes and in addition to that are involved in cancer development and progression [17,42]. When analyzing the stability against different cathepsins in vitro, cleavage sites at Asp-Phe and Met/Nle-Asp have been confirmed for different DOTA-conjugated MG analogs [14].

[^{177}Lu]Lu-**M2** and [^{177}Lu]Lu-**M5** with cleavage at Asp-1Nal and Gly-Trp, respectively, were present in all samples studies, indicating a major involvement of NEP in the metabolization of [^{177}Lu]Lu-**1** in vivo. The presence of [^{177}Lu]Lu-**M3** in almost all samples studies, as well as of [^{177}Lu]Lu-**M6** in the soluble phase of the homogenates from liver and kidney, suggests that also an ACE-dependent metabolism could occur. Kolenc-Peitl et al. have also suggested two different degradation pathways for MG analogs, one pathway via ACE-like enzyme activity with cleavage at Met-Asp, Gly-Trp, Tyr-Gly and Ala-Tyr, and another pathway directly releasing gastrin-6 with cleavage at Ala-Tyr [43]. Interestingly, hydrolysis of the C-terminal amide resulting in [^{177}Lu]Lu-**M1**, especially in blood and liver, was additionally observed, suggesting that other enzymes are also involved in the metabolism of [^{177}Lu]Lu-**1**, which have not yet been characterized. Our results are in contrast to Sauter et al. who investigated the susceptibility of different DOTA-conjugated MG analogs against various proteases in vitro [14]. Based on the radiometabolites found in the different tissue samples obtained from BALB/c mice injected with [^{177}Lu]Lu-**1**, we could not confirm ACE resistance in vivo, and also replacement of Met by Nle did not result in liability against neprylisins. However, the fact that the radiometabolites [^{177}Lu]Lu-**M2** and [^{177}Lu]Lu-**M3** were found in all investigated tissues supports a possible involvement of cathepsins in the degradation process. For [^{177}Lu]Lu-**1**, only minor amounts of [^{177}Lu]Lu-**M4** cleaved at (*N*-Me)Nle-Trp were observed in liver homogenate, indicating a stabilizing effect of the introduced *N*-methylated peptide bond. In a previous study, we could already show that single substitution with 1Nal in position 8 did not improve in vivo stability and only additional substitution with (N-Me)Nle in position 6 allowed to stabilize the linear peptide against enzymatic degradation [28,29]. When introducing 1,4-disubstituted 1,2,3-triazoles as metabolically stable bioisosteres in replacement of the amide bonds in Nle-substituted DOTA-MG11, Grob et al. also found the highest impact on the stability against proteases in human blood for the triazole insertion at Trp-Nle. The stability was not considerably further improved by additional triazole insertion at Ala-Tyr or Tyr-Gly. On the other hand, the insertion of triazoles at Tyr-Gly, Ala-Tyr, DGlu-Ala had a remarkable effect on tumor uptake, which was however clearly inferior to DOTA-MGS5 [28,30,44]. The introduction of an *N*-methylated peptide bond at Nle-Trp showed both effects of increased stability and improved receptor interaction. DOTA-MGS5 radiolabeled with different radiometals, showing the combined substitution of (*N*-Me)Nle and 1Nal in position 6 and 8, respectively, displayed a considerably higher cell uptake of >50%, when compared to ~25% observed for ^{111}In-labeled DOTA-MGS1 with single replacement of Phe by 1Nal [29]. The combination of increased stability and improved receptor interaction led to a considerable improvement in tumor uptake. In mice xenografted with A431-CCK2R cells a very high tumor uptake of more than 20%IA/g in combination with improved tumor-to-kidney ratio (4–6) was observed for DOTA-MGS5 labeled with indium-111, lutetium-177 or gallium-68 [28]. In the present study, very high receptor-mediated cell uptake in A431-CCK2R cells of >60% was also found for [^{177}Lu]Lu-**1** showing additional substitution with Pro in position 2. Interestingly, the radiometabolite [^{177}Lu]Lu-**M8** with cleavage at DGlu-Pro was not found in any of the tissues examined, indicating a possible additional stabilizing effect of the introduction of Pro. In a previous study analyzing the blood of patients injected with ^{111}In-labeled DOTA-MG11, high amounts of the short chain radiometabolites of DOTA-DGlu, DOTA-DGlu-Ala and DOTA-DGlu-Ala-Tyr were confirmed at 10 min after injection [12].

The preliminary biodistribution profile obtained for [^{177}Lu]Lu-**1** from the metabolic studies in mice confirmed a rapid clearance from blood and low unspecific uptake in most tissues, together with predominant renal excretion. The prolonged retention of radioactivity in CCK2R-expressing stomach and pancreas indicates that [^{177}Lu]Lu-**1** also shows high potential for targeting CCK2R-expressing tumors [45].

4. Materials and Methods

4.1. Materials

All commercially obtained chemicals were of analytical grade and used without further purification. No-carrier-added [^{177}Lu]$LuCl_3$ produced from highly enriched ^{176}Yb was purchased from Isotope Technologies (Garching, Germany). Dr. Luigi Aloj kindly provided the A431 human epidermoid carcinoma cell line stably transfected with the plasmid pCR3.1 containing the full coding sequence for the human CCK2R, as well as the same cell line transfected with the empty vector alone [46]. A431-CCK2R and A431-mock were cultured in Dulbecco's Modified Eagle Medium (DMEM) supplemented with 10% (*v/v*) fetal bovine serum and 5 mL of a 100× penicillin-streptomycin-glutamine mixture at 37 °C in a humidified 95% air/5% CO_2 atmosphere. Media and supplements were purchased from Invitrogen Corporation (Lofer, Austria). **1** was purchased from piCHEM (Raaba-Grambach, Austria).

4.2. Peptide Synthesis

The different metabolites of **1**, namely **M1–M6** shown in Table 1, were synthesized using 9-fluorenylmethoxycarbonyl (Fmoc) chemistry. The peptides were assembled on 2-chlorotritylchloride (2-CTC) resin with capacity 1.6 mmol/g (Iris Biotech GmbH, Marktredwitz, Germany). The reactive side chains of the amino acids were masked with the following protection groups: tert-butyl ester for Asp and DGlu, tert-butyl ether for Tyr, and tertbutyloxycarbonyl (BOC) for Trp. All coupling reactions were performed using a 5-fold excess of Fmoc-protected amino acids, 1-hydroxy-7-aza-benzotriazole (HOAt) and *O*-(7-Azabenzotriazole-1-yl)-*N,N,N′,N′*-tetramethyluronium hexa-fluorophosphate (HATU) in dimethtylformamide (DMF) and pH adjusted to 8 with *N,N′*-diisopropylethylamine (DIPEA). The resin was loaded with 30% of total capacity and the remaining binding sites were capped with methanol/DIPEA/dichloromethane (DCM) in a ratio of 200 μL/100 μL/2 mL for 30 min at room temperature. Between every conjugation step, the product was washed 6 times with DMF for 1 min. Removal of the Fmoc protecting groups was obtained by two consecutive treatments with 5 mL of 20% piperidin in DMF for 5 and 15 min each. For the coupling of DOTA, a 3-fold molar excess of DOTA-tris(tert-butyl ester), HOAt and HATU was used. Cleavage of the peptides from the resin with concomitant removal of acid-labile protecting groups was achieved by treatment with a mixture of trifluoroacetic acid (TFA), triisopropylsilane, and water in a ratio of 95/2.5/2.5 (*v/v/v*). The crude peptides were precipitated in ice-cold ether before HPLC purification and characterized by analytical HPLC (Dionex, Germering, Germany and matrix-assisted laser desorption/ionization time of flight mass spectroscopy (MALDI-TOF MS) (Bruker Daltonics, Bremen, Germany). Purification was performed by RP-HPLC on a GILSON 322 chromatography system with a GILSON UV/VIS-155D multi-wavelength UV detector, equipped with an Eurospher II 100-5 C18 A column, 250 × 8 mm (Knauer, Berlin, Germany) or an Eurosil Bioselect 300-5 C18 A column, Vertex Plus, 300 × 8 mm, combined with an Eurosil Bioselect 300-5 C18 precolumn, Vertex Plus A, 30 × 8 mm (Knauer, Berlin, Germany), using a water/ACN/0.1% TFA gradient.

Analytical HPLC was performed using an UltiMate 3000 chromatography system consisting of a variable UV-detector (UV-VIS at λ = 220 nm), a HPLC pump, an autosampler, a radiodetector (GabiStar, Raytest, Straubenhardt, Germany), equipped with a Phenomenex Jupiter 4 μm Proteo 90 Å C12 column, 250 × 4.6 mm (Phenomenex Ltd., Aschaffenburg, Germany) and analyzed with Chromeleon Dionex Software (Version 7.2.9.11323). The radiodetector was equipped with two different loops, a low-sensitivity loop of 5 μL and a high-sensitivity loop of 250 μL. Mass spectrometry was performed using a Bruker microflex benchtop MALDI-TOF MS with 200 shots per spot in reflector acquisition mode with a positive ion source. For mass determination, samples were prepared on α-cyano-4-hydroxycinnamic acid (HCCA) matrix using dried droplet procedure. Flex Analysis 2.4 software was used to analyze the recorded data. HPLC chromatograms and MS spectra are presented in the supporting information (Figures S1 and S2). The lyophilized peptide derivatives were stored at −20 °C.

M7 and **M8** were obtained using other strategies. For the synthesis of **M7**, the dipeptide DGlu-Pro was synthesized on 2-CTC resin following the synthesis protocol described above defining the amino acid sequence. After cleavage from the resin, precipitation in ice-cold ether, purification by HPLC and lyophilization, DGlu-Pro (0.015 g, 0.061 mmol) was transferred to a round-bottomed flask, dissolved in 1 mL ACN and pH adjusted to 8 with 25 μL DIPEA. DOTA-*N*-hydroxysuccinimid ester (0.023 g, 0.030 mmol) was added to the solution and the mixture was stirred at room temperature overnight. After evaporation of the solvent, the product was dissolved in 2 mL H_2O, purified by HPLC, lyophilized and characterized by HPLC and MALDI-TOF MS. For the synthesis of **M8**, to a solution of D-glutamic acid (0.30 g, 2.04 mmol) in 6.25 mL dry methanol, 0.89 mL distilled thionyl chloride (1.46 g, 12.23 mmol) was added at 0 °C over 30 min. Then, the reaction mixture was stirred for 12 h at room temperature. The solvent was evaporated under reduced pressure, diluted with saturated $NaHCO_3$, and extracted with dicloromethane (3 × 200 mL). The organic layer was washed with H_2O, followed by brine, dried over Na_2SO_4 and filtered. After evaporation of the solvent, the desired D-glutamic acid dimethyl ester was obtained as a yellowish oil (0.315 g, 1.79 mmol, 87% yield) [47]. NMR spectrum was measured on a 400 MHz Avance 4 Neo (Bruker) spectrometer. As solvents for NMR deuterated chloroform ($CDCl_3$) was used (Euriso-top®). The chemical shifts (δ) were referenced to tetramethylsilane or the solvent peak and were given in parts per million (ppm). Coupling constants (J) were reported in Hertz (Hz). The following descriptors for signals were used: s = singlet, t = triplet, q = quartet, m = multiplet. ^{1}H-NMR (400 MHz, $CDCL_3$ δ ppm): 3.66 (s, 3H, CH_3), 3.61 (s, 3H, CH_3), 3.43–3.39 (q, 1H, CH), 2.43–2.39 (t, 2H, CH_2), 2.04– 1.75 (m, 2H, CH_2) (Figure S3). To D-glutamic dimethylester (0.030 mg, 0.174 mmol) in a round-bottomed flask a mixture of DOTA-tris(tert-butylester) (0.05 g, 0.087 mmol), HATU (0.066 g, 0.174 mmol), HOAt (0.024 g, 0.174 mmol) in 3 mL DCM, adjusted to pH 8 with 35 μL DIPEA, was added and the solution was stirred overnight at room temperature. The reaction solution was evaporated and the protecting groups were removed by adding 3 mL 50% TFA in DCM at 60 °C for 24 h. After evaporation, the crude product was dissolved in 2 mL 50% ACN. After HPLC purification and lyophilization, the final product was characterized by HPLC and MALDI-TOF MS.

4.3. Radiolabeling and Characterization In Vitro

For labeling with lutetium-177, the DOTA-peptides (8–12 μg in 10 μL) were incubated with 10–30 μL [^{177}Lu]$LuCl_3$ solution (60–300 MBq in 0.05 M HCl) and a >1.2-fold volume of 0.4 M sodium acetate/0.24 M gentisic acid solution pH adjusted to 5 at 90 °C for 20 min. Radiochemical purity of the radiopeptides was analyzed using the analytical HPLC system described above using a flow rate of 1 mL/min together with the following water/ACN/0.1% TFA gradient: 0–3 min 10% ACN, 3–18 min 10–55% ACN, 18–20 min 55–80% ACN, 20–21 min 80–10% ACN, 21–25 min 10% ACN. [^{177}Lu]Lu-**1** used in animal studies was purified by solid phase extraction (SPE). For this purpose, a C18 SepPak tLight cartridge (Waters, Milford, MA) was pretreated with 5 mL 99% ethanol followed by 5 mL 0.9% saline. The radiolabeling solution was passed through the cartridge and washed with 5 mL 0.9% saline to elute hydrophilic impurities. The radiolabeled peptide was eluted with 50% ethanol from the cartridge and diluted with 0.9% saline.

For the determination of the distribution coefficient (LogD) in octanol/PBS, the radiolabeled DOTA-peptides (100 pmol) in 500 μL PBS (pH 7.4) were added to 500 μL octanol in an Eppendorf microcentrifuge tube (n = 8). The mixture was vigorously vortexed at room temperature over a period of 15 min using a small shaker (MS3 Basic, IKA, Staufen, Germany) with speed of 1500 rpm. After a waiting time of 10 min sufficient for the separation of the two phases, 100 μL aliquots of both layers were measured in a gamma counter (2480 Wizard2 3″, PerkinElmer Life Sciences and Analytical Sciences, formerly Wallac Oy, Turku, Finland) and the logD value was calculated.

For protein binding assessment, the radiolabeled DOTA-peptides were incubated in fresh human serum at 37 °C (500 pmol/mL, n = 2). After 1, 4 and 24 h of incubation, two samples were taken for each time point and analyzed by Sephadex G-50 size-exclusion chromatography (GE Healthcare Illustra, Little Chalfont, UK). The percentage of protein binding was determined by measuring the column and the eluate with the gamma-counter.

In vitro stability studies for the characterization of the metabolic stability of the radiolabeled peptide analogs in human serum were carried out with [^{177}Lu]Lu-**1**, [^{177}Lu]Lu-**M1**, [^{177}Lu]Lu-**M2**, [^{177}Lu]Lu-**M3** and [^{177}Lu]Lu-**M4** (n = 2). Additional stability studies in rat liver and kidney homogenates were performed for [^{177}Lu]Lu-**1** (n = 2). Tissue homogenates were prepared from dissected organs by homogenization for 1 min at RT (IKA-Werke, Staufen, Germany) in 20 mM HEPES buffer pH 7.3 (30% *v/v*). The radioligands were incubated in the different media at a concentration of 500 pmol/mL (corresponding to an activity of 6–11 MBq). After incubation at 37 °C at different time points for up to 24 h, a 100 µl sample was taken in duplicates and analyzed by HPLC. Samples obtained from human serum and rat homogenates were treated with ACN at a ratio of 1:1.5 (*v/v*) to precipitate proteins, centrifuged (14,000 rpm, 2 min, centrifuge 5424, Eppendorf AG, Germany) and diluted with water at a ratio of 1:1 (*v/v*). A 100 µl aliquot of this solution was injected into the radio HPLC system. The degradation of the radioligand was evaluated based on the radiochemical purity after radiolabeling and the percentage of intact radiopeptide observed during incubation in the different media.

4.4. Receptor Binding and Cell Uptake Studies

The binding affinity of **1**, pentagastrin and of the different metabolites for the CCK2R was tested in a competition assay against [^{125}I][3-iodo-Tyr^{12},Leu^{15}]gastrin-I. Radioiodination of gastrin-I was carried out using the chloramine-T method. [^{125}I][3-iodo-Tyr^{12},Leu^{15}]gastrin-I at high molar activity was obtained by HPLC purification and stored in aliquots at −25 °C. Binding assays were carried out using 96-well filter plates (MultiScreen$_{HTS}$-FB, Merck Group, Darmstadt, Germany) pre-treated with 10 mM TRIS/139 mM NaCl pH 7.4 (2 × 250 µL). For the assay, 200,000–400,000 A431-CCK2R cells per well were prepared in 20 mM HEPES buffer pH 7.4 containing 10 mM $MgCl_2$, 14 µM bacitracin and 0.5% BSA. The cells (n = 3) were incubated with increasing concentrations of the peptide conjugates (0.0003–30,000 nM) and [^{125}I][3-iodo-Tyr^{12},Leu^{15}]gastrin-I (50,000 cpm) for 1 h at RT. Incubation was interrupted by filtration of the medium and rapid rinsing with ice-cold 10 mM TRIS/139 mM NaCl pH 7.4 (2 × 200 µL) and the filters were measured in the gamma-counter. Half maximal inhibitory concentration (IC_{50}) values were calculated following nonlinear regression with Origin software (Microcal Origin 6.1, Northampton, MA, USA).

For internalization experiments with [^{177}Lu]Lu-**1** and its metabolites, A431-CCK2R and A431-mock cells were seeded at a density of 1.0×10^6 per well in 6-well plates (Greiner Labortechnik, Kremsmuenster, Austria)and grown for 48 h until reaching almost confluence. On the day of the experiment, the medium was replaced by 1.2 mL of fresh DMEM medium supplied with 1% (*v/v*) fetal bovine serum. The cells (n = 3) were incubated with ~25,000 cpm of the radioligand in 300 µL PBS/0.5% BSA in a total volume of 1.5 mL, corresponding to a final concentration of 0.4 nM of total peptide. After 1 h and 2 h incubation, the cell uptake was interrupted by removal of the medium and rapid rinsing with 2 × 1 mL PBS/0.5% BSA. Thereafter, the cells were incubated twice at ambient temperature with acid wash buffer (50 mM glycine buffer pH 2.8, 0.1 M NaCl) for 5 min, to remove the membrane-bound radioligand. Finally, the cells were lysed by treatment in 1 N NaOH and collected (internalized radioligand fraction). All fractions were counted in the gamma-counter and mean values were calculated. The internalized fraction was expressed in relation to the total radioactivity added to the cells. In an additional experiment, the cell uptake of [^{177}Lu]Lu-**M1**, [^{177}Lu]Lu-**M2**, [^{177}Lu]Lu-**M3** and [^{177}Lu]Lu-**M4** was investigated in A431-CCK2R cells for the time point of 1 and 2 h incubation.

4.5. Metabolic Biodistribution Studies and Characterization of the Radiometabolites In Vivo

Metabolic biodistribution studies were performed in accordance with the ethical standards of the institution and approved by the Austrian Ministry of Science (BMWFW-66.011/0072-V/3b/2019). These studies were carried out in 5- to 7-week-old female BALB/c mice (n = 3) injected with [^{177}Lu]Lu-**1** via a lateral tail vein. To allow monitoring of the metabolites by radio-HPLC, mice were injected with 30 MBq corresponding to 0.8 nmol total peptide. The mice were euthanized by cervical dislocation after different time points of 10 min, 30 min and 1 h p.i. and the urine and a venous blood sample were collected at the time of sacrifice. Liver and kidneys were dissected and homogenized in 20 mM HEPES buffer pH 7.3 at a ratio of 1:1 (*v/v*) with an Ultra-Turrax T8 homogenizer (IKA-Werke, Staufen, Germany) for 1 min at RT. Before radio-HPLC analysis, the samples (except urine) were treated with ACN at a ratio of 1:1 (*v/v*) to precipitate proteins, centrifuged at 2000× *g* for 2 min and diluted with water at a ratio of 1:1 (*v/v*). For HPLC analysis of the samples, the analytical HPLC system described above with a flow rate of 1 mL/min was used together with the following optimized water/ACN/0.1% TFA gradient to allow a better separation of [^{177}Lu]Lu-**1** and the different radiometabolites: 0–7 min 1–7% ACN, 7–8 min 7–10% ACN, 8–18 min 10–18% ACN, 18–20 min 18–32% ACN, 20–27 min 32% ACN, 27–37 min 32–0% ACN, 37–40 min 80% ACN, 40–40.1 min 80–1% ACN, 40.1–45 min 1% ACN. Urine was measured with the low-sensitivity loop of the radiodetector. The remaining samples were analyzed with the high-sensitivity loop due to the lower radioactivity present in blood, liver and kidney homogenate. To quantify the percentage of intact [^{177}Lu]Lu-**1** and [^{177}Lu]Lu-**M1** showing some overlap in the radiochromatogram when using the high-sensitivity loop of the radiodetector, the two radiopeptides were co-injected at different known ratios to enable a more accurate separation and integration of the two peaks (see supporting Figure S4). Mice from the metabolic biodistribution studies were subjected to a further dissection of all remaining tissues (blood, lung, heart, femur, spleen, muscle, intestine, pancreas and stomach). All organs, also including liver and kidneys, were weighed, and their radioactivity was measured in the gamma counter together with a standard and the residual body. To quantify the uptake of radioactivity in liver and kidneys, a part of liver and kidney homogenate was measured in the gamma counter to extrapolate the radioactivity for the whole organ.

5. Conclusions

Radiolabeled MG analogs with the modified receptor-specific C-terminal sequence Trp-(*N*-Me)Nle -Asp-1Nal-NH_2 are promising new candidates for diagnostic and therapeutic use in patients with advanced MTC and other CCK2R-expressing malignancies. [^{177}Lu]Lu-**1** with introduction of an additional tertiary peptide trough substitution with Pro in position 2, shows a highly improved stability against enzymatic degradation in vivo. From the radiometabolites identified in the blood of mice injected with [^{177}Lu]Lu-**1** hydrolysis of the C-terminal amide and cleavage of the peptide bonds of Asp-1Nal, (*N*-Me)Nle-Asp and Gly-Trp were found to occur in vivo. The high receptor-mediated cell uptake and favorable biodistribution profile in normal BALB/c mice support further studies evaluating the tumor targeting potential of [^{177}Lu]Lu-**1** and other alternative derivatives thereof.

Supplementary Materials: The following are available online, Figure S1: Representative UV-chromatogram of the metabolites **M1–M8**; Figure S2: MALDI-TOF-MS of the different synthesized metabolites **M1–M8**; Figure S3: 400 MHz 1H NMR of D-glutamic acid dimethyl ester; Figure S4. Radiochromatograms of [^{177}Lu]Lu-**1** and [^{177}Lu]Lu-**M1** co-analyzed in different ratios of approximately (a) 1:1, (b) 2:1 and (c) 10:1 (*v/v*) using the radiodetector equipped with the high sensitivity loop (250 μL)

Author Contributions: Conceptualization, Supervision, Project Administration and Funding Acquisition, E.v.G. and S.S.; Methodology and Investigation, A.A.H., M.K., M.R., R.G., N.H. and E.v.G.; Writing—Original Draft Preparation: A.A.H.; Writing—Review & Editing, E.v.G. and M.K. All authors have read and agreed to the published version of the manuscript.

Funding: This study was financially supported by the Austrian Science Fund (FWF), project P 27844.

Acknowledgments: Christine Rangger and Joachim Pfister are greatly acknowledged for technical assistance in cell culture and animal studies. Open Access Funding is supported by the Austrian Science Fund (FWF).

Conflicts of Interest: The authors declare no conflict of interest.

References

1. Fani, M.; Maecke, H.R.; Okarvi, S.M. Radiolabeled peptides: Valuable tools for the detection and treatment of cancer. *Theranostics* **2012**, *2*, 481–501. [CrossRef] [PubMed]
2. Bodei, L.; Cremonesi, M.; Grana, C.; Rocca, P.; Bartolomei, M.; Chinol, M.; Paganelli, G. Receptor radionuclide therapy with 90Y-[DOTA]0-Tyr3-octreotide (90Y-DOTATOC) in neuroendocrine tumours. *Eur. J. Nucl. Med. Mol. Imaging* **2004**, *31*, 1038–1046. [CrossRef] [PubMed]
3. Bison, S.M.; Konijnenberg, M.W.; Melis, M.; Pool, S.E.; Bernsen, M.R.; Teunissen, J.J.; Kwekkeboom, D.J.; de Jong, M. Peptide receptor radionuclide therapy using radiolabeled somatostatin analogs: Focus on future developments. *Clin. Transl. Imaging* **2014**, *2*, 55–66. [CrossRef] [PubMed]
4. Lutathera EMA. Available online: https://www.ema.europa.eu/en/medicines/human/EPAR/lutathera (accessed on 21 September 2020).
5. Lutathera FDA. Available online: https://www.accessdata.fda.gov/drugsatfda_docs/nda/2018/208700Orig1s000TOC.cfm (accessed on 21 September 2020).
6. Reubi, J.C. Somatostatin and other Peptide receptors as tools for tumor diagnosis and treatment. *Neuroendocrinology* **2004**, *80* (Suppl. 1), 51–56. [CrossRef]
7. Reubi, J.C.; Schaer, J.C.; Waser, B. Cholecystokinin(CCK)-A and CCK-B/gastrin receptors in human tumors. *Cancer Res.* **1997**, *57*, 1377–1386.
8. Klingler, M.; Hormann, A.A.; Guggenberg, E.V. Cholecystokinin-2 receptor targeting with radiolabeled peptides: Current status and future directions. *Curr. Med. Chem.* **2020**, *27*, 1–16. [CrossRef]
9. Behr, T.M.; Jenner, N.; Radetzky, S.; Behe, M.; Gratz, S.; Yucekent, S.; Raue, F.; Becker, W. Targeting of cholecystokinin-B/gastrin receptors in vivo: Preclinical and initial clinical evaluation of the diagnostic and therapeutic potential of radiolabelled gastrin. *Eur. J. Nucl. Med.* **1998**, *25*, 424–430. [CrossRef]
10. Béhé, M.; Becker, W.; Gotthardt, M.; Angerstein, C.; Behr, T.M. Improved kinetic stability of DTPA-DGlu as compared with conventional monofunctional DTPA in chelating indium and yttrium: Preclinical and initial clinical evaluation of radiometal labelled minigastrin derivatives. *Eur. J. Nucl. Med. Mol. Imaging* **2003**, *30*, 1140–1146. [CrossRef]
11. Good, S.; Walter, M.A.; Waser, B.; Wang, X.; Müller-Brand, J.; Béhé, M.P.; Reubi, J.C.; Maecke, H.R. Macrocyclic chelator-coupled gastrin-based radiopharmaceuticals for targeting of gastrin receptor-expressing tumours. *Eur. J. Nucl. Med. Mol. Imaging* **2008**, *35*, 1868–1877. [CrossRef]
12. Breeman, W.A.; Froberg, A.C.; de Blois, E.; van Gameren, A.; Melis, M.; de Jong, M.; Maina, T.; Nock, B.A.; Erion, J.L.; Macke, H.R.; et al. Optimised labeling, preclinical and initial clinical aspects of CCK-2 receptor-targeting with 3 radiolabeled peptides. *Nucl. Med. Biol.* **2008**, *35*, 839–849. [CrossRef]
13. Dubreuil, P.; Fulcrand, P.; Rodriguez, M.; Fulcrand, H.; Laur, J.; Martinez, J. Novel activity of angiotensin-converting enzyme. Hydrolysis of cholecystokinin and gastrin analogues with release of the amidated C-terminal dipeptide. *Biochem. J.* **1989**, *262*, 125–130. [CrossRef] [PubMed]
14. Sauter, A.W.; Mansi, R.; Hassiepen, U.; Muller, L.; Panigada, T.; Wiehr, S.; Wild, A.M.; Geistlich, S.; Behe, M.; Rottenburger, C.; et al. Targeting of the cholecystokinin-2 receptor with the minigastrin analog (177)Lu-DOTA-PP-F11N: Does the use of protease inhibitors further improve in vivo distribution? *J. Nucl. Med. Off. Publ. Soc. Nucl. Med.* **2019**, *60*, 393–399. [CrossRef]
15. Kaloudi, A.; Nock, B.A.; Lymperis, E.; Valkema, R.; Krenning, E.P.; de Jong, M.; Maina, T. Impact of clinically tested NEP/ACE inhibitors on tumor uptake of [(111)In-DOTA]MG11-first estimates for clinical translation. *EJNMMI Res.* **2016**, *6*, 15. [CrossRef] [PubMed]
16. Kaloudi, A.; Nock, B.A.; Krenning, E.P.; Maina, T.; De Jong, M. Radiolabeled gastrin/CCK analogs in tumor diagnosis: Towards higher stability and improved tumor targeting. *Q. J. Nucl. Med. Mol. Imaging* **2015**, *59*, 287–302. [PubMed]
17. Naqvi, S.A.R.; Matzow, T.; Finucane, C.; Nagra, S.A.; Ishfaq, M.M.; Mather, S.J.; Sosabowski, J. Insertion of a lysosomal enzyme cleavage site into the sequence of a radiolabeled neuropeptide influences cell trafficking in vitro and in vivo. *Cancer Biother. Radiopharm.* **2010**, *25*, 89–95. [CrossRef]

18. Roosenburg, S.; Laverman, P.; van Delft, F.L.; Boerman, O.C. Radiolabeled CCK/gastrin peptides for imaging and therapy of CCK2 receptor-expressing tumors. *Amino Acids* **2011**, *41*, 1049–1058. [CrossRef]
19. Laverman, P.; Joosten, L.; Eek, A.; Roosenburg, S.; Peitl, P.K.; Maina, T.; Mäcke, H.; Aloj, L.; Von Guggenber, E.; Sosabowski, J.K.; et al. Comparative biodistribution of 12 111In-labelled gastrin/CCK2 receptor-targeting peptides. *Eur. J. Nucl. Med. Mol. Imaging* **2011**, *38*, 1410–1416. [CrossRef]
20. Aloj, L.; Aurilio, M.; Rinaldi, V.; D'ambrosio, L.; Tesauro, D.; Peitl, P.K.; Maina, T.; Mansi, R.; Von Guggenberg, E.; Joosten, L.; et al. Comparison of the binding and internalization properties of 12 DOTA-coupled and 111In-labelled CCK2/gastrin receptor binding peptides: A collaborative project under COST Action BM0607. *Eur. J. Nucl. Med. Mol. Imaging* **2011**, *38*, 1417–1425. [CrossRef]
21. Ocak, M.; Helbok, A.; Rangger, C.; Peitl, P.K.; Nock, B.A.; Morelli, G.; Eek, A.; Sosabowski, J.K.; Breeman, W.A.; Reubi, J.C.; et al. Comparison of biological stability and metabolism of CCK2 receptor targeting peptides, a collaborative project under COST BM0607. *Eur. J. Nucl. Med. Mol. Imaging* **2011**, *38*, 1426–1435. [CrossRef]
22. Rottenburger, C.; Nicolas, G.P.; McDougall, L.; Kaul, F.; Cachovan, M.; Vija, A.H.; Schibli, R.; Geistlich, S.; Schumann, A.; Rau, T.; et al. Cholecystokinin 2 receptor agonist 177Lu-PP-F11N for radionuclide therapy of medullary thyroid carcinoma: Results of the lumed phase 0a study. *J. Nucl. Med.* **2020**, *61*, 520–526. [CrossRef]
23. Erba, P.A.; Maecke, H.; Mikolajczak, R.; Decristoforo, C.; Zaletel, K.; Maina-Nock, T.; Peitl, P.K.; Garnuszek, P.; Froberg, A.; Goebel, G.; et al. A novel CCK2/gastrin receptor-localizing radiolabeled peptide probe for personalized diagnosis and therapy of patients with progressive or metastatic medullary thyroid carcinoma: A multicenter phase I GRAN-T-MTC study. *Pol. Arch. Intern. Med.* **2018**, *128*, 791–795. [CrossRef] [PubMed]
24. Hubalewska-Dydejczyk, A.; Mikolajczak, R.; Decristoforo, C.; Kolenc-Peitl, P.; Erba, P.A.; Zaletel, K.; Maecke, H.; Maina, T.; Konijnenberg, M.; Garnuszek, P.; et al. Phase I clinical trial using a novel CCK2 receptor-localizing radiolabelled peptide probe for personalized diagnosis and therapy of patients with progressive or metastatic medullary thyroid carcinoma—Final results. *Eur. J. Nucl. Med. Mol. Imaging* **2019**, *46*, S339.
25. Kaloudi, A.; Nock, B.A.; Lymperis, E.; Krenning, E.P.; de Jong, M.; Maina, T. Improving the in vivo profile of minigastrin radiotracers: A comparative study involving the neutral endopeptidase inhibitor phosphoramidon. *Cancer Biother. Radiopharm.* **2016**, *31*, 20–28. [CrossRef] [PubMed]
26. McMurray, J.J.V. Neprilysin inhibition to treat heart failure: A tale of science, serendipity, and second chances. *Eur. J. Heart Fail.* **2015**, *17*, 242–247. [CrossRef] [PubMed]
27. Klingler, M.; Rangger, C.; Summer, D.; Kaeopookum, P.; Decristoforo, C.; von Guggenberg, E. Cholecystokinin-2 receptor targeting with novel c-terminally stabilized HYNIC-Minigastrin analogs radiolabeled with technetium-99m. *Pharmaceuticals* **2019**, *12*, 13. [CrossRef] [PubMed]
28. Klingler, M.; Summer, D.; Rangger, C.; Haubner, R.; Foster, J.; Sosabowski, J.; Decristoforo, C.; Virgolini, I.; von Guggenberg, E. DOTA-MGS5, a new cholecystokinin-2 receptor-targeting peptide analog with an optimized targeting profile for theranostic use. *J. Nucl. Med.* **2019**, *60*, 1010–1016. [CrossRef]
29. Klingler, M.; Decristoforo, C.; Rangger, C.; Summer, D.; Foster, J.; Sosabowski, J.K.; von Guggenberg, E. Site-specific stabilization of minigastrin analogs against enzymatic degradation for enhanced cholecystokinin-2 receptor targeting. *Theranostics* **2018**, *8*, 2896–2908. [CrossRef]
30. Grob, N.M.; Haussinger, D.; Deupi, X.; Schibli, R.; Behe, M.; Mindt, T.L. Triazolo-peptidomimetics: Novel radiolabeled minigastrin analogs for improved tumor targeting. *J. Med. Chem.* **2020**, *63*, 4484–4495. [CrossRef]
31. Li, S.-C.; Goto, N.K.; Williams, K.A.; Deber, C.M. Alpha-helical, but not beta-sheet, propensity of proline is determined by peptide environment. *Proc. Natl. Acad. Sci. USA* **1996**, *93*, 6676–6681. [CrossRef]
32. Peitl, P.K.; Tamma, M.; Kroselj, M.; Braun, F.; Waser, B.; Reubi, J.C.; Dolenc, M.S.; Maecke, H.R.; Mansi, R. Stereochemistry of amino acid spacers determines the pharmacokinetics of In-111 DOTA minigastrin analogues for targeting the CCK2/Gastrin receptor. *Bioconjug. Chem.* **2015**, *26*, 1113–1119. [CrossRef]
33. Langer, I.; Tikhonova, I.G.; Travers, M.A.; Archer-Lahlou, E.; Escrieut, C.; Maigret, B.; Fourmy, D. Evidence that interspecies polymorphism in the human and rat cholecystokinin receptor-2 affects structure of the binding site for the endogenous agonist cholecystokinin. *J. Biol. Chem.* **2005**, *280*, 22198–22204. [CrossRef] [PubMed]
34. Helbok, A.; Decristoforo, C.; Behe, M.; Rangger, C.; von Guggenberg, E. Preclinical evaluation of In-111 and Ga-68 labelled minigastrin analogues for CCK-2 receptor imaging. *Curr. Radiopharm.* **2009**, *2*, 304–310. [CrossRef]

35. Maina, T.; Konijnenberg, M.W.; KolencPeitl, P.; Garnuszek, P.; Nock, B.A.; Kaloudi, A.; Kroselj, M.; Zaletel, K.; Maecke, H.; Mansi, R.; et al. Preclinical pharmacokinetics, biodistribution, radiation dosimetry and toxicity studies required for regulatory approval of a phase I clinical trial with 111In-CP04 in medullary thyroid carcinoma patients. *Eur. J. Pharm. Sci.* **2016**, *91*, 236–242. [CrossRef] [PubMed]
36. Riordan, J.F. Angiotensin-I-converting enzyme and its relatives. *Genome. Biol.* **2003**, *4*, 225. [CrossRef]
37. Wilkins, M.R.; Unwin, R.J.; Kenny, A.J. Endopeptidase-24.11 and its inhibitors: Potential therapeutic agents for edematous disorders and hypertension. *Kidney Int.* **1993**, *43*, 273–285. [CrossRef]
38. Pauwels, S.; Najdovski, T.; Dimaline, R.; Lee, C.M.; Deschodtlanckman, M. Degradation of human gastrin and Cck by endopeptidase 24.11—Differential behavior of the sulfated and unsulfated peptides. *Biochim. Biophys. Acta* **1989**, *996*, 82–88. [CrossRef]
39. Deschodt-Lanckman, M.; Pauwels, S.; Najdovski, T.; Dimaline, R.; Dockray, G.J. In vitro and in vivo degradation of human gastrin by endopeptidase 24.11. *Gastroenterology* **1988**, *94*, 712–721. [CrossRef]
40. Wang, J.; Lin, Q.; Wu, Q.; Cooper, M.D. The enigmatic role of glutamyl aminopeptidase (BP-1/6C3 antigen) in immune system development. *Immunol. Rev.* **1998**, *161*, 71–77. [CrossRef]
41. Migaud, M.; Durieux, C.; Viereck, J.; SorocaLucas, E.; FournieZaluski, M.C.; Roques, B.P. The in vivo metabolism of cholecystokinin (CCK-8) is essentially ensured by aminopeptidase A. *Peptides* **1996**, *17*, 601–607. [CrossRef]
42. Mohamed, M.M.; Sloane, B.F. Cysteine cathepsins: Multifunctional enzymes in cancer. *Nat. Rev. Cancer* **2006**, *6*, 764–775. [CrossRef]
43. Kolenc-Peitl, P.; Mansi, R.; Tamma, M.; Gmeiner-Stopar, T.; Sollner-Dolenc, M.; Waser, B.; Baum, R.P.; Reubi, J.C.; Maecke, H.R. Highly improved metabolic stability and pharmacokinetics of indium-111-DOTA-gastrin conjugates for targeting of the gastrin receptor. *J. Med. Chem.* **2011**, *54*, 2602–2609. [CrossRef] [PubMed]
44. Grob, N.M.; Schmid, S.; Schibli, R.; Behe, M.; Mindt, T.L. Design of radiolabeled analogs of minigastrin by multiple amide-to-triazole substitutions. *J. Med. Chem.* **2020**, *63*, 4496–4505. [CrossRef] [PubMed]
45. Lay, J.M.; Jenkins, C.; Friis-Hansen, L.; Samuelson, L.C. Structure and developmental expression of the mouse CCK-B receptor gene. *Biochem. Biophys. Res. Commun.* **2000**, *272*, 837–842. [CrossRef] [PubMed]
46. Aloj, L.; Caraco, C.; Panico, M.; Zannetti, A.; Del Vecchio, S.; Tesauro, D.; De Luca, S.; Arra, C.; Pedone, C.; Morelli, G.; et al. In vitro and in vivo evaluation of In-111-DTPAGlu-G-CCK8 for cholecystokinin-B receptor imaging. *J. Nucl. Med.* **2004**, *45*, 485–494.
47. Varala, R.; Adapa, S.R. A practical and efficient synthesis of thalidomide via Na/liquid NH3 methodology. *Org. Process. Res. Dev.* **2005**, *9*, 853–856. [CrossRef]

Review

PET Radiotracers for CNS-Adrenergic Receptors: Developments and Perspectives

Santosh Reddy Alluri [1], Sung Won Kim [2], Nora D. Volkow [2,3,*] and Kun-Eek Kil [1,4,*]

[1] University of Missouri Research Reactor, University of Missouri, Columbia, MO 65211-5110, USA; srad3k@missouri.edu

[2] Laboratory of Neuroimaging, National Institute on Alcohol Abuse and Alcoholism, National Institutes of Health, Bethesda, MD 20892-1013, USA; kims8@mail.nih.gov

[3] National Institute on Drug Abuse, National Institutes of Health, Bethesda, MD 20892-1013, USA

[4] Department of Veterinary Medicine and Surgery, University of Missouri, Columbia, MO 65211, USA

* Correspondence: nvolkow@nida.nih.gov (N.D.V.); kilk@missouri.edu (K.-E.K.); Tel.: +1-(301)-443-6480 (N.D.V.); +1-(573)-884-7885 (K.-E.K.)

Academic Editor: Krishan Kumar

Received: 31 July 2020; Accepted: 1 September 2020; Published: 3 September 2020

Abstract: Epinephrine (E) and norepinephrine (NE) play diverse roles in our body's physiology. In addition to their role in the peripheral nervous system (PNS), E/NE systems including their receptors are critical to the central nervous system (CNS) and to mental health. Various antipsychotics, antidepressants, and psychostimulants exert their influence partially through different subtypes of adrenergic receptors (ARs). Despite the potential of pharmacological applications and long history of research related to E/NE systems, research efforts to identify the roles of ARs in the human brain taking advantage of imaging have been limited by the lack of subtype specific ligands for ARs and brain penetrability issues. This review provides an overview of the development of positron emission tomography (PET) radiotracers for in vivo imaging of AR system in the brain.

Keywords: adrenergic receptor; positron emission tomography; radiotracer

1. Introduction

Positron emission tomography (PET) is a noninvasive and highly sensitive in vivo imaging technique that uses small amounts of radiotracers to detect the concentration of relevant biomarkers in tissues such as receptors, enzymes, and transporters. These radiotracers can be used to characterize neurochemical changes in neuropsychiatric diseases and also to measure drug pharmacokinetics and pharmacodynamics directly in the human body including the brain [1,2]. PET technology requires positron-emitting radioisotopes, a radiotracer synthesis unit, a PET scanner, and data acquisition components. Among the PET isotopes, cyclotron produced carbon-11 (C-11, $t_{1/2}$ = 20.34 min), nitrogen-13 (N-13, $t_{1/2}$ = 9.96 min), and fluorine-18 (F-18, $t_{1/2}$ = 109.77 min) are frequently used for PET-neuroimaging. PET imaging is particularly valuable to characterize investigational drugs and their target proteins, providing a valuable tool for clinical neuroscience [3].

PET radiotracers for the central nervous system (CNS) should have proper pharmacokinetic profile in the brain and to achieve this property, they were designed to meet five molecular properties: (a) molecular weight <500 kDa, (b) Log $D_{7.4}$ between ~1 to 3 (lipophilicity factor), (c) number of hydrogen bond donors <5, (d) number of hydrogen bond acceptors <10, and (e) topological polar surface area <90 Å^2. Otherwise, they may either not cross the blood-brain barrier (BBB) or show lack of specific signal due to high nonspecific binding [3–5]. In addition, an ideal PET radiotracer should have, though unachievable, (a) high affinity (preferably subnanomolar range) for its target, (b) high selectivity between subtypes or void on off-targets, (c) high dynamic range in specific binding, (d) appropriate metabolic profile, (e) no adverse toxicology effects, and (f) kinetics suitable for mathematical modelling, which requires fast transfer to BBB and clearance in non-target tissues [3]. In general, a radiotracer with binding potential [ratio of target density (B_{max}) to ligand's affinity (K_d or K_i)] value of at least 10 is expected to provide a reliable specific signal in vivo [6,7]. Furthermore, there are some technical challenges associated with CNS PET radiotracer development, which include: limited synthesis time and/or complex syntheses to prepare a radiotracer, attainment of high molar activity, accurate radiometric metabolite analysis. In addition, the radiotracer ought to ideally address targets that are relevant to brain function and for the diagnosis and therapy of a disease [3,8]. Epinephrine (E) and norepinephrine (NE) were discovered in 1894 and 1907, respectively. They are both neurotransmitters and hormones that belong to the group of catecholamines crucial to the function of the body and brain [9–11]. Neither E nor NE crosses the BBB but they are synthetized in the brain. NE is synthesized in synapse from *L*-phenylalanine in four steps enzymatically through phenylalanine hydroxylase (PAH), tyrosine hydroxylase, dopa decarboxylase (DDC), and dopamine beta-hydroxylase (DBH), respectively. Further, NE is methylated to convert into E with phenylethanolamine-*N*-methyl transferase (PNMT) [12–14]. NE in synaptic cleft is removed either by reuptake via NE transporter (NET) or via metabolism by monoamine oxidase A (MAO-A) or catechol-*O*-methyl transferase (COMT) into various transitional metabolites. Though the noradrenergic neurons are confined to a few relatively small brain areas such as midbrain, pons, locus coeruleus, caudal ventrolateral nucleus, and medulla, they send extensive projections to most brain regions [15]. Table 1 summarizes the brain distribution (reported in mice) and function of various AR subtypes and their involvement in some brain disorders for CNS-ARs.

The effects of NE and E in the CNS and PNS is mediated mainly through two main classes of adrenergic receptors (ARs): alpha-ARs (α-ARs) and beta-ARs (β-ARs) [16]. The ARs were first identified in 1948 and pharmacological and molecular cloning techniques since then have identified various subclasses of ARs [17,18]. α-ARs are divided into α1 and α2 subclasses, wherein, each of these has three subtypes: α1A, α1B, α1D and α2A, α2B, α2C. The three subtypes of β-ARs are β1, β2, and β3 [19]. These receptors in the CNS are G-protein coupled receptors (GPCR) and are implicated in pathophysiology of various diseases, biochemical pathways, and biological functions [18].

Table 1. Brain distribution of AR subtypes and their associated brain disorders.

Receptor		Distribution	Distinct Functions and Associated Disorders	Ref
α1	α1A	High levels in olfactory system, hypothalamic nuclei, and brainstem. Moderate levels in amygdala, cerebral cortex, and cerebellum	Involved in neurotransmission of NE as well as γ-aminobutyric acid (GABA) and NMDA. May mediate effects of anti-depressants in treating depression and obsessive compulsive disorder (OCD)	[18,20–24]
	α1B	Thalamic nuclei, lateral nucleus of amygdala, cerebral cortex, some septal regions, brain stem regions	May play a role in behavioral activation. Associated with addiction, and neurodegenerative disorders (Multiple System Atrophy)	[18,20,21,24–27]
	α1D	Olfactory bulb, cerebral cortex, hippocampus, reticular thalamic nuclei, and amygdala	Mediates changes in locomotor behaviors. Associated with stress.	[18,20,23,28,29]
α2	α2A	Locus coeruleus, midbrain, hypothalamus, amygdala, cerebral cortex, and brain stem	Mediate functions of most of the α2-agonists used in sedation, antinociception, and behavioral actions. Associated with ADHD, anxiety	[18,23,30–35]
	α2B	Thalamus, hypothalamus, cerebellar Purkinje layer	Mediate antinociceptive action of nitrous oxide	[18,30,31]
	α2C	Hippocampus, striatum, olfactory tubercle, medulla, and basal ganglia	Involved in the neuronal release of NE as well as dopamine and serotonin. Potential therapeutic targets in depression & schizophrenia	[18,30,31,36–39]
β	β1	Homologous distribution. Expression was found (mostly β1 and β2) in frontal cortex, striatum, thalamus, putamen, amygdala, cerebellum, cerebral cortex and hippocampus.	Essential to motor learning, emotional memory storage and regulation of neuronal regeneration. Associated with mood disorders, aging, Alzheimer's disease, Parkinson's disease.	[16,18,40–48]
	β2			
	β3			

Given its biological significance, the adrenergic system has emerged as an important target for PET studies. Although ARs play major roles in the brain, most PET studies of ARs have focused on cardiac imaging [49,50] and PET studies of CNS-ARs are very limited. The other components of adrenergic system such as NET and MAO-A were studied using PET. Various C-11 and F-18 radiotracers of NET-selective anti-depressants (e.g., reboxetine) were examined using PET to monitor the function of noradrenergic system in CNS. Radiotracers, for example, (*S*,*S*)-[^{11}C]*O*-methyl-reboxetine ([^{11}C]MRB) and [^{18}F]FMeNER-D_2, were shown to exhibit desirable in vivo properties and their regional distribution in the brain is consistent with known distribution of NET in preclinical/clinical settings [51–54]. These radiotracers have been employed to monitor NET availability in different related diseases, including obesity, major depressive disorder, and Parkinson's disease. Likewise, radiotracers, for instance, [^{11}C]Harmine demonstrated clinical success for in vivo brain imaging of MAO-A, respectively [55–57]. [^{11}C]Harmine has been applied in several-PET neuroimaging studies to study the role of MAO-A in different pathological conditions, including nicotine/alcohol dependence, and Alzheimer's disease.

However, the metabolic functions of MAO-A do not necessarily reflect the activities of adrenergic system as they also metabolize other neurotransmitters, such as, dopamine and serotonin (5-HT). Therefore, the in vivo activities of adrenergic neurons, thus far, were exclusively examined by PET imaging based on NET. Since the activities of NET reflect only on presynaptic systems of adrenergic neurons, novel PET radiotracers that can observe the features of postsynaptic adrenergic system are still required. PET-neuroimaging of AR subtypes combined with NET can scrutinize the unique implications of adrenergic system in various pathophysiological conditions of the brain. Subtype specific PET radiotracers for CNS-ARs have the potential to help clarify the roles of ARs in brain pathophysiology and provide suggestions towards the diagnosis and treatment for diseases such as depression, attention deficit hyperactivity disorder (ADHD), substance use disorders, schizophrenia, and neurodegenerative diseases that involve ARs. This review, based on reports published through 2019, summarizes the development of PET radioligands for CNS-ARs in animal models and human subjects and presents suggestions for further development for CNS-AR radiotracers.

2. α1-AR PET Radiotracers

Three highly homologous subunits of α-1 ARs, α1A, α1B, and α1D, have been shown to have different amino acid sequence, pharmacological properties, and tissue distributions [58–60]. A detailed review of α1-AR pharmacology is given by Michael and Perez [58]. Binding of NE/E to any of the α1-AR subtypes is stimulatory and activates $G_{q/11}$-signalling pathway, which involves phospholipase C activation, generation of secondary messengers, inositol triphosphate and diacyl glycerol and intracellular calcium mobilization. In PNS, as well as in the brain's vascular system it results in smooth muscle contraction and vasoconstriction. While the signaling effects of α1-ARs in the cardiovascular system are well studied [61,62], the role of α1-ARs in CNS is complex and not completely understood. The Bmax values of α-1 ARs were measured from saturation assays using [^{3}H]prazosin (a selective α-1 blocker) with tissue homogenates from rats and the observed binding capacities (fmol/mg tissue) of prazosin in cortex, hippocampus, and cerebellum were 14.49 ± 0.38, 11.03 ± 0.39, and 7.72 ± 0.11, respectively [23,63]. The α1-ARs are postsynaptic receptors and can also modulate release of NE. In the human brain, α1-AR subtypes are localized in amygdala, cerebellum, thalamus, hippocampus, and to some extent in striatum [20,22,64].

The anti-depressant effects of noradrenergic enhancing drugs as well as their effects on anxiety and stress reactivity points to their relevance of α1A, α1D to these behaviors [21,28,65]. Furthermore, α1A-ARs regulate GABAergic and NMDAergic neurons [20]. A decrease in α1A-AR mRNA expression was observed in prefrontal cortex of subjects with dementia [66,67]. The α1B-ARs has also been shown to be crucial to brain function and disease [18]. For example, α1B-knockout (KO) mice study revealed that α1B-AR modulates behavior, showing increased reaction to novel situations [24,27]. In addition, the locomotor and rewarding effects of psychostimulants and opiates were decreased in α1B-KO mice, highlighting their role in the pharmacological effects of these drugs [24]. On the other hand, studies using the α1B-overexpression model suggest their involvement in neurodegenerative diseases [25].

Several α1-AR agonists and antagonists (α1-blockers) are available in the market as drugs to treat various heart and brain disorders [68,69]. Pharmacology of most of these drugs is complicated by the fact that they have strong affinities for other receptor systems, such as serotonin and dopamine receptors. The need to develop α1-AR selective/α1-AR subtype specific drugs is demanding. Undoubtedly, PET radiotracers selective for α1-AR are valuable to assess α1-ARs contribution to brain function and disease. In the late 1980s, prazosin, used for the treatment of hypertension, was labelled with carbon-11 to image α1-ARs with PET [49]. Following this, [^{11}C]prazosin analogous, [^{11}C]bunazosin and [^{11}C]GB67 (Figure 1A) were developed as PET radiotracers to image α1-ARs in the cardiovascular system [49,70]. These tracers were shown to have limited BBB permeability and were deemed to be not suitable for PET-neuroimaging. Efforts to develop PET radiotracers to image α1-ARs in the CNS was mainly based on antipsychotic drugs such as clozapine, sertindole, olanzapine, and risperidone that have mixed binding affinities for D_2, $5HT_{2A}$ receptors and α1-ARs in the nanomolar range. Their affinity for α1-ARs has been shown to contribute to antipsychotic efficacy uncovering their role in psychoses [71,72].

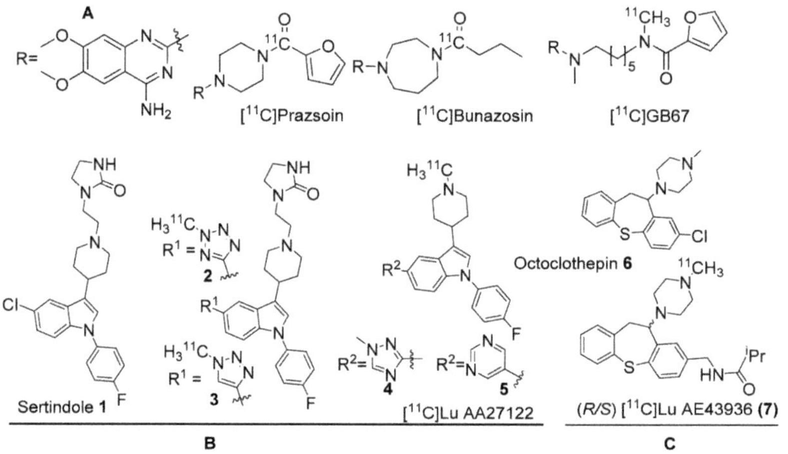

Figure 1. (**A**) Earlier PET radiotracers, [^{11}C]Prazosin, [^{11}C]Bunazosin, and [^{11}C]GB67 for cardiac α1-AR imaging. (**B**) Antagonist PET radiotracers based on sertindole. (**C**) Antagonist PET radiotracers based on octoclothepin for brain α1-AR imaging.

Two C-11 labelled analogs of the atypical antipsychotic drug sertindole were first reported by Balle et al. In the early 2000s (Figure 1B) [73,74]. Several other analogs with good affinities (K_i < 10 nM) for α1-ARs were subsequently reported by the same group. Two analogs, in which chlorine in sertindole (**1**) is replaced by a 2-methyl-tetrazol-5-yl (**2**) and a 1-methyl-1,2,3-triazol-4-yl (**3**), were labelled with C-11 [73,74]. The in vitro affinities (K_i) of **1**, **2** and **3** for α1-ARs are 1.4, 1.8 and 9.5 nM, respectively. Both [^{11}C]**2** and [^{11}C]**3** were prepared using ^{11}C-methylation with [^{11}C]methyl triflate from their corresponding *N*-desmethyl precursors. The molar activity of [^{11}C]**2** and [^{11}C]**3** was reported at 70 GBq/μmol and 15 GBq/μmol, respectively. Their brain distribution examined with PET in Cynomolgus monkeys showed that brain uptake of [^{11}C]**2** and [^{11}C]**3** was slow and low, with [^{11}C]**3** showing somewhat higher brain uptake than [^{11}C]**2**. Their brain distribution was homogenous and specific binding to α1-ARs could not be demonstrated. It was also concluded that these two radiotracers were not suitable to image α1-ARs in brain owing to rapid metabolism, substantial distribution to other organs, and substrates for active efflux mechanism.

In the early 2010s, further optimization of structure-activity relationship (SAR) studies led to the identification of two more sertindole analogs, **4** and **5** (Lu AA27122), with higher α1-AR selectivity over D_2 receptors (Figure 1B). Compared to the in vitro affinity for α1A-AR subtype (K_i = 0.16, 0.52 nM for **4** and **5**), the affinity for the α1B and α1D subtypes is 4 to 15 times less potent (Table 2). The $LogD_{7.4}$ values for **4** (2.7) and **5** (1.9) were in the optimal range for in vivo brain imaging [75–77]. Both [^{11}C]**4** and [^{11}C]**5** were prepared in a similar manner like above with >370 GBq/μmol molar activity for non-human primate studies [75]. Interestingly, while [^{11}C]**4** showed very poor brain uptake, [^{11}C]**5** had suitable brain uptake (4.6% ID/cc at 36 min). However, its binding was not blocked with a pharmacological dose of prazosin pretreatment, indicating lack of α1-AR specificity.

Table 2. In vitro affinities of compounds **1** to **7** for α1-ARs, D_2 and $5HT_{2C}$ receptors [73,75,76].

Compound	Receptor K_i (nM)				
	α1A	α1B	α1D	D_2	$5HT_{2C}$
1	0.37	0.33	0.66	0.45	0.55
2	0.23	1.1	2.0	140	330
3	3.0	6.0	8.6	310	1500
4	0.16	6.4	15	220	-
5	0.52	1.9	2.5	120	-
6	0.37	0.33	0.66	0.45	0.51
(*R*)-**7**	0.43	0.27	0.64	31	8.0
(*S*)-**7**	0.16	0.20	0.21	4.5	93

At the same time, another C-11 labelled radiotracer, [^{11}C]**7** ([^{11}C]Lu AE43936, Figure 1C) for brain α1-ARs was developed and evaluated by Risgaard et al. [76]. This radiotracer was based on the antipsychotic octoclothepin (**6**, Figure 1C), which belongs to the tricylic dibenzotheiepin group and has inverse agonist effects at dopamine, serotonin, and α1 receptor sites [78]. Two enantiomers of **7** (*R*/*S*) were radiolabelled on the basis of their varying selectivity and specificity for α1-AR subtypes (Table 2) and imaged with PET in female Danish Landrace pigs. The baseline PET imaging results indicated that neither of the radiolabelled isomers entered the pig's brain. Pretreatment with cyclosporin A (CsA) [79] increased the brain uptake of (*R*)-**7** in α1-AR rich cortex, thalamus (above 2 SUV), suggesting that (*R*)-**7** was a substrate for active efflux transporters. Further cell studies specified that (*R*)-**7** is a substrate for p-glycoprotein (Pgp).

So far, none of the reported radiotracers showed promising results for in vivo brain imaging of α1-ARs for they were limited by poor BBB penetration, being substrates of Pgp or lack of binding specificity. Therefore, novel α1-AR radiotracers that overcome brain permeability and display good affinity and subtype-selectivity are required to evaluate the role of α1-AR subtypes in brain pathophysiology. Several α1-AR subtype specific (high affinity for one subtype over the other) and nonspecific compounds have been reported over the years [26,29,80,81]. Either direct radiolabeling (if possible) or chemical modification and then radiolabeling of the most specific compounds could be an optimal approach for developing new α1-subtype specific radiotracer, considering the aforementioned CNS-PET radiotracer criteria.

3. α2-AR Subtype and Nonspecific PET Radiotracers

Unlike α1-ARs, α2-ARs decrease adenyl cyclase activity in association with G_i heterotrimeric G-protein and hence are inhibitory. They mediate many NE effects including cognition and readiness for action [11,30]. The brain distribution of the three subtypes, α2A, α2B, and α2C, was characterized by autoradiography and immunohistochemistry techniques. Among the three subtypes, α2A-ARs are the most abundant in the brain and localized in locus coeruleus, midbrain, hippocampus, hypothalamus, amygdala, cerebral cortex and brain stem. The α2B-ARs are located in thalamus and hypothalamus and α2C-ARs in cortex, hippocampus, olfactory tubercle and basal ganglia [51–53]. In the mouse brain, ~90% of α2-ARs are α2A-ARs and ~10% are α2C-ARs [30–32,82]. Notably, binding experiments using [^{3}H]2-methoxyidazoxan (a selective α-2 AR antagonist) with postmortem human brain detected 100% of α2A-ARs population in the hippocampus, cerebellum, and brainstem (B_{max} = 34–90 fmol/mg protein). In addition to this, α2A-AR (B_{max} = 53 fmol/mg of protein) and α2B/C-AR (B_{max} = 8 fmol/mg of protein) were detected in the frontal cortex [30]. An extensive array of agonists and antagonists for α2-ARs have been developed. The main limitation of these ligands is lack of subtype selectivity for α2-ARs and off-target binding to other receptors [83,84]. α2A-ARs are mostly presynaptic and agonists inhibit NE release from the terminals and are used to treat hypertension, drug withdrawal, and ADHD whereas antagonists increase NE release and are used as antidepressants. Development of α2-AR subtype selective PET tracers would facilitate medication development and help gain further understanding of their role in brain diseases.

During the 1980s, two research groups identified two potent α2-AR selective antagonists via radioligand binding assays. One WY-26703 belongs to the benzoquinolizine class, and the other MK-492 belongs to the benzo[b]furo-quinolizine class [85,86]. Based on these templates, Bylund's group developed two PET tracers: [^{11}C]WY-26703 (**8**) in 1992 and [^{11}C]MK-912 (**9**) in 1998 (Figure 2) [87,88]. Both radiotracers were prepared from their respective *N*-desmethyl precursors via ^{11}C-methylation with 30.71–34.41 GBq/μmol molar activity. The in vitro binding assays and ex vivo biodistribution studies (tissue dissection followed by γ-counting) in rodents indicated that both radiotracers crossed the BBB and **9** showed higher affinity and specific binding to α2-ARs than **8**. However, PET studies of **8** and **9** in Rhesus monkeys showed fast washout from brain and nonspecific binding; thus it was concluded that they were not appropriate for PET imaging of α2-ARs in brain.

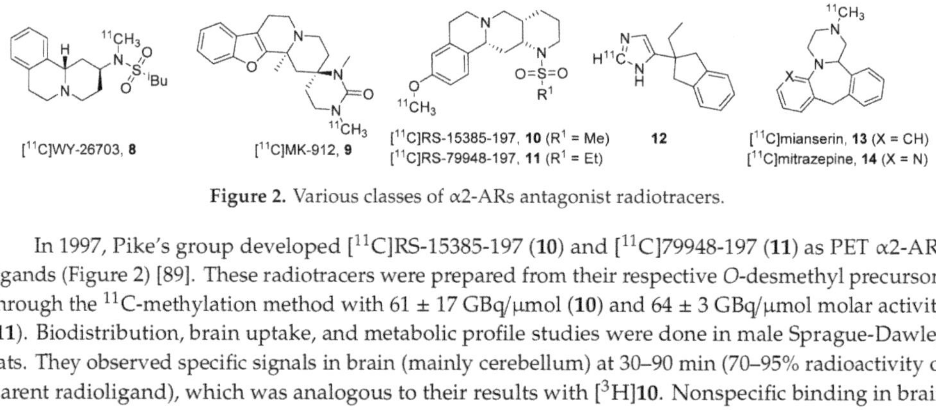

[^{11}C]WY-26703, **8** [^{11}C]MK-912, **9** [^{11}C]RS-15385-197, **10** (R^1 = Me) [^{11}C]RS-79948-197, **11** (R^1 = Et) **12** [^{11}C]mianserin, **13** (X = CH) [^{11}C]mitrazepine, **14** (X = N)

Figure 2. Various classes of α2-ARs antagonist radiotracers.

In 1997, Pike's group developed [^{11}C]RS-15385-197 (**10**) and [^{11}C]79948-197 (**11**) as PET α2-ARs ligands (Figure 2) [89]. These radiotracers were prepared from their respective *O*-desmethyl precursors through the ^{11}C-methylation method with 61 ± 17 GBq/μmol (**10**) and 64 ± 3 GBq/μmol molar activity (**11**). Biodistribution, brain uptake, and metabolic profile studies were done in male Sprague-Dawley rats. They observed specific signals in brain (mainly cerebellum) at 30–90 min (70–95% radioactivity of parent radioligand), which was analogous to their results with [^{3}H]**10**. Nonspecific binding in brain was **11** > **10**, which mostly likely reflected their differential metabolism (**11** > **10**). Thus, they chose **10** to quantify α2-ARs in the human brain using PET [90]. Studies in two volunteers with **10** revealed a low brain uptake index (BUI) due to high affinity to human plasma proteins. Consequently, **10** was not studied further.

In 2002, Crouzel's group chose atipamezole, an α2-AR selective antagonist, to develop [^{11}C]atipamezole, **12** (Figure 2) [91]. The radiotracer was prepared through an unique approach using 2-ethyl-2-oxoacetylindane, [^{11}C]formaldehyde ([^{11}C]HCHO) in the presence of zinc oxide and ammonium hydroxide (similar to Debus-Radziszewski imidazole synthesis) with an overall yield of 1.5%. However, no PET studies were reported with **12**.

Furthermore, in 2002, Smith's group developed two tetracyclic based anti-depressant radiotracers, mianserin (**13**) and mitrazepine (**14**), that have potent antagonist properties at α2-ARs and also at serotonin receptors ($5HT_{2A}$, $5HT_{2C}$) and labelled them with C-11 to prepare **13** and **14** (Figure 2) [92,93]. The radiotracers **13** and **14** were prepared from their respective *N*-desmethyl precursors via ^{11}C-methylation with ~40 GBq/μmol and 5–7 GBq/μmol molar activity, respectively. PET studies in female pigs with **13** showed limited binding potential in brain whereas **14** showed more favorable properties including slow metabolism, fast brain uptake and sufficient target-to-background ratio for pharmacokinetic parameters estimation.

Radiotracer **14** had higher binding in the frontal cortex, thalamus, and basal ganglia where pretreatment with unlabeled mitrazepine revealed that its binding was reversible, whereas, in the cerebellum and olfactory tubercle, it was not. Notably, using the α2-AR subtype KO mouse model they validated the receptor selectivity of **14**. In 2004 and 2009, this group conducted a clinical trial with volunteers using **14** to study its distribution, metabolism and pharmacokinetics [94,95]. The results revealed that **14** can serve as a PET radiotracer to image α2-ARs in the brain, though identification of its metabolites and its nonselective binding are limitations.

Two years later, Leysen's group developed a reversible, potent and selective α2-AR antagonist, viz. R107474 (**15**, Figure 3) [96]. They prepared $[^{11}C]$**15** through Pictet-Spengler condensation method using $[^{11}C]$HCHO and the respective secondary amine with 24–28 GBq/μmol molar activity at the end of bombardment (EOB). They carried out ex vivo autoradiography to measure in vivo α2A-ARs and α2C-ARs occupancy of **15** in rats. Biodistribution studies showed rapid uptake of **15** into brain and other tissues with the brain showing the highest uptake other than liver and kidneys. In the brain the highest uptake of **15** was in the septum (3.54 ± 0.52 ID/g) and entorhinal cortex (1.57 ± 0.50 ID/g) whereas the lowest was in the cerebellum, a region with very low density of α2-ARs. However, the potential of **15** was not investigated further.

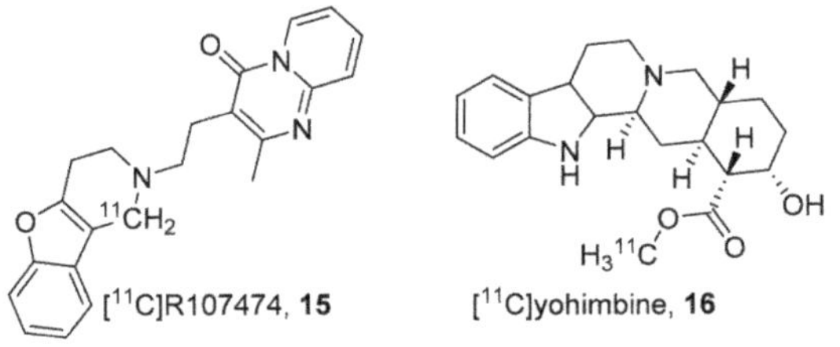

$[^{11}C]$R107474, **15**

$[^{11}C]$yohimbine, **16**

Figure 3. Anti-depressive & antihypertensive based α2-AR PET radiotracers.

In 2006, Jacobsen's group developed C-11 labelled yohimbine (**16**, Figure 3) [97], an antihypertensive agent. Yohimbine has potent antagonist properties at α2-ARs, but also interacts with α1 and 5-hydroxy tryptamine 1A receptors (5-HT_{1A}). The radiotracer **16** was prepared through ^{11}C-methylation of yohimbinic acid using C-11 methyl iodide ($[^{11}C]CH_3I$) and obtained with 40 GBq/μmol molar activity.

PET studies were performed in pigs to obtain whole-body and dosimetry recordings and for dynamic brain imaging. Interestingly, no radioactive metabolites of **16** were reported in pig plasma and binding of **16** was observed in α2-AR-rich regions where it was displaceable by co-injection of pharmacological doses of yohimbine or selective α2-AR antagonist (Figure 4). Later, **16** was used to image α2-ARs in the human brain (n = 6) using PET [98]. Highest binding of **16** was observed in cortex and hippocampus and the lowest in corpus callosum, which was used as a reference region to estimate the average total distribution (V_T) in other brain regions. The radiotracer **16** seems to be a suitable radiotracer to image α2-ARs but has similar issues as of **14,** which need to be addressed.

The concentration of α2-subtype receptors in the brain is low (5–90 fmol/mg range) increasing the challenge for their detection by PET [31]. Therefore, α2- subtype specific PET tracers (with subnanomolar affinities and >30-fold selectivity) still need to be developed for the quantification of α2-subtype receptors and to assess their role in brain diseases. A few research groups have developed α2-subtype specific PET radiotracers, but success has been limited.

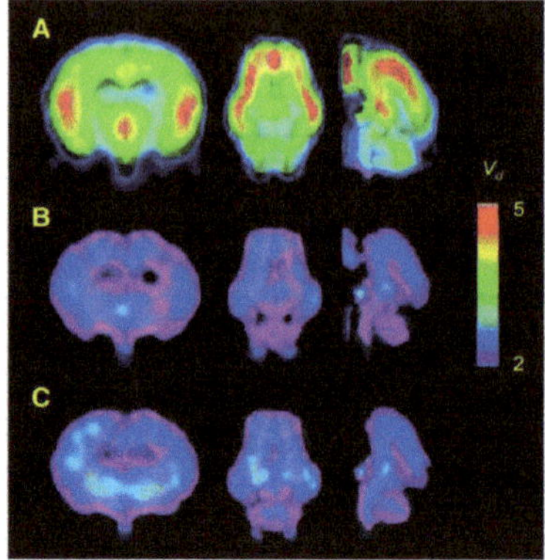

Figure 4. Parametric maps of **16** in living porcine brain. (**A**) Baseline study using **16** showed regional differences in its distribution. (**B**) Blocking experiment (yohimbine at 0.07 mg/kg) reduced the scale of distribution volume (V_d) to ~2 mL g^{-1} in all the α2-AR bound regions. (**C**) Increased dose of yohimbine (1.6 mg/kg) had no further significant effect in comparison to the low dose (n = 3) Maps are superimposed on the MR image. Adapted from JNM publication by Jacobsen S, Pedersen, K.; Smith, D.F.; Jensen, S.B.; Munk, O.L.; Cumming P [97]. Permission obtained from SNMMI.

3.1. α2A-Specific PET Radiotracers

Kumar's group in 2010 developed [^{11}C]MPTQ (**17**, Figure 5) for the quantification of α2A-ARs in vivo [35]. Compound **17** was shown to have blocking effects on α2A-ARs in vivo in brain and has stronger affinities for α2A-AR (K_i = 1.6 nM) than α2C-AR (K_i = 4.5 nM) and 5-HTT (serotonin transporter, K_i = 16 nM) [99]. They anticipated no binding of **17** to α2B and α2C-ARs since the densities of these receptors are lower than α2A-ARs. In addition, the 10-fold higher affinity of **17** α2A-ARs over 5-HTT is advantageous for α2A-ARs as both have similar B_{max} values. The radiosynthesis of **17** was accomplished through ^{11}C-methylation of its respective *N*-desmethyl precursor with 74–88.8 GBq/μmol molar activity at the end of synthesis (EOS). PET studies in baboons with **17** showed that it penetrated the BBB and accumulated in α2A-AR-rich brain areas. They ruled out binding of **17** to 5-HTT due to its low uptake in the hippocampus, temporal cortex, and occipital cortex, which are the brain regions with the highest binding of 5-HTT radiotracers. No further studies were reported using **17**.

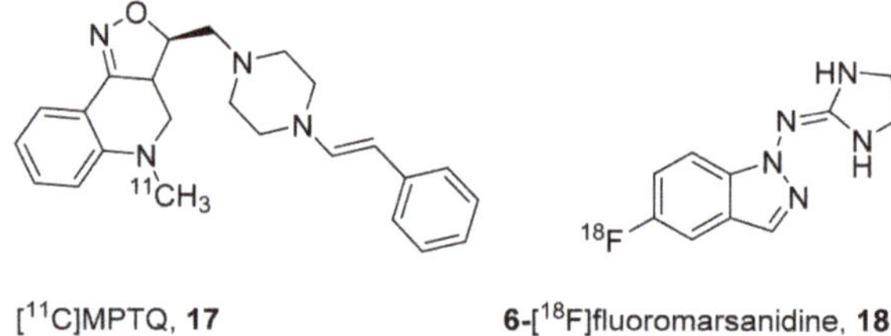

Figure 5. α2A-antagonist (**17**) and agonist (**18**) PET radiotracers.

In search of a selective agonist to α2-ARs, Lehmann's group identified 1-[(imidazolidin-2-yl) imino]indazole (marsanidine) [100] and later developed an α2A subtype specific ligand by introducing fluorine to marsanidine [33]. The reported binding affinity (K_d) of 6-fluoromarsanidine for α2A (33 nM) is higher than for α2B (72 nM) and α2C (600 nM). Solin et al., in 2019, prepared 6-[^{18}F]fluoromarsanidine (**18**, Figure 5) through electrophilic ^{18}F-radiofluorination using [^{18}F]selectfluor bis(triflate) and a corresponding precursor with 3–26 GBq/μmol molar activity at the EOS [34]. In vivo PET was performed in rats and α2A-KO mice, but the radiotracer was not continued further because of its rapid metabolism and high nonspecific uptake in rat and mouse brain.

3.2. α2C-Specific PET Radiotracers

Animal models, such as the forced swimming test (FST) and the prepulse inhibition (PPI) are used to screen for anti-depressants and anti-psychotics, respectively. The use of α2C-KO and α2C-overexpression (α2C-OE) mouse models in FST and PPI paradigms suggested that α2C-specific compounds may have therapeutic benefits for depression and schizophrenia [32,38]. In 2007, Orion pharma from Finland identified an acridine-based compound, JP-1302 [38], and a research group from Japan identified a methyl benzofuran based compound, MBF [101], as selective α2C antagonists. Both these ligands have high affinities for α2C (JP-1302 K_i = 28 nM, MBF K_i = 20 nM) than for α2A (JP-1302 K_i = 3500 nM, MBF K_i = 17,000 nM) and α2B (JP-1302 K_i = 1500 nM, MBF K_i = 750 nM). Based on these findings, Zhang's group, in 2010, synthesized [^{11}C]JP-1302 (**19**) and [^{11}C]MBF (**20**)(Figure 6) as PET probes to evaluate their BBB penetration and α2C selective binding in the brain [37]. The radiotracers **19** (molar activity 95 ± 24 GBq/μmol) and **20** (molar activity 62 ± 15 GBq/μmol) were prepared using ^{11}C-methylation from *N*-desmethyl and *O*-desmethyl precursors, respectively. PET studies were conducted in WT and Pgp, breast cancer resistance protein (BCRP) KO mice using both radiotracers.

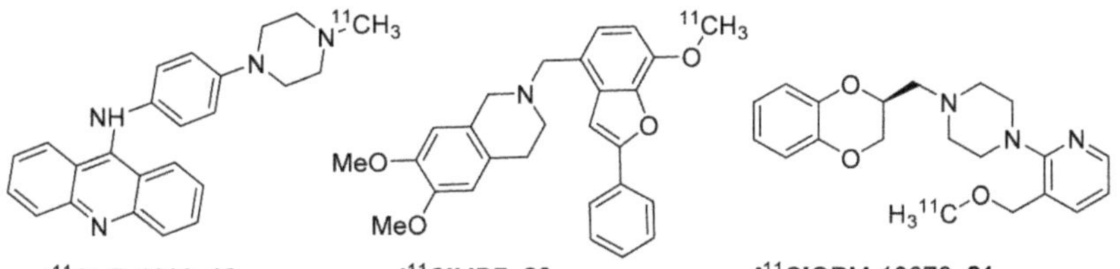

[^{11}C]JP-1302, **19** [^{11}C]MBF, **20** [^{11}C]ORM-13070, **21**

Figure 6. PET radiotracers for α2C-ARs.

This combined KO model is useful to evaluate whether brain penetration of PET probes is sensitive to Pgp and BCRP. After injection of the radiotracers, their levels in the brain were low in WT mice whereas they were higher in Pgp and BCRP KO mice. The regional binding of these radiotracers did not correspond with the regional brain distribution of α2C, so it was concluded that they were inadequate to evaluate α2C-ARs in brain with PET.

In 2014, researchers from Turku PET center and Orion Pharma reported the radiosynthesis of [^{11}C]ORM13070 (**21**, Figure 6) with molar activity 690 ± 340 GBq/μmol and its evaluation in rats and in α2A and α2AC KO mice with PET [36,102]. The binding affinities of **21** for α2C (3.8 nM) is higher than for α2A (109 nM) and α2B (23 nM).

The in vivo PET and ex vivo autoradiography of **21** in rat indicated that its brain distribution corresponds to the regional distribution of α2C in brain, with highest levels in striatum and olfactory tubercle. Pretreatment with atipamezole, a α2-sutype nonselective antagonist blocked the binding of **21** into these regions. Furthermore, by using α2A and α2AC KO model mice, they demonstrated α2C specificity of **21**. The brain uptake of **21** in α2A-KO and WT mice was similar whereas, negligible uptake occurred in α2AC KO (Figure 7, left). They represented time-activity curves for striatum and cerebellar cortex of three mice types (Figure 7, right) and the radioactivity ratios at 5–15 min for α2A, α2AC KO mice, and WT mice were 1.51–1.51, 1.06–1.09 and 1.51–1.57, respectively.

Accordingly, **21** was studied in healthy men to estimate its metabolism, pharmacokinetics, whole-body distribution and radiation dosimetry [39]. Good results were obtained in rodent and human PET studies with **21**, except for its fast washout from brain. Better pharmacokinetics, higher affinity, and specificity can potentially be enhanced by structural modifications to **21**. Given that the α2A-ARs are widely distributed in brain in contrast to α2C-ARs, a candidate with subnanomolar affinity for α2C-ARs (>50-fold affinity than α2A-ARs) is needed for a PET radiotracer. As α2C-ARs are of interest as therapeutic targets in brain diseases, the α2C-specifc PET radiotracers would facilitate their development as medications and help in investigations of α2C-ARs in the human brain.

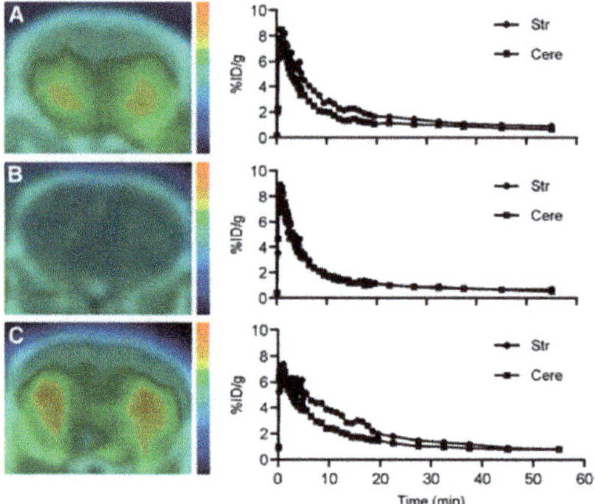

Figure 7. PET/CT images and time-activity curves of **21** for striatum and cerebellar cortex of (**A**) α2A KO (**B**) α2AC KO and (**C**) WT mice. Brain uptake of **21** in α2AC KO is negligible and is similar in α2A KO and WT mice with 7.8–8.1% ID/g at 1 min and 1.2% ID/g at 30 min after **21** injection. The striatum to cerebellar cortex radioactivity ratios (at 5–15 min) for α2AC KO mice did not differ and for α2A KO and WT mice are alike. Adapted from JNM publication by Arponen E.; Helin, S.; Marjamäki, P.; Grönroos, T.; Holm, P.; Löyttyniemi, E.; Någren, K.; Scheinin, M.; Haaparanta-Solin, M.; Sallinen, J.; [36]. Permission obtained from SNMMI.

4. β-ARs and Nonselective PET Radiotracers

β-ARs are associated with G_s-heterotrimeric G-protein and mediate intracellular signaling through adenyl cyclase activation and cyclic adenosine monophosphate (cAMP) production. β-ARs are classified into β1, β2, and β3 subtypes, in which, the former two have been much more explored [42,46]. Quite a lot of selective and nonselective β-AR agonists and antagonists (blockers) are available as drugs in the market to treat various cardiac and pulmonary disorders. In the brain, β-ARs are localized in the frontal cortex, striatum, thalamus, putamen, amygdala, cerebellum and hippocampus [48]. The density of β-ARs in brain is sensitive to brain pathophysiology. Notably, the density of β-ARs decrease with age [40]. Light microscopic autoradiography using [^{3}H]dihydroalprenolol (a selective β-blocker) with rat brain sections has shown a wide distribution of β-ARs in forebrain and cerebellum regions (B_{max} = 23 fmol/mg tissue) [103]. Similarly, B_{max} value of 18 fmol/mg protein was reported in pre-frontal cortex of subjects with Parkinson's disease [48,104]. By altering the Ca^{2+} levels through *N*-methyl-*D*-aspartate (NMDA) receptors in hippocampus, β-ARs modulate synaptic plasticity, including that needed for memory [44,45]. The blockade of β-ARs is associated with a small increased risk for Alzheimer's and Parkinson's disease [43,105]. In addition, abnormal function and densities of β-ARs have been reported in mood disorders and schizophrenia [41,47,106].

Several radioligands, mostly based on β-blockers, were validated for imaging of β-ARs in the heart [50]. The majority of β-blockers possess a hydroxyl propylamine moiety in their structures that is vital for binding to β-ARs and this moiety was maintained in most of these radioligands. PET radiotracers have succeeded in imaging and quantifying myocardial and pulmonary β-ARs in human [107,108], whereas, PET radiotracers for cerebral β-ARs have been more challenging. The clinical PET radiotracers for cardiac β-ARs have negative Log P values (<−2), which is not suitable for imaging the brain. Several lipophilic and high to moderate affinity β-AR nonselective antagonists were explored as PET radiotracers to image β-ARs in the brain.

During the 1980s, propranolol, a β-blocker drug was labelled with C-11 (**22**, Figure 8) but was unsuitable as a PET ligand for β-ARs because of high nonspecific binding in vivo [109,110]. Subsequently, Berridge's group described the synthesis of two isomers (*R*/*S*) of [^{18}F]fluorocarazolol

(**23**, Figure 8) through reductive amination using [^{18}F]fluoroacetone and desisopropylcarazolol with 18.5–37 GBq/µmol molar activity [111,112]. The radiotracer **23** has subnanomolar K_i values for β-ARs (β1 0.4 nM, β2 0.1 nM) and Log $P_{7.4}$ value of 2.19. The same group used *S*-**23** for PET imaging of the pig heart and lungs to validate the β-AR biding. In 1997, Waarde et al., employed *S*-**23** to image β-ARs in the human brain and obtained positive results [113]. They observed specific binding (blocked with pindolol) of *S*-**23** in β-AR rich areas, striatum and various cortical areas. However, the radiotracer was discontinued for further human studies as fluorocarazolol was positive for the Ames test i.e., mutagenic [114].

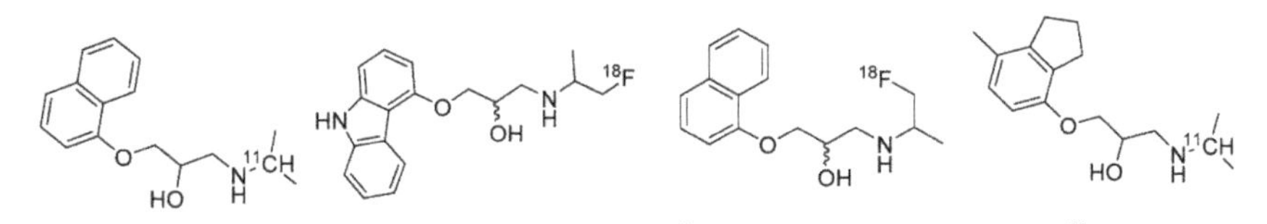

Figure 8. Early PET radiotracers for cerebral β-ARs.

Two research groups conducted biodistribution studies in rats using [^{18}F]fluoropropranolol (**24**) and [^{11}C]ICI 118,551 (**25**) (Figure 8), which failed because of their nonspecific binding [115,116]. In 2001, Fazio's group described two isomers (*R*/*S*) of C-11 labelled bisprolol (β1 K_i 1.6 nM, β2 K_i 100 nM and Log $P_{7.4}$ = −0.2) (**26**, Figure 9) to image β1-ARs in the brain [117]. The radiosynthesis of **26** was accomplished via reductive amination using [^{11}C]acetone and desisopropyl bisprolol precursor with 129.5 ± 37 GBq/µmol molar activity at the EOS. They observed little specific uptake of **26** in β1-AR rich regions in the rat's brain and also high nonspecific uptake in the pituitary (1.8 ± 0.3 ID at 30 min), a region with high β2-ARs levels. No further studies were reported using **26** to image β-ARs.

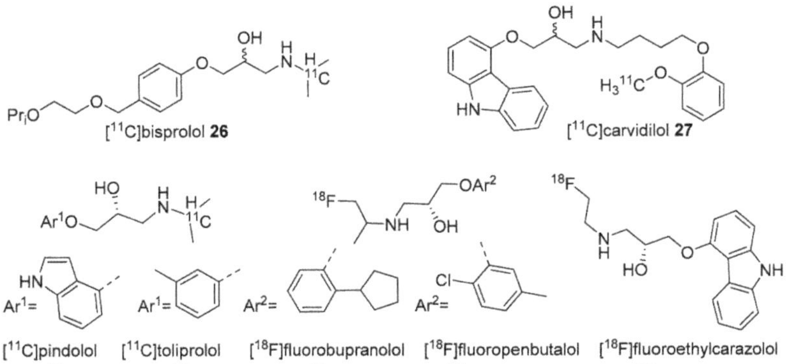

Figure 9. Radiotracers based on various β-AR blockers.

In 2002, Elsinga's group reported five different potent and lipophilic β-AR antagonists (carvedilol, pindolol, toliprolol, bupranolol, and penbutolol) as PET probes to image β-ARs in rat brain [118]. C-11 labelled carvedilol (**27**, Figure 9; molar activity 12.97–25.9 GBq/µmol) was prepared through ^{11}C-methylation using [^{11}C]CH_3I and its respective *O*-desmethyl precursor, whereas, [^{11}C]pindolol (**28**, molar activity 25.9–37 GBq/µmol) and [^{11}C]toliprolol (**29**, molar activity 22.2–25.9 GBq/µmol) were prepared via reductive amination using [^{11}C]acetone and the respective desisopropyl precursors. The F-18 tracers of bupranolol (**30**, molar activity 11.1–18.5 GBq/µmol) and penbutolol (**31**, 22.2–99.9 GBq/µmol) were also prepared by means of reductive amination but using [^{18}F]fluoroacetone and the respective des-fluoro-isopropyl precursors.

The five radiotracers had strong affinities (subnanomolar K_d) for β1 and β2-ARs. Although these radiotracers had sufficient affinity and lipophilicity for in vivo imaging, none showed good brain uptake. This group also evaluated S-[^{18}F]fluoroethylcarazolol (**32**, β1 K_i = 0.5 nM, β2 K_i = 0.4 nM and Log $P_{7.4}$ = 1.94) for in vivo imaging of β-ARs in rat brain [119]. The radiotracer **32** (Figure 9) was prepared via an epoxide ring-opening using [^{18}F]fluoroethylamine and the corresponding epoxide with >10 GBq/μmol molar activity. The radiotracer accumulated in brain with uptake reflecting cerebral β-ARs binding. However, no further PET imaging studies were reported using **32** probably because of its analogous nature to **23** which was shown to be positive Ames test [114,120].

In 2008, Elsinga's and Vasdev's groups chose exaprolol (β-AR K_d = 9–9.5 nM) and developed *S*-[^{11}C]exaprolol (**33**) and *S*-[^{18}F]fluoroexaprolol (**34**), respectively, to image β-ARs with PET (Figure 10) [120,121]. Radiotracer **33** was prepared via reductive amination using [^{11}C]acetone and desisopropylexaprolol precursor with >10 GBq/μmol molar activity and the radiotracer **34** was prepared through a nucleophilic substitution reaction using [^{18}F]fluoride and a corresponding tosylate precursor followed by reductive hydrolysis, with 34.29 GBq/μmol molar activity. Regardless of good binding and kinetic properties, both these radiotracers showed high nonspecific uptake in the brain and were found to be inadequate for PET imaging of β-ARs.

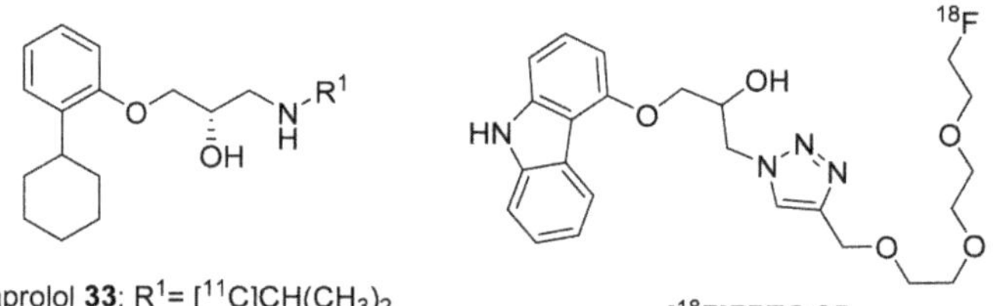

[^{11}C]exaprolol **33**; R^1= [^{11}C]$CH(CH_3)_2$
[^{18}F]fluoroexaprolol **34**; R^1= $CHCH_2{}^{18}FCH_3$

[^{18}F]FPTC **35**

Figure 10. Another set of latest β-AR PET radiotracers.

Again in 2014, Elsinga's group developed [^{18}F]FPTC (**35**, Figure 10) for PET imaging of β-ARs in brain [122]. The radiotracer **35** is a derivative of carazolol, wherein, isopropylamine group of carazolol was replaced by a PEGylated triazole group. It was prepared through Huisgen's 1,3-dipolar cycloaddition (click reaction) using F-18 labelled PEGylated alkyne and the corresponding azide with >120 GBq/μmol molar activity. Although **35** was shown to have appropriate $LogP_{7.4}$ (2.48) and specific binding in in vitro assays, it could not visualize β-ARs in the brain, lung or heart using micro-PET.

Thus, the development of PET radiotracers for neuroimaging of β-ARs remains a challenge and as of now, there are no β-AR subtype specific PET radiotracers. Such radiotracers are important to expand our understanding of the role of β-ARs in aging and memory formation and also to assess their function in behavioral disorders. Future research, as suggested by Elsinga and Waarde [48], should consider modifying the imaging agents used for myocardial β-ARs rather than radiolabeling existing β-blocker drugs. Alterations should optimize Log $P_{7.4}$ (2-3), high affinity and selectivity to β-ARs and no substrate affinity for Pgp.

5. Conclusions

Over the past four decades, significant efforts have been made to develop CNS-ARs PET ligands for brain imaging. Despite these efforts, very few PET radiotracers are available to selectively image AR subtypes in the brain. The development of specific radiotracers is hindered mainly by the low receptor densities of each AR subclass within the brain, which requires further optimization processes for highly potent and BBB permeable ligands. Though challenging, AR subtype specific agonist/antagonist PET radiotracers are needed to ascertain AR's role in brain pathophysiology and for medication development.

Funding: This research was funded by 'Tier 2 Research and Creative Works Strategic Investment Program Project of the University of Missouri (Internal Grant)' and the "Intramural Program of the National Institute on Alcohol Abuse and Alcoholism, grant number Y1AA-3009" and "The APC was funded by Y1AA-3009".

Conflicts of Interest: The authors declare no conflict of interest.

References

1. Scott, J.A. Positron Emission Tomography: Basic Science and Clinical Practice. *Am. J. Roentgenol.* **2004**, *182*, 418. [CrossRef]
2. Shukla, A.; Kumar, U. Positron emission tomography: An overview. *J. Med. Phys.* **2006**, *31*, 13–21. [CrossRef] [PubMed]
3. McCluskey, S.P.; Plisson, C.; Rabiner, E.A.; Howes, O. Advances in CNS PET: The state-of-the-art for new imaging targets for pathophysiology and drug development. *Eur. J. Nucl. Med. Mol. Imaging* **2020**, *47*, 451–489. [CrossRef]
4. Rankovic, Z. CNS Drug Design: Balancing Physicochemical Properties for Optimal Brain Exposure. *J. Med. Chem.* **2015**, *58*, 2584–2608. [CrossRef] [PubMed]
5. Adenot, M.; Lahana, R. Blood-Brain Barrier Permeation Models: Discriminating between Potential CNS and Non-CNS Drugs Including P-Glycoprotein Substrates. *J. Chem. Inf. Comput. Sci.* **2004**, *44*, 239–248. [CrossRef]
6. Agdeppa, E.D.; Spilker, M.E. A review of imaging agent development. *AAPS J.* **2009**, *11*, 286–299. [CrossRef]
7. Kumar, J.S.; Mann, J. PET Tracers for Serotonin Receptors and Their Applications. *Cent. Nerv. Syst. Agents Med. Chem.* **2014**, *14*, 96–112. [CrossRef]
8. Dunphy, M.P.S.; Lewis, J.S. Radiopharmaceuticals in preclinical and clinical development for monitoring of therapy with PET. *J. Nucl. Med.* **2009**, *50*, 106–122. [CrossRef]
9. Sneader, W. The discovery and synthesis of epinephrine. *Drug News Perspect.* **2001**, *14*, 491–494. [CrossRef]
10. Arthur, G. Epinephrine: A short history. *Lancet Respir. Med.* **2015**, *3*, 350–351. [CrossRef]
11. Schmidt, K.T.; Weinshenker, D. Adrenaline rush: The role of adrenergic receptors in stimulant-induced behaviors. *Mol. Pharmacol.* **2014**, *85*, 640–650. [CrossRef] [PubMed]
12. Eisenhofer, G.; Kopin, I.J.; Goldstein, D.S. Catecholamine metabolism: A contemporary view with implications for physiology and medicine. *Pharmacol. Rev.* **2004**, *56*, 331–349. [CrossRef] [PubMed]
13. Kirshner, N. Biosynthesis of adrenaline and noradrenaline. *Pharmacol. Rev.* **1959**, *11*, 350–357. [PubMed]
14. Flatmark, T. Catecholamine biosynthesis and physiological regulation in neuroendocrine cells. *Acta Physiol. Scand.* **2000**, *168*, 1–17. [CrossRef] [PubMed]
15. Ranjbar-Slamloo, Y.; Fazlali, Z. Dopamine and Noradrenaline in the Brain; Overlapping or Dissociate Functions? *Front. Mol. Neurosci.* **2020**, *12*, 1–8. [CrossRef]
16. Strosberg, A.D. Structure, function, and regulation of adrenergic receptors. *Protein Sci.* **1993**, *2*, 1198–1209. [CrossRef]
17. Ahlquist, R.P. A study of the adrenotropic receptors. *Am. J. Physiol.* **1948**, *153*, 586–600. [CrossRef]
18. Philipp, M.; Hein, L. Adrenergic receptor knockout mice: Distinct functions of 9 receptor subtypes. *Pharmacol. Ther.* **2004**, *101*, 65–74. [CrossRef]
19. Paul, J.; Bylund, B.; Lefkowitz, J.; Eikenburg, C.; Ruffolo, R.; Langer, Z.; Minneman, P. International Union of Pharmacology X. Recommendation for Nomenclature of adrenoreceptors: Consensus update. *Am. Soc. Pharmacol. Exp. Ther.* **1995**, *47*, 267–270.
20. Day, H.E.W.; Campeau, S.; Watson, S.J.; Akil, H. Distribution of α(1a)-, α(1b)- and α(1d)-adrenergic receptor mRNA in the rat brain and spinal cord. *J. Chem. Neuroanat.* **1997**, *13*, 115–139. [CrossRef]
21. Doze, V.A.; Handel, E.M.; Jensen, K.A.; Darsie, B.; Luger, E.J.; Haselton, J.R.; Talbot, J.N.; Rorabaugh, B.R. α1A- and α1B-adrenergic receptors differentially modulate antidepressant-like behavior in the mouse. *Brain Res.* **2009**, *1285*, 148–157. [CrossRef] [PubMed]
22. Leslie, A. Characterization of a1-Adrenergic Receptor Subtypes in Rat Brain: A Reevaluation of [3H] WB41 04 and [3H] Prazosin Binding. *Pharmacol. Exet. Am. Chem. Sciety Pharmacol. Experiemntal Ther.* **1986**, *29*, 321–330.
23. Reader, T.A.; Brière, R.; Grondin, L. Alpha-1 and alpha-2 adrenoceptor binding in cerebral cortex: Competition studies with [3H]prazosin and [3H]idazoxan. *J. Neural Transm.* **1987**, *68*, 79–95. [CrossRef] [PubMed]

24. Drouin, C.; Darracq, L.; Trovero, F.; Blanc, G.; Glowinski, J.; Cotecchia, S.; Tassin, J.P. α1b-Adrenergic Receptors Control Locomotor and Rewarding Effects of Psychostimulants and Opiates. *J. Neurosci.* **2002**, *22*, 2873–2884. [CrossRef]
25. Zuscik, M.J.; Sands, S.; Ross, S.A.; Waugh, D.J.J.; Gaivin, R.J.; Morilak, D.; Perez, D.M. Overexpression of the α(1B)-adrenergic receptor causes apoptotic neurodegeneration: Multiple system atrophy. *Nat. Med.* **2000**, *6*, 1388–1394. [CrossRef]
26. Hayashi, R.; Ohmori, E.; Isogaya, M.; Moriwaki, M.; Kumagai, H. Design and synthesis of selective α1B adrenoceptor antagonists. *Bioorg. Med. Chem. Lett.* **2006**, *16*, 4045–4047. [CrossRef]
27. Spreng, M.; Cotecchia, S.; Schenk, F. A behavioral study of alpha-1b adrenergic receptor knockout mice: Increased reaction to novelty and selectively reduced learning capacities. *Neurobiol. Learn. Mem.* **2001**, *75*, 214–229. [CrossRef]
28. Sadalge, A.; Coughlin, L.; Fu, H.; Wang, B.; Valladares, O.; Valentino, R.; Blendy, J.A. A1D Adrenoceptor Signaling Is Required for Stimulus Induced Locomotor Activity. *Mol. Psychiatry* **2003**, *8*, 664–672. [CrossRef]
29. Buckner, S.A.; Milicic, I.; Daza, A.; Lynch, J.J.; Kolasa, T.; Nakane, M.; Sullivan, J.P.; Brioni, J.D. A-315456: A selective α1D-adrenoceptor antagonist with minimal dopamine D2 and 5-HT1A receptor affinity. *Eur. J. Pharmacol.* **2001**, *433*, 123–127. [CrossRef]
30. Saunders, C.; Limbird, L.E. Localization and trafficking of α2-adrenergic receptor subtypes in cells and tissues. *Pharmacol. Ther.* **1999**, *84*, 193–205. [CrossRef]
31. Sastre, M.; García-Sevilla, J.A. α 2-Adrenoceptor Subtypes Identified by [3H]RX821002 Binding in the Human Brain: The Agonist Guanoxabenz Does Not Discriminate Different Forms of the Predominant α2A Subtype. *J. Neurochem.* **1994**, *63*, 1077–1085. [CrossRef] [PubMed]
32. Gilsbach, R.; Hein, L. Are the pharmacology and physiology of α 2adrenoceptors determined by α 2-heteroreceptors and autoreceptors respectively? *Br. J. Pharmacol.* **2012**, *165*, 90–102. [CrossRef] [PubMed]
33. Wasilewska, A.; Sączewski, F.; Hudson, A.L.; Ferdousi, M.; Scheinin, M.; Laurila, J.M.; Rybczyńska, A.; Boblewski, K.; Lehmann, A. Fluorinated analogues of marsanidine, a highly α2-AR/imidazoline I1 binding site-selective hypotensive agent. Synthesis and biological activities. *Eur. J. Med. Chem.* **2014**, *87*, 386–397. [CrossRef] [PubMed]
34. Krzyczmonik, A.; Keller, T.; López-Picón, F.R.; Forsback, S.; Kirjavainen, A.K.; Takkinen, J.S.; Wasilewska, A.; Scheinin, M.; Haaparanta-Solin, M.; Sączewski, F.; et al. Radiosynthesis and Preclinical Evaluation of an α2A-Adrenoceptor Tracer Candidate, 6-[18F]Fluoro-marsanidine. *Mol. Imaging Biol.* **2019**, *21*, 879–887. [CrossRef] [PubMed]
35. Prabhakaran, J.; Majo, V.J.; Milak, M.S.; Mali, P.; Savenkova, L.; Mann, J.J.; Parsey, R.V.; Kumar, J.S.D. Synthesis and in vivo evaluation of [11C]MPTQ: A potential PET tracer for alpha2A-adrenergic receptors. *Bioorg. Med. Chem. Lett.* **2010**, *20*, 3654–3657. [CrossRef]
36. Arponen, E.; Helin, S.; Marjamäki, P.; Grönroos, T.; Holm, P.; Löyttyniemi, E.; Någren, K.; Scheinin, M.; Haaparanta-Solin, M.; Sallinen, J.; et al. A PET tracer for brain α2C adrenoceptors, 11C-ORM-13070: Radiosynthesis and preclinical evaluation in rats and knockout mice. *J. Nucl. Med.* **2014**, *55*, 1171–1177. [CrossRef]
37. Kawamura, K.; Akiyama, M.; Yui, J.; Yamasaki, T.; Hatori, A.; Kumata, K.; Wakizaka, H.; Takei, M.; Nengaki, N.; Yanamoto, K.; et al. In vivo evaluation of limiting brain penetration of probes for α2C-adrenoceptor using small-animal positron emission tomography. *ACS Chem. Neurosci.* **2010**, *1*, 520–528. [CrossRef]
38. Sallinen, J.; Höglund, I.; Engström, M.; Lehtimäki, J.; Virtanen, R.; Sirviö, J.; Wurster, S.; Savola, J.M.; Haapalinna, A. Pharmacological characterization and CNS effects of a novel highly selective α2C-adrenoceptor antagonist JP-1302. *Br. J. Pharmacol.* **2007**, *150*, 391–402. [CrossRef]
39. Luoto, P.; Suilamo, S.; Oikonen, V.; Arponen, E.; Helin, S.; Herttuainen, J.; Hietamäki, J.; Holopainen, A.; Kailajärvi, M.; Peltonen, J.M.; et al. 11C-ORM-13070, a novel PET ligand for brain α2C-adrenoceptors: Radiometabolism, plasma pharmacokinetics, whole-body distribution and radiation dosimetry in healthy men. *Eur. J. Nucl. Med. Mol. Imaging* **2014**, *41*, 1947–1956. [CrossRef]
40. Sscarpace, P.J.; Abrass, I.B. Alpha- and beta-adrenergic receptor function in the brain during senescence. *Neurobiol. Aging* **1988**, *9*, 53–58. [CrossRef]
41. Pandey, S.C.; Ren, X.; Sagen, J.; Pandey, G.N. β-Adrenergic receptor subtypes in stress-induced behavioral depression. *Pharmacol. Biochem. Behav.* **1995**, *51*, 339–344. [CrossRef]
42. Wallukat, G. The β-adrenergic receptors. *Herz* **2002**, *27*, 683–690. [CrossRef] [PubMed]

43. Lu'O'Ng, K.V.Q.; Nguyen, L.T.H. The role of beta-adrenergic receptor blockers in Alzheimer's disease: Potential genetic and cellular signaling mechanisms. *Am. J. Alzheimers. Dis. Demen.* **2013**, *28*, 427–439. [CrossRef] [PubMed]
44. O'Dell, T.J.; Connor, S.A.; Guglietta, R.; Nguyen, P.V. β-Adrenergic receptor signaling and modulation of long-term potentiation in the mammalian hippocampus. *Learn. Mem.* **2015**, *22*, 461–471. [CrossRef]
45. Gelinas, J.; Nguyen, P. Neuromodulation of Hippocampal Synaptic Plasticity, Learning, and Memory by Noradrenaline. *Cent. Nerv. Syst. Agents Med. Chem.* **2008**, *7*, 17–33. [CrossRef]
46. Wachter, S.B.; Gilbert, E.M. Beta-adrenergic receptors, from their discovery and characterization through their manipulation to beneficial clinical application. *Cardiology* **2012**, *122*, 104–112. [CrossRef]
47. Joyce, J.N.; Lexow, N.; Kim, S.J.; Artymyshyn, R.; Senzon, S.; Lawerence, D.; Cassanova, M.F.; Kleinman, J.E.; Bird, E.D.; Winokur, A. Distribution of beta-adrenergic receptor subtypes in human post-mortem brain: Alterations in limbic regions of schizophrenics. *Synapse* **1992**, *10*, 228–246. [CrossRef]
48. Waarde, A.; Vaalburg, W.; Doze, P.; Bosker, F.; Elsinga, P. PET Imaging of Beta-Adrenoceptors in Human Brain: A Realistic Goal or a Mirage? *Curr. Pharm. Des.* **2005**, *10*, 1519–1536. [CrossRef]
49. Pike, V.W.; Law, M.P.; Osman, S.; Davenport, R.J.; Rimoldi, O.; Giardinà, D.; Camici, P.G. Selection, design and evaluation of new radioligands for PET studies of cardiac adrenoceptors. *Pharm. Acta Helv.* **2000**, *74*, 191–200. [CrossRef]
50. Chen, X.; Werner, R.A.; Javadi, M.S.; Maya, Y.; Decker, M.; Lapa, C.; Herrmann, K.; Higuchi, T. Radionuclide imaging of neurohormonal system of the heart. *Theranostics* **2015**, *5*, 545–558. [CrossRef]
51. Logan, J.; Wang, G.J.; Telang, F.; Fowler, J.S.; Alexoff, D.; Zabroski, J.; Jayne, M.; Hubbard, B.; King, P.; Carter, P.; et al. Imaging the norepinephrine transporter in humans with (S,S)-[11C]O-methyl reboxetine and PET: Problems and progress. *Nucl. Med. Biol.* **2007**, *34*, 667–679. [CrossRef] [PubMed]
52. Chen, X.; Kudo, T.; Lapa, C.; Buck, A.; Higuchi, T. Recent advances in radiotracers targeting norepinephrine transporter: Structural development and radiolabeling improvements. *J. Neural Transm.* **2020**, *127*, 851–873. [CrossRef] [PubMed]
53. Zeng, F.; Mun, J.; Jarkas, N.; Stehouwer, J.S.; Voll, R.J.; Tamagnan, G.D.; Howell, L.; Votaw, J.R.; Kilts, C.D.; Nemeroff, C.B.; et al. Synthesis, Radiosynthesis, and Biological Evaluation of Carbon-11 and Fluorine-18 Labeled Reboxetine Analogs: Potential Positron Emission Tomography Radioligands for in Vivo Imaging of the Norepinephrine Transporter. *J. Med. Chem.* **2009**, *52*, 62–73. [CrossRef] [PubMed]
54. Moriguchi, S.; Kimura, Y.; Ichise, M.; Arakawa, R.; Takano, H.; Seki, C.; Ikoma, Y.; Takahata, K.; Nagashima, T.; Yamada, M.; et al. PET quantification of the norepinephrine transporter in human brain with (S,S)-18F-FMeNER-D2. *J. Nucl. Med.* **2017**, *58*, 1140–1145. [CrossRef]
55. Narayanaswami, V.; Drake, L.R.; Brooks, A.F.; Meyer, J.H.; Houle, S.; Kilbourn, M.R.; Scott, P.J.H.; Vasdev, N. Classics in Neuroimaging: Development of PET Tracers for Imaging Monoamine Oxidases. *ACS Chem. Neurosci.* **2019**, *10*, 1867–1871. [CrossRef]
56. Fowler, J.S.; Logan, J.; Shumay, E.; Alia-Klein, N.; Wang, G.J.; Volkow, N.D. Monoamine oxidase: Radiotracer chemistry and human studies. *J. Label. Compd. Radiopharm.* **2015**, *58*, 51–64. [CrossRef]
57. Narayanaswami, V.; Dahl, K.; Bernard-Gauthier, V.; Josephson, L.; Cumming, P.; Vasdev, N. Emerging PET Radiotracers and Targets for Imaging of Neuroinflammation in Neurodegenerative Diseases: Outlook Beyond TSPO. *Mol. Imaging* **2018**, *17*, 1–25. [CrossRef]
58. Piascik, M.T.; Perez, D.M. α1-Adrenergic receptors: New insights and directions. *J. Pharmacol. Exp. Ther.* **2001**, *298*, 403–410.
59. Graham, R.M.; Perez, D.M.; Hwa, J.; Piascik, M.T. α1 -Adrenergic Receptor Subtypes. *Circ. Res.* **1996**, *78*, 737–749. [CrossRef]
60. Chen, Z.J.; Minneman, K.P. Recent progress in α1-adrenergic receptor research. *Acta Pharmacol. Sin.* **2005**, *26*, 1281–1287. [CrossRef]
61. O'Connell, T.D.; Jensen, B.C.; Baker, A.J.; Simpson, P.C. Cardiac alpha1-adrenergic receptors: Novel aspects of expression, signaling mechanisms, physiologic function, and clinical importance. *Pharmacol. Rev.* **2014**, *66*, 308–333. [CrossRef] [PubMed]
62. Jensen, B.C.; O'Connell, T.D.; Simpson, P.C. Alpha-1-adrenergic receptors in heart failure: The adaptive arm of the cardiac response to chronic catecholamine stimulation. *J. Cardiovasc. Pharmacol.* **2014**, *63*, 291–301. [CrossRef] [PubMed]

63. Morrow, A.L.; Creese, I. Characterization of Alpha-1-Adrenergic Receptor Subtypes in Rat-Brain—A Reevaluation of [H-3] Wb4104 and [H-3] Prazosin Binding. *Mol. Pharmacol.* **1986**, *29*, 321–330. [PubMed]
64. Romero-grimaldi, C.; Moreno-lo, B. Age-Dependent Effect of Nitric Oxide on Subventricular Zone and Olfactory Bulb. *J. Comp. Neurol.* **2008**, *346*, 339–346. [CrossRef] [PubMed]
65. Campeau, S.; Nyhuis, T.J.; Kryskow, E.M.; Masini, C.V.; Babb, J.A.; Sasse, S.K.; Greenwood, B.N.; Fleshner, M.; Day, H.E.W. Stress rapidly increases alpha 1d adrenergic receptor mRNA in the rat dentate gyrus. *Brain Res.* **2010**, *1323*, 109–118. [CrossRef]
66. Szot, P.; White, S.S.; Greenup, J.L.; Leverenz, J.B.; Peskind, E.R.; Raskind, M.A. Changes in Adrenoreceptors in The Prefrontal Cortex Of Subjects With Dementia: Evidence Of Compensatory Changes. *Neuroscience* **2007**, *146*, 471–480. [CrossRef]
67. Perez, D.M.; Doze, V.A. Cardiac and neuroprotection regulated by alpha1-AR subtypes. *J. Recept Signal. Transduct Res.* **2011**, *31*, 98–110. [CrossRef]
68. Sica, D. Alpha-1 adrenergic Blockers: Current Usage Considerations. *J. Clin. Hypertens.* **2005**, *7*, 757–762. [CrossRef]
69. Miriam, H.; Hein, L. α1-Adrenozeptor-Antagonisten. Bei BPS und Hypertonie. *Pharm. Unserer Zeit* **2008**, *37*, 290–295. [CrossRef]
70. Law, M.P.; Osman, S.; Pike, V.W.; Davenport, R.J.; Cunningham, V.J.; Rimoldi, O.; Rhodes, C.G.; Giardinà, D.; Camici, P.G. Evaluation of [11C]GB67, a novel radioligand for imaging myocardial (α1-adrenoceptors with positron emission tomography. *Eur. J. Nucl. Med.* **2000**, *27*, 7–17. [CrossRef]
71. Nyberg, S.; Eriksson, B.; Oxenstierna, G.; Halldin, C.; Farde, L. Suggested minimal effective dose of risperidone based on PET-measured D2 and 5-HT(2A) receptor occupancy in schizophrenic patients. *Am. J. Psychiatry* **1999**, *156*, 869–875. [CrossRef] [PubMed]
72. Nyberg, S.; Olsson, H.; Nilsson, U.; Maehlum, E.; Halldin, C.; Farde, L. Low striatal and extra-striatal D2 receptor occupancy during treatment with the atypical antipsychotic sertindole. *Psychopharmacology (Berl.)* **2002**, *162*, 37–41. [CrossRef]
73. Balle, T.; Perregaard, J.; Ramirez, M.T.; Larsen, A.K.; Søby, K.K.; Liljefors, T.; Andersen, K. Synthesis and structure-affinity relationship investigations of 5-heteroaryl-substituted analogues of the antipsychotic sertindole. A new class of highly selective α1 adrenoceptor antagonists. *J. Med. Chem.* **2003**, *46*, 265–283. [CrossRef] [PubMed]
74. Balle, T.; Halldin, C.; Andersen, L.; Alifrangis, L.H.; Badolo, L.; Jensen, K.G.; Chou, Y.W.; Andersen, K.; Perregaard, J.; Farde, L. New α1-adrenoceptor antagonists derived from the antipsychotic sertindole—Carbon-11 labelling and pet examination of brain uptake in the cynomolgus monkey. *Nucl. Med. Biol.* **2004**, *31*, 327–336. [CrossRef] [PubMed]
75. Airaksinen, A.J.; Finnema, S.J.; Balle, T.; Varnäs, K.; Bang-Andersen, B.; Gulyás, B.; Farde, L.; Halldin, C. Radiosynthesis and evaluation of new α1-adrenoceptor antagonists as PET radioligands for brain imaging. *Nucl. Med. Biol.* **2013**, *40*, 747–754. [CrossRef]
76. Risgaard, R.; Ettrup, A.; Balle, T.; Dyssegaard, A.; Hansen, H.D.; Lehel, S.; Madsen, J.; Pedersen, H.; Püschl, A.; Badolo, L.; et al. Radiolabelling and PET brain imaging of the α1-adrenoceptor antagonist Lu AE43936. *Nucl. Med. Biol.* **2013**, *40*, 135–140. [CrossRef]
77. Jorgensen, M.; Jorgensen, P.N.; Christoffersen, C.T.; Jensen, K.G.; Balle, T.; Bang-Andersen, B. Discovery of novel α1-adrenoceptor ligands based on the antipsychotic sertindole suitable for labeling as PET ligands. *Bioorg. Med. Chem.* **2013**, *21*, 196–204. [CrossRef]
78. Bøgesø, K.P.; Arnt, J.; Hyttel, J.; Pedersen, H.; Liljefors, T. Octoclothepin Enantiomers. A Reinvestigation of Their Biochemical and Pharmacological Activity in Relation to a New Receptor-Interaction Model for Dopamine D-2 Receptor Antagonists. *J. Med. Chem.* **1991**, *34*, 2023–2030. [CrossRef]
79. Liow, J.S.; Lu, S.; McCarron, J.A.; Hong, J.; Musachio, J.L.; Pike, V.W.; Innis, R.B.; Zoghbi, S.S. Effect of a P-Glycoprotein Inhibitor, Cyclosporin A, on the Disposition in Rodent Brain and Blood of the 5-HT1A Receptor Radioligand, [11C](R)-(—)-RWAY. *Synapse* **2007**, *61*, 96–105. [CrossRef]
80. Roberts, L.R.; Bryans, J.; Conlon, K.; McMurray, G.; Stobie, A.; Whitlock, G.A. Novel 2-imidazoles as potent, selective and CNS penetrant α1A adrenoceptor partial agonists. *Bioorg. Med. Chem. Lett.* **2008**, *18*, 6437–6440. [CrossRef]
81. Whitlock, G.A.; Brennan, P.E.; Roberts, L.R.; Stobie, A. Potent and selective α1A adrenoceptor partial agonists-Novel imidazole frameworks. *Bioorg. Med. Chem. Lett.* **2009**, *19*, 3118–3121. [CrossRef] [PubMed]

82. Bücheler, M.M.; Hadamek, K.; Hein, L. Two α2-adrenergic receptor subtypes, α2A and α2C, inhibit transmitter release in the brain of gene-targeted mice. *Neuroscience* **2002**, *109*, 819–826. [CrossRef]
83. Nguyen, V.; Tiemann, D.; Park, E.; Salehi, A. Alpha-2 Agonists. *Anesthesiol. Clin.* **2017**, *35*, 233–245. [CrossRef] [PubMed]
84. Khan, Z.P.; Ferguson, C.N.; Jones, R.M. Alpha-2 and imidazoline receptor agonists. Their pharmacology and therapeutic role. *Anaesthesia* **1999**, *54*, 146–165. [CrossRef]
85. Paciorek, P.M.; Pierce, V.; Shepperson, N.B.; Waterfall, J.F. An investigation into the selectivity of a novel series of benzoquinolizines for α2-adrenoceptors in vivo. *Br. J. Pharmacol.* **1984**, *82*, 127–134. [CrossRef]
86. Pettibone, D.J.; Flagg, S.D.; Totarol, J.A.; Clineschmidt, B.V.; Huff, J.R.; Young, S.D.; Chen, R. [3H]L-657, 743 (MK-912): A new, high affinity, selective radioligand for brain α-2 adrenoceptors. *Life Sci.* **1989**, *44*, 459–467. [CrossRef]
87. Pleus, R.C.; Shiue, C.Y.; Shiue, G.G.; Rysavy, J.A.; Huang, H.; Cornish, K.G.; Sunderland, J.J.; Bylund, D.B. Synthesis and biodistribution of the alpha 2-adrenergic receptor antagonist (11C)WY26703. Use as a radioligand for positron emission tomography. *Receptor* **1992**, *2*, 241–252.
88. Shiue, C.Y.; Pleus, R.C.; Shiue, G.G.; Rysavy, J.A.; Sunderland, J.J.; Cornish, K.G.; Young, S.D.; Bylund, D.B. Synthesis and biological evaluation of [11C]MK-912 as an α2- adrenergic receptor radioligand for PET studies. *Nucl. Med. Biol.* **1998**, *25*, 127–133. [CrossRef]
89. BK, F.D.; Terrazzino, S.; Tavitian, B.; Hinnen, F.; Vaufrey, F.; Crouzel, C. XIIth international symposium on radiopharmaceutical chemistry: Abstracts and programme. *J. Label. Compd. Radiopharm.* **1997**, *40*, 1–72. [CrossRef]
90. Hume, S.P.; Hirani, E.; Opacka-Juffry, J.; Osman, S.; Myers, R.; Gunn, R.N.; McCarron, J.A.; Clark, R.D.; Melichar, J.; Nutt, D.J.; et al. Evaluation of [O-methyl-11C]RS.15385-197 as a positron emission tomography radioligand for central α2-adrenoceptors. *Eur. J. Nucl. Med.* **2000**, *27*, 475–484. [CrossRef]
91. Roeda, D.; Sipil, H.T.; Bramoull, Y.; Enas, J.D.; Vaufrey, F.; Doll, F.; Crouzel, C. Synthesis of [11C]atipamezole, a potential PET ligand for the α2-adrenergic receptor in the brain. *J. Label. Compd. Radiopharm.* **2002**, *45*, 37–47. [CrossRef]
92. Marthi, K.; Bender, D.; Watanabe, H.; Smith, D.F. PET evaluation of a tetracyclic, atypical antidepressant, [N-methyl-11C]mianserin, in the living porcine brain. *Nucl. Med. Biol.* **2002**, *29*, 317–319. [CrossRef]
93. Marthi, K.; Bender, D.; Gjedde, A.; Smith, D.F. [11C]Mirtazapine for PET neuroimaging: Radiosynthesis and initial evaluation in the living porcine brain. *Eur. Neuropsychopharmacol.* **2002**, *12*, 427–432. [CrossRef]
94. Marthi, K.; Jakobsen, S.; Bender, D.; Hansen, S.B.; Smith, S.B.; Hermansen, F.; Rosenberg, R.; Smith, D.F. [N-methyl-11C]Mirtazapine for positron emission tomography neuroimaging of antidepressant actions in humans. *Psychopharmacology (Berl.)* **2004**, *174*, 260–265. [CrossRef] [PubMed]
95. Smith, D.F.; Stork, B.S.; Wegener, G.; Ashkanian, M.; Jakobsen, S.; Bender, D.; Audrain, H.; Vase, K.H.; Hansen, S.B.; Videbech, P.; et al. [11C]mirtazapine binding in depressed antidepressant nonresponders studied by PET neuroimaging. *Psychopharmacology (Berl.)* **2009**, *206*, 133–140. [CrossRef] [PubMed]
96. Van der Mey, M.; Windhorst, A.D.; Klok, R.P.; Herscheid, J.D.M.; Kennis, L.E.; Bischoff, F.; Bakker, M.; Langlois, X.; Heylen, L.; Jurzak, M.; et al. Synthesis and biodistribution of [11C]R107474, a new radiolabeled α2-adrenoceptor antagonist. *Bioorg. Med. Chem.* **2006**, *14*, 4526–4534. [CrossRef]
97. Jakobsen, S.; Pedersen, K.; Smith, D.F.; Jensen, S.B.; Munk, O.L.; Cumming, P. Detection of a 2 -Adrenergic Receptors in Brain of Living Pig with 11 C-Yohimbine. *J. Nucl. Med.* **2006**, *47*, 2008–2015.
98. Nahimi, A.; Jakobsen, S.; Munk, O.L.; Vang, K.; Phan, J.A.; Rodell, A.; Gjedde, A. Mapping α2 adrenoceptors of the human brain with11C-yohimbine. *J. Nucl. Med.* **2015**, *56*, 392–398. [CrossRef]
99. Andrés, J.I.; Alcázar, J.; Alonso, J.M.; Alvarez, R.M.; Bakker, M.H.; Biesmans, I.; Cid, J.M.; De Lucas, A.I.; Fernández, J.; Font, L.M.; et al. Discovery of a new series of centrally active tricyclic isoxazoles combining serotonin (5-HT) reuptake inhibition with α2- adrenoceptor blocking activity. *J. Med. Chem.* **2005**, *48*, 2054–2071. [CrossRef]
100. Sączewski, F.; Kornicka, A.; Rybczyńska, A.; Hudson, A.L.; Shu, S.M.; Gdaniec, M.; Boblewski, K.; Lehmann, A. 1-[(imidazolidin-2-yl)imino]indazole. Highly α2/I 1 selective agonist: Synthesis, X-ray structure, and biological activity. *J. Med. Chem.* **2008**, *51*, 3599–3608. [CrossRef]
101. Hagihara, K.; Kashima, H.; Iida, K.; Enokizono, J.; Uchida, S.I.; Nonaka, H.; Kurokawa, M.; Shimada, J. Novel 4-(6,7-dimethoxy-1,2,3,4-tetrahydroisoquinolin-2-yl)methylbenzofuran derivatives as selective α2C-adrenergic receptor antagonists. *Bioorg. Med. Chem. Lett.* **2007**, *17*, 1616–1621. [CrossRef] [PubMed]

102. Finnema, S.J.; Hughes, Z.A.; Haaparanta-Solin, M.; Stepanov, V.; Nakao, R.; Varnäs, K.; Varrone, A.; Arponen, E.; Marjamäki, P.; Pohjanoksa, K.; et al. Amphetamine decreases α2C-adrenoceptor binding of [11C]ORM-13070: A PET study in the primate brain. *Int. J. Neuropsychopharmacol.* **2015**, *18*, 1–10. [CrossRef] [PubMed]
103. Palacios, J.M.; Kuhar, M.J. Beta adrenergic receptor localization in rat brain by light microscopic autoradiography. *Neurochem. Int.* **1982**, *4*, 473–490. [CrossRef]
104. Cash, R.; Ruberg, M.; Raisman, R.; Agid, Y. Adrenergic receptors in Parkinson's disease. *Brain Res.* **1984**, *322*, 269–275. [CrossRef]
105. Hopfner, F.; Höglinger, G.U.; Kuhlenbäumer, G.; Pottegård, A.; Wod, M.; Christensen, K.; Tanner, C.M.; Deuschl, G. β-adrenoreceptors and the risk of Parkinson's disease. *Lancet Neurol.* **2020**, *19*, 247–254. [CrossRef]
106. Klimek, V.; Rajkowska, G.; Luker, S.N.; Dilley, G.; Meltzer, H.Y.; Overholser, J.C.; Stockmeier, C.A.; Ordway, G.A. Brain noradrenergic receptors in major depression and schizophrenia. *Neuropsychopharmacology* **1999**, *21*, 69–81. [CrossRef]
107. Ueki, J.; Rhodes, C.G.; Hughes, J.M.; De Silva, R.; Lefroy, D.C.; Ind, P.W.; Qing, F.; Brady, F.; Luthra, S.K.; Steel, C.J. In vivo quantification of pulmonary B-adrenoceptor density in humans with (*S*)-[11C]CGP-12177 and PET. *Am. Physiol. Soc.* **1993**, *75*, 559–565. [CrossRef] [PubMed]
108. Elsinga, P.H.; Doze, P.; Van Waarde, A.; Pieterman, R.M.; Blanksma, P.K.; Willemsen, A.T.M.; Vaalburg, W. Imaging of β-adrenoceptors in the human thorax using (*S*)-[11C]CGP12388 and positron emission tomography. *Eur. J. Pharmacol.* **2001**, *433*, 173–176. [CrossRef]
109. Berger, G.; Maziere, M.; Prenant, C.; Sastre, J.; Syrota, A.; Comar, D. Synthesis of 11 C propranolol. *J. Radioanal. Chem.* **1982**, *74*, 301–306. [CrossRef]
110. Antoni, G.; Ulin, J.; Långström, B. Synthesis of the 11C-labelled β-adrenergic receptor ligands atenolol, metoprolol and propranolol. *Int. J. Radiat. Appl. Instrum. Part A* **1989**, *40*, 561–564. [CrossRef]
111. Berridge, M.S.; Nelson, A.D.; Zheng, L.; Leisure, G.P.; Miraldi, F. Specific beta-adrenergic receptor binding of carazolol measured with PET. *J. Nucl. Med.* **1994**, *35*, 1665–1676.
112. Zheng, L.; Berridge, M.S.; Ernsberger, P. Synthesis, Binding Properties, and 18F Labeling of Fluorocarazolol, a High-Affinity β-Adrenergic Receptor Antagonist. *J. Med. Chem.* **1994**, *37*, 3219–3230. [CrossRef] [PubMed]
113. Doze, P.; Van Waarde, A.; Elsinga, P.H.; Van-Loenen Weemaes, A.M.A.; Willemsen, A.T.M.; Vaalburg, W. Validation of S-1′-[18F]fluorocarazolol for in vivo imaging and quantification of cerebral β-adrenoceptors. *Eur. J. Pharmacol.* **1998**, *353*, 215–226. [CrossRef]
114. Doze, P.; Elsinga, P.H.; De Vries, E.F.J.; Van Waarde, A.; Vaalburg, W. Mutagenic activity of a fluorinated analog of the beta-adrenoceptor ligand carazolol in the Ames test. *Nucl. Med. Biol.* **2000**, *27*, 315–319. [CrossRef]
115. Tewson, T.J.; Stekhova, S.; Kinsey, B.; Chen, L.; Wiens, L.; Barber, R. Synthesis and biodistribution of R- and S-isomers of [18F]- fluoropropranolol, a lipophilic ligand for the β-adrenergic receptor. *Nucl. Med. Biol.* **1999**, *26*, 891–896. [CrossRef]
116. Moresco, R.M.; Matarrese, M.; Soloviev, D.; Simonelli, P.; Rigamonti, M.; Gobbo, C.; Todde, S.; Carpinelli, A.; Galli Kienle, M.; Fazio, F. Synthesis and in vivo evaluation of [11C]ICI 118551 as a putative subtype selective β2-adrenergic radioligand. *Int. J. Pharm.* **2000**, *204*, 101–109. [CrossRef]
117. Soloviev, D.V.; Matarrese, M.; Moresco, R.M.; Todde, S.; Bonasera, T.A.; Sudati, F.; Simonelli, P.; Magni, F.; Colombo, D.; Carpinelli, A.; et al. Asymmetric synthesis and preliminary evaluation of (*R*)- and (*S*)-[11C]bisoprolol, a putative β1-selective adrenoceptor radioligand. *Neurochem. Int.* **2001**, *38*, 169–180. [CrossRef]
118. Doze, P.; Elsinga, P.H.; Maas, B.; Van Waarde, A.; Wegman, T.; Vaalburg, W. Synthesis and evaluation of radiolabeled antagonists for imaging of β-adrenoceptors in the brain with PET. *Neurochem. Int.* **2002**, *40*, 145–155. [CrossRef]
119. Doze, P.; Van Waarde, A.; Tewson, T.J.; Vaalburg, W.; Elsinga, P.H. Synthesis and evaluation of (*S*)-[18F]-fluoroethylcarazolol for in vivo β-adrenoceptor imaging in the brain. *Neurochem. Int.* **2002**, *41*, 17–27. [CrossRef]
120. van Waarde, A.; Doorduin, J.; de Jong, J.R.; Dierckx, R.A.; Elsinga, P.H. Synthesis and preliminary evaluation of (*S*)-[11C]-exaprolol, a novel β-adrenoceptor ligand for PET. *Neurochem. Int.* **2008**, *52*, 729–733. [CrossRef]

121. Stephenson, K.A.; van Oosten, E.M.; Wilson, A.A.; Meyer, J.H.; Houle, S.; Vasdev, N. Synthesis and preliminary evaluation of [18F]-fluoro-(2S)-Exaprolol for imaging cerebral β-adrenergic receptors with PET. *Neurochem. Int.* **2008**, *53*, 173–179. [CrossRef] [PubMed]
122. Mirfeizi, L.; Rybczynska, A.A.; van Waarde, A.; Campbell-Verduyn, L.; Feringa, B.L.; Dierckx, R.A.J.O.; Elsinga, P.H. [18F]-(fluoromethoxy)ethoxy)methyl)-1H-1,2,3-triazol-1-yl)propan-2-ol ([18F FPTC) a novel PET-ligand for cerebral beta-adrenoceptors. *Nucl. Med. Biol.* **2014**, *41*, 203–209. [CrossRef] [PubMed]

Review

Overview of Radiolabeled Somatostatin Analogs for Cancer Imaging and Therapy

Romain Eychenne [1,2,3], Christelle Bouvry [4,5], Mickael Bourgeois [2,3], Pascal Loyer [6], Eric Benoist [1] and Nicolas Lepareur [4,6,*]

1 UPS, CNRS, SPCMIB (Laboratoire de Synthèse et Physico-Chimie de Molécules d'Intérêt Biologique)—UMR 5068, Université de Toulouse, F-31062 Toulouse, France; eychenne@arronax-nantes.fr (R.E.); benoist@chimie.ups-tlse.fr (E.B.)
2 Groupement d'Intérêt Public ARRONAX, 1 Rue Aronnax, F-44817 Saint Herblain, France; mickael.bourgeois@univ-nantes.fr
3 CNRS, CRCINA (Centre de Recherche en Cancérologie et Immunologie Nantes—Angers)—UMR 1232, ERL 6001, Inserm, Université de Nantes, F-44000 Nantes, France
4 Comprehensive Cancer Center Eugène Marquis, Rennes, F-35000, France; c.bouvry@rennes.unicancer.fr
5 CNRS, ISCR (Institut des Sciences Chimiques de Rennes)—UMR 6226, Univ Rennes, F-35000 Rennes, France
6 INRAE, Institut NUMECAN (Nutrition, Métabolismes et Cancer)—UMR_A 1341, UMR_S 1241, Inserm, Univ Rennes, F-35000 Rennes, France; pascal.loyer@univ-rennes1.fr
* Correspondence: n.lepareur@rennes.unicancer.fr; Tel.: +33-029-925-3144

Received: 28 July 2020; Accepted: 1 September 2020; Published: 2 September 2020

Abstract: Identified in 1973, somatostatin (SST) is a cyclic hormone peptide with a short biological half-life. Somatostatin receptors (SSTRs) are widely expressed in the whole body, with five subtypes described. The interaction between SST and its receptors leads to the internalization of the ligand–receptor complex and triggers different cellular signaling pathways. Interestingly, the expression of SSTRs is significantly enhanced in many solid tumors, especially gastro-entero-pancreatic neuroendocrine tumors (GEP-NET). Thus, somatostatin analogs (SSAs) have been developed to improve the stability of the endogenous ligand and so extend its half-life. Radiolabeled analogs have been developed with several radioelements such as indium-111, technetium-99 m, and recently gallium-68, fluorine-18, and copper-64, to visualize the distribution of receptor overexpression in tumors. Internal metabolic radiotherapy is also used as a therapeutic strategy (e.g., using yttrium-90, lutetium-177, and actinium-225). With some radiopharmaceuticals now used in clinical practice, somatostatin analogs developed for imaging and therapy are an example of the concept of personalized medicine with a theranostic approach. Here, we review the development of these analogs, from the well-established and authorized ones to the most recently developed radiotracers, which have better pharmacokinetic properties and demonstrate increased efficacy and safety, as well as the search for new clinical indications.

Keywords: somatostatin analogs; radiolabeling; radiopharmaceuticals; radionuclide therapy; imaging

1. Introduction

Somatostatin (SST), also called somatotropin release inhibiting factor (SRIF), is a cyclic peptide hormone, first isolated in 1968 from an ovine hypothalamus, and actually identified in 1973 [1]. It was originally discovered as a growth hormone inhibitor, but is now known to be involved in the inhibition of numerous metabolic processes relating to neurotransmitters, endocrine secretions (e.g., growth hormone, insulin, glucagon, and gastrin) but also modulating exocrine secretions (e.g., gastric acid and pancreatic enzymes). In the body, its synthesis takes place in the form of an inactive precursor of 116 amino acids (AA), preprosomatostatin, which is then converted by the action of proteases into prosomatostatin (96 AA). Depending on where it is produced in the body, enzymes do not cleave the pro-peptide on the same amino acid motif, resulting in two distinct active forms, SRIF-28 and SRIF-14. Although SRIF-14 is predominant in the central nervous system and SRIF-28 in the digestive tract, the distribution of these two biologically active forms is similar.

In the early 1990s, concomitantly to studies on the binding properties and mechanisms of action of somatostatin, five receptor subtypes were discovered (SSTR1 to SSTR5) [2]. These subtypes belong to the family of receptors coupled to G-proteins, and their length varies from 364 to 418 AA. They all exhibit seven α helices with transmembrane domains and most of the differences between subtypes are found in the extracellular (N-terminal) and intracellular (C-terminal) ends. SSTR-1, -3, -4, and -5 have a single subtype, while two variants exist for SSTR2, called SSTR2A and SSTR2B. SSTR1 to 4 link SRIF-14 and -28 with a very high affinity (in the nanomolar order), whereas SSTR5 shows an affinity 5 to 10 times higher, but for SRIF-28 only.

Somatostatin receptors are widely distributed in healthy tissues, with distinct expression throughout the body (Figure 1). It is quite possible to find several subtypes in the same tissue. Each of the SSTRs is involved in the regulation of the various processes: (i) SSTR1 is involved in the antisecretory effects of growth hormone, prolactin (a peptide hormone involved in lactation, reproduction, growth, and immunity) and calcitonin (regulation of calcemia); (ii) SSTR2 also inhibits the secretion of growth hormone and adrenocorticotropin (hormone that stimulates the adrenal glands), glucagon, insulin, interferon-γ (protein produced by immune cells), and stomach acid; (iii) SSTR5 has the same inhibiting effect on growth hormone, adrenocorticotropin, insulin, and inhibits the secretion of amylase (digestive enzyme constituting saliva and pancreatic juice); (iv) SSTR3 reduces cell proliferation and causes cell apoptosis; (v) the functions of SSTR4 are not yet well defined [3].

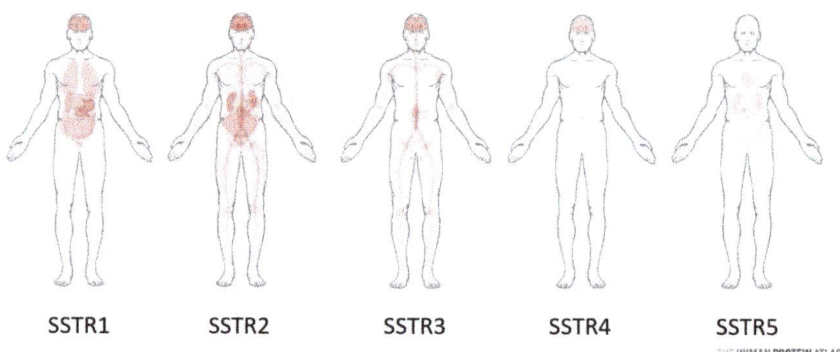

Figure 1. Somatostatin receptors (SSTRs) biodistribution in the body (from The Human Protein Atlas https://www.proteinatlas.org/).

The effects of somatostatin are expressed through different signaling pathways [4,5]. After a cascade of reactions, this leads on the one hand to the inhibition of tumor growth (action on the secretion of hormones) and blocking proliferation via the activation of different tyrosine phosphatases (anti-proliferative and pro-apoptotic action), but also to the inhibition of the secretion of growth factors such as growth hormone or IGF-1 having a major role in the inhibition of tumor growth (anti-angiogenic) (Figure 2) [6,7].

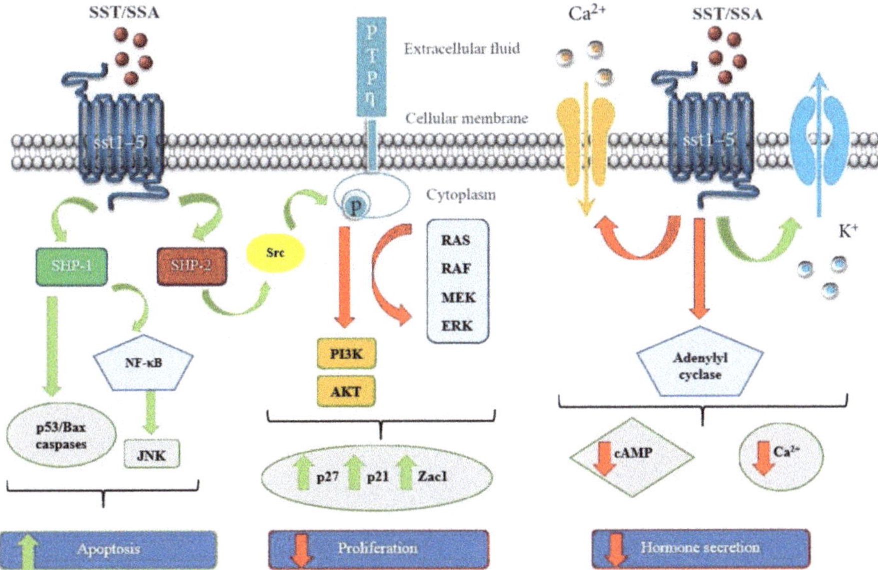

Figure 2. Schematic representation of the signaling pathways induced by somatostatin receptors activation. Green arrows: activated pathways; red arrows: inhibited pathways. Adapted from [8].

Over the past 20 years, our understanding of the phenomena due to the activation of SSTRs has increased thanks to numerous translational and clinical studies, leading to the development of new therapeutic options [3]. The use of SST analogs has demonstrated real effectiveness in the treatment of various pathologies: acromegaly (production of an excess of growth hormone), pancreatitis, complications linked to diabetes and obesity (e.g., retinopathy or nephropathy), action on inflammation and pain in some cases [5,9]. However, SSTRs and SST analogs are mainly known for their presence and role in the detection and treatment of some solid tumors. Tumor cells and peritumoral vessels express receptor subtypes whose density depends on the type of tumors (Table 1) [10–13]. For those overexpressing SSTRs, such as pituitary adenomas, gastroentero-pancreatic neuroendocrine tumors (GEP-NET), or other cancers (e.g., lymphomas, small cell lung cancers, etc.), targeting with SST analogs becomes possible [14]. Many therapeutic protocols based on these analogs (classic octreotide or with a longer release time (octreotide LAR), Lanreotide, Vapreotide, Pasireotide, etc.) have been the subject of phase II and III clinical trials. The majority of results were generally disappointing and did not provide clear evidence of a significant antitumor effect on solid tumors, probably due to the existence of other pathways of tumor progression [15,16].

Table 1. SSTRs expression in different tumor types.

Tumor Type	SSTR Expression	Ref
Astrocytoma	+	[17]
Breast carcinoma	+ (SSTR2)	[11]
Cholangiocarcinoma	+ (SSTR2)	[18]
Colorectal carcinoma	-	[17]
Endometrial carcinoma	-	[17]
Ependymoma	+ (SSTR1, SSTR5)	[11]
Esophageal carcinoma	-	[17]
Ewing sarcoma	-	[17]
Exocrine pancreatic tumor	-	[17]
Gastric carcinoma	+ (SSTR1 > SSTR2, SSTR5)	[11]
Gastrinoma	+ (SSTR2)	[17]
Glioblastoma	-	[17]

Table 1. *Cont.*

Tumor Type	SSTR Expression	Ref
Growth hormone-producing pituitary adenoma	**+** (SSTR2, SSTR5)	[17]
Gut carcinoid	**+** (SSTR2 > SSTR1, SSTR5)	[17]
Hepatocellular carcinoma	+ (SSTR2, SSTR5)	[19]
Insulinoma	+ (SSTR1, SSTR2, SSTR3)	[20]
Leiomyoma	+	[17]
Lymphoma	+ (SSTR2)	[11]
Medullary thyroid carcinoma	+ (SSTR2)	[11]
Medulloblastoma	**+** (SSTR2)	[17]
Meningioma	**+** (SSTR2)	[17]
Neuroblastoma	**+** (SSTR2)	[17]
Non-functioning pituitary adenoma	+ (SSTR3 > SSTR2)	[17]
Non-small cell lung cancer	-	[17]
Ovarian carcinoma	+	[17]
Paraganglioma	**+** (SSTR2)	[17]
Pheochromocytoma	**+** (SSTR1, SSTR2)	[17]
Prostate carcinoma	+ (SSTR1)	[17]
Renal cell carcinoma	+ (SSTR2)	[11]
Small cell lung cancer	+ (SSTR2)	[17]
Urinary bladder carcinoma	-	[17]

Bold +, receptors with particularly high density and incidence. Subtypes preferentially expressed are listed in parentheses, only when compelling evidence is available (immunohistochemistry or autoradiography). Adapted from [11] and [17].

For example, regarding liver tumors, such as hepatocellular carcinoma (HCC), in vitro studies clearly demonstrated (i) the lack of SSTRs expression in healthy liver cells; (ii) overexpression in tumors and metastases of HCC, even though their density is less than in neuroendocrine tumors [21,22]. On the other hand, the results show a heterogeneous expression and strong inter-individual differences. In fact, according to studies, HCCs express high levels of SSTR2 [21,23,24] or SSTR5 [13,19], or even SSTR1 [22] or SSTR3 [25]. In general, around 40% of HCCs studied express somatostatin receptors. These differences could be due to the different methodologies used during the measurements, by studying different stages of the disease or even by heterogeneous behaviors of HCC. Further studies have also found a correlation between the density of SSTRs expression, disease aggressiveness [26], and the rate of tumor recurrence after treatment with octreotide LAR [27]. In a study by Nguyen-Khac et al. [23], 41.2% of extrahepatic metastases express SSTR2. Preclinical tests on HCC cell lines have shown an antiproliferative effect of SST analogs [25,28]. In addition, a real decrease in invasion and cell migration of HCC cells after stimulation of SSTR1 by a specific agonist has also been demonstrated [22]. This action has also been confirmed in vivo [29], with the demonstration of a similar effect on metastatic dissemination [23,30]. These initial results paved the way for clinical trials on patients with HCC, but their conclusions are quite contradictory, [31] showing rather positive effects in the advanced stages [32,33] and others quite negative [34,35]. These outcome discrepancies could come from heterogeneity in the choice of patients, but available data are still insufficient to truly conclude on the effectiveness of analogs of SST alone in the control of HCC tumors [6,31,36]. Cholangiocarcinoma, the other main primary liver tumor, might also be a potential target [18,37].

On the other hand, in certain cases, and in particular for neuroendocrine tumors (a category of tumors where SSTRs are the most expressed), a benefit has been proven via two Phase III studies, which have greatly contributed to the fact that SST analogs are now used in clinical routine [38,39].

2. Somatostatin Analogs

Somatostatin has a short half-life in the body (between one and three minutes), because it is rapidly degraded by peptidases found in plasma and tissues [40]. Therefore, the amount present in the bloodstream is extremely low (between 14 and 32.5 pg/mL). This very short half-life has been considered a limiting factor for possible clinical applications, thus many analogs with better metabolic properties (longer half-life between 1.5 h and 12 h) have been rapidly developed [2,5,9]. These are most often hexapeptide or octapeptide molecules which incorporate the biologically active core of native somatostatin (see some examples in Figure 3). Indeed, studies on the structure–activity correlation have shown that the Phe^7, Trp^8, Lys^9, and Thr^{10} sequence in the form of a β-sheet is necessary for biological activity. The residues Trp^8 and Lys^9 are essential for this activity, whereas Phe^7 and Thr^{10} may undergo some substitutions. Among somatostatin analogs, there are two main categories: the agonists (substances capable of activating somatostatin receptors) and the antagonists (molecules that interact with somatostatin receptors and block or reduce the physiological effect of an agonist). It is also important to note that somatostatin analogs have different affinities for the different receptor subtypes [2].

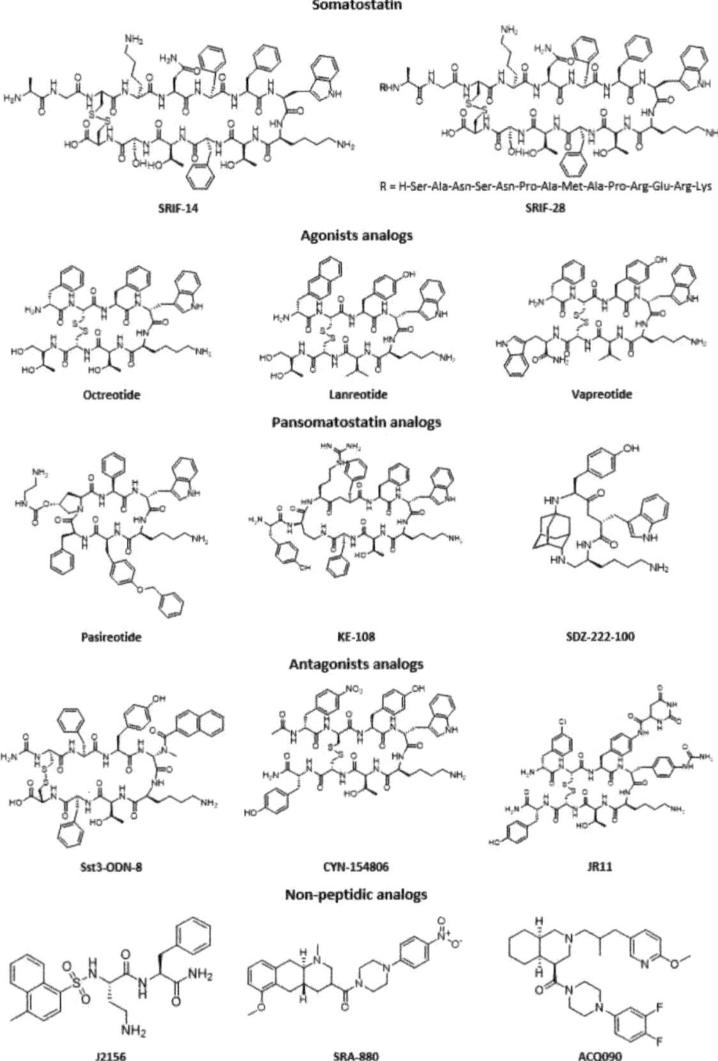

Figure 3. Chemical structures of SRIF-14, SRIF-28, and selected examples of somatostatin analogs.

The first agonist peptide analog to be approved by the FDA was octreotide (SMS 201-995), marketed under the name Sandostatin®. From a structural point of view, it has a D-Trp and a D-Phe, to stabilize the β-sheet and a disulfide bridge closer to the active core, for a better metabolic stability. Its pharmacodynamics is highly similar to native SST, which has made it widely used in clinical trials for the treatment of GEP (gastro-entero-pancreatic) tumors [41,42]. Next, Lanreotide (BIM 23014, tradename Somatuline®), whose structure is similar to that of octreotide (Phe and Thr having been replaced by Tyr and Val respectively), showed comparable characteristics and is also widely used in the treatment of neuroendocrine tumors [43]. In 2005, another analog, Vapreotide (RC160), was marketed under the name Sanvar®, with properties close to those of the two previous analogs, and is also used for the treatment of esophageal varices. More recently, Pasireotide (SOM-230 or Signifor®) was one of the first analogs to show a strong affinity for most of the somatostatin receptor subtypes (pansomatostatin analog). Marketed by Novartis, it is used for the treatment of Cushing's disease [44]. Many other analogs have been developed, from "ultra-short" peptides, such as SDZ 222-100 (an adamantine cyclopeptide), to longer ones, such as KE-108 or CH-275 [5]. Regarding antagonist peptide analogs, the wide variety of compounds that the octapeptide model can offer has allowed the discovery of several structures that can block this kind of receptors. The first antagonist that has been described in the literature is CYN-154806, followed by PRL-2970, sst3-ODN-8 or even non-cyclic models such as BIM-23056 and BIM-23627. New non-peptide compounds have also emerged [45]. These agonists and antagonists (selective or not) constitute a very promising field in the chemistry of somatostatin analogs, in particular because of their pharmacological, pharmacokinetic, and physicochemical properties. This type of compound may have a stronger affinity and/or selectivity for certain subtypes of somatostatin receptors than the majority of peptide analogs. They can thus provide additional information on the exact role of each of these subtypes [5,9].

3. Targeting of Somatostatin Receptors with Radiopharmaceuticals

In the field of medicine, much research is focused on finding methods to achieve earlier detection of pathologies to allow treatment at early stages of the disease, to increase the chances of total recovery. For this purpose, nuclear medicine, through the use of radiopharmaceuticals, is a very powerful tool. Its application can have two different aims: imaging, with the visualization of a radioactive element's distribution in the body, or therapy, with specific irradiation of abnormal cells, thereby reducing damages to nearby healthy tissue. Having a broad range of potential biological targets and desirable pharmacokinetic characteristics—such as high uptake in target tissue and fast blood and non-target tissue clearance—peptides can also be easily chemically modified for incorporation into a radiopharmaceutical, making them a very potent targeting vector for nuclear medicine. Research in that domain has thus gained widespread interest [46–49]. These compounds can be directly labeled with a radionuclide, such as a halogen radioisotope, but they are generally based on a triple structure involving: (i) a radiometal, the radiation of which allows either the localization (γ and β^+ emitters) or the destruction (β^-, α or Auger electron emitters) of the targeted cells; (ii) a bifunctional chelating agent (BFCA), the dual role of which is not only to bind the radiometal in a very stable manner to minimize its dissociation in vivo, but also to allow its conjugation with targeting moiety (or vector) via a functionalized arm; (iii) a targeting moiety (the peptide analog), which aims to convey this set in a specific way to a well-defined target. To limit the influence of the chelating moiety, a linker (or spacer) is usually inserted between the BFCA and the biomolecule (Figure 4).

The choice of the radiometal is crucial, since it deeply influences the design of the chelating structure [50–53]. Several criteria govern the choice of radionuclide: (i) the nature of the radiation emitted, depending on the intended application (diagnosis or therapy); (ii) the half-life, which must be long enough to allow effective fixation of the radiotracer on the target cells, but relatively short to avoid irradiation of the organism (neighboring healthy tissues) and more specifically non-targeted organs; (iii) the isotope decay profile. By emitting its radiation, the nuclide disintegrates into a daughter nuclide, which must be non-radioactive to avoid any additional harmfulness to the organism; (iv) the means of production. Most of the radioelements used in nuclear medicine are artificial. They can be produced in three different ways: from a nuclear reactor, a cyclotron or via a generator. Generator production remains the most convenient way for clinical application, as it can provide in-house radionuclides when a cyclotron is not available nearby, but cyclotron production still remains the cheapest and most used. As an example, Table 2 shows some of the characteristics of radioactive nuclides among the most used today for the radiolabeling of peptides.

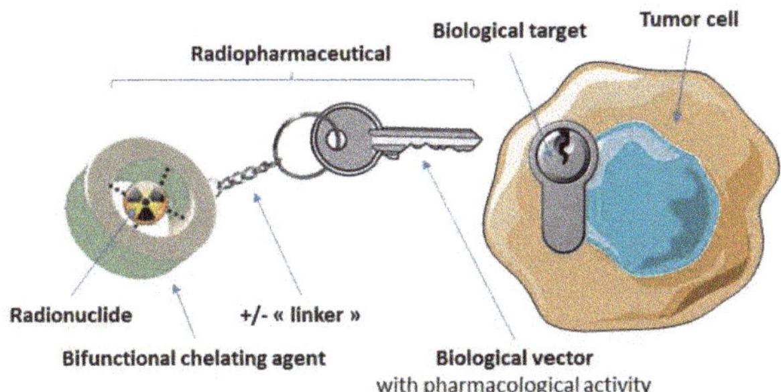

Figure 4. Schematic design of a radiometallated bioconjugate.

Table 2. Some of the main radionuclides studied for imaging and therapy (SPECT—Single-Photon Emission Computed Tomography; PET—Positron Emission Tomography).

Radionuclide	Half-Life (h)	Type of Emission	Energy of Emitted Radiation (keV)	Source	Application
^{99m}Tc	6.01	γ	140	Generator	SPECT imaging
^{111}In	67.4	γ	172, 245	Cyclotron	SPECT imaging
^{18}F	1.83	β^+	634	Cyclotron	PET imaging
^{64}Cu	12.7	$\beta^+/\gamma/\beta^-$	653	Cyclotron	PET imaging
^{68}Ga	1.1	β^+	1190	Generator/Cyclotron	PET imaging
^{90}Y	64.1	β^-	2284	Generator	Therapy
^{177}Lu	160.8	β^-/γ	497	Cyclotron	Therapy
^{188}Re	17	β^-/γ	2118	Generator	Therapy
^{211}At	7.2	α	5870	Cyclotron	Therapy
^{225}Ac	238	α	5830	Generator	Therapy

From a structural point of view, each radiometal has its own properties such as polarizability, degree of oxidation, or coordination number. These features have a direct impact on the choice of the bifunctional chelating agent, in particular in terms of denticity and nature of the donor atoms (most often *O*-, *N*-, or *S*-donors) [54,55]. The BFCA makes it possible to link the biomolecule and the radiometal; its choice is a crucial step in the construction of a radiopharmaceutical. As indicated above, this structure plays a double role: the first is to complex the radioelement in a very stable manner. Several criteria can be evaluated to truly attest to the stability of the complex formed. First of all, the formed radiocomplex must be thermodynamically stable, i.e., the metal-ligand affinity must be as strong as possible. Then it must be kinetically inert. Many metalation reactions take place in the body and the complex formed must be stable enough to avoid any in vivo degradation (e.g., demetallation or transchelation). In addition, radiolabeling conditions with low concentrations are required, ideally with efficient complexation kinetics (high labeling yield) and fast and mild reaction conditions. Beside chemistry considerations, the radiotracer must have: (i) a strong affinity for the target receptor; (ii) a high accumulation for the target and low for the non-target organs; (iii) relatively rapid clearance in the organism; (iv) preferably a mainly renal route of excretion.

Chelating ligands used for the design of radiotracers are usually classified into two categories: macrocyclic and acyclic compounds (Figure 5). Generally, acyclic ligands are less kinetically inert than macrocycles, although some may have shown very good characteristics. On the other hand, these ligands generally have faster metal-chelate binding kinetics compared to macrocyclic analogs, which represents a huge advantage for working with isotopes that have a short lifespan. Despite the coordination properties specific to each metal, some chelating agents—such as polyaminopolycarboxylic acids—are considered to be 'universal' because they can complex different radiometals.

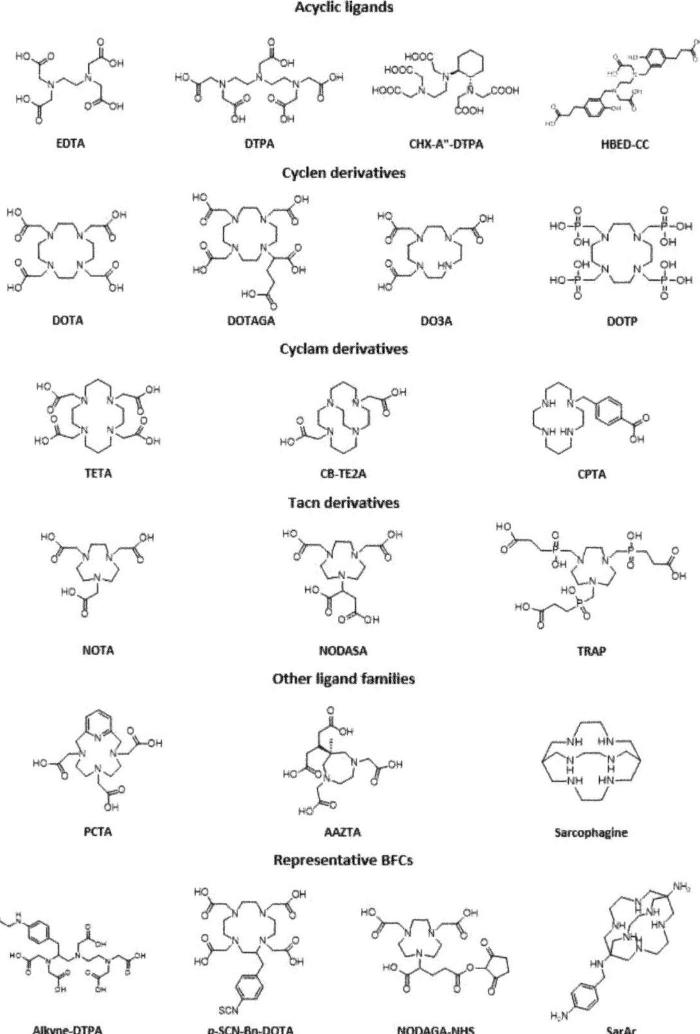

Figure 5. Representative (but not exhaustive) examples of acyclic and macrocyclic polyamino and polyaminocarboxylic chelator families and their derivatives.

Among acyclic ligands, the first BFCAs developed were EDTA (ethylenediaminetetraacetic acid) and DTPA (diethylenetriaminepentaacetic acid). They have been widely used in the chemistry of radiopharmaceuticals, in particular with radioelements such as ^{111}In, ^{90}Y or ^{177}Lu, and even ^{99m}Tc [54]. Later on, DTPA derivatives such as CHX-A''-DTPA with a cyclohexyl moiety bringing more rigidity to the DTPA backbone (allowing a pre-organization of the system) showed better kinetic inertia [56]. Regarding cyclic compounds, cyclen derivatives such as DOTA (1,4,7,10-tetraazacyclododecane-1,4,7,10-tetraacetic acid) and triaza analogs—NOTA (1,4,7-triazacyclononane-1,4,7-triacetic acid)—are among the most studied ligands. NOTA has the smaller chelating cavity of the two, and is generally used for Ga (III) or Cu (II) because it has a particular attraction for these metals, which results in mild radiolabeling conditions and good in vivo stability of the complexes formed. DOTA (which is considered as the gold standard chelator) and its derivatives play an important role in clinical applications because they form very stable complexes with a wide range of trivalent radiometals such as Ga (III), Y (III), In (III), Lu (III), or even divalent such as Cu (II) [57,58]. For DOTA or NOTA, the introduction of a functionalized arm offers the possibility of coupling a biomolecule (NODASA/NODAGA and DOTASA/DOTAGA). Similarly, TETA (1,4,8,11-tetraazacyclotetradecane-1,4,8,11-tetraacetic acid), has mainly been studied with Cu (II) and have shown a stability similar to DOTA [59].

Whether on the side of macrocyclic ligands, derivatives or variations of DOTA (e.g., p-SCN-Bn-DOTA, DOTAGA, CB-DO2A, TCMC ...), NOTA (e.g., p-SCN-Bn-NOTA, NETA ...), or TETA (e.g., CB-TE2A, p-NH2-Bn-TE3A ...), or on the side of acyclic ligands, derivatives or variations of DTPA (e.g., CHX-A″-DTPA ...), a large number of ligands have been developed so far. A wide choice of ligands is available for the design of new agents, and numerous journals have described and carefully classified all the structures that can be used in the design of a radiopharmaceutical, whatever the intended application [46,51,53,54,57,60].

BFCA's second role is to allow the conjugation of the complex with a biomolecule. The nature of this link is very important, because it is essential for it to be stable, and above all, for it to not interfere in any way with the binding to the receiver. The slightest structural modification of the ligand and/or of the biomolecule can have a very marked effect on the affinity to the targeted receptors. To minimize this impact as much as possible, that sometimes a 'spacer' or 'linker' can be used between these two entities. Biomolecules are often functionalized through a primary amine, which provides an ideal conjugation site for a coupling reaction, most often with peptide or thiourea type links. Other links based on thioether, triazole, oxime, or more recently via a copper-free click-chemistry with tetrazine/cyclooctyne may prove to be interesting, in particular, because they have very good stability in vivo [51,54,61].

Many somatostatin analogs have already been labeled with various radioelements, whether for imaging, with probes used today in clinical applications, or for therapy, with many compounds in clinical studies [17,61–63]. These analogs were obtained from modifications in the sequence of amino acids that make up the peptide. For example, replacing Phe^3 in octreotide (OC) with Tyr^3 (TOC) improves the affinity for SSTRs (in particular SSTR2) and introduction of a Thr (TATE) instead of Thr(ol) (TOC) further improves this. By following this procedure, many analogs have been developed and studied, often with the same chelating cavity to be able to compare their properties (Table 3) [64,65].

Table 3. Peptidic sequences of the main somatostatin agonist analogs. Differences towards Octreotide (OC) are highlighted in red.

Peptide	Peptidic Sequence
OC Octreotide	D-Phe^1-cyclo(Cys^2-Phe^3-D-Trp^4-Lys^5-Thr^6-Cys^7)$Thr(ol)^8$
LAN Lanreotide	β-D-Nal^1-cyclo(Cys^2-Tyr^3-D-Trp^4-Lys^5-Val^6-Cys^7)Thr^8-NH_2
VAP Vapreotide	D-Phe^1-cyclo(Cys^2-Phe^3-D-Trp^4-Lys^5-Val^6-Cys^7)Trp^8-NH_2
TOC [Tyr^3]-Octreotide	D-Phe^1-cyclo(Cys^2-Tyr^3-D-Trp^4-Lys^5-Thr^6-Cys^7)$Thr(ol)^8$
TATE [Tyr^3]-Octreotate	D-Phe^1-cyclo(Cys^2-Tyr^3-D-Trp^4-Lys^5-Thr^6-Cys^7)Thr^8
NOC [1-Nal^3]-Octreotide	D-Phe^1-cyclo(Cys^2-1-Nal^3-D-Trp^4-Lys^5-Thr^6-Cys^7)$Thr(ol)^8$
NOC-ATE [1-Nal^3, Thr^8]-Octreotide	D-Phe^1-cyclo(Cys^2-1-Nal^3-D-Trp^4-Lys^5-Thr^6-Cys^7)Thr^8
BOC [$BzThi^3$]-Octreotide	D-Phe^1-cyclo(Cys^2-$BzThi^3$-D-Trp^4-Lys^5-Thr^6-Cys^7)$Thr(ol)^8$
BOC-ATE [$BzThi^3$, Thr^8]-Octreotide	D-Phe^1-cyclo(Cys^2-$BzThi^3$-D-Trp^4-Lys^5-Thr^6-Cys^7)Thr^8

3.1. Radiolabeled Somatostatin Analogs for Imaging

The very first proof of concept for the visualization of tumors expressing SSTRs was carried out with [^{123}I-Tyr^{3}]-octreotide, obtained from an iodination reaction (electrophilic substitution) of tyrosine [66,67]. This compound demonstrated biological activity and an affinity for receptors similar to those of native SST [68]. Despite the obvious interest of this probe, several factors such as the difficult radiolabeling procedure, the significant cost, and particularly, the clearance via the liver and the hepatobiliary system (which makes it difficult to interpret the obtained images) were the main drawbacks of its application [67]. To overcome all of these disadvantages, iodine-123 has been replaced with indium-111, which, through the chelating agent DTPA, has been coupled to octreotide (Figure 6) [69]. In vivo studies of [^{111}In-DTPA^{0}]-octreotide ([^{111}In]-pentetreotide) have shown that it is possible to visualize tumors expressing SSTRs and their metastases, even 24 h after injection. In comparison with the compounds coupled to antibodies, this reveals a relatively rapid clearance via the kidneys, which represents a huge advantage compared to [^{123}I-Tyr^{3}]-octreotide [70,71]. This compound was the first radiopharmaceutical targeting SSTRs to be approved by the FDA (Octreoscan® marketed in 1994). It has been widely used, and has long been considered a 'gold standard' for the visualization of neuroendocrine tumors. It still has a few limits: in fact, it requires a high tumor/noise intensity ratio, shows low spatial resolution, has a moderate affinity for receptors and finally, and possesses a high γ energy which results in a high dose of radioactivity received by the patient. For all these reasons, research in the field of radiopharmaceuticals has focused on other radioelements such as technetium-99m for SPECT and gallium-68 for PET. In addition to having excellent physical properties, these two elements are available from a commercial clinical-grade generator, an important advantage for clinical applications.

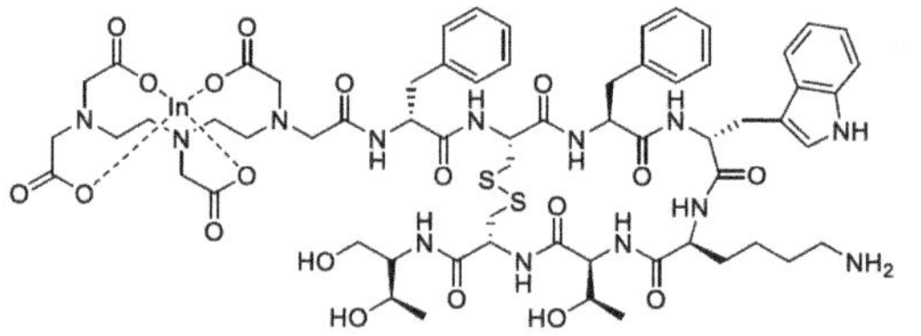

Figure 6. Structure of [^{111}In]-pentetreotide (Octreoscan®).

3.1.1. Gallium-68 and Indium-111

DOTATOC analog was the first to be radiolabeled with indium-111, and its comparative study with Octreoscan® showed similar diagnostic accuracy, but with better biodistribution and clearance [72]. Although DOTATATE alone showed better affinity for SSTRs, the two analogs [^{111}In]-DOTATOC and [^{111}In]-DOTATATE showed relatively similar pharmacokinetic properties [73]. SSTR2 receptors—and to a lesser extent, SSTR5—are most often overexpressed in tumors. Consequently, the majority of the radiotracers described have a strong affinity for these two SSTRs subtypes. Systems such as DOTANOC were designed to develop a probe capable of targeting all subtypes. Compared to DOTATOC and DOTATATE, it has a similar affinity for SSTR2 and SSTR5 subtypes, but a much higher affinity towards SSTR3. Their high internalization rate results in interesting biodistribution data, with a greater accumulation of the probe in the tumor and in the organs or tissues expressing SSTRs (e.g., pancreas and adrenal glands), ending with excretion mainly by kidneys [74].

These three systems, similarly labeled with gallium-68 (Figure 7), have proven to be very good radiotracers, and are currently routinely used in clinical applications [75]. These three radiopharmaceuticals have slightly different pharmacokinetic properties, but their diversity is mainly due to the variation in affinity for certain subtypes. This feature is even more marked depending on the radioelement chosen (^{68}Ga or ^{111}In). This can be explained by the differences in the geometry of the complexes. [^{68}Ga]-DOTATOC is very affine for SSTR2 and more moderate for SSTR5, [^{68}Ga]-DOTATATE is specific to SSTR2 and finally, [^{68}Ga]-DOTANOC binds with great affinity to SSTR2, SSTR3, and SSTR5 [76–78].

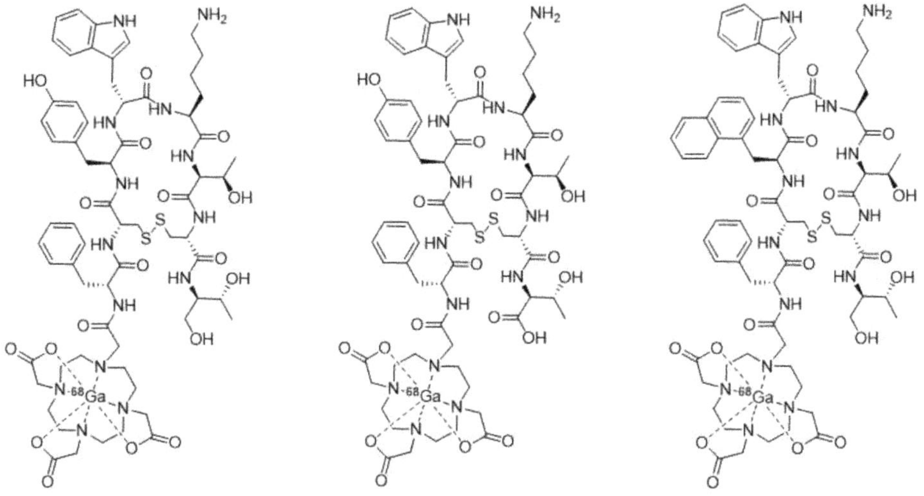

Figure 7. Structures of the three main systems radiolabeled with gallium-68.

A study with DOTANOC aimed at determining the impact of the introduction of a spacer on the pharmacokinetic properties of the formed radiotracer. The aim was to insert polyethyleneglycol (PEG) moieties or sugars between the chelating cavity (DOTA) and the biomolecule (NOC), which resulted in the modification of the lipophilicity or the charge of the final compound. As a result, the hydrophilicity of the system seems to be involved only in the affinity phenomenon towards the receptor, and the overall charge of the compound influences the excretion profile [79].

DOTA is not the only macrocycle to have been coupled to somatostatin analogs. Knowing the attraction of Ga (III) for NOTA, the latter has been the subject of comparative studies. Conjugated with octreotide (NODAGATOC), the compound showed a strong affinity for SSTR2 (similar to that of DOTATOC). Once marked with ^{111}In, affinity was even stronger for SSTR2, with even a gain on SSTR3 and SSTR5 (compared to ^{68}Ga-NODAGATOC), which confirms the influence that the geometry of the complex can have on affinity. In terms of stability, as expected, that of [^{68}Ga]-NODAGATOC was higher than that of [^{111}In]-NODAGATOC. The biodistribution of [^{68}Ga]-NODAGATOC was similar to that of [^{68}Ga]-DOTATOC, but showed a better accumulation in the tumor than [^{111}In]-DOTATOC. This is probably due to the strong agonist character, and the high rate of internalization of the NODAGATOC derivative [80].

A large variety of derivatives have also been investigated, such as DOTALAN, DOTABOC, DOTAGA [81], DOTANOCATE or DOTABOCATE (all derivatives of DOTANOC) [82,83], or THP-TATE (comparison of the overall behavior of the tris chelating system (hydroxypyridinone) with DOTATATE) [84]. New generation analogs with broader affinity profiles or pan-somatostatin analogs have been developed. For instance, AM3 (DOTA-Tyr-cyclo(DAB-Arg-cyclo(Cys-Phe-D-Trp-Lys-Thr-Cys))), a bicyclic somatostatin analog demonstrated affinity to SSTR2, 3, and 5, when labeled with ^{68}Ga. It showed a fast background clearance coupled with a high tumor/non-tumor ratio. [85] KE108 was coupled with DOTA and labeled with ^{111}In and ^{68}Ga, giving [^{111}In/^{68}Ga]-KE88 (DOTA-D-Dab-Arg-Phe-Phe-D-Trp-Lys-Thr-Phe), which bound to all five SSTRs with high affinity. [86] However, in an in vitro study, it had a low SSTR2 uptake, but was very effective for SSTR3-expressing tumors. More recently, a Pasireotide derivative, DOTA-PA1

(DOTA-cyclo-[HyPro-Phe-D-Trp-Lys-Tyr(Bzl)-Phe]) was labeled with ^{68}Ga and was investigated in three human lung cancer models, where it demonstrated superiority compared to [^{68}Ga]-DOTATATE [87]. In parallel, the group from Demokritos Institute, in Athens, developed pansomatostatin radiopeptides based on native somatostatin (SRIF-14 and SRIF-28). Both were derivatized with DOTA chelator and labeled with ^{111}In. Subsequent radiotracers exhibited high affinity and internalization profiles. SRIF-14 derivatives unfortunately demonstrated low in vivo stability. [^{111}In]-DOTA-LTT-SS28, on the contrary, demonstrated a much higher stability and showed more promise [88,89].

3.1.2. Technetium-99m

A wide range of chelating agents have been used to prepare somatostatin analogs labeled with technetium-99m: peptide moieties [90,91], propyleneaminooxime [92], tetraamines [93,94] or a cyclopentadienyl group [95]. Macrocyclic ligands have also been investigated [96]. Three systems stand out for the radiolabeling of somatostatin analogs: HYNIC-TOC and Demotate scaffolds, and P829 (Figure 8).

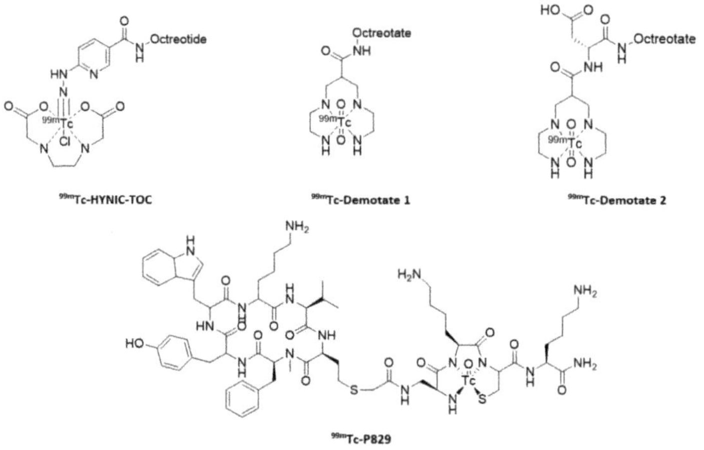

Figure 8. [^{99m}Tc]-labeled somatostatin analogs.

Initially, the HYNIC core (hydrazinonicotinamide) was designed for the radiolabeling of antibodies and proteins with technetium-99m [97], then this was transposed to peptides and more specifically to octreotide. This ligand can complex the metal in a monodentate or bidentate way, therefore, it is necessary to use one or more co-ligands to complete the coordination of the [^{99m}Tc]-HYNIC core. Among the most commonly used co-ligands are tricin, nicotinic acid, or EDDA (ethylenediaminodiacetic acid). Each co-ligand has its own influences on the properties of the complex obtained (e.g., lipophilicity and biodistribution) [98]. The first studies were carried out using tricin as a co-ligand ([^{99m}Tc]-HYNIC-TOC), but quickly EDDA demonstrated a very favorable influence on the pharmacokinetics of the radiotracers [99]. Compared to Octreoscan®, [^{99m}Tc]-EDDA/HYNIC-TOC showed better accumulation in the tumor and a weaker accumulation in the kidneys. The improved spatial resolution, the reduction in the radiation dose received by the patient and the better availability of ^{99m}Tc made it a possible alternative to Octreoscan® [99,100]. Finally, its conjugation with the octreotate analog ([^{99m}Tc]-EDDA/HYNIC-TATE) has shown significantly similar behavior to its octreotide counterpart [101]. [^{99m}Tc]-EDDA/HYNIC-TOC (Tektrotyd®) was granted marketing authorization in Europe in adult patients with gastro-enteropancreatic neuroendocrine tumors (GEP-NET) for localizing primary tumors and their metastases.

The second radiotracer, based on the tetraamine motif 6-R-1,4,8,11-tetraazaundecane, is available in a series with [^{99m}Tc]-Demotate 1 ([^{99m}Tc-$N_4{}^0$, Tyr^3]-octreotate) and ^{99m}Tc-Demotate 2 ([^{99m}Tc-$N_4{}^{0-1}$,

Asp^0, Tyr^3]-octreotate). The first version of this probe demonstrated excellent pharmacokinetic properties, including faster accumulation in the tumor compared to Octreoscan® [102]. The objective of the second version was to improve the qualities of [^{99m}Tc]-Demotate 1, by modifying the overall charge of the complex and adding an Asp residue. In the end, [^{99m}Tc]-Demotate 2 showed overall behavior similar to [^{111}In]-DOTATATE, even if the latter has a faster clearance and a better retention time in the tumor [103]. The last of the main analogs based on technetium-99m is [^{99m}Tc]-P829 (^{99m}Tc-Depreotide), marketed in 2000 by the company CISBio International under the name of NeoSpect®, but recently withdrawn from the market. The P829 peptide (directly radiolabeled with ^{99m}Tc) showed results similar to the other SST analogs [104]. Its use for the detection of neuroendocrine tumors appeared to be less precise than with Octreoscan® [105]. On the other hand, its affinity for SSTR3, subtype which may be the origin of cross-competition from other types of receptors (notably VIP receptors), gave it the ability to bind to a larger number of primary tumors [104]. In particular, it was used clinically for the diagnosis of malignant lung tumors [106–108], for which it got its market authorization [109], and also demonstrated some interest in breast cancer, but it was never confirmed in a larger series of patients [110].

The question that now remains to be answered is that of the clinical interest of a SPECT tracer among the wide choice of PET SSTRs imaging agents [111,112].

3.1.3. Copper-64

Due to the short half-life of ^{68}Ga ($T_{1/2}$ = 67.7 min.) each center willing to perform ^{68}Ga PET imaging must purchase a currently expensive ^{68}Ge/^{68}Ga generator and a specifically shielded hot-cell. For this reason and despite the FDA and EMA market authorizations for [^{68}Ga]-DOTATATE and [^{68}Ga]-DOTATOC and the better diagnostic performances for these two radiopharmaceuticals products, the use of ^{68}Ga appears to be under the dependence of an economic choice for many hospitals and only a few large centers are making the financial investment to perform ^{68}Ga-radiolabeling. In this context, the use of a PET-emitter with a longer half-life such as copper-64 ($T_{1/2}$ = 12.7 h) appears to be an interesting alternative to remove the financial hindrance of gallium-68 [113]. This physical parameter allows for a centralized radiolabeling site with a large multicentric supply of ready-to-use ^{64}Cu-radiolabeled compounds. The chemistry of copper is also well known, which is a real asset in the design of new radiotracers. Many systems already presented before, such as DOTATOC/TATE or NODAGATOC/TATE, or others more copper-specific BFCAs, such as TETA (1,4,8,11-tetraazacyclotetradecane-*N*,*N*′,*N*′′,*N*′′′-tetraacetic acid) [114], and its more stable derivatives such as cross-bridge CB-TE2A (4,11-bis(carboxymethyl)-1,4,8,11-tetraazabicyclo [6.6.2]hexadecane) [115], and CPTA (4-[(1,4,8,11-tetraazacyclotetradec-1-yl)methyl]benzoic acid]) [116] or sarcophagine derivatives [117] have been studied. A review on the development of copper radiolabeled somatostatin analogs was recently published by Marciniak et al. [118].

To validate the clinical interest of [^{64}Cu]-somatostatin analogs, various clinical studies have been conducted around the world. Among the different somatostatin analogs, [^{64}Cu]-DOTATATE was one of the first used. In 2015, [^{64}Cu]-DOTATATE was compared head-to-head to [^{111}In]-DTPA-octreotide in 112 patients and showed that the PET ^{64}Cu-compound was far superior to SPECT ^{111}In compound performances [119]. In 2017, [^{64}Cu]-DOTATATE was challenged to [^{68}Ga]-DOTATOC according to an identical PET/CT imaging modality [120]. The results of this study, where 59 patients were injected with [^{68}Ga]-DOTATOC followed by an injection of [^{64}Cu]-DOTATATE one week later, concluded that the two radiopharmaceuticals had the same sensitivity. Nevertheless, in this cohort of neuroendocrine tumors, [^{64}Cu]-DOTATATE had a substantially better lesion detection rate. The patient follow-up revealed that these additional lesions detected by [^{64}Cu]-DOTATATE were true positives. To evaluate the benefits of this better detection of lesions with [^{64}Cu]-DOTATATE than with [^{68}Ga]-DOTATOC, the correlation between PET image [^{64}Cu]-DOTATATE uptake (expressed in maximal standardized uptake value - SUV_{max}) and overall (OS)/progression free survival (PFS) was studied during 24 months after [^{64}Cu]-DOTATATE PET/CT acquisition. The conclusion of this study claimed a good correlation/prognostic between SUV_{max} and PFS but not with OS [121]. The major drawback of these preliminary human studies consist of the affinities differences for the five SSTRs subtypes between DOTATOC and DOTATATE compounds. To circumvent these discrepancies, an in vitro study in a mouse model was conducted and compared [^{64}Cu]-DOTATATE to [^{68}Ga]-DOTATATE. The results showed a similar pharmacokinetic and absolute uptake between both compounds 1 h post-injection [122]. In Europe, where the PET radiopharmaceutical approved is [^{68}Ga]-DOTATOC, it could be interesting to perform some PET imaging with [^{64}Cu]-DOTATOC to compare the performance of the two tracers. A first-in-human retrospective study was recently conducted and seems to present same results than [^{64}Cu]-DOTATATE with high detection rate of suspected lesion associated to a high target-to-background contrast [123]. A recent first-in-human study also demonstrated potential interest for [^{64}Cu]-SARTATE analog [124].

In conclusion, despite a higher dosimetric impact for copper-64 (only 17.6% of radioactive decay lead to positron emission), copper-64 somatostatin analogs appear to be an advantageous alternative to gallium-68 radiopharmaceuticals. Compared to ^{68}Ga, in addition to economic advantages, ^{64}Cu has a lower positron range which leads to a better PET intrinsic resolution and a higher half-life which allows for a more flexible scanning window. The better patient care management and outcomes remain to be proven and the work is in progress to establish these points [121,125]. In parallel, at present, a radiopharmaceutical industrial company submitted a market authorization from FDA for [^{64}Cu]-DOTATATE and thus confirms the interest of copper-64 in SSTRs imaging.

3.1.4. Other Radiometals

Other radionuclides have also been investigated for SSTRs imaging. Cobalt-55 seems to be a possible alternative to gallium-68 and copper-64 compounds, with similar behavior and lifespan (17.5 h vs. 12.7 h) to the latter, but with a higher positron yield (75.9% vs. 17.6%). Preliminary complexation tests of DOTATOC with the isotope ^{57}Co as a surrogate for ^{55}Co showed a higher affinity for SSTR2 than [^{68}Ga]-DOTATOC, implying a rate of internalization among the highest of all derivatives of SST and thus, a strong accumulation in targeted tissues. Despite similar structures, the analogs of cobalt and gallium have different biological behaviors. This confirms the fact that the physical characteristics of radioactive elements influence the affinity, biodistribution, and pharmacokinetics of radiolabeled peptides [126]. The properties of cobalt-based compounds have been further investigated with the comprehensive evaluation of other octreotide analogs such as DOTANOC and DOTATATE [127]. Furthermore, [^{55}Co]-DOTATATE compared favorably with [^{68}Ga]-DOTATATE and [^{64}Cu]-DOTATATE in an animal model [122]. Associated with the Auger-emitting ^{58m}Co, it could represent a potentially interesting theranostic pair [128].

Scandium and terbium are two metals that recently emerged as possibly useful for theranostic applications, as both possess imaging and therapeutic radionuclides [129]. DOTATOC was radiolabeled with scandium-44 ($T_{1/2}$ = 3.97 h, $E_{\beta+}$ = 632 keV) [130] and terbium-152 ($T_{1/2}$ = 17.5 h, $E_{\beta+}$ = 1140 keV) [131] and rapidly injected in patients in proof-of-concept studies [132,133]. No adverse effects were observed during follow-up periods and images proved suitable for diagnosis. With DOTATATE, it seems the affinity to SSTR2 receptors is lower with scandium than with gallium, thus limiting its interest [134]. In a study comparing the labeling and stability of DOTANOC and NODAGANOC with ^{44}Sc and ^{68}Ga, it was observed that [^{44}Sc]-NODAGANOC labeling was more challenging and less stable than [^{44}Sc]-DOTANOC [135]. The opposite was observed with ^{68}Ga. Recently, a new chelator was proposed, AAZTA (1,4-bis (carboxymethyl)-6-[bis (carboxymethyl)]amino-6-methylperhydro-1,4-diazepine), which enables fast and easy labeling at room temperature. AAZTA-TOC labeled with ^{44}Sc demonstrated high in vitro stability [136]. Affinity tests are now necessary to assess its potential utility. DOTATATE has also been labeled with ^{155}Tb ($T_{1/2}$ = 5.32 days, E_{γ} = 87 keV (32%), 105 keV (25%)) for SPECT imaging [137]. Though a potentially promising radionuclide for theranostic applications, availability of ^{155}Tb is currently the main limitation for further development.

At the turn of the millennium, yttrium-86 ($T_{1/2}$ = 14.74 h, 32% β^+) was thought to be a potential radionuclide of interest, particularly for pretherapeutic dosimetry of ^{90}Y-radiotracers, and notably ^{90}Y-labeled somatostatin analogs [138]. Thus, several octreotide analogs were developed [139,140]. [^{86}Y]-DOTATOC even reached the clinics [139,141]; however, ^{86}Y properties are less than optimal, and availability is limited, so interest soon faded out.

3.1.5. Fluorine-18

Radiometals' production is currently still limited, even for the most advanced ones [142–144]. Fluorine-18, on the contrary, can be mass-produced and distributed daily, thanks to a worldwide network of cyclotrons. Because of this availability, and favorable decay characteristics ($T_{1/2}$ = 110 min, 97% β^+), it thus should be noted that some radiotracers based on fluorine-18 have been described (Figure 9) [145]. The first generations such as 2-[^{18}F]fluoropropionyl-D-Phe^1-octreotide [146] or 4-[^{18}F]fluorobenzoyl-D-Phe^1-octreotide [147] generally showed unfavorable biokinetic properties (low accumulation and low retention in the tumor). The probes developed subsequently contained hydrophilic or charged moieties to reduce the lipophilicity of the radiotracer. In particular, several carbohydrate derivatives of octreotide/octreotate have been developed [148,149]. A disadvantage of fluorine-labeling compared to radiometal labeling is the use of generally long and tedious multi-step procedures. To circumvent this, innovative strategies, enabling fast and purification-less labeling, have been developed, such as the formation of ^{18}F-boron or ^{18}F-silicon bonds, or the use of click-chemistry [150–152]. Another elegant method to label somatostatin analogs is the use of [^{18}F]-aluminum fluoride with radiotracers previously developed for radiometals, such as NOTATOC [153]. These new generation analogs demonstrated general properties (affinity for the targeted receptors, metabolic stability, biodistribution and clearance) which are much more interesting, and some of them have been investigated in patients, where they gave results comparable to [^{68}Ga]-DOTATOC [154,155]. In addition, [^{18}F]F-FET-βAG-TOCA and [^{18}F]-IMP466 ([Al^{18}F]-NOTATOC) are currently being evaluated in phase I clinical trials (EudraCT number 2013-003152-20 and NCT03511768, respectively). Recently published results with [^{18}F]-IMP466 demonstrated it was safe and well-tolerated, with a physiologic uptake pattern similar to [^{68}Ga]-DOTATATE [156]. Besides cost and availability, another advantage of fluorine-18 is its shorter positron range compared with gallium-68, leading to an improved spatial resolution, and thus, better quantification of uptake [157].

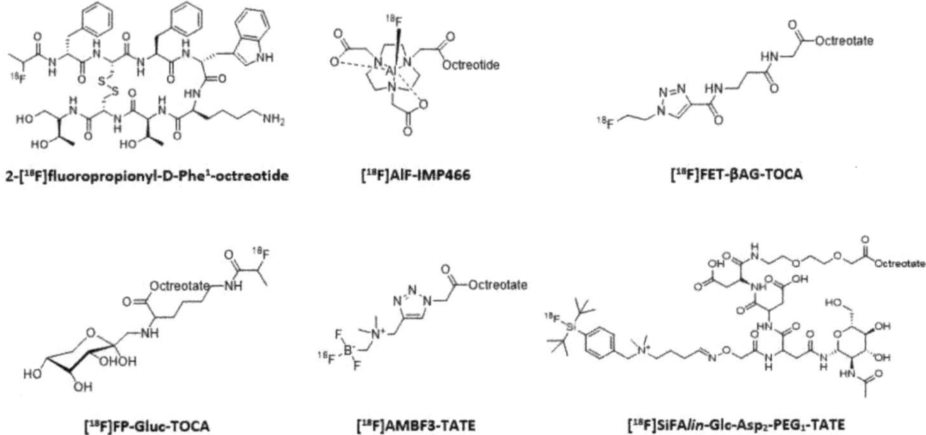

Figure 9. [^{18}F]-labeled somatostatin analogs.

3.2. Radiolabeled Somatostatin Analogs for Therapy

Concerning radionuclide therapy and more particularly peptide receptor radionuclide therapy (PRRT), radioactivity is used to destroy the targeted cells. Radiopharmaceuticals used in therapy are designed in the same way as those used in imaging, only the nature of the radioelement being modified. Contrary to imaging, which uses radioelements having very penetrating but little ionizing radiations, PRRT privileges the use of radionuclides that have little penetrating and more energetic and thus more ionizing radiations. Brought directly to the cancer cell, the radiation emitted by the radioactive decay causes irreversible ionization of the cell's DNA, which induces its apoptosis. The main isotopes used today are iodine-131, yttrium-90, lutetium-177 and, to a lesser extent, rhenium-188 [158]. As mentioned earlier, the purpose of the DOTA-SSA design was to work with a chelating cavity capable of complexing radioelements for imaging or therapy. Consequently, most of the platforms discussed above have been transposed for therapeutic application via the use of β^- emitters [64,74,81,82].

3.2.1. Yttrium-90 and Lutetium-177

Yttrium-90, a pure high energy β^- emitter ($T_{1/2}$ = 64 h, $E_{\beta max}$ = 2.28 MeV), and lutetium-177, a medium energy β^- emitter ($T_{1/2}$ = 6.7 d, $E_{\beta max}$ = 0.5 MeV) with a γ component (208 keV), are currently the most used in PRRT. Each of these two elements has its own advantages for targeted therapy. The particles emitted by ^{90}Y are more energetic and more penetrating; they are able to diffuse on a thicker layer of cells, which is an advantage for the treatment of large tumors. However, even if high energy radiation allows a more uniform irradiation of the tumor, the risk of imposing an excessive dose of radiation on the adjacent tissues is very present. For its part, the ^{177}Lu emits less energetic radiation, more suited to small tumors. In addition, the energy of its γ radiation is sufficient to allow detection by scintigraphy and establish dosimetry during the therapy sequences [159].

The first analog to be studied was [^{90}Y]-DOTATOC (Octreother®), and the first treatment sessions quickly showed good results, stopping the progression of the tumor [72,160,161]. Many studies on this long-used treatment have made it possible to observe a good tolerance for this radiotracer, with fairly mild side effects (fatigue) and in very rare cases a little more severe ones (nausea). However, it also showed some toxicity for the kidneys and the bones, these two aspects being the dose-limiting factors for the patient. In vitro, a greater affinity for SSTR2 has been demonstrated for [^{90}Y]-DOTATATE compared to [^{90}Y]-DOTATOC [64]. However, for the diagnosis in humans, a better contrast between the kidneys and the tumor was found for [^{111}In]-DOTATOC compared to [^{111}In]-DOTATATE [73], which may explain the wider use of DOTATOC analog. Despite this, these two analogs have relatively similar properties and have proven to be effective treatment methods that improve survival in some patients with neuroendocrine tumors (approximately 50 months vs. 18 months without treatment) [162]. In a Phase IIA study with [^{90}Y]-DOTALAN (MAURITIUS trial), this one demonstrated lower tumor uptake in neuroendocrine tumors compared to ^{90}Y-DOTATOC, but could be of potential interest for other tumors, such as HCC or lung cancers [163]. With the perspective of several years of clinical use, PRRT with ^{90}Y-labeled somatostatin analogs appears to be well-tolerated with favorable long-term outcome. Unfortunately, Phase III studies are still lacking [164,165].

The same analogs have also been radiolabeled with lutetium-177. Initially, [^{177}Lu]-DOTATOC was used in cases of relapse of neuroendocrine tumors after treatment with [^{90}Y]-DOTATOC. Despite satisfactory results [166], its subsequently developed analog [^{177}Lu]-DOTATATE has shown more promise, mainly due to a more significant retention time in the tumor. For this reason, octreotate analog (TATE) is being preferred to octreotide (TOC) for labeling with lutetium [164,167]. It is also important to note that, unlike ^{90}Y, no cases of nephrotoxicity after treatment with ^{177}Lu have been reported. In 2005, the possibility of combining these two β^- emitters for therapy in cases where tumors of variable sizes are detected, was demonstrated [168]. From there, different treatment combinations between the four main systems ([^{90}Y]-DOTATOC, [^{90}Y]-DOTATATE, [^{177}Lu]-DOTATOC, and [^{177}Lu]-DOTATATE) have proven to be interesting and sometimes even more effective than using a single treatment modality [169,170]. Similarly, combination treatments with non-labeled somatostatin analogs, chemotherapy, targeted therapy, and/or radiosensitizers might further improve the efficacy and/or tolerability [171,172]. [^{177}Lu]-DOTATATE has been investigated in a phase III trial, in well-differentiated, unresectable or metastatic, progressive midgut neuroendocrine tumors (Netter 1 trial). Treatment with [^{177}Lu]-DOTATATE resulted in a significant tumor response rate of 18% compared with 3% in the high-dose octreotide LAR group, coupled with a 79% risk reduction for disease progression or death [173]. Following these positive findings, [^{177}Lu]-DOTATATE was granted marketing authorization in this indication, both in Europe and in the US (Lutathera®) [174]. Coupled with ^{68}Ga-imaging (Figure 10), it represents a powerful theranostic tool for the management of neuroendocrine tumors (NETs) [175]. Current research with [^{177}Lu]-DOTATATE aims to improve the safety and efficacy of this procedure, enlarge possible indication, notably in advanced, poorly-differentiated, GEP-NETs, [176,177] or other NETs, such as pheomochromocytoma or paraganglioma [178,179].

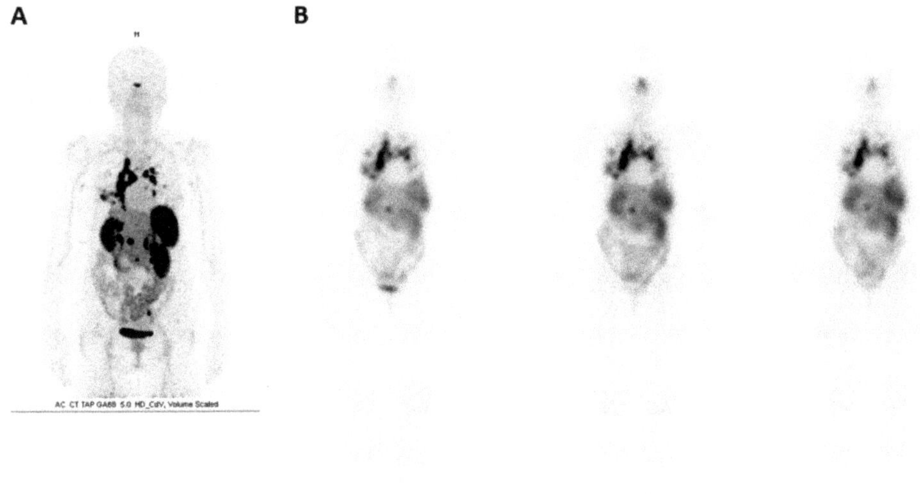

Figure 10. (**A**) [^{68}Ga]-DOTATOC (Somakit®) and (**B**) [^{177}Lu]-DOTATATE (Lutathera®, cures 1, 2, and 3) imaging of a patient treated for progressive metastatic midgut NET (images courtesy of Centre Eugene Marquis, Rennes, France).

3.2.2. Rhenium-188 and Other β-Emitting Radionuclides

Despite equally interesting characteristics, rhenium-188 remains widely less used than ^{90}Y and ^{177}Lu [180]. This is mainly due to more difficult chemistry and the unavailability of a pharmaceutical-grade ^{188}W/^{188}Re generator, as compared to the other two. Vapreotide and Lanreotide analogs have been described in the literature with ^{188}Re. They have been investigated in experimental cancer models (e.g., pancreas, colorectal, lungs and cervical) to reduce tumor growth [181–184]. [^{188}Re]-Lanreotide notably demonstrated favorable pharmacokinetics and distribution profiles (tumor-to-liver ratio) in HCC-bearing rats compared to healthy ones [185]. Another example is an equivalent to Depreotide (P829). After the development of ^{99m}Tc-Depreotide for imaging, the idea was to label this compound with ^{188}Re, to assess its potential in vivo. Although the radiolabeling proceeded successfully, the study showed unacceptable toxicity to non-target organs. To improve its properties, structural modifications of the peptide sequences close to the chelating moiety were tested. This optimization led to P2045, which showed better accumulation in the tumor, weaker retention in the kidneys, and faster urinary excretion than [^{99m}Tc]-depreotide [186]. This new rhenium-based analog of depreotide, [^{188}Re]-P2045 (Figure 11), went up to phase I in therapy for small cell lung cancer [187] and has shown promising in vivo results in the treatment of pancreatic tumors in mice [188]. To the best of our knowledge, no HYNIC-TOC/TATE or demotate derivatives have yet been radiolabeled with rhenium. Recent research with rhenium isotopes has been focusing on tricarbonyl core derivatives for the labeling of NOTA-SSAs [96].

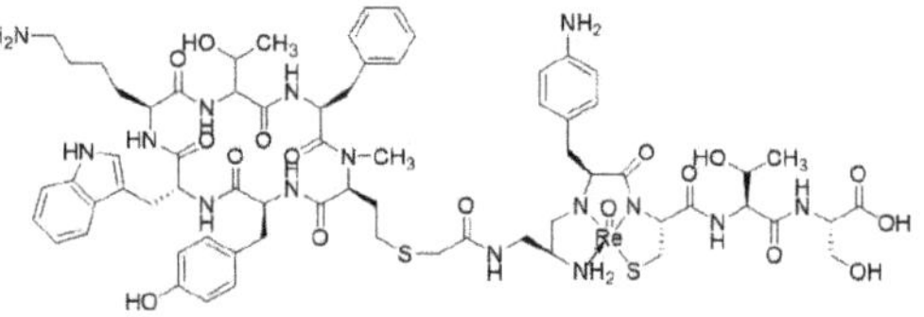

Figure 11. [^{188}Re]-P2045.

In a theranostic perspective, other β-emitting nuclides could have a potential interest—such as ^{47}Sc ($T_{1/2}$ = 3.35 d, $E_{\beta max}$ = 600.8 keV), ^{67}Cu ($T_{1/2}$ = 2.58 d, $E_{\beta max}$ = 577 keV), and ^{161}Tb ($T_{1/2}$ = 6.91 d, $E_{\beta max}$ = 593 keV)—to be coupled with ^{44}Sc, ^{64}Cu, and ^{152}Tb/^{155}Tb respectively [129,158,189]. To date, no ^{67}Cu-labeled somatostatin analogs have been described so far, and only very preliminary studies have been described with [^{161}Tb]-DTPA-Octreotide and [^{47}Sc]-DOTATOC [190,191].

3.2.3. Alpha and Auger Emitters

Recently, alpha emitters have attracted particular attention for radionuclide therapy. Long confined to hematological tumors, they are now being considered for the potential treatment of solid tumors [192]. In vitro, α-labeled somatostatin analogs (DOTATOC and DOTATATE) demonstrated a significantly higher killing effect compared to ^{177}Lu [193–195]. [^{213}Bi]- and [^{225}Ac]-labeled DOTATOC (^{213}Bi: $T_{1/2}$ = 45.6 min, E_α = 5.88 MeV; ^{225}Ac: $T_{1/2}$ = 9.92 d, E_α = 5.83 MeV) have demonstrated promising therapeutic effects in pre-clinical animal studies [196,197]; whereas [^{213}Bi]-DOTATATE, investigated in human small cell lung carcinoma and rat pancreatic tumor models, demonstrated a great therapeutic effect in both small (50 mm^3) and large (200 mm^3) tumors, but with a higher probability for stable disease in small tumors [198]. First, and, to date, the only clinical experience with [^{213}Bi]-DOTATOC, was published by Kratochwil et al., and included seven patients with advanced NETs with liver metastases refractory to treatment with [^{90}Y]-DOTATOC or [^{177}Lu]-DOTATOC [199]. It demonstrated specific tumor binding, lower toxicity than with β-irradiation and partial remission of metastases. Two years after intra-arterial injection of [^{213}Bi]-DOTATOC, all seven patients were still alive. Regarding ^{225}Ac, a first-in-human study included 10 patients with progressive NETs after β-PRRT. As with ^{213}Bi, [^{225}Ac]-DOTATOC was well tolerated and effective [200]. A recent study with [^{225}Ac]-DOTATATE confirmed the potential of these radiotracers as an additional, and valuable, treatment option for patients who are refractory to [^{177}Lu]-DOTATATE therapy. 32 patients with previous [^{177}Lu]-DOTATATE therapy were treated with [^{225}Ac]-DOTATATE (100 kBq/kg body weight). The response was assessed in 24 patients, with 9 stabilized diseases and 15 partial remissions [201].

Though not stricto sensu an α-emitter, lead-212 ($T_{1/2}$ = 10.6 h) eventually decays to stable 2^{08}Pb through a cascade chain with two α-emissions of potential therapeutic interest. A somatostatin analog, DOTAMTATE (Figure 12), has been labeled with ^{212}Pb and investigated in a murine model of neuroendocrine tumor. Results showed a promising safety index with a 3.2-fold increase in median survival and one-third of the animals being tumor-free. A combination with 5-FU (Fluorouracyl) was able to durably cure approximately 80% of the animals. [202] Given these promising outcomes, a Phase I dose-escalation clinical trial has recently been started with [^{212}Pb]-DOTAMTATE (AlphaMedix™) including 50 patients with unresectable or metastatic neuroendocrine tumors (NCT03466216). Preliminary results (nine patients enrolled) demonstrated a favorable safety profile at the tested doses [203].

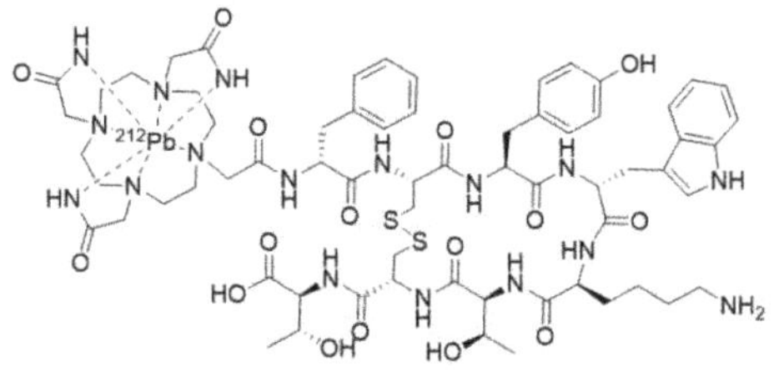

Figure 12. [^{212}Pb]-DOTAMTATE.

Cyclotron-produced astatine-211 ($T_{1/2}$ = 7.2 h, E_{α} = 5.87 MeV) is another very promising α-emitting radionuclide. Astatine is the heaviest halogen with a behavior somehow similar to iodine, but, in certain circumstances, it also displays significant metallic characteristics [204]. Direct astatination of somatostatin analogs is feasible, through tyrosine residues, but it led to poor stability of the resulting analogs, therefore different prosthetic groups have been developed [205–207]. Although *N*-(3-[^{211}At]astato-4-guanidinomethylbenzoyl)-Phe^{1}-octreotate ([^{211}At]-AGMBO) and N^{α}-(1-deoxy-D-fructosyl)-N^{ε}-(3-[^{211}At]astatobenzoyl)-Lys^{0}-octreotate ([^{211}At]-GABLO) showed disappointing biodistribution results, with poor tumor uptake, [^{211}At]-SPC-octreotide displayed a more favorable biodistribution profile, and a dose-dependent apoptosis in an NSCLC murine model.

Auger electron emitters are also very potent for specific tumor cell killing, sparing surrounding cells, with a highly localized energy deposition. Indium-111 emits Auger electrons (E_{Ae-} = 19 keV, 16%), and, as such, has been investigated for therapy. Several clinical trials have been undertaken with high doses of [^{111}In]-Pentetreotide. A first study with 20 patients that had neuroendocrine progressive tumors demonstrated stabilization of the disease in 5 patients, and tumor shrinkage in 5 others. All of them had received a cumulated dose higher than 20 GBq [208]. In a study with 50 SSTR-positive patients treated with cumulated doses from 20 to 160 GBq, of which 40 were evaluable, there was a stabilization in 14 patients, minor remission in 6 and partial remission in 1, with mild bone marrow toxicity [209]. However, half of the patients receiving more than 100 GBq developed a myelodysplastic syndrome or leukemia. A dose of 100 GBq was thus considered the maximal tolerated activity. Another study with 27 patients with GEP-NETs found that two doses of 6.6 GBq (180 mCi) were safe and well-tolerated, demonstrating a clinical benefit in 62% of patients [210]. Benefit of ^{111}In-Pentetreotide treatment was shown to last at least 6 months for 70% of patients, while only 31% of them still had sustained benefit after 18 months [211]. Efficacy in large tumors and end-stage patients is limited, mainly because of heterogeneous radiopharmaceutical uptake due to poor tumor vascularity and central necrosis [212]. This has been demonstrated by Capello et al. in a rat tumor model, with different sizes of tumors [213]. Effects were much more pronounced in small ($\leq$ 1 cm^2) tumors than in large ($\geq$8 cm^2). They also found a significant increase in tumor receptor density after tumor regrowth, indicating repeated injections would probably be more efficient than single-dose treatment. It could also be worth using PRRT with Auger emitters in an adjuvant setting after surgery, to destroy occult metastases. A final example is [^{58m}Co]-DOTATOC. This radiotracer presented for potential use in Auger-based therapy, particularly for disseminated tumor cells and micrometastases, appears to have more beneficial in vitro properties than those of [^{177}Lu]-DOTATATE, with a significantly more efficient cell killing effect per cumulated decay, which has to be confirmed in vivo [127].

4. Antagonists vs. Agonists

Pharmacomodulation around the synthetic somatostatin analogs has led to a change of chirality in the first amino-acid (from D to L form) and in cysteine number 2 (from L to D form). These modifications have given a new class of SSTR specific compounds with antagonist effects (Table 4). From a pharmacological point of view, the biological and molecular mechanisms responsible for their targeting effectiveness in vivo are completely different. After binding to an SST receptor, an agonist analog is internalized into the cell as a ligand-receptor complex. This internalization allows it to accumulate in the cell, and to increase the amount of radiation emitted. This very powerful and specific internalization mechanism enables efficient in vivo targeting of receptors. This phenomenon does not occur (or very little) for somatostatin antagonists, and they do not stimulate the G-protein coupled to the SSTR with an associate blockage of the agonist-induced activity. Surprisingly, it has been shown that targeting receptors can also be effective without internalization of the ligand-receptor complex, and some antagonist analogs can sometimes behave better than agonists (e.g., better accumulation in tumor, poor kidney retention, and rapid clearance) [214,215]. This high tumor uptake appears to be a consequence of a greater number of target binding sites for antagonists and a more slowly dissociation than for agonists, which allows for a longer accumulation of radiation [216,217]. The hypothesis of a ligand rebinding mechanism has been put forward, but this still requires some investigation before it can be validated. These first results were confirmed by preclinical studies and by preliminary clinical trials and seems to show superior results for antagonist-based tracers than agonists [218–221]. The first comparative study of antagonists with Octreoscan® confirmed the good characteristics of the [^{111}In]-DOTA-BASS analog, and better accumulation at the level of the tumor and better visualization of metastases. It was truly the first proof of the concept of antagonist SSTRs imaging [222].

Table 4. Main somatostatin antagonist analogs. Differences towards octreotide (OC) are highlighted in red.

Antagonist Peptide	*Peptidic Sequence*
Sst2-ANT (BASS)	p-NO_2-Phe^1-cyclo(D-Cys^2-Tyr^3-D-Trp^4-Lys^5-Thr^6-Cys^7)D-Tyr^8-NH_2
LM3	p-Cl-Phe^1-cyclo(D-Cys^2-Tyr^3-D-Aph^4(Cbm)-Lys^5-Thr^6-Cys^7)D-Tyr^8-NH_2
JR10	p-NO_2-Phe^1-cyclo(D-Cys^2-Tyr^3-D-Aph^4(Cbm)-Lys^5-Thr^6-Cys^7)D-Tyr^8-NH_2
JR11 (Satoreotide)	p-Cl-Phe^1-cyclo(D-Cys^2-Aph^3(Hor)-D-Aph^4(Cbm)-Lys^5-Thr^6-Cys^7)D-Tyr^8-NH_2

Concerning the affinity for each SSTR subtype, it turned out that the nature of the chelator and the radiometal is of great importance for the in vivo pharmacokinetic fate (mainly for the tumor uptake and retention time) [223]. Ultimately, copper-64 based radiotracers seem to be more interesting, especially when comparing their contrast ratio between the tumor and normal tissues which increases over time—a direct consequence of their higher half-life. The influence of radiometals (^{111}In, ^{90}Y, ^{177}Lu, ^{64}Cu, and ^{68}Ga) and chelates (DOTA and NODAGA) on three antagonist families (LM3, JR10, and JR11) were also studied. On the radiometric side, the overall affinity of [^{68}Ga]-DOTA was found to be much lower than for the other elements, which is the opposite of the results obtained with the agonists. For the chelate, the substitution of DOTA by NODAGA seems to greatly improve the affinity of the antagonist analogs. During this study, two particularly promising platforms emerged, DOTA-JR11 and NODAGA-JR11 [224]. Another example highlighting the influence of the chelate is 406-040-15 (cyclo (2–11) H-Cpa-DCys-Asn-Phe-Phe-DTrp-Lys-Thr-Phe-Thr-Cys-2NalNH_2), a pansomatostatin analog, with an SSTR3 antagonist behavior. Chelation to DOTA turned this analog to an agonist [225]. Note that the first antagonist labeled via a [^{99m}Tc]-tricarbonyl core has been described. ^{99m}TcL-sst2-ANT (with L = tridentate ligand type N, S, N) has shown very promising in vivo behavior, but requires some modifications to improve its pharmacokinetics [226].

As for imaging, antagonists are also an interesting alternative for therapy. As discussed above, the first proof of the feasibility of imaging using antagonists was highlighted by comparing Octreoscan® and [^{111}In]-DOTA-BASS. However, this analog has shown only a very modest affinity for the SSTR2 receptor subtype targeted in the therapy of neuroendocrine tumors [214]. To overcome this problem, the second generation of somatostatin antagonists was synthesized to improve affinity for this receptor. DOTA-JR11 showed the highest affinity for SSTR2 and was selected for use in targeted therapy [218]. A pilot study to assess the possibility of treatment with [^{177}Lu]-DOTA-JR11, by comparing it to [^{177}Lu]-DOTATATE, was carried out. This new antagonist has shown favorable properties, such as better accumulation in the tumor and a higher dose received by the tumor, thanks to a longer retention time [227]. Further developments led to a theranostic pair with JR11: one with a NODAGA chelator (satoreotide trizoxetan, OPS-202) and one with DOTA chelator (satoreotide tetraxetan, OPS-201) [228,229]. Satoreotide trizoxetan is currently radiolabeled with ^{68}Ga and used in PET imaging clinical trials (Figure 13) [230,231]. Satoreotide tetraxetan radiolabeled with ^{177}Lu has been evaluated in a therapeutic clinical trial [232]. First clinical results for this somatostatin antagonist theranostic pair seem to be promising with high sensitivity for neuroendocrine tumors and require further studies in larger patient population.

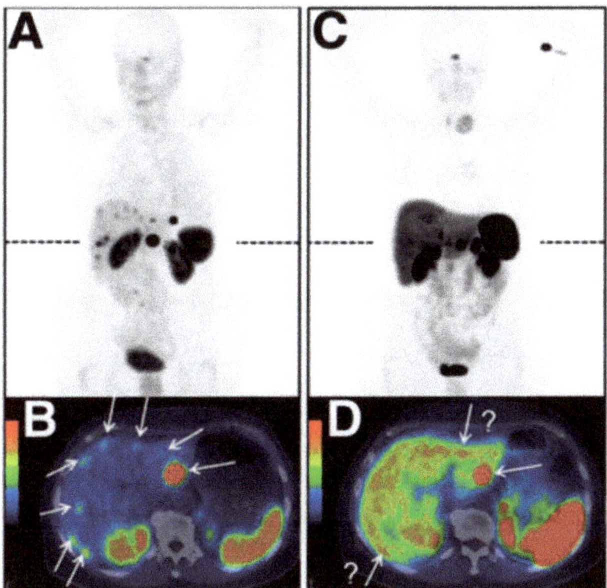

Figure 13. Comparison between [^{68}Ga]-OPS202 (**A**,**B**) and [^{68}Ga]-DOTATOC (**C**,**D**) PET/CT images of the same patient with ileal neuroendocrine tumours, showing bilobar liver metastases (from Rangger et al. [233]).

5. Future Prospects

Regarding clinically established somatostatin analogs, the development of kit-based ^{68}Ga radiotracers, as well as cyclotron production of gallium-68 should improve their availability and worldwide dissemination. Further clinical translation of ^{64}Cu- and ^{18}F-based somatostatin SSAs could also represent an attractive alternative. For therapy, current research focuses on optimizing the dose received by the tumor while sparing healthy tissues. Fractionation, as well as combination of ^{90}Y and ^{177}Lu, have demonstrated their interest [168,234]. The same approach with other treatment modalities, such as external-beam radiotherapy or chemotherapy could enhance treatment response [235,236]. Targeted α-therapy also seems to hold promises and is currently attracting much interest, notably from the industry.

Recent developments showed a switch from agonist to antagonist derivatives, demonstrating higher efficacy. With the advent of new promising radionuclides and somatostatin analogs with better pharmacokinetic properties and binding profiles, the future looks bright for radiolabeled somatostatin analogs, expanding their use for wider indications, than just GEP-NETs. With peptide derivatives with improved targeting, tumors with lower SSTR expression might nonetheless be clinically relevant. In this context, as already demonstrated with some analogs, use of somatostatin-based radiopharmaceuticals might be of interest in pulmonary or hepatic cancers, warranting further studies. The development of bivalent radiotracers to target several receptors concomitantly expressed could be of interest to improve targeting [237]. Similarly, improved detection and sensitivity could be achieved using bimodal agents [238]. Besides, the clinical success for radiolabeled somatostatin analogs both with diagnostic and therapeutic radionuclides paved the way for new promising peptide derivatives, such as bombesin, neurotensin, or CXCR4 ligands, and, in a similar way, PSMA ligands, for cancer theranostics [49,233,239,240].

Author Contributions: All authors contributed to the writing of the manuscript. All authors have read and agreed to the published version of the manuscript.

Funding: This work was partly supported with a funding from Ligue Contre le Cancer (R.E.), a grant from Ligue 22 Contre le Cancer, and Labex IRON (grant no. ANR-11-LABX-0018).

Acknowledgments: The authors thank Sophie Laffont for providing the [^{68}Ga]-DOTATOC and [^{177}Lu]-DOTATATE pictures.

Conflicts of Interest: The authors declare no conflict of interest.

References

1. Brazeau, P.; Vale, W.; Burgus, R.; Ling, N.; Butcher, M.; Rivier, J.; Guillemin, R. Hypothalamic polypeptide that inhibits the secretion of immunoreactive pituitary growth hormone. *Science* **1973**, *179*, 77–79. [CrossRef] [PubMed]
2. Patel, Y.C.; Greenwood, M.T.; Panetta, R.; Demchyshyn, L.; Niznik, H.; Srikant, C.B. The somatostatin receptor family. *Life Sci.* **1995**, *57*, 1249–1265. [CrossRef]
3. Günther, T.; Tulipano, G.; Dournaud, P.; Bousquet, C.; Csaba, Z.; Kreienkamp, H.J.; Lupp, A.; Korbonits, M.; Castaño, J.P.; Wester, H.J.; et al. International Union of Basic and Clinical Pharmacology. CV. Somatostatin Receptors: Structure, Function, Ligands, and New Nomenclature. *Pharmacol. Rev.* **2018**, *70*, 763–835. [CrossRef]
4. Patel, Y.C. Somatostatin and Its Receptor Family. *Front. Neuroendocrinol.* **1999**, *20*, 157–198. [CrossRef] [PubMed]
5. Weckbecker, G.; Lewis, I.; Albert, R.; Schmid, H.A.; Hoyer, D.; Bruns, C. Opportunities in somatostatin research: Biological, chemical and therapeutic aspects. *Nat. Rev. Drug Discov.* **2003**, *2*, 999–1017. [CrossRef] [PubMed]
6. Abdel-Rahman, O.; Lamarca, A.; Valle, J.W.; Hubner, R.A. Somatostatin receptor expression in hepatocellular carcinoma: Prognostic and therapeutic considerations. *Endocr. Relat. Cancer* **2014**, *21*, R485–R493. [CrossRef]
7. Pyronnet, S.; Bousquet, C.; Najib, S.; Azar, R.; Laklai, H.; Susini, C. Antitumor effects of somatostatin. *Mol. Cell. Endocrinol.* **2008**, *286*, 230–237. [CrossRef]
8. Barbieri, F.; Bajetto, A.; Pattarozzi, A.; Gatti, M.; Würth, R.; Thellung, S.; Corsaro, A.; Villa, V.; Nizzari, M.; Florio, T. Peptide receptor targeting in cancer: The somatostatin paradigm. *Int. J. Pept.* **2013**, *2013*, 926295. [CrossRef]
9. Rai, U.; Thrimawithana, T.R.; Valery, C.; Young, S.A. Therapeutic uses of somatostatin and its analogues: Current view and potential applications. *Pharmacol. Ther.* **2015**, *152*, 98–110. [CrossRef]
10. Reubi, J.C.; Schaer, J.C.; Laissue, J.A.; Waser, B. Somatostatin receptors and their subtypes in human tumors and in peritumoral vessels. *Metabolism* **1996**, *45*, 39–41. [CrossRef]
11. Reubi, J.C.; Waser, B.; Schaer, J.C.; Laissue, J.A. Somatostatin receptor sst1-sst5 expression in normal and neoplastic human tissues using receptor autoradiography with subtype-selective ligands. *Eur. J. Nucl. Med.* **2001**, *28*, 836–846. [CrossRef] [PubMed]
12. Guillermet-Guibert, J.; Lahlou, H.; Pyronnet, S.; Bousquet, C.; Susini, C. Somatostatin receptors as tools for diagnosis and therapy: Molecular aspects. *Best Pract. Res. Clin. Gastroenterol.* **2005**, *19*, 535551. [CrossRef]
13. Hasskarl, J.; Kaufmann, M.; Schmid, H.A. Somatostatin receptors in non-neuroendocrine malignancies: The potential role of somatostatin analogs in solid tumors. *Future Oncol.* **2011**, *7*, 895–913. [CrossRef]
14. Gomes-Porras, M.; Cárdenas-Salas, J.; Álvarez-Escolá, C. Somatostatin Analogs in Clinical Practice: A Review. *Int. J. Mol. Sci.* **2020**, *21*, 1682. [CrossRef]
15. Hejna, M.; Schmidinger, M.; Raderer, M. The clinical role of somatostatin analogues as antineoplastic agents: Much ado about nothing? *Ann. Oncol.* **2002**, *13*, 653–668. [CrossRef]

16. Keskin, O.; Yalcin, S. A review of the use of somatostatin analogs in oncology. *OncoTargets Ther.* **2013**, *6*, 471–483. [CrossRef]
17. Reubi, J.C. Peptide receptors as molecular targets for cancer diagnosis and therapy. *Endocr. Rev.* **2003**, *24*, 389–427. [CrossRef]
18. Zhao, B.; Zhao, H.; Zhao, N.; Zhu, X.G. Cholangiocarcinoma cells express somatostatin receptor subtype 2 and respond to octreotide treatment. *J. Hepatobiliary Pancreat. Surg.* **2002**, *9*, 497–502. [CrossRef]
19. Bläker, M.; Schmitz, M.; Gocht, A.; Burghardt, S.; Schulz, M.; Bröring, D.C.; Pace, A.; Greten, H.; De Weerth, A. Differential expression of somatostatin receptor subtypes in hepatocellular carcinomas. *J. Hepatol.* **2004**, *41*, 112–118. [CrossRef]
20. Reubi, J.C.; Waser, B. Concomitant expression of several peptide receptors in neuroendocrine tumours: Molecular basis for in vivo multireceptor tumour targeting. *Eur. J. Nucl. Med. Mol. Imaging* **2003**, *30*, 781–793. [CrossRef]
21. Reubi, J.C.; Zimmermann, A.; Jonas, S.; Waser, B.; Neuhaus, P.; Läderach, U.; Wiedenmann, B. Regulatory peptide receptors in human hepatocellular carcinomas. *Gut* **1999**, *45*, 766–774. [CrossRef] [PubMed]
22. Reynaert, H.; Rombouts, K.; Vandermonde, A.; Urbain, D.; Kumar, U.; Bioulac-Sage, P.; Pinzani, M.; Rosenbaum, J.; Geerts, A. Expression of somatostatin receptors in normal and cirrhotic human liver and in hepatocellular carcinoma. *Gut* **2004**, *53*, 1180–1189. [CrossRef] [PubMed]
23. Nguyen-Khac, E.; Ollivier, I.; Aparicio, T.; Moullart, V.; Hugentobler, A.; Lebtahi, R.; Lobry, C.; Susini, C.; Duhamel, C.; Hommel, S.; et al. Somatostatin receptor scintigraphy screening in advanced hepatocarcinoma: A multi-center French study. *Cancer Biol. Ther.* **2009**, *8*, 2033–2039. [CrossRef] [PubMed]
24. Verhoef, C.; van Dekken, H.; Hofland, L.J.; Zondervan, P.E.; de Wilt, J.H.; van Marion, R.; de Man, R.A.; IJzermans, J.N.; van Eijck, C.H. Somatostatin receptors in human hepatocellular carcinomas: Biological, patient and tumor characteristics. *Dig. Surg.* **2008**, *25*, 21–26. [CrossRef] [PubMed]
25. Liu, H.L.; Huo, L.; Wang, L. Octreotide inhibits proliferation and induces apoptosis of hepatocellular carcinoma cells. *Acta Pharmacol. Sin.* **2004**, *25*, 1380–1386.
26. Li, S.; Liu, Y.; Shen, Z. Characterization of somatostatin receptor 2 and 5 expression in operable hepatocellular carcinomas. *Hepatogastroenterology* **2012**, *59*, 2054–2058. [CrossRef]
27. Liu, Y.; Jiang, L.; Mu, Y. Somatostatin receptor subtypes 2 and 5 are associated with better survival in operable hepatitis B-related hepatocellular carcinoma following octreotide long-acting release treatment. *Oncol. Lett.* **2013**, *6*, 821–828. [CrossRef]
28. Huang, C.Z.; Huang, A.M.; Liu, J.F.; Wang, B.; Lin, K.C.; Ye, Y.B. Somatostatin Octapeptide Inhibits Cell Invasion and Metastasis in Hepatocellular Carcinoma Through PEBP1. *Cell Physiol. Biochem.* **2018**, *47*, 2340–2349. [CrossRef]
29. Hua, Y.P.; Huang, J.F.; Liang, L.J.; Li, S.Q.; Lai, J.M.; Liang, H.Z. The study of inhibition effect of octreotide on the growth of hepatocellular carcinoma xenografts in situ in nude mice. *Chin. J. Surg.* **2005**, *43*, 721–725.
30. Jia, W.D.; Xu, G.L.; Wang, W.; Wang, Z.H.; Li, J.S.; Ma, J.L.; Ren, W.H.; Ge, Y.S.; Yu, J.H.; Liu, W.B. A somatostatin analogue, octreotide, inhibits the occurrence of second primary tumors and lung metastasis after resection of hepatocellular carcinoma in mice. *Tohoku J. Exp. Med.* **2009**, *218*, 155160. [CrossRef]
31. Reynaert, H.; Colle, I. Treatment of Advanced Hepatocellular Carcinoma with Somatostatin Analogues: A Review of the Literature. *Int. J. Mol. Sci.* **2019**, *20*, 4811. [CrossRef] [PubMed]
32. Kouroumalis, E.; Skordilis, P.; Thermos, K.; Vasilaki, A.; Moschandrea, J.; Manousos, O.N. Treatment of hepatocellular carcinoma with octreotide: A randomised controlled study. *Gut* **1998**, *42*, 442–447. [CrossRef] [PubMed]
33. Dimitroulopoulos, D.; Xinopoulos, D.; Tsamakidis, K.; Zisimopoulos, A.; Andriotis, E.; Panagiotakos, D.; Fotopoulou, A.; Chrysohoou, C.; Bazinis, A.; Daskalopoulou, D.; et al. Long acting octreotide in the treatment of advanced hepatocellular cancer and overexpression of somatostatin receptors: Randomized placebo-controlled trial. *World J. Gastroenterol.* **2007**, *13*, 3164–3170. [CrossRef] [PubMed]
34. Becker, G.; Allgaier, H.P.; Olschewski, M.; Zähringer, A.; Blum, H.E.; HECTOR Study Group. Long-acting octreotide versus placebo for treatment of advanced HCC: A randomized controlled double-blind study. *Hepatology* **2007**, *45*, 9–15. [CrossRef]

35. Barbare, J.C.; Bouché, O.; Bonnetain, F.; Dahan, L.; Lombard-Bohas, C.; Faroux, R.; Raoul, J.L.; Cattan, S.; Lemoine, A.; Blanc, J.F.; et al. Treatment of advanced hepatocellular carcinoma with long-acting octreotide: A phase III multicenter, randomised, double blind placebo-controlled study. *Eur. J. Cancer* **2009**, *45*, 1788–1797. [CrossRef]
36. Samonakis, D.N.; Notas, G.; Christodoulakis, N.; Kouroumalis, E.A. Mechanisms of action and resistance of somatostatin analogues for the treatment of hepatocellular carcinoma: A message not well taken. *Dig. Dis. Sci.* **2008**, *53*, 2359–2365. [CrossRef]
37. Kaemmerer, D.; Schindler, R.; Mußbach, F.; Dahmen, U.; Altendorf-Hofmann, A.; Dirsch, O.; Sänger, J.; Schulz, S.; Lupp, A. Somatostatin and CXCR4 chemokine receptor expression in hepatocellular and cholangiocellular carcinomas: Tumor capillaries as promising targets. *BMC Cancer* **2017**, *17*, 896. [CrossRef]
38. Rinke, A.; Müller, H.H.; Schade-Brittinger, C.; Klose, K.J.; Barth, P.; Wied, M.; Mayer, C.; Aminossadati, B.; Pape, U.F.; Bläker, M.; et al. Placebo-controlled, double-blind, prospective, randomized study on the effect of octreotide LAR in the control of tumor growth in patients with metastatic neuroendocrine midgut tumors: A report from the PROMID study group. *J. Clin. Oncol.* **2009**, *27*, 4656–4663. [CrossRef]
39. Caplin, M.E.; Pavel, M.; Ćwikła, J.B.; Phan, A.T.; Raderer, M.; Sedláčková, E.; Cadiot, G.; Wolin, E.M.; Capdevila, J.; Wall, L.; et al. Anti-tumour effects of lanreotide for pancreatic and intestinal neuroendocrine tumours: The CLARINET open-label extension study. *Endocr. Relat. Cancer* **2016**, *23*, 191–199. [CrossRef]
40. Sheppard, M.; Shapiro, B.; Pimstone, B.; Kronheim, S.; Berelowitz, M.; Gregory, M. Metabolic clearance and plasma half-disappearance time of exogenous somatostatin in man. *J. Clin. Endocrinol. Metab.* **1979**, *48*, 50–53. [CrossRef]
41. Lowell, A.; Freda, P.U. From somatostatin to octreotide LAR: Evolution of a somatostatin analogue. *Curr. Med. Res. Opin.* **2009**, *25*, 2989–2999. [CrossRef]
42. Öberg, K.; Lamberts, S.W.J. Somatostatin analogues in acromegaly and gastroenteropancreatic neuroendocrine tumours: Past, present and future. *Endocr. Relat. Cancer* **2016**, *23*, R551–R566. [CrossRef] [PubMed]
43. Ryan, P.; McBride, A.; Ray, D.; Pulgar, S.; Ramirez, R.A.; Elquza, E.; Favaro, J.P.; Dranitsaris, G. Lanreotide vs. octreotide LAR for patients with advanced gastroenteropancreatic neuroendocrine tumors: An observational time and motion analysis. *J. Oncol. Pharm. Pract.* **2019**, *25*, 1425–1433. [CrossRef] [PubMed]
44. Feelders, R.A.; Yasothan, U.; Kirkpatrick, P. Pasireotide. *Nat. Rev. Drug Discov.* **2012**, *11*, 597–598. [CrossRef]
45. Crider, A.M. Recent Advances in the Development of Nonpeptide Somatostatin Receptor Ligands. *Mini Rev. Med. Chem.* **2002**, *2*, 507–517. [CrossRef]
46. Correia, J.D.; Paulo, A.; Raposinho, P.D.; Santos, I. Radiometallated peptides for molecular imaging and targeted therapy. *Dalton Trans.* **2011**, *40*, 6144–6167. [CrossRef]
47. Jamous, M.; Haberkorn, U.; Mier, W. Synthesis of Peptide Radiopharmaceuticals for the Therapy and Diagnosis of Tumor Diseases. *Molecules* **2013**, *18*, 3379–3409. [CrossRef]
48. Tornesello, A.L.; Buonaguro, L.; Tornesello, M.L.; Buonaguro, F.M. New Insights in the Design of Bioactive Peptides and Chelating Agents for Imaging and Therapy in Oncology. *Molecules* **2017**, *22*, 1282. [CrossRef]
49. Tornesello, A.L.; Tornesello, M.L.; Buonaguro, F.M. An Overview of Bioactive Peptides for in vivo Imaging and Therapy in Human Diseases. *Mini Rev. Med. Chem.* **2017**, *17*, 758–770. [CrossRef]
50. Cutler, C.S.; Hennkens, H.M.; Sisay, N.; Huclier-Markai, S.; Jurisson, S.S. Radiometals for Combined Imaging and Therapy. *Chem. Rev.* **2013**, *113*, 858–883. [CrossRef]
51. Ramogida, C.F.; Orvig, C. Tumour Targeting with Radiometals for Diagnosis and Therapy. *Chem. Commun.* **2013**, *49*, 4720–4739. [CrossRef] [PubMed]
52. Blower, P.J. A nuclear chocolate box: The periodic table of nuclear medicine. *Dalton Trans.* **2015**, *44*, 4819–4844. [CrossRef] [PubMed]
53. Kostelnik, T.I.; Orvig, C. Radioactive Main Group and Rare Earth Metals for Imaging and Therapy. *Chem. Rev.* **2019**, *119*, 902–956. [CrossRef] [PubMed]
54. Price, E.W.; Orvig, C. Matching Chelators to Radiometals for Radiopharmaceuticals. *Chem. Soc. Rev.* **2014**, *43*, 260–290. [CrossRef] [PubMed]
55. Boros, E.; Packard, A.B. Radioactive Transition Metals for Imaging and Therapy. *Chem. Rev.* **2019**, *119*, 870–901. [CrossRef]
56. Hancock, R.D.; Martell, A.E. Ligand design for selective complexation of metal ions in aqueous solution. *Chem. Rev.* **1989**, *89*, 1875–1914. [CrossRef]

57. Stasiuk, G.J.; Long, N.J. The ubiquitous DOTA and its derivatives: The impact of 1,4,7,10-tetraazacyclododecane-1,4,7,10-tetraacetic acid on biomedical imaging. *Chem. Commun.* **2013**, *49*, 2732–2746. [CrossRef]
58. Baranyai, Z.; Tircsó, G.; Rösch, F. The Use of the Macrocyclic Chelator DOTA in Radiochemical Separations. *Eur. J. Inorg. Chem.* **2020**, 36–56. [CrossRef]
59. Sun, X.; Wuest, M.; Weisman, G.R.; Wong, E.H.; Reed, D.P.; Boswell, C.A.; Motekaitis, R.; Martell, A.E.; Welch, M.J.; Anderson, C.J. Radiolabeling and in vivo behavior of copper-64-labeled cross-bridged cyclam ligands. *J. Med. Chem.* **2002**, *45*, 469–477. [CrossRef]
60. Bhattacharyya, S.; Dixit, M. Metallic radionuclides in the development of diagnostic and therapeutic radiopharmaceuticals. *Dalton Trans.* **2011**, *40*, 6112–6128. [CrossRef]
61. Mushtaq, S.; Yun, S.J.; Jeon, J. Recent Advances in Bioorthogonal Click Chemistry for Efficient Synthesis of Radiotracers and Radiopharmaceuticals. *Molecules* **2019**, *24*, 3567. [CrossRef] [PubMed]
62. Reubi, J.C.; Maecke, H.R. Peptide-based probes for cancer imaging. *J. Nucl. Med.* **2008**, *49*, 1735–1738. [CrossRef] [PubMed]
63. Maecke, H.R.; Reubi, J.C. Somatostatin receptors as targets for nuclear medicine imaging and radionuclide treatment. *J. Nucl. Med.* **2011**, *52*, 841–844. [CrossRef] [PubMed]
64. Reubi, J.C.; Schär, J.C.; Waser, B.; Wenger, S.; Heppeler, A.; Schmitt, J.S.; Mäcke, H.R. Affinity profiles for human somatostatin receptor subtypes SST1-SST5 of somatostatin radiotracers selected for scintigraphic and radiotherapeutic use. *Eur. J. Nucl. Med.* **2000**, *27*, 273–282. [CrossRef]
65. Fani, M.; Maecke, H.R. Radiopharmaceutical development of radiolabeled peptides. *Eur. J. Nucl. Med. Mol. Imaging* **2012**, *39*, S11–S30. [CrossRef]
66. Lamberts, S.W.; Bakker, W.H.; Reubi, J.C.; Krenning, E.P. Somatostatin-receptor imaging in the localization of endocrine tumors. *N. Engl. J. Med.* **1990**, *323*, 1246–1249. [CrossRef]
67. Bakker, W.H.; Krenning, E.P.; Breeman, W.A.; Kooij, P.P.; Reubi, J.C.; Koper, J.W.; de Jong, M.; Laméris, J.S.; Visser, T.J.; Lamberts, S.W. In vivo use of a radioiodinated somatostatin analogue: Dynamics, metabolism, and binding to somatostatin receptor-positive tumors in man. *J. Nucl. Med.* **1991**, *32*, 1184–1189.
68. Bakker, W.H.; Krenning, E.P.; Breeman, W.A.; Koper, J.W.; Kooij, P.P.; Reubi, J.C.; Klijn, J.G.; Visser, T.J.; Docter, R.; Lamberts, S.W. Receptor scintigraphy with a radioiodinated somatostatin analogue: Radiolabeling, purification, biologic activity, and in vivo application in animals. *J. Nucl. Med.* **1990**, *31*, 1501–1509.
69. Bakker, W.H.; Albert, R.; Bruns, C.; Breeman, W.A.; Hofland, L.J.; Marbach, P.; Pless, J.; Pralet, D.; Stolz, B.; Koper, J.W.; et al. [^{111}In-DTPA-D-Phe^{1}]-octreotide, a potential radiopharmaceutical for imaging of somatostatin receptor-positive tumors: Synthesis, radiolabeling and in vitro validation. *Life Sci.* **1991**, *49*, 1583–1591. [CrossRef]
70. Bakker, W.H.; Krenning, E.P.; Reubi, J.C.; Breeman, W.A.; Setyono-Han, B.; de Jong, M.; Kooij, P.P.; Bruns, C.; van Hagen, P.M.; Marbach, P.; et al. In vivo application of [^{111}In-DTPA-D-Phe^{1}]-octreotide for detection of somatostatin receptor-positive tumors in rats. *Life Sci.* **1991**, *49*, 1593–1601. [CrossRef]
71. Krenning, E.P.; Bakker, W.H.; Kooij, P.P.; Breeman, W.A.; Oei, H.Y.; de Jong, M.; Reubi, J.C.; Visser, T.J.; Bruns, C.; Kwekkeboom, D.J.; et al. Somatostatin receptor scintigraphy with indium-111-DTPA-D-Phe-1-octreotide in man: Metabolism, dosimetry and comparison with iodine-123-Tyr-3-octreotide. *J. Nucl. Med.* **1992**, *33*, 652–658. [PubMed]
72. Otte, A.; Jermann, E.; Behe, M.; Goetze, M.; Bucher, H.C.; Roser, H.W.; Heppeler, A.; Mueller-Brand, J.; Maecke, H.R. DOTATOC: A powerful new tool for receptor-mediated radionuclide therapy. *Eur. J. Nucl. Med.* **1997**, *24*, 792–795. [CrossRef] [PubMed]
73. Forrer, F.; Uusijärvi, H.; Waldherr, C.; Cremonesi, M.; Bernhardt, P.; Mueller-Brand, J.; Maecke, H.R. A comparison of ^{111}In-DOTATOC and ^{111}In-DOTATATE: Biodistribution and dosimetry in the same patients with metastatic neuroendocrine tumours. *Eur. J. Nucl. Med. Mol. Imaging* **2004**, *31*, 1257–1262. [CrossRef] [PubMed]
74. Wild, D.; Schmitt, J.S.; Ginj, M.; Mäcke, H.R.; Bernard, B.F.; Krenning, E.; De Jong, M.; Wenger, S.; Reubi, J.C. DOTA-NOC, a high-affinity ligand of somatostatin receptor subtypes 2, 3 and 5 for labeling with various radiometals. *Eur. J. Nucl. Med. Mol. Imaging* **2003**, *30*, 1338–1347. [CrossRef]
75. Virgolini, I.; Ambrosini, V.; Bomanji, J.B.; Baum, R.P.; Fanti, S.; Gabriel, M.; Papathanasiou, N.D.; Pepe, G.; Oyen, W.; De Cristoforo, C.; et al. Procedure guidelines for PET/CT tumour imaging with ^{68}Ga-DOTA-conjugated peptides: ^{68}Ga-DOTA-TOC, ^{68}Ga-DOTA-NOC, ^{68}Ga-DOTA-TATE. *Eur. J. Nucl. Med. Mol. Imaging* **2010**, *37*, 2004–2010. [CrossRef]

76. Hofmann, M.; Maecke, H.; Börner, R.; Weckesser, E.; Schöffski, P.; Oei, L.; Schumacher, J.; Henze, M.; Heppeler, A.; Meyer, J.; et al. Biokinetics and imaging with the somatostatin receptor PET radioligand ^{68}Ga-DOTATOC: Preliminary data. *Eur. J. Nucl. Med.* **2001**, *28*, 1751–1757. [CrossRef]
77. Hofman, M.S.; Lau, W.F.; Hicks, R.J. Somatostatin receptor imaging with ^{68}Ga DOTATATE PET/CT: Clinical utility, normal patterns, pearls, and pitfalls in interpretation. *Radiographics* **2015**, *35*, 500–516. [CrossRef]
78. Wild, D.; Mäcke, H.R.; Waser, B.; Reubi, J.C.; Ginj, M.; Rasch, H.; Müller-Brand, J.; Hofmann, M. ^{68}Ga-DOTANOC: A first compound for PET imaging with high affinity for somatostatin receptor subtypes 2 and 5. *Eur. J. Nucl. Med. Mol. Imaging* **2005**, *32*, 724. [CrossRef]
79. Antunes, P.; Ginj, M.; Walter, M.A.; Chen, J.; Reubi, J.C.; Maecke, H.R. Influence of different spacers on the biological profile of a DOTA-Somatostatin analogue. *Bioconjugate Chem.* **2007**, *18*, 84–92. [CrossRef]
80. Eisenwiener, K.P.; Prata, M.I.; Buschmann, I.; Zhang, H.W.; Santos, A.C.; Wenger, S.; Reubi, J.C.; Mäcke, H.R. NODAGATOC, a new chelator-coupled somatostatin analogue labeled with [$^{67/68}$Ga] and [^{111}In] for SPECT, PET, and targeted therapeutic applications of somatostatin receptor (hsst2) expressing tumors. *Bioconjugate Chem.* **2002**, *13*, 530–541. [CrossRef]
81. Laznickova, A.; Laznicek, M.; Trejtnar, F.; Maecke, H.R.; Eisenwiener, K.P.; Reubi, J.C. Biodistribution of two octreotate analogs radiolabeled with indium and yttrium in rats. *Anticancer Res.* **2010**, *30*, 2177–2184. [PubMed]
82. Ginj, M.; Chen, J.; Walter, M.A.; Eltschinger, V.; Reubi, J.C.; Maecke, H.R. Preclinical evaluation of new and highly potent analogues of octreotide for predictive imaging and targeted radiotherapy. *Clin. Cancer Res.* **2005**, *11*, 1136–1145. [PubMed]
83. Boubaker, A.; Prior, J.O.; Willi, J.P.; Champendal, M.; Kosinski, M.; Bischof-Delaloye, A.; Maecke, H.R.; Ginj, M.; Baechler, S.; Buchegger, F. Biokinetics and dosimetry of ^{111}In-DOTA-NOC-ATE compared with ^{111}In-DTPA-octreotide. *Eur. J. Nucl. Med. Mol. Imaging* **2012**, *39*, 1868–1875. [CrossRef] [PubMed]
84. Ma, M.T.; Cullinane, C.; Waldeck, K.; Roselt, P.; Hicks, R.J.; Blower, P.J. Rapid kit-based ^{68}Ga-labeling and PET imaging with THP-Tyr^{3}-octreotate: A preliminary comparison with DOTA-Tyr3octreotate. *EJNMMI Res.* **2015**, *5*, 52. [CrossRef] [PubMed]
85. Fani, M.; Mueller, A.; Tamma, M.L.; Nicolas, G.; Rink, H.R.; Cescato, R.; Reubi, J.C.; Maecke, H.R. Radiolabeled bicyclic somatostatin-based analogs: A novel class of potential radiotracers for SPECT/PET of neuroendocrine tumors. *J. Nucl. Med.* **2010**, *51*, 1771–1779. [CrossRef]
86. Ginj, M.; Zhang, H.; Eisenwiener, K.P.; Wild, D.; Schulz, S.; Rink, H.; Cescato, R.; Reubi, J.C.; Maecke, H.R. New pansomatostatin ligands and their chelated versions: Affinity profile, agonist activity, internalization, and tumor targeting. *Clin. Cancer Res.* **2008**, *14*, 2019–2027. [CrossRef]
87. Liu, F.; Liu, T.; Xu, X.; Guo, X.; Li, N.; Xiong, C.; Li, C.; Zhu, H.; Yang, Z. Design, Synthesis, and Biological Evaluation of ^{68}Ga-DOTA-PA1 for Lung Cancer: A Novel PET Tracer for Multiple Somatostatin Receptor Imaging. *Mol. Pharm.* **2018**, *15*, 619–628. [CrossRef]
88. Tatsi, A.; Maina, T.; Cescato, R.; Waser, B.; Krenning, E.P.; de Jong, M.; Cordopatis, P.; Reubi, J.C.; Nock, B.A. [^{111}In-DOTA]Somatostatin-14 analogs as potential pansomatostatin-like radiotracers—First results of a preclinical study. *EJNMMI Res.* **2012**, *2*, 25. [CrossRef]
89. Maina, T.; Cescato, R.; Waser, B.; Tatsi, A.; Kaloudi, A.; Krenning, E.P.; de Jong, M.; Nock, B.A.; Reubi, J.C. [^{111}In-DOTA]LTT-SS28, a first pansomatostatin radioligand for in vivo targeting of somatostatin receptor-positive tumors. *J. Med. Chem.* **2014**, *57*, 6564–6571. [CrossRef]
90. Pearson, D.A.; Lister-James, J.; McBride, W.J.; Wilson, D.M.; Martel, L.J.; Civitello, E.R.; Taylor, J.E.; Moyer, B.R.; Dean, R.T. Somatostatin receptor-binding peptides labeled with technetium-99m: Chemistry and initial biological studies. *J. Med. Chem.* **1996**, *39*, 1361–1371. [CrossRef]
91. Decristoforo, C.; Mather, S.J. Preparation, ^{99m}Tc-labeling and in vitro characterization of HYNIC and N3S modified RC-160 and [Tyr^{3}]octreotide. *Bioconjugate Chem.* **1999**, *10*, 431–438. [CrossRef] [PubMed]
92. Maina, T.; Stolz, B.; Albert, R.; Bruns, C.; Koch, P.; Mäcke, H. Synthesis, radiochemistry and biological evaluation of a new somatostatin analogue (SDZ 219-387) labeled with technetium-99m. *Eur. J. Nucl. Med.* **1994**, *21*, 437–444. [CrossRef] [PubMed]
93. Thakur, M.L.; Kolan, H.; Li, J.; Wiaderkiewicz, R.; Pallela, V.R.; Duggaraju, R.; Schally, A.V. Radiolabeled somatostatin analogues in prostate cancer. *Nucl. Med. Biol.* **1997**, *24*, 105–113. [CrossRef]

94. Abiraj, K.; Ursillo, S.; Tamma, M.L.; Rylova, S.N.; Waser, B.; Constable, E.C.; Fani, M.; Nicolas, G.P.; Reubi, J.C.; Maecke, H.R. The tetraamine chelator outperforms HYNIC in a new technetium-99m-labeled somatostatin receptor 2 antagonist. *EJNMMI Res.* **2018**, *8*, 75. [CrossRef]
95. Spradau, T.W.; Edwards, W.B.; Anderson, C.J.; Welch, M.J.; Katzenellenbogen, J.A. Synthesis and biological evaluation of Tc-99m-cyclopentadienyltricarbonyltechnetium-labeled octreotide. *Nucl. Med. Biol.* **1999**, *26*, 1–7. [CrossRef]
96. Makris, G.; Kuchuk, M.; Gallazzi, F.; Jurisson, S.S.; Smith, C.J.; Hennkens, H.M. Somatostatin receptor targeting with hydrophilic [^{99m}Tc/^{186}Re]Tc/Re-tricarbonyl NODAGA and NOTA complexes. *Nucl. Med. Biol.* **2019**, *71*, 39–46. [CrossRef]
97. Abrams, M.J.; Juweid, M.; tenKate, C.I.; Schwartz, D.A.; Hauser, M.M.; Gaul, F.E.; Fuccello, A.J.; Rubin, R.H.; Strauss, H.W.; Fischman, A.J. Technetium-99m-human polyclonal IgG radiolabeled via the hydrazino nicotinamide derivative for imaging focal sites of infection in rats. *J. Nucl. Med.* **1990**, *31*, 2022–2028.
98. Mikolajczak, R.; Maecke, H.R. Radiopharmaceuticals for somatostatin receptor imaging. *Nucl. Med. Rev. Cent. East Eur.* **2016**, *19*, 126–132. [CrossRef]
99. Decristoforo, C.; Melendez-Alafort, L.; Sosabowski, J.K.; Mather, S.J. ^{99m}Tc-HYNIC-[Tyr^3]octreotide for imaging somatostatin receptor positive tumors: Preclinical evaluation and comparison with ^{111}In-octreotide. *J. Nucl. Med.* **2000**, *41*, 1114–1119.
100. Gabriel, M.; Decristoforo, C.; Donnemiller, E.; Ulmer, H.; Watfah Rychlinski, C.; Mather, S.J.; Moncayo, R. An intrapatient comparison of ^{99m}Tc-EDDA/HYNIC-TOC with ^{111}In-DTPA-Octreotide for diagnosis of somatostatin receptor expressing tumors. *J. Nucl. Med.* **2003**, *44*, 708–716.
101. Cwikla, J.B.; Mikolajczak, R.; Pawlak, D.; Buscombe, J.R.; Nasierowska-Guttmejer, A.; Bator, A.; Maecke, H.R.; Walecki, J. Initial direct comparison of ^{99m}Tc-TOC and ^{99m}Tc-TATE in identifying sites of disease in patients with proven GEP NETs. *J. Nucl. Med.* **2008**, *49*, 1060–1065. [CrossRef] [PubMed]
102. Decristoforo, C.; Maina, T.; Nock, B.; Gabriel, M.; Cordopatis, P.; Moncayo, R. ^{99m}Tc-Demotate 1: First data in tumour patients-results of a pilot/phase I study. *Eur. J. Nucl. Med. Mol. Imaging* **2003**, *30*, 1211–1219. [CrossRef] [PubMed]
103. Maina, T.; Nock, B.A.; Cordopatis, P.; Bernard, B.F.; Breeman, W.A.; van Gameren, A.; van den Berg, R.; Reubi, J.C.; Krenning, E.P.; de Jong, M. [^{99m}Tc]Demotate 2 in the detection of sst2-positive tumours: A preclinical comparison with [^{111}In]DOTA-tate. *Eur. J. Nucl. Med. Mol. Imaging* **2006**, *33*, 831–840. [CrossRef] [PubMed]
104. Virgolini, I.; Leimer, M.; Handmaker, H.; Lastoria, S.; Bischof, C.; Muto, P.; Pangerl, T.; Gludovacz, D.; Peck-Radosavljevic, M.; Lister-James, J.; et al. Somatostatin receptor subtype specificity and in vivo binding of a novel tumor tracer, ^{99m}Tc-P829. *Cancer Res.* **1998**, *58*, 1850–1859.
105. Lebtahi, R.; Le Cloirec, J.; Houzard, C.; Daou, D.; Sobhani, I.; Sassolas, G.; Mignon, M.; Bourguet, P.; Le Guludec, D. Detection of neuroendocrine tumors: ^{99m}Tc-P829 scintigraphy compared with ^{111}In-pentetreotide scintigraphy. *J. Nucl. Med.* **2002**, *43*, 889–895.
106. Blum, J.E.; Handmaker, H.; Rinne, N.A. The utility of a somatostatin-type receptor binding peptide radiopharmaceutical (P829) in the evaluation of solitary pulmonary nodules. *Chest* **1999**, *115*, 224–232. [CrossRef]
107. Bååth, M.; Kolbeck, K.G.; Danielsson, R. Somatostatin receptor scintigraphy with ^{99m}Tc-Depreotide (NeoSpect) in discriminating between malignant and benign lesions in the diagnosis of lung cancer: A pilot study. *Acta Radiol.* **2004**, *45*, 833–839. [CrossRef]
108. Axelsson, R.; Herlin, G.; Bååth, M.; Aspelin, P.; Kolbeck, K.G. Role of scintigraphy with technetium-99m depreotide in the diagnosis and management of patients with suspected lung cancer. *Acta Radiol.* **2008**, *49*, 295–302. [CrossRef]
109. Menda, Y.; Kahn, D. Somatostatin receptor imaging of non-small cell lung cancer with ^{99m}Tc depreotide. *Semin. Nucl. Med.* **2002**, *32*, 92–96. [CrossRef]
110. Van Den Bossche, B.; D'haeninck, E.; Bacher, K.; Thierens, H.; Van Belle, S.; Dierckx, R.A.; Van de Wiele, C. Biodistribution and dosimetry of ^{99m}Tc-depreotide (P829) in patients suffering from breast carcinoma. *Cancer Biother. Radiopharm.* **2004**, *19*, 776–783. [CrossRef]
111. Briganti, V.; Cuccurullo, V.; Di Stasio, G.D.; Mansi, L. Gamma Emitters in Pancreatic Endocrine Tumors Imaging in the PET Era: Is there a Clinical Space for ^{99m}Tc-peptides? *Curr. Radiopharm.* **2019**, *12*, 156–170. [CrossRef] [PubMed]

112. Boschi, A.; Uccelli, L.; Martini, P. A Picture of Modern Tc-99m Radiopharmaceuticals: Production, Chemistry, and Applications in Molecular Imaging. *Appl. Sci.* **2019**, *9*, 2526. [CrossRef]
113. Boschi, A.; Martini, P.; Janevik-Ivanovska, E.; Duatti, A. The emerging role of copper-64 radiopharmaceuticals as cancer theranostics. *Drug Discov. Today* **2018**, *23*, 1489–1501. [CrossRef] [PubMed]
114. Anderson, C.J.; Dehdashti, F.; Cutler, P.D.; Schwarz, S.W.; Laforest, R.; Bass, L.A.; Lewis, J.S.; McCarthy, D.W. ^{64}Cu-TETA-octreotide as a PET imaging agent for patients with neuroendocrine tumors. *J. Nucl. Med.* **2001**, *42*, 213–221.
115. Sprague, J.E.; Peng, Y.; Sun, X.; Weisman, G.R.; Wong, E.H.; Achilefu, S.; Anderson, C.J. Preparation and biological evaluation of copper-64-labeled Tyr^3-octreotate using a cross-bridged macrocyclic chelator. *Clin. Cancer Res.* **2004**, *10*, 8674–8682. [CrossRef]
116. Edwards, W.B.; Fields, C.G.; Anderson, C.J.; Pajeau, T.S.; Welch, M.J.; Fields, G.B. Generally Applicable, Convenient Solid-Phase Synthesis and Receptor Affinities of Octreotide Analogs. *J. Med. Chem.* **1994**, *37*, 3749–3757. [CrossRef]
117. Paterson, B.M.; Roselt, P.; Denoyer, D.; Cullinane, C.; Binns, D.; Noonan, W.; Jeffery, C.M.; Price, R.I.; White, J.M.; Hicks, R.J.; et al. PET imaging of tumours with a ^{64}Cu labeled macrobicyclic cage amine ligand tethered to Tyr^3-octreotate. *Dalton Trans.* **2014**, *43*, 1386–1396. [CrossRef]
118. Marciniak, A.; Brasuń, J. Somatostatin analogues labeled with copper radioisotopes: Currrent status. *J. Radioanal. Nucl. Chem.* **2017**, *313*, 279–289. [CrossRef]
119. Pfeifer, A.; Knigge, U.; Binderup, T.; Mortensen, J.; Oturai, P.; Loft, A.; Berthelsen, A.K.; Langer, S.W.; Rasmussen, P.; Elema, D.; et al. ^{64}Cu-DOTATATE PET for Neuroendocrine Tumors: A Prospective Head-to-Head Comparison with ^{111}In-DTPA-Octreotide in 112 Patients. *J. Nucl. Med.* **2015**, *56*, 847–854. [CrossRef]
120. Johnbeck, C.B.; Knigge, U.; Loft, A.; Berthelsen, A.K.; Mortensen, J.; Oturai, P.; Langer, S.W.; Elema, D.R.; Kjaer, A. Head-to-head comparison of ^{64}Cu-DOTATATE and ^{68}Ga-DOTATOC PET/CT: A prospective study of 59 patients with neuroendocrine tumors. *J. Nucl. Med.* **2017**, *58*, 451–458. [CrossRef]
121. Carlsen, E.A.; Johnbeck, C.B.; Binderup, T.; Loft, M.; Pfeifer, A.; Mortensen, J.; Oturai, P.; Loft, A.; Berthelsen, A.K.; Langer, S.W.; et al. ^{64}Cu-DOTATATE PET/CT and prediction of overall and progression-free survival in patients with neuroendocrine neoplasms. *J. Nucl. Med.* **2020**. [CrossRef]
122. Andersen, T.L.; Baun, C.; Olsen, B.B.; Dam, J.H.; Thisgaard, H. Improving Contrast and Detectability: Imaging with [^{55}Co]Co-DOTATATE in Comparison with [^{64}Cu]Cu-DOTATATE and [^{68}Ga]Ga-DOTATATE. *J. Nucl. Med.* **2020**, *61*, 228–233. [CrossRef] [PubMed]
123. Mirzaei, S.; Revheim, M.; Raynor, W.; Zehetner, W.; Knoll, P.; Zandieh, S.; Alavi, A. ^{64}Cu-DOTATOC PET-CT in Patients with Neuroendocrine Tumors. *Oncol. Ther.* **2019**, *8*, 125–131. [CrossRef]
124. Hicks, R.J.; Jackson, P.; Kong, G.; Ware, R.E.; Hofman, M.S.; Pattison, D.A.; Akhurst, T.A.; Drummond, E.; Roselt, P.; Callahan, J.; et al. ^{64}Cu-SARTATE PET Imaging of Patients with Neuroendocrine Tumors Demonstrates High Tumor Uptake and Retention, Potentially Allowing Prospective Dosimetry for Peptide Receptor Radionuclide Therapy. *J. Nucl. Med.* **2019**, *60*, 777–785. [CrossRef] [PubMed]
125. Delpassand, E.S.; Ranganathan, D.; Wagh, N.; Shafie, A.; Gaber, A.; Abbasi, A.; Kjaer, A.; Tworowska, I.; Núñez, R. ^{64}Cu-DOTATATE PET/CT for Imaging Patients with Known or Suspected Somatostatin Receptor-Positive Neuroendocrine Tumors: Results of the First US Prospective, Reader-Blinded Clinical Trial. *J. Nucl. Med.* **2020**, *61*, 890–896. [CrossRef] [PubMed]
126. Heppeler, A.; André, J.P.; Buschmann, I.; Wang, X.; Reubi, J.C.; Hennig, M.; Kaden, T.A.; Maecke, H.R. Metal-ion-dependent biological properties of a chelator-derived somatostatin analogue for tumour targeting. *Chem. Eur. J.* **2008**, *14*, 3026–3034. [CrossRef] [PubMed]
127. Thisgaard, H.; Olsen, B.B.; Dam, J.H.; Bollen, P.; Mollenhauer, J.; Høilund-Carlsen, P.F. Evaluation of cobalt-labeled octreotide analogs for molecular imaging and Auger electron-based radionuclide therapy. *J. Nucl. Med.* **2014**, *55*, 1311–1316. [CrossRef]
128. Thisgaard, H.; Olesen, M.; Dam, J.H. Radiosynthesis of Co-55- and Co-58m-labeled DOTATOC for positron emission tomography imaging and targeted radionuclide therapy. *J. Label. Compd. Radiopharm.* **2011**, *54*, 758–762. [CrossRef]
129. Müller, C.; Domnanich, K.A.; Umbricht, C.A.; van der Meulen, N.P. Scandium and terbium radionuclides for radiotheranostics: Current state of development towards clinical application. *Br. J. Radiol.* **2018**, *91*, 20180074. [CrossRef]

130. Pruszyński, M.; Majkowska-Pilip, A.; Loktionova, N.S.; Eppard, E.; Roesch, F. Radiolabeling of DOTATOC with the long-lived positron emitter ^{44}Sc. *Appl. Radiat. Isot.* **2012**, *70*, 974–979. [CrossRef]
131. Müller, C.; Vermeulen, C.; Johnston, K.; Köster, U.; Schmid, R.; Türler, A.; van der Meulen, N.P. Preclinical in vivo application of ^{152}Tb-DOTANOC: A radiolanthanide for PET imaging. *EJNMMI Res.* **2016**, *6*, 35. [CrossRef] [PubMed]
132. Singh, A.; van der Meulen, N.P.; Muller, C.; Klette, I.; Kulkarni, H.R.; Türler, A.; Schibli, R.; Baum, R.P. First-in-human PET/CT imaging of metastatic neuroendocrine neoplasms with cyclotron-produced ^{44}Sc-DOTATOC: A proof-of-concept study. *Cancer Biother. Radiopharm.* **2017**, *32*, 124–132. [CrossRef] [PubMed]
133. Baum, R.P.; Singh, A.; Benešová, M.; Vermeulen, C.; Gnesin, S.; Köster, U.; Johnston, K.; Müller, D.; Senftleben, S.; Kulkarni, H.R.; et al. Clinical evaluation of the radiolanthanide terbium-152: First-in-human PET/CT with ^{152}Tb-DOTATOC. *Dalton Trans.* **2017**, *46*, 14638–14646. [CrossRef] [PubMed]
134. Koumarianou, E.; Pawlak, D.; Korsak, A.; Mikolajczak, R. Comparison of receptor affinity of ^{nat}Sc-DOTA-TATE versus ^{nat}Ga-DOTA-TATE. *Nucl. Med. Rev. Cent. East Eur.* **2011**, *14*, 85–89. [CrossRef]
135. Domnanich, K.A.; Müller, C.; Farkas, R.; Schmid, R.M.; Ponsard, B.; Schibli, R.; Türler, A.; van der Meulen, N.P. ^{44}Sc for labeling of DOTA- and NODAGA-functionalized peptides: Preclinical in vitro and in vivo investigations. *EJNMMI Radiopharm. Chem.* **2017**, *1*, 8. [CrossRef]
136. Sinnes, J.P.; Nagel, J.; Rösch, F. AAZTA5/AAZTA5-TOC: Synthesis and radiochemical evaluation with ^{68}Ga, ^{44}Sc and ^{177}Lu. *EJNMMI Radiopharm. Chem.* **2019**, *4*, 18. [CrossRef]
137. Müller, C.; Fischer, E.; Behe, M.; Köster, U.; Dorrer, H.; Reber, J.; Haller, S.; Cohrs, S.; Blanc, A.; Grünberg, J.; et al. Future prospects for SPECT imaging using the radiolanthanide terbium-155—Production and preclinical evaluation in tumor-bearing mice. *Nucl. Med. Biol.* **2014**, *41*, e58–e65. [CrossRef]
138. Walrand, S.; Jamar, F.; Mathieu, I.; De Camps, J.; Lonneux, M.; Sibomana, M.; Labar, D.; Michel, C.; Pauwels, S. Quantitation in PET using isotopes emitting prompt single gammas: Application to yttrium-86. *Eur. J. Nucl. Med. Mol. Imaging* **2003**, *30*, 354–361. [CrossRef]
139. Helisch, A.; Förster, G.J.; Reber, H.; Buchholz, H.G.; Arnold, R.; Göke, B.; Weber, M.M.; Wiedenmann, B.; Pauwels, S.; Haus, U.; et al. Pre-therapeutic dosimetry and biodistribution of ^{86}Y-DOTA-Phe^{1}-Tyr^{3}-octreotide versus ^{111}In-pentetreotide in patients with advanced neuroendocrine tumours. *Eur. J. Nucl. Med. Mol. Imaging* **2004**, *31*, 1386–1392. [CrossRef]
140. Clifford, T.; Boswell, C.A.; Biddlecombe, G.B.; Lewis, J.S.; Brechbiel, M.W. Validation of a novel CHX-A "derivative suitable for peptide conjugation: Small animal PET/CT imaging using yttrium-86-CHX-A"-octreotide. *J. Med. Chem.* **2006**, *49*, 4297–4304. [CrossRef]
141. Jamar, F.; Barone, R.; Mathieu, I.; Walrand, S.; Labar, D.; Carlier, P.; de Camps, J.; Schran, H.; Chen, T.; Smith, M.C.; et al. ^{86}Y-DOTA^{0}-D-Phe^{1}-Tyr^{3}-octreotide (SMT487)—A phase 1 clinical study: Pharmacokinetics, biodistribution and renal protective effect of different regimens of amino acid co-infusion. *Eur. J. Nucl. Med. Mol. Imaging* **2003**, *30*, 510–518. [CrossRef] [PubMed]
142. Mikolajczak, R.; van der Meulen, N.P.; Lapi, S.E. Radiometals for imaging and theranostics, current production, and future perspectives. *J. Label. Compd. Radiopharm.* **2019**, *62*, 615–634. [CrossRef] [PubMed]
143. do Carmo, S.J.C.; Scott, P.J.H.; Alves, F. Production of radiometals in liquid targets. *EJNMMI Radiopharm. Chem.* **2020**, *5*, 2. [CrossRef] [PubMed]
144. Talip, Z.; Favaretto, C.; Geistlich, S.; Meulen, N.P.V. A Step-by-Step Guide for the Novel Radiometal Production for Medical Applications: Case Studies with ^{68}Ga, ^{44}Sc, ^{177}Lu and ^{161}Tb. *Molecules* **2020**, *25*, 966. [CrossRef]
145. Waldmann, C.M.; Stuparu, A.D.; van Dam, R.M.; Slavik, R. The search for an alternative to [^{68}Ga]Ga-DOTA-TATE in neuroendocrine tumor theranostics: Current state of ^{18}F-labeled somatostatin analog development. *Theranostics* **2019**, *9*, 1336–1347. [CrossRef]
146. Guhlke, S.; Wester, H.J.; Bruns, C.; Stöcklin, G. (2-[^{18}F]Fluoropropionyl-(D)phe^{1})-octreotide, a potential radiopharmaceutical for quantitative somatostatin receptor imaging with PET: Synthesis, Radiolabeling, in vitro validation and biodistribution in mice. *Nucl. Med. Biol.* **1994**, *21*, 819–825. [CrossRef]
147. Hostetler, E.D.; Edwards, W.B.; Anderson, C.J.; Welch, M.J. Synthesis of 4-[^{18}F]fluorobenzoyl octreotide and biodistribution in tumour-bearing Lewis rats. *J. Label. Compd Radiopharm.* **1999**, *42*, S720–S722.

148. Wester, H.J.; Schottelius, M.; Poethko, T.; Bruus-Jensen, K.; Schwaiger, M. Radiolabeled Carbohydrated Somatostatin Analogs: A Review of the Current Status. *Cancer Biother. Radiopharm.* **2004**, *19*, 231–244. [CrossRef]
149. Maschauer, S.; Heilmann, M.; Wängler, C.; Schirrmacher, R.; Prante, O. Radiosynthesis and preclinical evaluation of ^{18}F-fluoroglycosylated octreotate for somatostatin receptor imaging. *Bioconjugate Chem.* **2016**, *27*, 2707–2714. [CrossRef]
150. Liu, Z.; Pourghiasian, M.; Bénard, F.; Pan, J.; Lin, K.S.; Perrin, D.M. Preclinical evaluation of a high-affinity ^{18}F-trifluoroborate octreotate derivative for somatostatin receptor imaging. *J. Nucl. Med.* **2014**, *55*, 1499–1505. [CrossRef]
151. Niedermoser, S.; Chin, J.; Wängler, C.; Kostikov, A.; Bernard-Gauthier, V.; Vogler, N.; Soucy, J.P.; McEwan, A.J.; Schirrmacher, R.; Wängler, B. In vivo evaluation of ^{18}F-SiFAlin-Modified TATE: A potential challenge for ^{68}Ga-DOTATATE, the clinical gold standard for somatostatin receptor imaging with PET. *J. Nucl. Med.* **2015**, *56*, 1100–1105. [CrossRef]
152. Allott, L.; Dubash, S.; Aboagye, E.O. [^{18}F]FET-βAG-TOCA: The Design, Evaluation and Clinical Translation of a Fluorinated Octreotide. *Cancers* **2020**, *12*, 865. [CrossRef] [PubMed]
153. Laverman, P.; McBride, W.J.; Sharkey, R.M.; Eek, A.; Joosten, L.; Oyen, W.J.G.; Goldenberg, D.M.; Boerman, O.C. A novel facile method of labeling octreotide with ^{18}F-fluorine. *J. Nucl. Med.* **2010**, *51*, 454–461. [CrossRef] [PubMed]
154. Meisetschlaeger, G.; Poethko, T.; Stahl, A.; Wolf, I.; Scheidhauer, K.; Schottelius, M.; Herz, M.; Wester, H.J.; Schwaiger, M. Gluc-Lys([^{18}F]FP)-TOCA PET in patients with SSTR-positive tumors: Biodistribution and diagnostic evaluation compared with [^{111}In]DTPA-octreotide. *J. Nucl. Med.* **2006**, *47*, 566–573.
155. Ilhan, H.; Lindner, S.; Todica, A.; Cyran, C.C.; Tiling, R.; Auernhammer, C.J.; Spitzweg, C.; Boeck, S.; Unterrainer, M.; Gildehaus, F.J.; et al. Biodistribution and first clinical results of ^{18}F-SiFAlin-TATE PET: A novel ^{18}F-labeled somatostatin analog for imaging of neuroendocrine tumors. *Eur. J. Nucl. Med. Mol. Imaging* **2020**, *47*, 870–880. [CrossRef] [PubMed]
156. Pauwels, E.; Cleeren, F.; Tshibangu, T.; Koole, M.; Serdons, K.; Dekervel, J.; Van Cutsem, E.; Verslype, C.; Van Laere, K.; Bormans, G.; et al. [^{18}F]AlF-NOTA-octreotide PET imaging: Biodistribution, dosimetry and first comparison with [^{68}Ga]Ga-DOTATATE in neuroendocrine tumour patients. *Eur. J. Nucl. Med. Mol. Imaging* **2020**. [CrossRef]
157. Kemerink, G.J.; Visser, M.G.; Franssen, R.; Beijer, E.; Zamburlini, M.; Halders, S.G.; Brans, B.; Mottaghy, F.M.; Teule, G.J. Effect of the positron range of ^{18}F, ^{68}Ga and ^{124}I on PET/CT in lung-equivalent materials. *Eur. J. Nucl. Med. Mol. Imaging* **2011**, *38*, 940–948. [CrossRef]
158. Uccelli, L.; Martini, P.; Cittanti, C.; Carnevale, A.; Missiroli, L.; Giganti, M.; Bartolomei, M.; Boschi, A. Therapeutic Radiometals: Worldwide Scientific Literature Trend Analysis (2008–2018). *Molecules* **2019**, *24*, 640. [CrossRef]
159. Cremonesi, M.; Ferrari, M.E.; Bodei, L.; Chiesa, C.; Sarnelli, A.; Garibaldi, C.; Pacilio, M.; Strigari, L.; Summers, P.E.; Orecchia, R.; et al. Correlation of dose with toxicity and tumour response to ^{90}Y- and ^{177}Lu-PRRT provides the basis for optimization through individualized treatment planning. *Eur. J. Nucl. Med. Mol. Imaging* **2018**, *45*, 2426–2441. [CrossRef]
160. Otte, A.; Mueller-Brand, J.; Dellas, S.; Nitzsche, E.U.; Herrmann, R.; Maecke, H.R. Yttrium-90 labeled somatostatin-analogue for cancer treatment. *Lancet* **1998**, *351*, 417–418. [CrossRef]
161. Otte, A.; Herrmann, R.; Heppeler, A.; Behe, M.; Jermann, E.; Powell, P.; Maecke, H.R.; Muller, J. Yttrium-90 DOTATOC: First clinical results. *Eur. J. Nucl. Med.* **1999**, *26*, 1439–1447. [CrossRef] [PubMed]
162. Vinjamuri, S.; Gilbert, T.M.; Banks, M.; McKane, G.; Maltby, P.; Poston, G.; Weissman, H.; Palmer, D.H.; Vora, J.; Pritchard, D.M.; et al. Peptide Receptor Radionuclide Therapy With ^{90}Y-DOTATATE/^{90}Y-DOTATOC in Patients with Progressive Metastatic Neuroendocrine Tumours: Assessment of Response, Survival and Toxicity. *Br. J. Cancer* **2013**, *108*, 1440–1448. [CrossRef] [PubMed]
163. Virgolini, I.; Britton, K.; Buscombe, J.; Moncayo, R.; Paganelli, G.; Riva, P. In- and Y-DOTA-lanreotide: Results and implications of the MAURITIUS trial. *Semin. Nucl. Med.* **2002**, *32*, 148–155. [CrossRef] [PubMed]
164. Bodei, L.; Cremonesi, M.; Grana, C.M.; Chinol, M.; Baio, S.M.; Severi, S.; Paganelli, G. Yttrium-labeled peptides for therapy of NET. *Eur. J. Nucl. Med. Mol. Imaging* **2012**, *39*, S93–S102. [CrossRef]
165. Gabriel, M.; Nilica, B.; Kaiser, B.; Virgolini, I.J. Twelve-Year Follow-up After Peptide Receptor Radionuclide Therapy. *J. Nucl. Med.* **2019**, *60*, 524–529. [CrossRef]

166. Baum, R.P.; Kluge, A.W.; Kulkarni, H.; Schorr-Neufing, U.; Niepsch, K.; Bitterlich, N.; van Echteld, C.J. [^{177}Lu-DOTA]^0-D-Phe^1-Tyr^3-Octreotide (^{177}Lu-DOTATOC) For Peptide Receptor Radiotherapy in Patients with Advanced Neuroendocrine Tumours: A Phase-II Study. *Theranostics* **2016**, *6*, 501–510. [CrossRef]
167. Esser, J.P.; Krenning, E.P.; Teunissen, J.J.; Kooij, P.P.; van Gameren, A.L.; Bakker, W.H.; Kwekkeboom, D.J. Comparison of [^{177}Lu-DOTA^0,Tyr^3]octreotate and [^{177}Lu-DOTA^0,Tyr^3]octreotide: Which peptide is preferable for PRRT? *Eur. J. Nucl. Med. Mol. Imaging* **2006**, *33*, 1346–1351. [CrossRef]
168. de Jong, M.; Breeman, W.A.; Valkema, R.; Bernard, B.F.; Krenning, E.P. Combination radionuclide therapy using ^{177}Lu- and ^{90}Y-labeled somatostatin analogs. *J. Nucl. Med.* **2005**, *46*, 13S–17S.
169. Kunikowska, J.; Królicki, L.; Hubalewska-Dydejczyk, A.; Mikołajczak, R.; Sowa-Staszczak, A.; Pawlak, D. Clinical results of radionuclide therapy of neuroendocrine tumours with ^{90}Y-DOTATATE and tandem ^{90}Y/^{177}Lu-DOTATATE: Which is a better therapy option? *Eur. J. Nucl. Med. Mol. Imaging* **2011**, *38*, 1788–1797. [CrossRef]
170. Kunikowska, J.; Zemczak, A.; Kołodziej, M.; Gut, P.; Łoń, I.; Pawlak, D.; Mikołajczak, R.; Kamiński, G.; Ruchała, M.; Kos-Kudła, B.; et al. Tandem peptide receptor radionuclide therapy using ^{90}Y/^{177}Lu-DOTATATE for neuroendocrine tumors efficacy and side-effects—Polish multicenter experience. *Eur. J. Nucl. Med. Mol. Imaging* **2020**, *47*, 922–933. [CrossRef]
171. Brabander, T.; Nonnekens, J.; Hofland, J. The next generation of peptide receptor radionuclide therapy. *Endocr. Relat. Cancer* **2019**, *26*, C7–C11. [CrossRef] [PubMed]
172. Adant, S.; Shah, G.M.; Beauregard, J.M. Combination treatments to enhance peptide receptor radionuclide therapy of neuroendocrine tumours. *Eur. J. Nucl. Med. Mol. Imaging* **2020**, *47*, 907–921. [CrossRef] [PubMed]
173. Strosberg, J.; El-Haddad, G.; Wolin, E.; Hendifar, A.; Yao, J.; Chasen, B.; Mittra, E.; Kunz, P.L.; Kulke, M.H.; Jacene, H.; et al. Phase 3 Trial of 177Lu-Dotatate for Midgut Neuroendocrine Tumors. *N. Engl. J. Med.* **2017**, *376*, 125–135. [CrossRef] [PubMed]
174. Hennrich, U.; Kopka, K. Lutathera®: The First FDA- and EMA-Approved Radiopharmaceutical for Peptide Receptor Radionuclide Therapy. *Pharmaceuticals* **2019**, *12*, 114. [CrossRef]
175. Werner, R.A.; Bluemel, C.; Allen-Auerbach, M.S.; Higuchi, T.; Herrmann, K. ^{68}Gallium- and ^{90}Yttrium-/^{177}Lutetium: "theranostic twins" for diagnosis and treatment of NETs. *Ann. Nucl. Med.* **2015**, *29*, 1–7. [CrossRef]
176. Waseem, N.; Aparici, C.M.; Kunz, P.L. Evaluating the Role of Theranostics in Grade 3 Neuroendocrine Neoplasms. *J. Nucl. Med.* **2019**, *60*, 882–891. [CrossRef]
177. Sorbye, H.; Kong, G.; Grozinsky-Glasberg, S. PRRT in high-grade gastroenteropancreatic neuroendocrine neoplasms (WHO G3). *Endocr. Relat. Cancer* **2020**, *27*, R67–R77. [CrossRef]
178. Mak, I.Y.F.; Hayes, A.R.; Khoo, B.; Grossman, A. Peptide Receptor Radionuclide Therapy as a Novel Treatment for Metastatic and Invasive Phaeochromocytoma and Paraganglioma. *Neuroendocrinology* **2019**, *109*, 287–298. [CrossRef]
179. Vyakaranam, A.R.; Crona, J.; Norlén, O.; Granberg, D.; Garske-Román, U.; Sandström, M.; Fröss-Baron, K.; Thiis-Evensen, E.; Hellman, P.; Sundin, A. Favorable Outcome in Patients with Pheochromocytoma and Paraganglioma Treated with ^{177}Lu-DOTATATE. *Cancers* **2019**, *11*, 909. [CrossRef]
180. Lepareur, N.; Lacœuille, F.; Bouvry, C.; Hindré, F.; Garcion, E.; Chérel, M.; Noiret, N.; Garin, E.; Knapp, F.F.R., Jr. Rhenium-188 Labeled Radiopharmaceuticals: Current Clinical Applications in Oncology and Promising Perspectives. *Front. Med.* **2019**, *6*, 132. [CrossRef]
181. Zamora, P.O.; Gulhke, S.; Bender, H.; Diekmann, D.; Rhodes, B.A.; Biersack, H.J.; Knapp, F.F., Jr. Experimental radiotherapy of receptor-positive human prostate adenocarcinoma with ^{188}Re-RC-160, a directly-radiolabeled somatostatin analogue. *Int. J. Cancer* **1996**, *65*, 214–220. [CrossRef]
182. Zamora, P.O.; Bender, H.; Gulhke, S.; Marek, M.J.; Knapp, F.F., Jr.; Rhodes, B.A.; Biersack, H.J. Pre-clinical experience with Re-188-RC-160, a radiolabeled somatostatin analog for use in peptide-targeted radiotherapy. *Anticancer Res.* **1997**, *17*, 1803–1808. [PubMed]
183. Arteaga de Murphy, C.; Pedraza-López, M.; Ferro-Flores, G.; Murphy-Stack, E.; Chávez-Mercado, L.; Ascencio, J.A.; García-Salinas, L.; Hernández-Gutiérrez, S. Uptake of ^{188}Re-beta-naphthyl-peptide in cervical carcinoma tumours in athymic mice. *Nucl. Med. Biol.* **2001**, *28*, 319–326. [CrossRef]
184. Molina-Trinidad, E.M.; de Murphy, C.A.; Ferro-Flores, G.; Murphy-Stack, E.; Jung-Cook, H. Radiopharmacokinetic and dosimetric parameters of ^{188}Re-lanreotide in athymic mice with induced human cancer tumors. *Int. J. Pharm.* **2006**, *310*, 125–130. [CrossRef]

185. Molina-Trinidad, E.M.; de Murphy, C.A.; Jung-Cook, H.; Stack, E.M.; Pedraza-Lopez, M.; Morales-Marquez, J.L.; Serrano, G.V. Therapeutic ^{188}Re-lanreotide: Determination of radiopharmacokinetic parameters in rats. *J. Pharm. Pharmacol.* **2010**, *62*, 456–461. [CrossRef]
186. Cyr, J.E.; Pearson, D.A.; Wilson, D.M.; Nelson, C.A.; Guaraldi, M.; Azure, M.T.; Lister-James, J.; Dinkelborg, L.M.; Dean, R.T. Somatostatin receptor-binding peptides suitable for tumour radiotherapy with Re-188 or Re-186. Chemistry and initial biological studies. *J. Med. Chem.* **2007**, *50*, 1354–1364. [CrossRef]
187. Edelman, M.J.; Clamon, G.; Kahn, D.; Magram, M.; Lister-James, J.; Line, B.R. Targeted radiopharmaceutical therapy for advanced lung cancer: Phase I trial of rhenium Re188 P2045, a somatostatin analog. *J. Thorac. Oncol.* **2009**, *4*, 1550–1554. [CrossRef]
188. Nelson, C.A.; Azure, M.T.; Adams, C.T.; Zinn, K.R. The somatostatin analog ^{188}Re-P2045 inhibits the growth of AR42J pancreatic tumor xenografts. *J. Nucl. Med.* **2014**, *55*, 2020–2025. [CrossRef]
189. Champion, C.; Quinto, M.A.; Morgat, C.; Zanotti-Fregonara, P.; Hindié, E. Comparison between Three Promising ß-emitting Radionuclides, ^{67}Cu, ^{47}Sc and ^{161}Tb, with Emphasis on Doses Delivered to Minimal Residual Disease. *Theranostics* **2016**, *6*, 1611–1618. [CrossRef]
190. De Jong, M.; Breeman, W.A.; Bernard, B.F.; Rolleman, E.J.; Hofland, L.J.; Visser, T.J.; Setyono-Han, B.; Bakker, W.H.; van der Pluijm, M.E.; Krenning, E.P. Evaluation in vitro and in rats of ^{161}Tb-DTPA-octreotide, a somatostatin analogue with potential for intraoperative scanning and radiotherapy. *Eur. J. Nucl. Med.* **1995**, *22*, 608–616. [CrossRef]
191. Loveless, C.S.; Radford, L.L.; Ferran, S.J.; Queern, S.L.; Shepherd, M.R.; Lapi, S.E. Photonuclear production, chemistry, and in vitro evaluation of the theranostic radionuclide ^{47}Sc. *EJNMMI Res.* **2019**, *9*, 42. [CrossRef] [PubMed]
192. Tafreshi, N.K.; Doligalski, M.L.; Tichacek, C.J.; Pandya, D.N.; Budzevich, M.M.; El-Haddad, G.; Khushalani, N.I.; Moros, E.G.; McLaughlin, M.L.; Wadas, T.J.; et al. Development of Targeted Alpha Particle Therapy for Solid Tumors. *Molecules* **2019**, *24*, 4314. [CrossRef] [PubMed]
193. Nayak, T.; Norenberg, J.; Anderson, T.; Atcher, R. A comparison of high- versus low-linear energy transfer somatostatin receptor targeted radionuclide therapy in vitro. *Cancer Biother. Radiopharm.* **2005**, *20*, 52–57. [CrossRef] [PubMed]
194. Nayak, T.K.; Norenberg, J.P.; Anderson, T.L.; Prossnitz, E.R.; Stabin, M.G.; Atcher, R.W. Somatostatin-receptor-targeted α-emitting ^{213}Bi is therapeutically more effective than β^{-}-emitting ^{177}Lu in human pancreatic adenocarcinoma cells. *Nucl. Med. Biol.* **2007**, *34*, 185–193. [CrossRef] [PubMed]
195. Chan, H.S.; de Blois, E.; Morgenstern, A.; Bruchertseifer, F.; de Jong, M.; Breeman, W.; Konijnenberg, M. In Vitro comparison of ^{213}Bi- and ^{177}Lu-radiation for peptide receptor radionuclide therapy. *PLoS ONE* **2017**, *12*, e0181473. [CrossRef]
196. Norenberg, J.P.; Krenning, B.J.; Konings, I.R.H.M.; Kusewitt, D.F.; Nayak, T.K.; Anderson, T.L.; de Jong, M.; Garmestani, K.; Brechbiel, M.W.; Kvols, L.K. ^{213}Bi-[DOTA^{0}, Tyr^{3}]Octreotide Peptide Receptor Radionuclide Therapy of Pancreatic Tumors in a Preclinical Animal Model. *Clin. Cancer Res.* **2006**, *12*, 897–903. [CrossRef]
197. Miederer, M.; Henriksen, G.; Alke, A.; Mossbrugger, I.; Quintanilla-Martinez, L.; Senekowitsch-Schmidtke, R.; Essler, M. Preclinical evaluation of the alpha-particle generator nuclide ^{225}Ac for somatostatin receptor radiotherapy of neuroendocrine tumors. *Clin. Cancer Res.* **2008**, *14*, 3555–3561. [CrossRef]
198. Chan, H.S.; Konijnenberg, M.W.; de Blois, E.; Koelewijn, S.; Baum, R.P.; Morgenstern, A.; Bruchertseifer, F.; Breeman, W.A.; de Jong, M. Influence of tumour size on the efficacy of targeted alpha therapy with ^{213}Bi-[DOTA^{0},Tyr^{3}]-octreotate. *EJNMMI Res.* **2016**, *6*, 6. [CrossRef]
199. Kratochwil, C.; Giesel, F.L.; Bruchertseifer, F.; Mier, W.; Apostolidis, C.; Boll, R.; Murphy, K.; Haberkorn, U.; Morgenstern, A. ^{213}Bi-DOTATOC receptor-targeted alpha-radionuclide therapy induces remission in neuroendocrine tumours refractory to beta radiation: A first-in-human experience. *Eur. J. Nucl. Med. Mol. Imaging* **2014**, *41*, 2106–2119. [CrossRef]
200. Zhang, J.; Singh, A.; Kulkarni, H.R.; Schuchardt, C.; Müller, D.; Wester, H.J.; Maina, T.; Rösch, F.; van der Meulen, N.P.; Müller, C.; et al. From Bench to Bedside-The Bad Berka Experience with First-in-Human Studies. *Semin. Nucl. Med.* **2019**, *49*, 422–437. [CrossRef]
201. Ballal, S.; Yadav, M.P.; Bal, C.; Sahoo, R.K.; Tripathi, M. Broadening horizons with ^{225}Ac-DOTATATE targeted alpha therapy for gastroenteropancreatic neuroendocrine tumour patients stable or refractory to ^{177}Lu-DOTATATE PRRT: First clinical experience on the efficacy and safety. *Eur. J. Nucl. Med. Mol. Imaging* **2020**, *47*, 934–946. [CrossRef] [PubMed]

202. Stallons, T.A.R.; Saidi, A.; Tworowska, I.; Delpassand, E.S.; Torgue, J.J. Preclinical Investigation of ^{212}Pb-DOTAMTATE for Peptide Receptor Radionuclide Therapy in a Neuroendocrine Tumor Model. *Mol. Cancer Ther.* **2019**, *18*, 1012–1021. [CrossRef] [PubMed]
203. Tworowska, I.; Delpassand, E.S.; Bolek, L.; Shanoon, F.; Sgouros, G.; Frey, E.; He, B.; Muzammil, A.; Ghaly, M.; Stallons, T.; et al. Targeted Alpha-emitter Therapy of Neuroendocrine Tumors using ^{212}Pb-octreotate (AlphaMedix™). *J. Med. Imaging Radiat. Sci.* **2019**, *50*, S34. [CrossRef]
204. Guérard, F.; Gestin, J.F.; Brechbiel, M.W. Production of [^{211}At]-astatinated radiopharmaceuticals and applications in targeted α-particle therapy. *Cancer Biother. Radiopharm.* **2013**, *28*, 1–20. [CrossRef]
205. Vaidyanathan, G.; Boskovitz, A.; Shankar, S.; Zalutsky, M.R. Radioiodine and ^{211}At-labeled guanidinomethyl halobenzoyl octreotate conjugates: Potential peptide radiotherapeutics for somatostatin receptor-positive cancers. *Peptides* **2004**, *25*, 2087–2097. [CrossRef] [PubMed]
206. Vaidyanathan, G.; Affleck, D.J.; Schottelius, M.; Wester, H.; Friedman, H.S.; Zalutsky, M.R. Synthesis and evaluation of glycosylated octreotate analogues labeled with radioiodine and ^{211}At via a tin precursor. *Bioconjugate Chem.* **2006**, *17*, 195–203. [CrossRef]
207. Zhao, B.; Qin, S.; Chai, L.; Lu, G.; Yang, Y.; Cai, H.; Yuan, X.; Fan, S.; Huang, Q.; Yu, F. Evaluation of astatine-211-labeled octreotide as a potential radiotherapeutic agent for NSCLC treatment. *Bioorg. Med. Chem.* **2018**, *26*, 1086–1091. [CrossRef]
208. Krenning, E.P.; Valkema, R.; Kooij, P.P.; Breeman, W.A.; Bakker, W.H.; de Herder, W.W.; van Eijck, C.H.; Kwekkeboom, D.J.; de Jong, M.; Pauwels, S. Scintigraphy and radionuclide therapy with [indium-111-labeled-diethyltriamine penta-acetic acid-D-Phe^1]-octreotide. *Ital. J. Gastroenterol. Hepatol.* **1999**, *31*, S219–S223.
209. Valkema, R.; De Jong, M.; Bakker, W.H.; Breeman, W.A.; Kooij, P.P.; Lugtenburg, P.J.; De Jong, F.H.; Christiansen, A.; Kam, B.L.; De Herder, W.W.; et al. Phase I study of peptide receptor radionuclide therapy with [In-DTPA]octreotide: The Rotterdam experience. *Semin. Nucl. Med.* **2002**, *32*, 110–122. [CrossRef]
210. Anthony, L.B.; Woltering, E.A.; Espenan, G.D.; Cronin, M.D.; Maloney, T.J.; McCarthy, K.E. Indium-111-pentetreotide prolongs survival in gastroenteropancreatic malignancies. *Semin. Nucl. Med.* **2002**, *32*, 123–132. [CrossRef]
211. Buscombe, J.R.; Caplin, M.E.; Hilson, A.J. Long-term efficacy of high-activity ^{111}In-pentetreotide therapy in patients with disseminated neuroendocrine tumors. *J. Nucl. Med.* **2003**, *44*, 1–6. [PubMed]
212. Lewington, V.J. Targeted radionuclide therapy for neuroendocrine tumours. *Endocr. Relat. Cancer* **2003**, *10*, 497–501. [CrossRef] [PubMed]
213. Capello, A.; Krenning, E.; Bernard, B.; Reubi, J.C.; Breeman, W.; de Jong, M. ^{111}In-labeled somatostatin analogues in a rat tumour model: Somatostatin receptor status and effects of peptide receptor radionuclide therapy. *Eur. J. Nucl. Med. Mol. Imaging* **2005**, *32*, 1288–1295. [CrossRef]
214. Ginj, M.; Zhang, H.; Waser, B.; Cescato, R.; Wild, D.; Wang, X.; Erchegyi, J.; Rivier, J.; Mäcke, H.R.; Reubi, J.C. Radiolabeled somatostatin receptor antagonists are preferable to agonists for in vivo peptide receptor targeting of tumors. *Proc. Natl. Acad. Sci. USA* **2006**, *103*, 16436–16441. [CrossRef] [PubMed]
215. Waser, B.; Tamma, M.L.; Cescato, R.; Maecke, H.R.; Reubi, J.C. Highly efficient in vivo agonist induced internalization of sst2 receptors in somatostatin target tissues. *J. Nucl. Med.* **2009**, *50*, 936941. [CrossRef] [PubMed]
216. Cescato, R.; Schulz, S.; Waser, B.; Eltschinger, V.; Rivier, J.E.; Wester, H.J.; Culler, M.; Ginj, M.; Liu, Q.; Schonbrunn, A.; et al. Internalization of sst2, sst3, and sst5 receptors: Effects of somatostatin agonists and antagonists. *J. Nucl. Med.* **2006**, *47*, 502–511. [PubMed]
217. Fani, M.; Peitl, P.K.; Velikyan, I. Current status of radiopharmaceuticals for the theranostics of neuroendocrine neoplasms. *Pharmaceuticals* **2017**, *10*, 30. [CrossRef] [PubMed]
218. Reubi, J.C.; Waser, B.; Mäcke, H.; Rivier, J. Highly Increased ^{125}I-JR11 Antagonist Binding In Vitro Reveals Novel Indications for sst2 Targeting in Human Cancers. *J. Nucl. Med.* **2017**, *58*, 300–306. [CrossRef]
219. Dude, I.; Zhang, Z.; Rousseau, J.; Hundal-Jabal, N.; Colpo, N.; Merkens, H.; Lin, K.S.; Bénard, F. Evaluation of agonist and antagonist radioligands for somatostatin receptor imaging of breast cancer using positron emission tomography. *EJNMMI Radiopharm. Chem.* **2017**, *2*, 4. [CrossRef]
220. Rylova, S.N.; Stoykow, C.; Del Pozzo, L.; Abiraj, K.; Tamma, M.L.; Kiefer, Y.; Fani, M.; Maecke, H.R. The somatostatin receptor 2 antagonist ^{64}Cu-NODAGA-JR11 outperforms ^{64}Cu-DOTA-TATE in a mouse xenograft model. *PLoS ONE* **2018**, *13*, e0195802. [CrossRef]

221. Krebs, S.; Pandit-Taskar, N.; Reidy, D.; Beattie, B.J.; Lyashchenko, S.K.; Lewis, J.S.; Bodei, L.; Weber, W.A.; O'Donoghue, J.A. Biodistribution and radiation dose estimates for ^{68}Ga-DOTA-JR11 in patients with metastatic neuroendocrine tumors. *Eur. J. Nucl. Med. Mol. Imaging* **2019**, *46*, 677–685. [CrossRef] [PubMed]
222. Wild, D.; Fani, M.; Behe, M.; Brink, I.; Rivier, J.E.; Reubi, J.C.; Maecke, H.R.; Weber, W.A. First clinical evidence that imaging with somatostatin receptor antagonists is feasible. *J. Nucl. Med.* **2011**, *52*, 1412–1417. [CrossRef] [PubMed]
223. Fani, M.; Del Pozzo, L.; Abiraj, K.; Mansi, R.; Tamma, M.L.; Cescato, R.; Waser, B.; Weber, W.A.; Reubi, J.C.; Maecke, H.R. PET of somatostatin receptor-positive tumors using ^{64}Cu- and ^{68}Ga-somatostatin antagonists: The chelate makes the difference. *J. Nucl. Med.* **2011**, *52*, 1110–1118. [CrossRef] [PubMed]
224. Fani, M.; Braun, F.; Waser, B.; Beetschen, K.; Cescato, R.; Erchegyi, J.; Rivier, J.E.; Weber, W.A.; Maecke, H.R.; Reubi, J.C. Unexpected sensitivity of sst2 antagonists to N-terminal radiometal modifications. *J. Nucl. Med.* **2012**, *53*, 1481–1489. [CrossRef]
225. Reubi, J.C.; Erchegyi, J.; Cescato, R.; Waser, B.; Rivier, J.E. Switch from antagonist to agonist after addition of a DOTA chelator to a somatostatin analog. *Eur. J. Nucl. Med. Mol. Imaging* **2010**, *37*, 15511558. [CrossRef]
226. Radford, L.; Gallazzi, F.; Watkinson, L.; Carmack, T.; Berendzen, A.; Lewis, M.R.; Jurisson, S.S.; Papagiannopoulou, D.; Hennkens, H.M. Synthesis and evaluation of a ^{99m}Tc tricarbonyl-labeled somatostatin receptor-targeting antagonist peptide for imaging of neuroendocrine tumors. *Nucl. Med. Biol.* **2017**, *47*, 4–9. [CrossRef]
227. Wild, D.; Fani, M.; Fischer, R.; Del Pozzo, L.; Kaul, F.; Krebs, S.; Fischer, R.; Rivier, J.E.; Reubi, J.C.; Maecke, H.R.; et al. Comparison of somatostatin receptor agonist and antagonist for peptide receptor radionuclide therapy: A pilot study. *J. Nucl. Med.* **2014**, *55*, 1248–1252. [CrossRef]
228. Fani, M.; Nicolas, G.P.; Wild, D. Somatostatin Receptor Antagonists for Imaging and Therapy. *J. Nucl. Med.* **2017**, *58*, 61S–66S. [CrossRef]
229. Mansi, R.; Fani, M. Design and development of the theranostic pair ^{177}Lu-OPS201/^{68}Ga-OPS202 for targeting somatostatin receptor expressing tumors. *J. Label. Compd. Radiopharm.* **2019**, *62*, 635–645. [CrossRef]
230. Nicolas, G.P.; Beykan, S.; Bouterfa, H.; Kaufmann, J.; Bauman, A.; Lassmann, M.; Reubi, J.C.; Rivier, J.E.F.; Maecke, H.R.; Fani, M.; et al. Safety, Biodistribution, and Radiation Dosimetry of ^{68}Ga-OPS202 in Patients with Gastroenteropancreatic Neuroendocrine Tumors: A Prospective Phase I Imaging Study. *J. Nucl. Med.* **2018**, *59*, 909–914. [CrossRef]
231. Nicolas, G.P.; Schreiter, N.; Kaul, F.; Uiters, J.; Bouterfa, H.; Kaufmann, J.; Erlanger, T.E.; Cathomas, R.; Christ, E.; Fani, M.; et al. Sensitivity Comparison of ^{68}Ga-OPS202 and ^{68}Ga-DOTATOC PET/CT in Patients with Gastroenteropancreatic Neuroendocrine Tumors: A Prospective Phase II Imaging Study. *J. Nucl. Med.* **2018**, *59*, 915–921. [CrossRef] [PubMed]
232. Reidy-Lagunes, D.; Pandit-Taskar, N.; O'Donoghue, J.A.; Krebs, S.; Staton, K.D.; Lyashchenko, S.K.; Lewis, J.S.; Raj, N.; Gönen, M.; Lohrmann, C.; et al. Phase I Trial of Well-Differentiated Neuroendocrine Tumors (NETs) with Radiolabeled Somatostatin Antagonist ^{177}Lu-Satoreotide Tetraxetan. *Clin. Cancer Res.* **2019**, *25*, 6939–6947. [CrossRef] [PubMed]
233. Rangger, C.; Haubner, R. Radiolabeled Peptides for Positron Emission Tomography and Endoradiotherapy in Oncology. *Pharmaceuticals* **2020**, *13*, 22. [CrossRef] [PubMed]
234. Villard, L.; Romer, A.; Marincek, N.; Brunner, P.; Koller, M.T.; Schindler, C.; Ng, Q.K.T.; Mäcke, H.R.; Müller-Brand, J.; Rochlitz, C.; et al. Cohort study of somatostatin-based radiopeptide therapy with [^{90}Y-DOTA]-TOC versus [^{90}Y-DOTA]-TOC plus [^{177}Lu-DOTA]-TOC in neuroendocrine cancers. *J. Clin. Oncol.* **2012**, *30*, 1100–1106. [CrossRef]
235. Gill, M.R.; Falzone, N.; Du, Y.; Vallis, K.A. Targeted radionuclide therapy in combined-modality regimens. *Lancet Oncol.* **2017**, *18*, e414–e423. [CrossRef]
236. Hartrampf, P.E.; Hänscheid, H.; Kertels, O.; Schirbel, A.; Kreissl, M.C.; Flentje, M.; Sweeney, R.A.; Buck, A.K.; Polat, B.; Lapa, C. Long-term results of multimodal peptide receptor radionuclide therapy and fractionated external beam radiotherapy for treatment of advanced symptomatic meningioma. *Clin. Transl. Radiat. Oncol.* **2020**, *22*, 29–32. [CrossRef]
237. Reubi, J.C.; Maecke, H.R. Approaches to multireceptor targeting: Hybrid radioligands, radioligand cocktails, and sequential radioligand applications. *J. Nucl. Med.* **2017**, *58*, 10S–16S. [CrossRef]

238. Ghosh, S.C.; Rodriguez, M.; Carmon, K.S.; Voss, J.; Wilganowski, N.L.; Schonbrunn, A.; Azhdarinia, A. A Modular Dual-Labeling Scaffold That Retains Agonistic Properties for Somatostatin Receptor Targeting. *J. Nucl. Med.* **2017**, *58*, 1858–1864. [CrossRef]
239. Langbein, T.; Weber, W.A.; Eiber, M. Future of Theranostics: An Outlook on Precision Oncology in Nuclear Medicine. *J. Nucl. Med.* **2019**, *60*, 13S–19S. [CrossRef]
240. Jones, W.; Griffiths, K.; Barata, P.C.; Paller, C.J. PSMA Theranostics: Review of the Current Status of PSMA-Targeted Imaging and Radioligand Therapy. *Cancers* **2020**, *12*, 1367. [CrossRef]

Article

Evaluation of Met-Val-Lys as a Renal Brush Border Enzyme-Cleavable Linker to Reduce Kidney Uptake of ^{68}Ga-Labeled DOTA-Conjugated Peptides and Peptidomimetics

Shreya Bendre [1], Zhengxing Zhang [1], Hsiou-Ting Kuo [1], Julie Rousseau [1], Chengcheng Zhang [1], Helen Merkens [1], Áron Roxin [1], François Bénard [1,2,3] and Kuo-Shyan Lin [1,2,3,*]

1 Department of Molecular Oncology, BC Cancer, Vancouver, BC V5Z 1L3, Canada; sbendre@bccrc.ca (S.B.); zzhang@bccrc.ca (Z.Z.); hkuo@bccrc.ca (H.-T.K.); jrousseau@bccrc.ca (J.R.); cczhang@bccrc.ca (C.Z.); hmerkens@bccrc.ca (H.M.); aroxin@bccrc.ca (Á.R.); fbenard@bccrc.ca (F.B.)
2 Department of Functional Imaging, BC Cancer, Vancouver, BC V5Z 4E6, Canada
3 Department of Radiology, University of British Columbia, Vancouver, BC V5Z 1M9, Canada
* Correspondence: klin@bccrc.ca

Academic Editor: Krishan Kumar
Received: 29 July 2020; Accepted: 21 August 2020; Published: 25 August 2020

Abstract: High kidney uptake is a common feature of peptide-based radiopharmaceuticals, leading to reduced detection sensitivity for lesions adjacent to kidneys and lower maximum tolerated therapeutic dose. In this study, we evaluated if the Met-Val-Lys (MVK) linker could be used to lower kidney uptake of ^{68}Ga-labeled DOTA-conjugated peptides and peptidomimetics. A model compound, [^{68}Ga]Ga-DOTA-AmBz-MVK(Ac)-OH (AmBz: aminomethylbenzoyl), and its derivative, [^{68}Ga]Ga-DOTA-AmBz-MVK(HTK01166)-OH, coupled with the PSMA (prostate-specific membrane antigen)-targeting motif of the previously reported HTK01166 were synthesized and evaluated to determine if they could be recognized and cleaved by the renal brush border enzymes. Additionally, positron emission tomography (PET) imaging, ex vivo biodistribution and in vivo stability studies were conducted in mice to evaluate their pharmacokinetics. [^{68}Ga]Ga-DOTA-AmBz-MVK(Ac)-OH was effectively cleaved specifically by neutral endopeptidase (NEP) of renal brush border enzymes at the Met-Val amide bond, and the radio-metabolite [^{68}Ga]Ga-DOTA-AmBz-Met-OH was rapidly excreted via the renal pathway with minimal kidney retention. [^{68}Ga]Ga-DOTA-AmBz-MVK(HTK01166)-OH retained its PSMA-targeting capability and was also cleaved by NEP, although less effectively when compared to [^{68}Ga]Ga-DOTA-AmBz-MVK(Ac)-OH. The kidney uptake of [^{68}Ga]Ga-DOTA-AmBz-MVK(HTK01166)-OH was 30% less compared to that of [^{68}Ga]Ga-HTK01166. Our data demonstrated that derivatives of [^{68}Ga]Ga-DOTA-AmBz-MVK-OH can be cleaved specifically by NEP, and therefore, MVK can be a promising cleavable linker for use to reduce kidney uptake of radiolabeled DOTA-conjugated peptides and peptidomimetics.

Keywords: radiopharmaceuticals; kidney uptake; cleavable linkers; neutral endopeptidase (NEP); renal brush border enzymes; prostate-specific membrane antigen (PSMA); cancer imaging and therapy

1. Introduction

The use of low molecular weight radiolabeled peptides and antibody fragments for applications in oncology is rapidly gaining momentum [1–5]. Such a site-directed radiation delivery involves targeting of certain specific receptors overexpressed on the surface of cancer cells for the purpose of targeted imaging and radionuclide therapy [6]. While these oncophilic molecules serve as biological targeting vectors, they commonly exhibit very high and sustained renal uptake [7]. This is caused by either a high renal expression of cancer markers targeted by these oncophilic molecules, megalin-cubilin mediated endocytosis and transcellular transport, or lysosomal proteolysis following glomerular filtration and renal reabsorption [7–9]. This reduces detection sensitivity for lesions adjacent to kidneys and lowers the maximum tolerated dose for radiotherapy.

Arano et al. reported an effective strategy to reduce kidney uptake of radiopharmaceuticals by incorporating specific cleavable linkages into them, with the Met-Val-Lys (MVK) sequence found to be the most effective thus far [1]. Use of this strategy is attributable to recognition and cleavage of MVK between Met-Val residues by a metalloendopeptidase enzyme called neutral endopeptidase or neprilysin (NEP) [10]. This enzyme is found in abundance on the renal brush border membrane lining, particularly of the proximal convoluted tubules [11]. Arano et al. exploited this renal brush border enzyme and designed a NOTA(1,4,7-triazacyclononane-1,4,7-triacetic acid)-conjugated $^{67/68}$Ga-labeled antibody fragment bearing this MVK linker sequence to successfully lower renal uptake by 80% at 3 h post-injection (p.i.) without loss of tumor uptake [12].

By adopting this design strategy Zhang et al. recently reported the comparison of two radiolabeled Exendin 4 derivatives for imaging the expression of the glucagon-like peptide-1 receptor (GLP-1R) with positron emission tomography (PET) [13]. GLP-1R is highly expressed in insulinomas. However, the very high and sustained uptake of radiolabeled GLP-1R-targeting Exendin 4 derivatives in kidneys hinders the application of these tracers for detecting insulinomas. Compared with [^{68}Ga]Ga-NOTA-Cys^{40}-Leu^{14}-Exendin 4, the derivative bearing the MVK linker ([^{68}Ga]Ga-NOTA-MVK-Cys^{40}-Leu^{14}-Exendin 4) had similar tumor uptake values but only one third of kidney uptake at 2 h p.i., greatly enhancing the tumor-to-kidney contrast and detection sensitivity [13].

Despite successful application of this strategy to lower kidney uptake of radiolabeled peptides, the use of NOTA as a radiometal chelator unfortunately excludes application for using therapeutic isotopes like ^{177}Lu [14]. ^{177}Lu emits both β- and γ-radiation and is widely used as a theranostic pair with the positron-emitter ^{68}Ga. The preferred radiometal chelator for potential theranostic applications is 1,4,7,10-tetraazacyclododecane-1,4,7,10-tetraacetic acid (DOTA), a macrocyclic bifunctional chelating agent which forms stable complexes with a variety of radiometals including ^{68}Ga and ^{177}Lu.

In this study, we evaluated the MVK sequence as a cleavable tripeptide linker to reduce renal uptake of radiolabeled DOTA-conjugated peptides and peptidomimetics. We modified the original design, NOTA-MVK(Targeting vector)-OH, reported by Uehara et al. [12], and replaced the NOTA chelator with DOTA for potential radiolabeling with both imaging and radiotherapeutic isotopes (Figure 1). We also inserted an aminomethylbenzoyl (AmBz) group between DOTA and the MVK sequence to mimic the aminobenzyl group in the original design. The maleimide-thiol linkage at the Lys side chain was replaced with an amide linkage to enable the facile synthesis of the DOTA-conjugated peptides on solid phase.

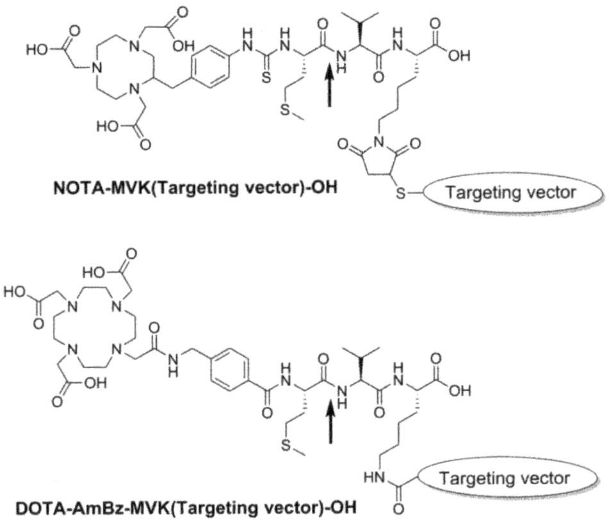

Figure 1. DOTA-AmBz-MVK(Targeting vector)-OH inspired by the reported design of NOTA-MVK (Targeting vector)-OH. The amide bond between Met-Val (pointed by an arrow) is recognized and cleaved by NEP.

We first synthesized the model compound, [^{68}Ga]Ga-DOTA-AmBz-MVK(Ac)-OH (Figure 2), and confirmed that it could be recognized by the renal brush border enzymes and cleaved at the Met-Val amide bond. We conducted PET imaging and in vivo stability studies in mice and confirmed that the expected radio-metabolite [^{68}Ga]Ga-DOTA-AmBz-Met-OH was rapidly excreted through the urinary pathway with minimal uptake in kidneys. We then replaced the acetyl group of [^{68}Ga]Ga-DOTA-AmBz-MVK(Ac)-OH with the pharmacophore of HTK01166 (Figure 2), a peptidomimetic targeting the prostate-specific membrane antigen (PSMA), which is highly expressed in prostate cancer [15]. The targeting pharmacophore of [^{68}Ga]Ga-HTK01166 was selected as it was shown previously to have very high renal uptake (147 %ID/g, 1 h p.i.) in mice. Since Met is prone to oxidation to generate methionine sulfoxide (Met(O) or M(O)) [16] and the resulting M(O)VK sequence might not be recognized and cleaved by the renal brush border enzyme, we also synthesized [^{68}Ga]Ga-DOTA-AmBz-M(O)VK(HTK01166)-OH for comparison. Here we report the syntheses of [^{68}Ga]Ga-DOTA-AmBz-MVK(Ac)-OH, [^{68}Ga]Ga-DOTA-AmBz-MVK(HTK01166)-OH and [^{68}Ga]Ga-DOTA-AmBz-M(O)VK(HTK01166)-OH, and the results of enzyme assay, PET imaging, ex vivo biodistribution, and in vivo stability studies to evaluate the applicability of using the MVK linker to reduce the renal uptake of radiolabeled DOTA-conjugated peptides and peptidomimetics.

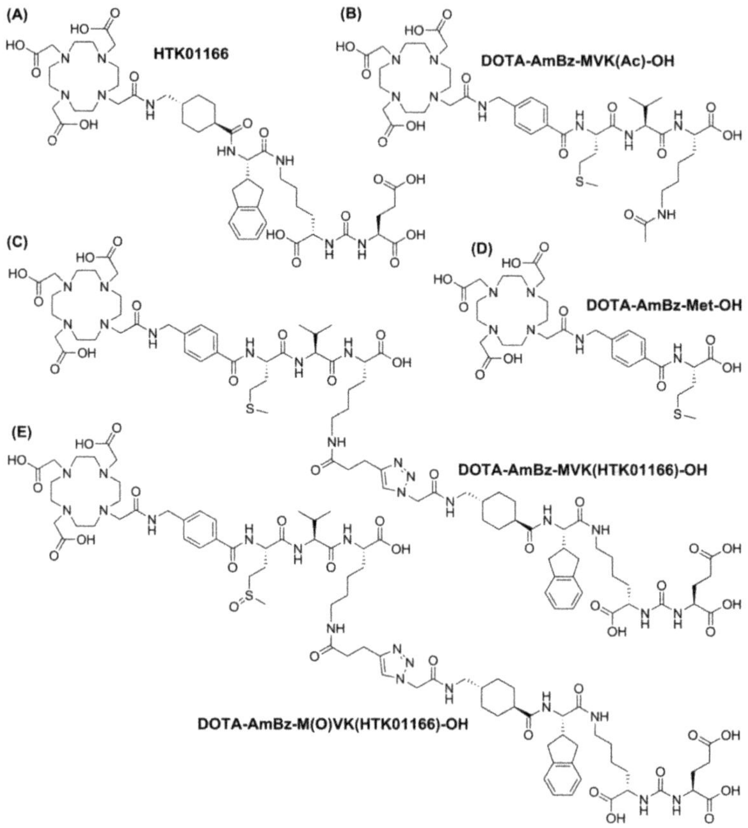

Figure 2. Chemical structures of (**A**) HTK01166, (**B**) DOTA-AmBz-MVK(Ac)-OH, (**C**) DOTA-AmBz-MVK (HTK01166)-OH, (**D**) DOTA-AmBz-Met-OH, and (**E**) DOTA-AmBz-M(O)VK(HTK01166)-OH.

2. Results

2.1. Chemistry and Radiochemistry

Synthesis of Fmoc-Lys(pentynoyl)-O*t*Bu (**2**) is shown in Scheme 1. Fmoc-Lys(pentynoyl)-OH (**1**) prepared from literature procedures [17] was reacted with 2,2,2-trichloroacetimidate (2.2 equiv) in CH_2Cl_2. After overnight incubation at room temperature and purification by flash column chromatography, the desired product **2** was obtained in 79% yield.

The DOTA-conjugated peptides and peptidomimetics including DOTA-AmBz-MVK(Ac)-OH, DOTA-AmBz-MVK(HTK01166)-OH, DOTA-AmBz-Met-OH, and DOTA-AmBz-M(O)VK(HTK01166)-OH (Figure 2) were assembled on solid phase (Schemes 2 and 3). After cleavage/deprotection with TFA and HPLC purification, these DOTA-conjugated peptides and peptidomimetics were obtained in 1–35% isolated yields. Their nonradioactive Ga-complexed standards were obtained by incubating the DOTA-conjugated peptides and peptidomimetics with excess $GaCl_3$ in acetate buffer (0.1 M, pH 4.2) at 80 °C. After HPLC purification, the nonradioactive Ga-complexed standards were obtained in 16–78% isolated yields. Detailed HPLC conditions and retention times for the purification of the DOTA-conjugated peptides and peptidomimetics and their nonradioactive Ga-complexed standards are provided in Supplemental Tables S1 and S2 (see the Supplemental Materials). The identities of the DOTA-conjugated peptides and peptidomimetics and their nonradioactive Ga-complexed standards were confirmed by MS analysis.

^{68}Ga labeling of the DOTA-conjugated peptides and peptidomimetics was conducted in HEPES buffer (2 M, pH 5.0) with microwave heating for 1 min. After HPLC purification, [^{68}Ga]Ga-DOTA-AmBz-MVK(Ac)-OH, [^{68}Ga]Ga-DOTA-AmBz-MVK(HTK01166)-OH and [^{68}Ga]Ga-DOTA-AmBz-M(O)VK(HTK01166)-OH were obtained in 40–78% decay-corrected radiochemical yields with >2 GBq/μmol molar activity and >93% radiochemical purity. Detailed HPLC conditions and retention times for the purification and quality control of the ^{68}Ga-labeled DOTA-conjugated peptides and peptidomimetics are provided in Supplemental Table S3 (see the Supplemental Materials).

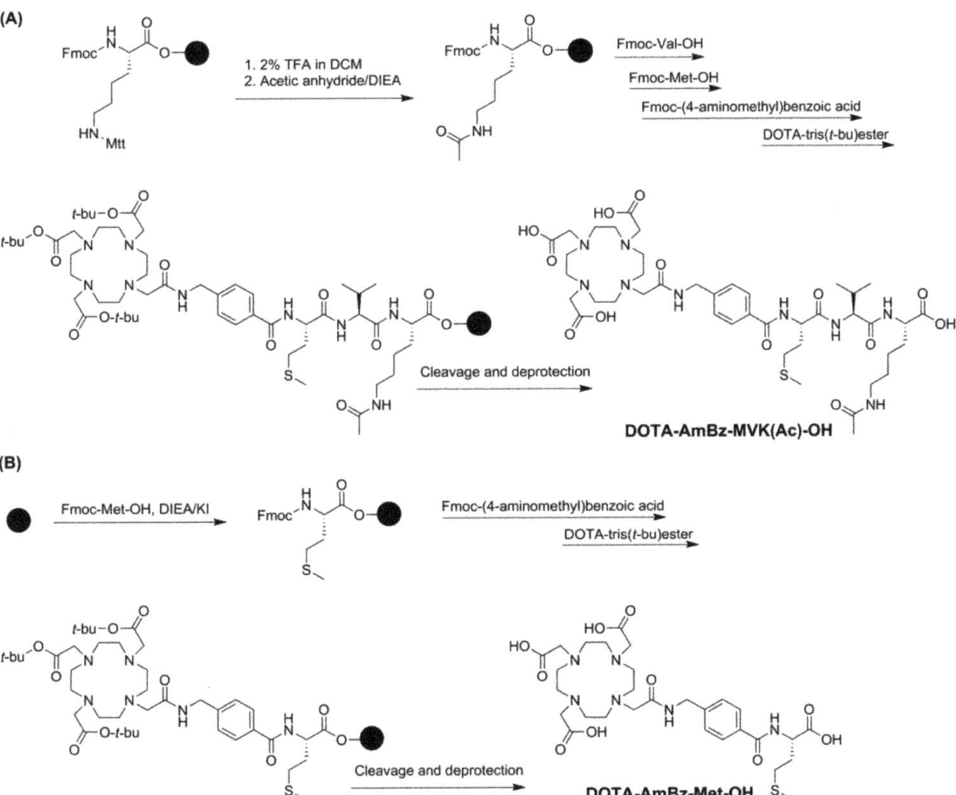

Scheme 1. Synthesis of Fmoc-L-Lys(pentynoyl)-O*t*Bu (**2**).

Scheme 2. Synthesis of (**A**) DOTA-AmBz-MVK(Ac)-OH and (**B**) DOTA-AmBz-Met-OH. • = resin.

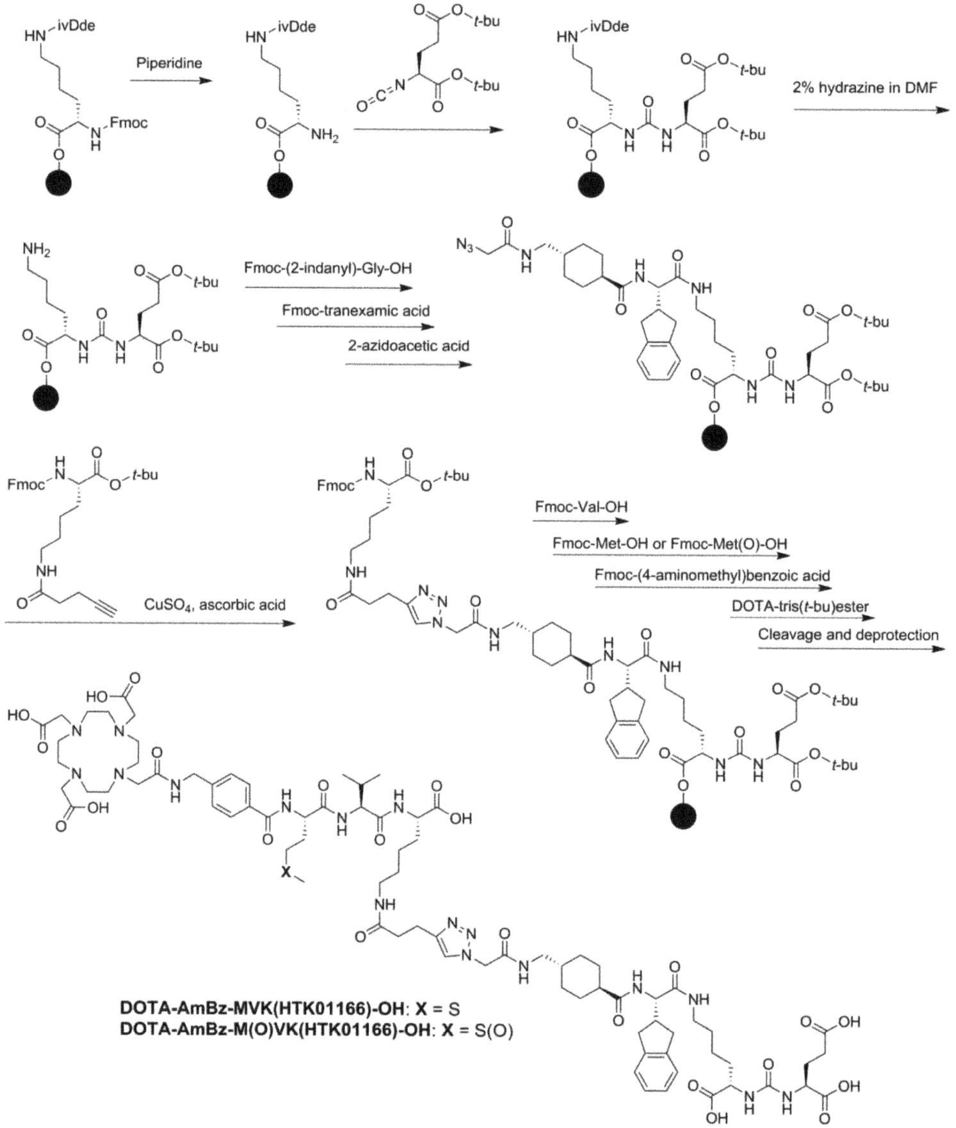

Scheme 3. Synthesis of DOTA-AmBz-MVK(HTK01166)-OH and DOTA-AmBz-M(O)VK(HTK01166)-OH. • = resin.

2.2. In Vitro Enzyme Assays

In vitro enzyme assays revealed very efficient cleavage (>95%) of the ^{68}Ga-labeled DOTA-conjugated linker, [^{68}Ga]Ga-DOTA-AmBz-MVK(Ac)-OH, with the expected fragment [^{68}Ga]Ga-DOTA-AmBz-Met-OH as the dominant radio-metabolite (>90%) (Figure 3A). The presence of the NEP inhibitor phosphoramidon significantly inhibited cleavage of [^{68}Ga]Ga-DOTA-AmBz-MVK(Ac)-OH, with 85% of the recovered radioactivity as the intact tracer (Figure 3B).

[^{68}Ga]Ga-DOTA-AmBz-MVK(HTK01166)-OH, a derivative of [^{68}Ga]Ga-DOTA-AmBz-MVK(Ac)-OH by replacing the acetyl group with the PSMA-targeting motif Lys-urea-Glu and the lipophilic linker of HTK01166, was cleaved less effectively under the same assay conditions. After 1-h incubation, ~80% of [^{68}Ga]Ga-DOTA-AmBz-MVK(HTK01166)-OH remained intact and ~16% of the recovered radioactivity was present as the expected radio-metabolite [^{68}Ga]Ga-DOTA-AmBz-Met-OH (Figure 4A). An unidentified radio-metabolite with retention time at ~4.3 min accounted for ~4% of the recovered radioactivity. In the presence of phosphoramidon, the intact [^{68}Ga]Ga-DOTA-AmBz-MVK(HTK01166)-OH fraction increased to ~95% and the formation of [^{68}Ga]Ga-DOTA-AmBz-Met-OH was completely inhibited (Figure 4B).

[^{68}Ga]Ga-DOTA-AmBz-M(O)VK(HTK01166)-OH, the oxidized version of [^{68}Ga]Ga-DOTA-AmBz-MVK(HTK01166)-OH was fairly stable under the same assay conditions with 94% and 99.5% remaining intact without and with the presence of phosphoramidon, respectively (Figure 5).

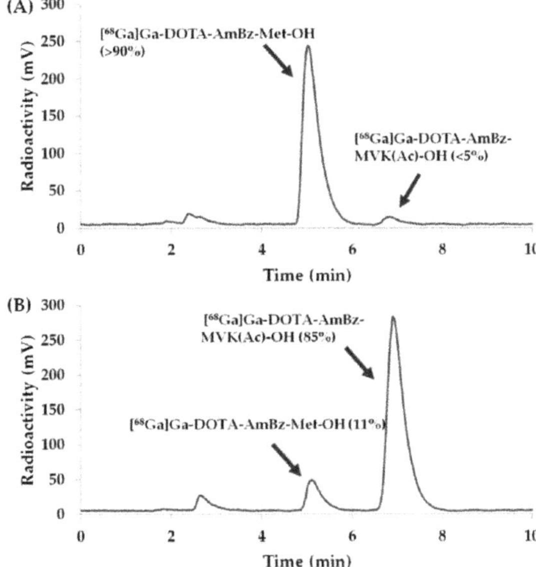

Figure 3. Radio-HPLC chromatograms of in vitro enzyme assay samples of [^{68}Ga]Ga-DOTA-AmBz-MVK(Ac)-OH (**A**) without and (**B**) with the presence of phosphoramidon. HPLC conditions were 82/18 A/B at a flow rate of 2 mL/min; A: H_2O containing 0.1% TFA; B: CH_3CN containing 0.1% TFA; HPLC column: Luna C18, 5 μm particle size, 100 Å pore size, 250 × 4.6 mm.

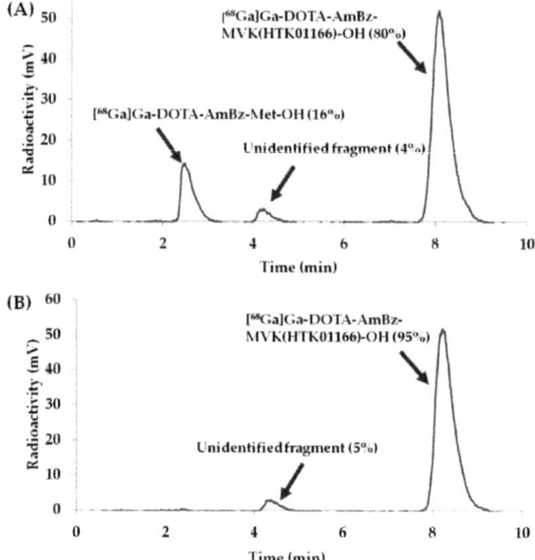

Figure 4. Radio-HPLC chromatograms of in vitro enzyme assay samples of [^{68}Ga]Ga-DOTA-AmBz-MVK(HTK01166)-OH (**A**) without and (**B**) with the presence of phosphoramidon. HPLC conditions were 72/28 A/B at a flow rate of 2 mL/min; A: H_2O containing 0.1% TFA; B: CH_3CN containing 0.1% TFA; HPLC column: Luna C18, 5 µm particle size, 100 Å pore size, 250 × 4.6 mm.

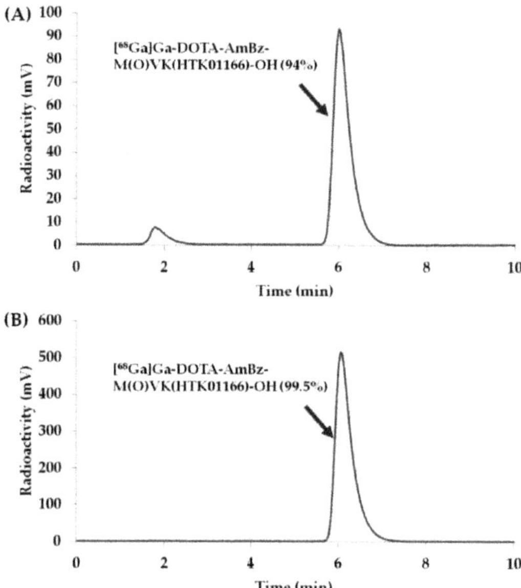

Figure 5. Radio-HPLC chromatograms of in vitro enzyme assay samples of [^{68}Ga]Ga-DOTA-AmBz-M(O)VK(HTK01166)-OH (**A**) without and (**B**) with the presence of phosphoramidon. HPLC conditions were 76/24 A/B at a flow rate of 2 mL/min; A: H_2O containing 0.1% TFA; B: CH_3CN containing 0.1% TFA; HPLC column: Luna C18, 5 µm particle size, 100 Å pore size, 250 × 4.6 mm.

2.3. PET/CT Imaging and Ex Vivo Biodistribution Studies

The pharmacokinetics of [^{68}Ga]Ga-DOTA-AmBz-MVK(Ac)-OH was first evaluated in mice via PET/CT imaging studies. As shown in Figure 6, radioactivity from the injected [^{68}Ga]Ga-DOTA-AmBz-MVK(Ac)-OH was excreted rapidly from blood pool and all background organs/tissues predominately via the renal pathway. At 1 h p.i., only kidneys and urinary bladder were clearly visualized in PET images, with low radioactivity (<2.5 %ID/g) retained in kidneys.

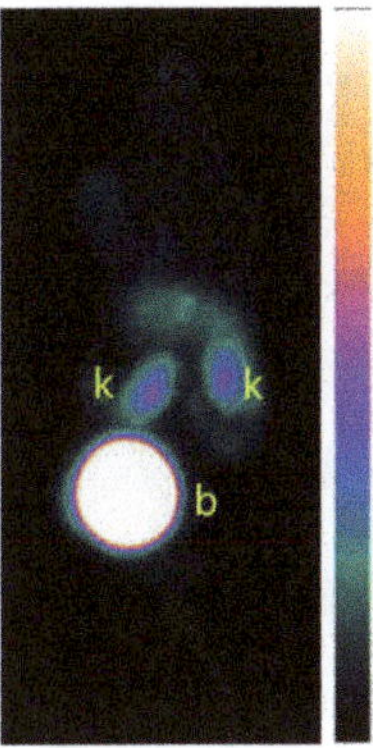

Figure 6. A representative maximum-intensity-projection PET image of [^{68}Ga]Ga-DOTA-AmBz-MVK(Ac)-OH showing its rapid excretion predominantly via the renal pathway with minimal kidney retention (<2.5 %ID/g). The range of color bar is 0–5 %ID/g. k: kidney; b: bladder.

Next, we conducted the PET/CT imaging study of [^{68}Ga]Ga-DOTA-AmBz-MVK(HTK01166)-OH in mice bearing PSMA-expressing LNCaP tumor xenografts to evaluate its pharmacokinetics and PSMA-targeting capability. As shown in Figure 7A, [^{68}Ga]Ga-DOTA-AmBz-MVK(HTK01166)-OH was excreted quickly from background organs/tissues predominately via the renal pathway. At 1 h p.i., only urinary bladder and the PSMA-expressing kidneys and LNCaP tumor were clearly visualized in the PET images (Figure 7A). Co-injection of the PSMA inhibitor, 2-PMPA (2-(phosphonomethyl)-pentanedioic acid), blocked most of the uptake into the tumor and kidneys (Figure 7B).

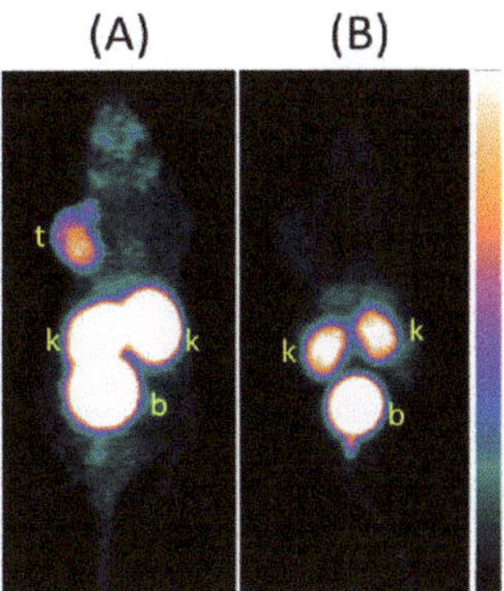

Figure 7. Representative maximum-intensity-projection PET images of [^{68}Ga]Ga-DOTA-AmBz-MVK(HTK01166)-OH acquired at 1 h p.i. from LNCaP tumor-bearing mice (**A**) without and (**B**) with the co-injection of 2-PMPA (0.2 mg). The range of color bar is 0–5 %ID/g. t: tumor; k: kidney; b: bladder.

For comparison, the PET/CT imaging study was also conducted using [^{68}Ga]Ga-DOTA-AmBz-M(O)VK(HTK01166)-OH, the sulfoxide analog of [^{68}Ga]Ga-DOTA-AmBz-MVK(HTK01166)-OH. Similar to [^{68}Ga]Ga-DOTA-AmBz-MVK(HTK01166)-OH, the excretion of [^{68}Ga]Ga-DOTA-AmBz-M(O)VK(HTK01166)-OH was fast and predominately via the renal pathway (Figure 8A). At 1 h p.i., only urinary bladder, LNCaP tumor xenograft and kidneys were clearly visualized in PET images. Co-injection of 2-PMPA blocked most of the uptake in tumors and kidneys (Figure 8B), demonstrating the uptake was PSMA-mediated.

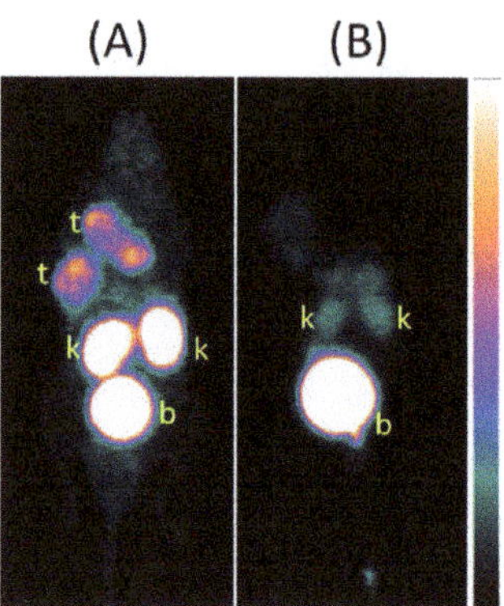

Figure 8. Representative maximum-intensity-projection PET/CT images of [^{68}Ga]Ga-DOTA-AmBz-M(O)VK(HTK01166)-OH acquired at 1 h p.i. from LNCaP tumor-bearing mice (**A**) without and (**B**) with the co-injection of 2-PMPA (0.2 mg). The range of color bar is 0–5 %ID/g. t: tumor; k: kidney; b: bladder.

The ex vivo biodistribution studies were conducted for [^{68}Ga]Ga-DOTA-AmBz-MVK(HTK01166)-OH and [^{68}Ga]Ga-DOTA-AmBz-M(O)VK(HTK01166)-OH in LNCaP tumor-bearing mice, and the results are shown in Table 1. The ex vivo biodistribution data acquired at 1 h p.i. were consistent with the observations from their PET images (Figures 7A and 8A, and Table 1) with low background in blood pool and non-target tissues/organs, good uptake in LNCaP tumor xenografts, and high uptake in kidneys. [^{68}Ga]Ga-DOTA-AmBz-MVK(HTK01166)-OH and [^{68}Ga]Ga-DOTA-AmBz-M(O)VK(HTK01166)-OH had similar uptake values (%ID/g) for the collected tissues/organs (blood: 0.55 ± 0.09 vs. 0.56 ± 0.13; pancreas: 0.60 ± 0.22 vs. 0.56 ± 0.24; spleen: 8.87 ± 2.63 vs. 11.5 ± 2.70; kidneys: 104 ± 12.2 vs. 89.7 ± 19.9; heart: 0.25 ± 0.12 vs. 0.23 ± 0.05; lung: 0.96 ± 0.12 vs. 0.88 ± 0.10; muscle: 0.21 ± 0.05 vs. 0.22 ± 0.05) and comparable tumor-to-background (blood, muscle and kidney) contrast ratios. Compared with the previously reported [^{68}Ga]Ga-HTK01166 [15], [^{68}Ga]Ga-DOTA-AmBz-M(O)VK (HTK01166)-OH showed 39% reduction in average kidney uptake (147 vs. 89.7 %ID/g, p = 0.025). [^{68}Ga]Ga-DOTA-AmBz-MVK(HTK01166)-OH also showed 30% reduction in average kidney uptake (147 vs. 104 %ID/g), but the difference is not statistically significant (p = 0.055).

Table 1. Biodistribution data and uptake ratios of ^{68}Ga-labeled PSMA-targeted tracers in LNCaP tumor-bearing mice acquired at 1 h p.i. (*** $p < 0.001$).

Tissue (%ID/g)	[^{68}Ga]Ga-DOTA-AmBz-MVK(HTK01166)-OH (n = 6)	[^{68}Ga]Ga-DOTA-AmBz-M(O)VK(HTK01166)-OH (n = 6)
Blood	0.55 ± 0.09	0.56 ± 0.13
Urine	187 ± 53.8	221 ± 56.2
Fat	0.72 ± 0.33	0.52 ± 0.13
Testes	0.36 ± 0.06	0.31 ± 0.07
Intestines	0.40 ± 0.10	0.29 ± 0.10
Stomach	0.11 ± 0.02	0.13 ± 0.05
Pancreas	0.60 ± 0.22	0.56 ± 0.24
Spleen	8.87 ± 2.63	11.48 ± 2.70
Adrenal Glands	5.59 ± 4.02	4.33 ± 2.07
Kidneys	104 ± 12.2	89.7 ± 19.3
Liver	0.58 ± 0.08	0.32 ± 0.06 ***
Heart	0.25 ± 0.12	0.23 ± 0.05
Lungs	0.96 ± 0.12	0.88 ± 0.10
LNCaP Tumor	3.98 ± 0.87	4.06 ± 1.27
Muscle	0.21 ± 0.05	0.22 ± 0.05
Bone	0.16 ± 0.04	0.11 ± 0.05
Brain	0.02 ± 0.00	0.03 ± 0.01
Tail	1.03 ± 0.82	1.86 ± 2.44
Tumor/Muscle	19.9 ± 5.41	17.9 ± 7.57
Tumor/Blood	7.24 ± 1.16	6.83 ± 2.81
Tumor/Kidney	0.04 ± 0.01	0.04 ± 0.02

2.4. *Quantification of Radio-Metabolites in Blood and Urine*

Radio-metabolites of [^{68}Ga]Ga-DOTA-AmBz-MVK(Ac)-OH and [^{68}Ga]Ga-DOTA-AmBz-MVK (HTK01166)-OH in mouse blood and urine samples were analyzed by HPLC. The blood samples were collected at 5 min p.i. due to the fast excretion nature of the small radio-metabolites, whereas the urine samples were collected at 15 min p.i. to allow sufficient time for the accumulation of enough radio-metabolites for analysis.

As shown in Figure 9A, there were mainly unidentified polar metabolites (retention time 1.5–3.0 min) present in the blood samples, but no intact [^{68}Ga]Ga-DOTA-AmBz-MVK(Ac)-OH and only minimal amounts of the expected metabolite [^{68}Ga]Ga-DOTA-AmBz-Met-OH (<5%) was detected. On the contrary, >90% of the radioactivity presented in the urine samples was [^{68}Ga]Ga-DOTA-AmBz-Met-OH (Figure 9B).

Analysis of blood samples from mice injected with [^{68}Ga]Ga-DOTA-AmBz-MVK(HTK01166)-OH showed only 9% remaining intact and 35% of the tracer was metabolized to the expected fragment [^{68}Ga]Ga-DOTA-AmBz-Met-OH (Figure 10A). However, a major unidentified radio-metabolite, accounting for 56% of the recovered radioactivity, was also observed. Analysis of the urine samples revealed that only 20% of the recovered radioactivity was presented as the intact tracer and the remaining 80% was the expected fragment [^{68}Ga]Ga-DOTA-AmBz-Met-OH (Figure 10B).

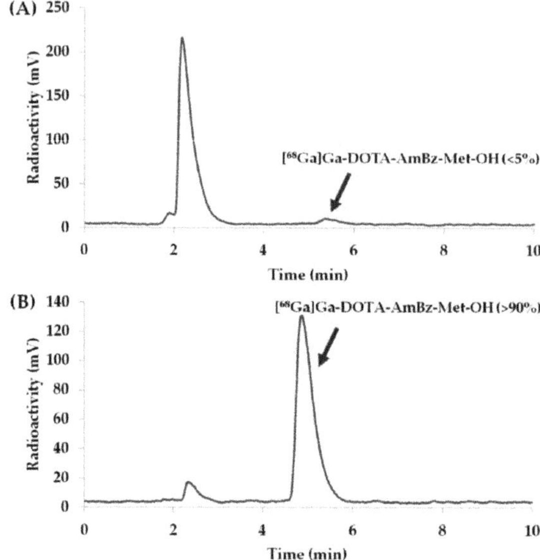

Figure 9. Representative radio-HPLC chromatograms of [^{68}Ga]Ga-DOTA-AmBz-MVK(Ac)-OH obtained from mouse (**A**) blood and (**B**) urine samples collected at 5 and 15 min p.i., respectively. HPLC conditions were 82/18 A/B at a flow rate of 2 mL/min; A: H_2O containing 0.1% TFA; B: CH_3CN containing 0.1% TFA; HPLC column: Luna C18, 5 μm particle size, 100 Å pore size, 250 × 4.6 mm.

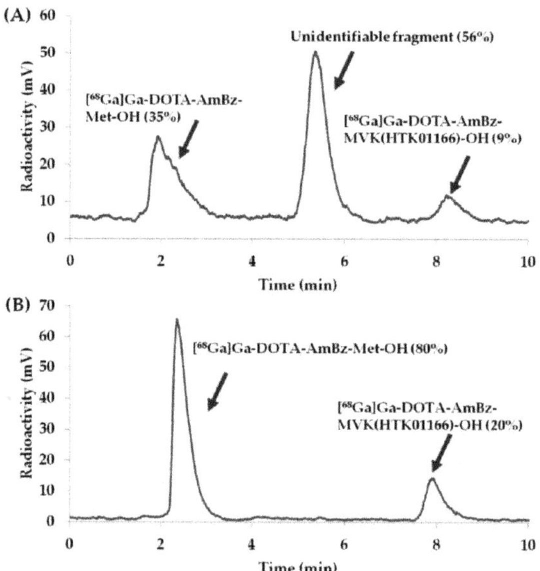

Figure 10. Representative radio-HPLC chromatograms of [^{68}Ga]Ga-DOTA-AmBz-MVK(HTK01166)-OH obtained from mouse (**A**) blood and (**B**) urine samples collected at 5 and 15 min p.i., respectively. HPLC conditions were 72/28 A/B at a flow rate of 2 mL/min; A: H_2O containing 0.1% TFA; B: CH_3CN containing 0.1% TFA; HPLC column: Luna C18, 5 μm particle size, 100 Å pore size, 250 × 4.6 mm.

3. Discussion

The potential of NEP to recognize and cleave specific sequences has been exploited by various groups in the recent years to reduce kidney uptake of radiopharmaceuticals [1,13,18]. This is attributable to the abundant expression of NEP in the brush border membrane lining primarily of the proximal convoluted tubules of the juxtamedullary nephrons [11]. The mechanism underlying degradation of certain specific linker sequences has been well elucidated [1]. The type of amino acid in the radio-metabolite(s) as well as the radiometal chelate used plays a critical role in deciding the kidney residence time of the generated radio-metabolite(s).

NEP is a type-II integral membrane glycoprotein with metalloendopeptidase activity, and also presents an even better carboxydipeptidase activity when the two situations are possible [19]. To apply this strategy to reduce the kidney uptake of radiolabeled peptides and peptidomimetics, we modified the original design as shown in Figure 2. We followed the same design with the targeting vector conjugated to the cleavable linker at the Lys side chain. Such design preserves the free carboxylic group of Lys and enhances the cleavage of the MVK linker by NEP via its carboxydipeptidase activity.

We first synthesized the model tracer [^{68}Ga]Ga-DOTA-AmBz-MVK(Ac)-OH with an acetyl group coupled to the ε-amino group of Lys to provide the amide linkage which would be present when a targeting vector is coupled to this linker. Enzyme assays using the brush border membrane vesicles (BBMVs) extracted from mouse kidneys confirmed that [^{68}Ga]Ga-DOTA-AmBz-MVK(Ac)-OH can be efficiently cleaved (>95%) by renal brush border enzymes (Figure 3), and the expected [^{68}Ga]Ga-DOTA-AmBz-Met-OH was identified as the major radio-metabolite (>90%). Co-incubation with the NEP inhibitor phosphoramidon greatly enhanced the stability of [^{68}Ga]Ga-DOTA-AmBz-MVK(Ac)-OH against renal brush border enzymes (~85% remaining intact), indicating the cleavage of [^{68}Ga]Ga-DOTA-AmBz-MVK(Ac)-OH into [^{68}Ga]Ga-DOTA-AmBz-Met-OH was mediated by NEP. These data confirmed the success of our modifications on the original design (NOTA→DOTA, aminobenzyl→aminomethylbenzoyl, and maleimide-thiol linkage→amide linkage, Figure 1), and the resulting [^{68}Ga]Ga-DOTA-AmBz-MVK(Ac)-OH was still recognized and cleaved by NEP at the same Met-Val amide bond.

We then conducted PET imaging and in vivo stability studies in mice to confirm that [^{68}Ga]Ga-DOTA-AmBz-MVK(Ac)-OH can be metabolized in vivo into [^{68}Ga]Ga-DOTA-Met-OH, and low retention of [^{68}Ga]Ga-DOTA-AmBz-Met-OH in kidneys. This is vital for the success of this modified strategy as it depends on the generation of the expected radio-metabolite [^{68}Ga]Ga-DOTA-AmBz-Met-OH and most importantly [^{68}Ga]Ga-DOTA-AmBz-Met-OH needs to have low kidney retention too. As shown in Figure 6, [^{68}Ga]Ga-DOTA-AmBz-MVK(Ac)-OH was excreted rapidly and predominately via the renal pathway with low kidney retention (<2.5 %ID/g at 1 h p.i.). The blood samples of [^{68}Ga]Ga-DOTA-AmBz-MVK(Ac)-OH revealed no presence of the intact tracer and a small fraction of the expected radio-metabolite [^{68}Ga]Ga-DOTA-AmBz-Met-OH (<5%, Figure 9A). This indicates that [^{68}Ga]Ga-DOTA-AmBz-MVK(Ac)-OH was cleared rapidly from the blood pool either as its intact form or as radio-metabolite(s). Besides kidneys, NEP is also present in some tissues although in a much lower expression level. Therefore, the cleavage of [^{68}Ga]Ga-DOTA-AmBz-MVK(Ac)-OH by NEP present in other tissues cannot be ruled out. Analysis of the urine samples revealed no presence of [^{68}Ga]Ga-DOTA-AmBz-MVK(Ac)-OH and [^{68}Ga]Ga-DOTA-AmBz-Met-OH was presented as the major radio-metabolite (>90%). These data suggest that [^{68}Ga]Ga-DOTA-AmBz-Met-OH had low retention in kidneys and any intact [^{68}Ga]Ga-DOTA-AmBz-MVK(Ac)-OH excreted through kidneys was metabolized presumably by renal brush border enzymes mainly into [^{68}Ga]Ga-DOTA-AmBz-Met-OH.

After confirming the in vivo cleavage of [^{68}Ga]Ga-DOTA-AmBz-MVK(Ac)-OH into [^{68}Ga]Ga-DOTA-AmBz-Met-OH, and the low kidney retention of the [^{68}Ga]Ga-DOTA-AmBz-Met-OH, we next tested this modified design with a PSMA-targeting vector coupled to the Lys side chain. PSMA has become a very promising imaging and therapeutic target for the management of prostate cancer. However, due to a high expression level of PSMA in kidneys, high and sustained renal uptake of PSMA-targeting radioligands are constantly observed, leading to suboptimal detection sensitivity for lesions adjacent to kidneys, and concerns for renal toxicity when radiotherapeutic agents are used. The idea of incorporating the MVK cleavable linker to a radiolabeled DOTA-conjugated PSMA-targeting radioligand is to have the radioligand cleaved by the renal brush border enzymes when it is excreted through kidneys. Since the expected radiometal-complexed DOTA-AmBz-Met-OH is not retained in kidneys, the overall renal uptake of the PSMA-targeting radioligand containing the MVK cleavable linker will be reduced.

Enzyme assays revealed that unlike the instability of [^{68}Ga]Ga-DOTA-AmBz-MVK(Ac)-OH against renal brush border enzymes, [^{68}Ga]Ga-DOTA-AmBz-MVK(HTK01116)-OH was relatively stable with 80% remaining intact and 16% converted to the expected radio-metabolite [^{68}Ga]Ga-DOTA-AmBz-Met-OH under the same assay conditions (Figure 4). The enhanced stability could be due to the steric hindrance introduced by replacing the acetyl group with the much bulkier HTK01166 motif. Further enhancement in stability achieved by co-incubation with the NEP inhibitor phosphoramidon indicates that [^{68}Ga]Ga-DOTA-AmBz-MVK(HTK01116)-OH was cleaved mainly by NEP into [^{68}Ga]Ga-DOTA-AmBz-Met-OH.

PET imaging studies showed that [^{68}Ga]Ga-DOTA-AmBz-MVK(HTK01166)-OH and its oxidized version [^{68}Ga]Ga-DOTA-AmBz-M(O)VK(HTK01166)-OH had very similar distribution patterns: low background, high kidney retention and good tumor visualization (Figures 7 and 8). The observed PET images were consistent with the ex vivo biodistribution data showing comparable uptake for all collected organs/tissues (Table 1). The brain uptake (≤0.03 %ID/g) of both tracers was negligible indicating that they cannot freely cross the blood-brain barrier. Their fast blood clearance (~0.55 %ID/g at 1 h p.i.) suggests that the [^{68}Ga]Ga-DOTA complex was stable as free ^{68}Ga would be captured by transferrins leading to prolonged retention in blood pool [20]. Minimal uptake (≤0.60 %ID/g) of both tracers in liver and intestines indicates that they were excreted mainly by the renal pathway. Higher uptake was observed in PSMA-expressing LNCaP tumors (~4 %ID/g), kidneys (90–104 %ID/g) and spleen (8.9–11.5 %ID/g) suggesting that the uptake of both tracers in these tissues was PSMA-mediated [15]. This was further confirmed by PET imaging studies as co-injection of the PSMA inhibitor, 2-PMPA (0.2 mg), reduced the tumor uptake of both tracers to the background level (Figures 7 and 8).

The average kidney uptake of [^{68}Ga]Ga-DOTA-AmBz-MVK(HTK01166)-OH was lower than that of the previously reported [^{68}Ga]Ga-HTK01166 (104 vs. 147 %ID/g), suggesting the insertion of the cleavable MVK linker might be a useful strategy to reduce kidney uptake of radiopharmaceuticals. However, although statistically not significant (p = 0.15), a slightly higher kidney retention was observed for mice injected with [^{68}Ga]Ga-DOTA-AmBz-MVK(HTK01166)-OH (104 ± 12.2 %ID/g) than for mice injected with [^{68}Ga]Ga-DOTA-AmBz-MV(O)K(HTK01166)-OH (89.7 ± 19.3 %ID/g). Since [^{68}Ga]Ga-DOTA-AmBz-MVK(HTK01166)-OH contains a cleavable MVK linker, whereas [^{68}Ga]Ga-DOTA-AmBz-M(O)VK(HTK01166)-OH contains a non-cleavable M(O)VK linker, we would expect a much lower kidney uptake from [^{68}Ga]Ga-DOTA-AmBz-MVK(HTK01166)-OH. These discrepant data suggest that there might be some unknown mechanism(s) causing the higher kidney retention of [^{68}Ga]Ga-DOTA-AmBz-MVK(HTK01166)-OH. In addition, we noticed that while the kidney retention (1.3 %ID/g) in the mouse injected with [^{68}Ga]Ga-DOTA-AmBz-M(O)VK(HTK01166)-OH and 2-PMPA was minimal (Figure 8B), there was significantly higher kidney uptake (4.2 %ID/g) in the mouse injected with [^{68}Ga]Ga-DOTA-AmBz-MVK(HTK01166)-OH and 2-PMPA (Figure 7B).

The retained kidney uptake was unlikely to have resulted from the intact [^{68}Ga]Ga-DOTA-AmBz-MVK(HTK01166)-OH, as the mouse was co-injected with excess 2-PMPA to block PSMA. The retained uptake was unlikely to have resulted from the expected radio-metabolite [^{68}Ga]Ga-DOTA-AmBz-Met-OH either, as we have shown that no significant kidney retention was observed in the mouse injected with [^{68}Ga]Ga-DOTA-AmBz-MVK(Ac)-OH (Figure 6). The only possibility for the kidney retention would be unidentified radio-metabolite(s), which cannot bind PSMA but can be retained in kidneys. Therefore, the in vivo stability study was further conducted for [^{68}Ga]Ga-DOTA-AmBz-MVK(HTK01166)-OH to discern the cause of its higher kidney retention in mice.

As shown in Figure 10A, unlike the good stability observed in enzyme assays, [^{68}Ga]Ga-DOTA-AmBz-MVK(HTK01166)-OH was rapidly metabolized in vivo with only 9% of the tracer remaining intact in the blood at 5 min p.i. However, the expected radio-metabolite [^{68}Ga]Ga-DOTA-AmBz-Met-OH was accounted for only 35% of the recovered radioactivity, while 56% consists of an unidentified radio-metabolite. Interestingly, in urine samples, only [^{68}Ga]Ga-DOTA-AmBz-MVK(HTK01166)-OH (20%) and [^{68}Ga]Ga-DOTA-AmBz-Met-OH (80%) were detected, and there was no presence of the unidentified radio-metabolite (Figure 10B). The absence of the unidentified radio-metabolite in urine samples suggests that it might be retained in kidneys. This would explain the higher kidney uptake of [^{68}Ga]Ga-DOTA-AmBz-MVK(HTK01166)-OH than [^{68}Ga]Ga-DOTA-AmBz-M(O)VK(HTK01166)-OH (Table 1), as well as its higher kidney retention when co-injected with 2-PMPA (Figures 7 and 8). This would also explain the insufficient reduction (~30%) in kidney uptake when compared [^{68}Ga]Ga-DOTA-AmBz-MVK(HTK01166)-OH and [^{68}Ga]Ga-HTK01166 [15], which is far less than the ~80% and ~67% reduction in kidney uptake reported by Uehara et al. [12] and Zhang et al. [13], respectively, using the radiolabeled NOTA-MVK(Targeting vector)-OH design. The insufficient reduction reported here using the DOTA-AmBz-MVK(Targeting vector)-OH design is likely caused by the unidentified radio-metabolite that could be trapped in kidneys. Therefore, further optimization of the design of DOTA-conjugated cleavable linkers should avoid the generation of radio-metabolites that could be trapped in kidneys and cause high and sustained kidney uptake.

The identity of the unidentified fragment remains unknown (Figure 10). This is because the core structure of HTK01166 contains no amide bonds formed by two natural amino acids. Moreover, the data from our [^{68}Ga]Ga-DOTA-AmBz-MVK(Ac)-OH, [^{68}Ga]Ga-DOTA-AmBz-M(O)VK(HTK01166)-OH and the previously reported [^{68}Ga]Ga-NOTA- MVK-conjugated antibody fragment [12] and Exendin 4 [13], did not suggest the cleavage of the Val-Lys amide bond. Therefore, we did not expect cleavage of [^{68}Ga]Ga-DOTA-AmBz-MVK(HTK01166)-OH at locations other than the Met-Val amide bond, and more investigations are needed to verify the identity of this unknown radio-metabolite.

To conclude, we showed that replacing NOTA and the aminobenzyl group in the reported NOTA-MVK linker with DOTA and AmBz, respectively, generated the model compound [^{68}Ga]Ga-DOTA-AmBz-MVK(Ac)-OH, which was still recognized and specifically cleaved at the Met-Val amide bond by NEP. Coupling a bulkier PSMA-targeting vector to the side chain of Lys in DOTA-AmBz-MVK enhanced its stability against NEP, but possibly also rendered its vulnerability against other enzyme(s) as evident by the formation of an unidentified radio-metabolite that could be retained in kidneys. Nevertheless, the renal uptake of the resulting [^{68}Ga]Ga-DOTA-AmBz-MVK(HTK01166)-OH was still lower than that of [^{68}Ga]Ga-HTK01166. These data demonstrated that MVK could be a promising cleavable linker for use to reduce renal uptake of radiolabeled DOTA-conjugated tumor-targeting peptides and peptidomimetics. This strategy can be used to enhance detection sensitivity of the imaging agents for lesions adjacent to kidneys, and improve the tumor-to-kidney absorbed dose ratio for the radiotherapeutic agents.

4. Materials and Methods

4.1. General Methods

Fmoc-L-Lys(pentynoyl)-OH (**1**) was synthesized according to the literature procedures [17]. Brush border membrane vesicles (BBMVs) were extracted from mouse kidneys following literature procedures [21]. All other chemicals were procured from commercial sources and used without further purification. All peptides and peptidomimetics were synthesized either on an AAPPTec (Louisville, KY, USA) Endeavor 90 peptide synthesizer or a CEM (Matthews, NC, USA) Liberty Blue™ automated microwave peptide synthesizer. Purification and quality control of radiolabeling precursor, nonradioactive Ga-complexed standards and ^{68}Ga-labeled peptides and peptidomimetics were performed on Agilent (Santa Clara, CA, USA) HPLC systems equipped with a model 1200 quaternary pump, a model 1200 UV absorbance detector (set at 220 nm), and a Bioscan (Washington, DC, USA) NaI scintillation detector. The operation of Agilent HPLC systems was controlled using the Agilent ChemStation software. HPLC columns used were a semipreparative column (Luna C18, 5 μm particle size, 100 Å pore size, 250 × 10 mm) and an analytical column (Luna C18, 5 μm particle size, 100 Å pore size, 250 × 4.6 mm) from Phenomenex (Torrance, CA, USA). The HPLC solvents were A: H_2O containing 0.1% TFA; B: CH_3CN containing 0.1% TFA. The collected HPLC eluates containing the desired peptides were lyophilized using a Labconco (Kansas City, MO, USA) FreeZone 4.5 Plus freeze drier. ^{1}H-NMR spectrum was acquired using an AVANCE Bruker 400 MHz NMR spectrometer equipped with BBI probe with Z gradients. Mass analyses were performed using an AB SCIEX (Framingham, MA, USA) 4000 QTRAP mass spectrometer system with an ESI ion source. C18 Sep-Pak cartridges (1 cm^3, 50 mg) were obtained from Waters (Milford, MA, USA). ^{68}Ga was eluted from an iThemba Laboratories (Somerset West, South Africa) generator and purified according to the previously published procedures using a DGA resin column from Eichrom Technologies LLC (Lisle, IL, USA) [22,23]. Radioactivity of ^{68}Ga-labeled peptides and peptidomimetics was measured using a Capintec (Ramsey, NJ, USA) CRC-25R/W dose calibrator. PET/CT imaging was performed using a Siemens Inveon (Knoxville, TN, USA) micro PET/CT scanner. The radioactivity of mouse tissues collected from biodistribution studies was counted using a PerkinElmer (Waltham, MA, USA) Wizard2 2480 automatic gamma counter.

4.2. *Synthesis of Fmoc-L-Lys(pentynoyl)-OtBu (2)*

Fmoc-L-Lys(pentynoyl)-OH (**1**) (2.05 g, 4.6 mmol) in CH_2Cl_2 (20 mL) was added *t*-butyl 2,2,2-trichloroacetimidate (2.18 g, 10 mmol). The resulting mixture was stirred at RT for 22 h, and purified by flash column chromatography eluted with 1:1 ethyl acetate/hexane to 100% ethyl acetate. The desired product **2** was obtained as a white solid (1.82 g, 79%). ^{1}H-NMR (400 MHz, $CDCl_3$) δ ppm: 1.39–2.08 (m, 7H), 1.47 (s, 9H), 2.37 (t, J = 7.0 Hz, 2H), 2.52 (dt, J = 2.2, 7.0 Hz, 2H). 3.28 (m, 2H), 4.22 (t, J = 7.0 Hz, 2H), 4.39 (m, 2H), 5.42 (d, J = 8.0 Hz, 1H), 5.90 (bs, 1H), 7.31 (dt, J = 1.0, 7.4 Hz, 2H), 7.40 (t, J = 7.5 Hz, 2H), 7.60 (d, J = 7.4 Hz, 2H), 7.77 (d, J = 7.5 Hz, 2H). ESI-MS: calculated $[M + H]^+$ for Fmoc-L-Lys(pentynoyl)-OtBu $C_{30}H_{36}N_2O_5$ 505.3; found 506.0.

4.3. *Synthesis of DOTA-Conjugated Precursors*

4.3.1. Synthesis of DOTA-AmBz-MVK(Ac)-OH

For synthesizing the DOTA-conjugated linker DOTA-AmBz-MVK(Ac)-OH, Fmoc-Lys(Mtt) wang resin (0.05 mmol scale, 0.5–0.8 mmol/g loading) was first swollen using DMF. The Mtt protecting group was removed using 2% TFA in DCM for 30 min (5 min, 6 times) and neutralized using DIEA in DMF. The free amine was acetylated using acetic anhydride (20 equiv)/DIEA (20 equiv)/DMF. Fmoc on Lys was then deprotected using 20% piperidine in DMF and coupled with Fmoc-Val-OH (5 equiv) in the presence of activators HATU/HOAt (5 equiv) and DIEA (10 equiv) in DMF. Further elongation of the peptide chain was carried out by repeating the Fmoc deprotection and coupling steps. Fmoc-Met-OH, Fmoc-(4-aminomethyl)benzoic acid, and finally DOTA-tris(*t*-bu)ester (tri-*t*-butyl 1,4,7,10-tetraazacyclododecane-1,4,7,10-tetraacetate) were coupled in that order. The DOTA conjugated peptide was cleaved off the resin and simultaneously deprotected using TFA:H_2O:triisopropylsilane (TIS):2,2′-(ethylenedioxy)diethanethiol (DODT) for 1.5 h at RT, in a 92.5:2.5:2.5:2.5 ratio. The cleaved peptide was filtered and precipitated with cold diethyl ether before purification using the semipreparative HPLC column. HPLC conditions were 84/16 A/B at a flow rate of 4.5 mL/min. The retention time of DOTA-AmBz-MVK(Ac)-OH was 14.1 min. Eluates containing the desired peptide were collected, pooled, and lyophilized. The isolated yield was 12%. ESI-MS: calculated $[M + H]^+$ for DOTA-AmBz-MVK(Ac)-OH $C_{42}H_{67}N_9O_{13}S$ 938.5; found 938.4.

4.3.2. Synthesis of DOTA-AmBz-MVK(HTK01166)-OH

For synthesizing the PSMA-targeting DOTA-AmBz-MVK(HTK01166)-OH the PSMA-targeting motif (Lys-urea-Glu) and the lipophilic linker ((2-indanyl)-Gly-tranexamic acid) of HTK01166 were first assembled on solid phase following our previously published procedures [15]. Briefly, Fmoc-Lys(ivDde) wang resin (0.1 mmol scale, 0.58 mmol/g loading) was swollen using DMF. The isocyanate derivative of di-*tert*-butyl ester of glutamate (5 equiv) was prepared according to literature procedures [24]. The isocyante derivative was then added to Fmoc-deprotected Lys(ivDde) wang resin and the reaction mixture was allowed to shake overnight. Next, the ivDde protecting group was removed using 2% hydrazine in DMF (5 mL, five times, 5 min). Subsequent couplings of Fmoc-(2-indanyl)-Gly-OH, Fmoc-tranexamic acid and 2-azidoacetic acid were conducted using standard Fmoc chemistry. Next, Fmoc-Lys(pentynoyl)-O*t*Bu (**2**), (0.5 mmol) was clicked onto the azido group using $CuSO_4$ (0.05 mmol) and ascorbic acid (0.25 mmol). After the click reaction, further peptide elongation was continued using Fmoc-Val-OH, Fmoc-Met-OH, Fmoc-(4-aminomethyl)benzoic acid, and finally, DOTA-tris(*t*-bu)ester. Peptide cleavage and deprotection was performed using TFA:H_2O:TIS:DODT for 1.5 h at RT in a 92.5:2.5:2.5:2.5 ratio. The cleaved peptide was filtered and precipitated with cold diethyl ether before purification using the semipreparative HPLC column. HPLC conditions were 73/27 A/B at a flow rate of 4.5 mL/min. The retention time of DOTA-AmBz-MVK(HTK01166)-OH was 12.6 min. Eluates containing the desired peptide were collected, pooled, and lyophilized. The isolated yield was 1%. ESI-MS: calculated $[M + H]^+$ for DOTA-AmBz-MVK(HTK01166)-OH $C_{78}H_{115}N_{17}O_{23}S$ 1690.8; found 1690.9.

4.3.3. Synthesis of DOTA-AmBz-M(O)VK(HTK01166)-OH

DOTA-AmBz-M(O)VK(HTK01166)-OH, the oxidized (sulfoxide) version of DOTA-AmBz-MVK (HTK01166)-OH was synthesized as a control using a similar synthetic and purification strategy as for DOTA-AmBz-MVK(HTK01166)-OH. For this purpose, Fmoc-Met-OH was replaced with Fmoc-Met(O)-OH. The HPLC conditions were 74/26 A/B at a flow rate of 4.5 mL/min. The retention time and isolated yield were 8.3 min and 2%, respectively. ESI-MS: calculated $[M + H]^+$ for DOTA-AmBz-M(O)VK(HTK01166)-OH $C_{78}H_{115}N_{17}O_{24}S$ 1706.8; found 1706.9.

4.3.4. Synthesis of DOTA-AmBz-Met-OH

For synthesizing the expected cleaved fragment DOTA-AmBz-Met-OH, ClTCP(Cl)ProTide resin (0.1 mmol scale, 0.48 mmol/g loading) was swollen in DMF. Fmoc-Met-OH (5 equiv) in DMF containing 1 M DIEA/0.125M KI was coupled to the resin. After elongation with Fmoc-(4-aminomethyl)benzoic acid and DOTA-tris(*t*-bu)ester, the peptide was cleaved off the resin using TFA:H_2O:TIS:DODT for 1.5 h at RT in a 92.5:2.5:2.5:2.5 ratio and purified using the semipreparative HPLC column. The HPLC conditions were 83/17 A/B at a flow rate of 4.5 mL/min. The retention time was 10.2 min, and the isolated yield was 35%. ESI-MS: calculated $[M + H]^+$ for DOTA-AmBz-Met-OH $C_{29}H_{44}N_6O_{10}S$ 669.3; found 669.0.

4.4. Synthesis of Nonradioactive Ga-Complexed Standards

Ga-DOTA-AmBz-MVK(Ac)-OH, Ga-DOTA-AmBz-Met-OH, Ga-DOTA-AmBz-MVK(HTK01166)-OH and Ga-DOTA-AmBz-M(O)VK(HTK01166)-OH were prepared by incubating the DOTA-conjugated peptides and peptidomimetics with $GaCl_3$ (5 equiv) in NaOAc buffer (0.1 M, 300–500 μL, pH 4.2) at 80 °C for 15 min. The reaction mixtures were directly purified using the semipreparative HPLC column. The eluates containing the desired product were collected, pooled, and lyophilized.

For Ga-DOTA-AmBz-MVK(Ac)-OH the HPLC conditions were 82/18 A/B at a flow rate of 4.5 mL/min. The retention time and isolated yield were 11.2 min and 78%, respectively. ESI-MS: calculated $[M + H]^+$ for Ga-DOTA-AmBz-MVK(Ac)-OH $C_{42}GaH_{65}N_9O_{13}S$ 1005.4; found 1004.4.

For Ga-DOTA-AmBz-MVK(HTK01166)-OH the HPLC conditions were 74/26 A/B at a flow rate of 4.5 mL/min. The retention time and isolated yield were 8.0 min and 16%, respectively. ESI-MS: calculated $[M + H]^+$ for Ga-DOTA-AmBz-MVK(HTK01166)-OH $C_{78}GaH_{113}N_{17}O_{23}S$ 1757.7; found 1758.4.

For Ga-DOTA-AmBz-M(O)VK(HTK01166)-OH the HPLC conditions were 74/26 A/B at a flow rate of 4.5 mL/min. The retention time and isolated yield were 10.8 min and 35%, respectively. ESI-MS: calculated $[M + 2H]^{2+}$ for Ga-DOTA-AmBz-M(O)VK(HTK01166)-OH $C_{78}GaH_{113}N_{17}O_{24}S$ 887.4; found 887.5.

For Ga-DOTA-AmBz-Met-OH, the HPLC conditions were 83/17 at a flow rate of 4.5 mL/min. The retention time and the isolated yield were 10.5 min and 72%, respectively. ESI-MS: calculated $[M + H]^+$ for Ga-DOTA-AmBz-Met-OH $C_{29}GaH_{42}N_6O_{10}S$ 736.2; found 736.1.

4.5. Synthesis of ^{68}Ga-Labeled Peptides and Peptidomimetics

Purification of ^{68}Ga eluated from ^{68}Ge/^{68}Ga generator and labeling experiments were performed following our previously published procedures [22,23]. Purified ^{68}Ga in 0.5 mL of water was added into a 4-mL glass vial preloaded with 0.7 mL of HEPES buffer (2 M, pH 5.0) and 25 μg of the precursor. The radiolabeling reaction was carried out under microwave heating for 1 min. The reaction mixtures were purified using the semipreparative HPLC column and eluted with 84/16 A/B for [^{68}Ga]Ga-DOTA-AmBz-MVK(Ac)-OH, 74/26 A/B for [^{68}Ga]Ga-DOTA-AmBz-MVK(HTK01166)-OH, and 23% CH_3CN and 0.1% HCOOH in H_2O for [^{68}Ga]Ga-DOTA-AmBz-M(O)VK(HTK01166)-OH at a flow rate of 4.5 mL/min. The retention times for [^{68}Ga]Ga-DOTA-AmBz-MVK(Ac)-OH, [^{68}Ga]Ga-DOTA-AmBz-MVK(HTK01166)-OH and [^{68}Ga]Ga-DOTA-AmBz-M(O)VK(HTK01166)-OH were 24.3, 26.1 and 15.7 min, respectively. The eluate fraction containing the radiolabeled product was collected, diluted with water (50 mL), and passed through a C18 Sep-Pak cartridge that was prewashed with ethanol (10 mL) and water (10 mL). After washing the C18 Sep-Pak cartridge with water (10 mL), the ^{68}Ga-labeled product was eluted off the cartridge with ethanol (0.4 mL) and diluted with saline for all the in vitro enzyme assays, PET imaging, ex vivo biodistribution and in vivo stability studies.

4.6. In Vitro Enzyme Assay

The enzymatic recognition of synthesized peptides and peptidomimetics was determined by incubating the ^{68}Ga-labeled peptides and peptidomimetics with the extracted BBMVs at 37 °C for 1 h. For the assay, aliquots (25 μL) of the enzyme solution (1.52 mg/mL) and enzyme buffer (250 mM NaCl, 57.5 mM Tris-base; adjusted to a pH 7.5) were mixed in a 96-well clear bottom plate and incubated at 37 °C for 10 min. The mixtures also contained 100 ppm ascorbic acid to prevent oxidation of the Met residue in the tested peptides and peptidomimetics during the assay. The radiolabeled peptide (50 μL, ~3.7 MBq) was then added to the test well. The control well contained phosphoramidon, a potent NEP inhibitor, at a final concentration of 1 mmol/L in addition to the contents of the test well. After 1 h incubation, all reactions were quenched using equal volume of CH_3CN and centrifuged at 13,000 rpm for 10 min. The resulting supernatant was collected and analyzed using the analytical HPLC column to identify and quantify radio-metabolite(s). The assay was performed in duplicates.

4.7. Cell Culture

Human prostate cancer LNCaP cells obtained from ATCC (Manassas, VA, USA) were cultured in RPMI 1640 medium, supplemented with 10% FBS, penicillin (100 U/mL), and streptomycin (100 μg/mL) at 37 °C in a Panasonic Healthcare (Tokyo, Japan) MCO-19AIC humidified incubator containing 5% CO_2. At 80–90% confluency, cells were washed with sterile phosphate-buffered saline, trypsinized and pooled before they were manually counted using a Bal Supply (Sylvania, OH, USA) 202C counter.

4.8. PET/CT Imaging and Ex Vivo Biodistribution in Tumor-Bearing Mice

All imaging and biodistribution studies were performed using male NOD-*scid* IL2Rg^{null} (NSG) mice and conducted according to the guidelines established by the Canadian Council on Animal

Care and approved by Animal Ethics Committee of the University of British Columbia. For tumor inoculations, mice were anesthetized by inhalation with 2% isoflurane in oxygen and implanted subcutaneously with 5×10^6 LNCaP cells below the left shoulder. Imaging and biodistribution studies were performed only after tumors grew to 5–8 mm in diameter over a period of 5–7 weeks.

For PET/CT imaging studies, ~3–6 MBq of the ^{68}Ga-labeled tracer was injected through the tail vein. For the blocking study, 2-PMPA (0.2 mg) was co-injected with the tracer. Mice were allowed to recover and roam freely in the cages after injecting the tracer. At 45 min p.i., mice were sedated again and positioned on the scanner. First, a 10 min CT scan was conducted for localization and attenuation correction for reconstruction of PET images, before a 10 min PET image was acquired. Heating pads were used during the entire procedure to keep the mice warm.

For ex vivo biodistribution studies, mice were injected with ~1.5–3 MBq of the ^{68}Ga-labeled tracer. At 1 h p.i., mice were euthanized, blood was drawn from heart, and organs/tissues of interest were collected, rinsed with PBS, blotted dry, weighed, and counted using an automated gamma counter. The uptake in each organ/tissue was normalized to the injected dose and expressed as the percentage of the injected dose per gram of tissue (%ID/g).

4.9. Quantification of Radio-Metabolites in Blood and Urine

Male NSG mice were injected with 3–17 MBq of the ^{68}Ga-labeled peptide. For blood profiling, mice were anesthetized with 2% isoflurane in O_2 and euthanized by CO_2 inhalation at 5 min p.i. Blood draw was then performed by cardiac puncture and blood was collected in an eppendorf tube with equal volume of CH_3CN. Each tube was then centrifuged at RT for 10–15 min and the resulting supernatant was collected and analyzed using the analytical HPLC column to identify and quantify radio-metabolite(s) in blood. For the purpose of urine profiling, urine was collected after euthanizing the mice at 15 min p.i. The urine samples were also collected and analyzed using the analytical HPLC column to identify and quantify radio-metabolite(s) in urine.

4.10. Statistical Analysis

Statistical analyses were performed by Student's *t*-test using the Microsoft (Redmond, WA, USA) Excel software. The unpaired, two-tailed test was used to compare tissue uptake and tumor-to-background (muscle, blood and kidney) contrast ratios of [^{68}Ga]Ga-DOTA-AmBz-MVK(HTK01166)-OH and [^{68}Ga]Ga-DOTA-AmBz-M(O)VK(HTK01166)-OH. The unpaired, one-tailed test was used to compare kidney uptake of [^{68}Ga]Ga-DOTA-AmBz-MVK(HTK01166)-OH (or [^{68}Ga]Ga-DOTA-AmBz-M(O)VK(HTK01166)-OH) with that of the previously reported [^{68}Ga]Ga-HTK01166 [15]. The difference was considered statistically significant when the p value was <0.05.

Supplementary Materials: The following are available online at http://www.mdpi.com/1420-3049/25/17/3854/s1, Table S1: HPLC conditions and retention times for the purification of DOTA-conjugated peptides and peptidomimetics using the semipreparative column (Luna C18, 5 μm particle size, 100 Å pore size, 250 × 10 mm), Table S2: HPLC conditions and retention times for the purification of nonradioactive Ga-complexed DOTA-conjugated peptides and peptidomimetics using the semipreparative column (Luna C18, 5 μm particle size, 100 Å pore size, 250 × 10 mm), and Table S3: HPLC conditions and retention times for the purification/QC of 68Ga-labeled DOTA-conjugated peptides and peptidomimetics using the semipreparative column—Luna C18, 5 μm particle size, 100 Å pore size, 250 × 10 mm; the analytical (QC) column—Luna C18, 5 μm particle size, 100 Å pore size, 250 × 4.6 mm.

Author Contributions: Conceptualization, K.-S.L.; methodology, S.B., Z.Z. and K.-S.L.; formal analysis, S.B., Z.Z. and K.-S.L.; investigation, S.B., Z.Z., H.-T.K., J.R., C.Z., Á.R. and H.M.; resources, F.B. and K.-S.L.; data curation, S.B., Z.Z., and C.Z.; writing—original draft preparation, S.B., Z.Z., and K.-S.L.; writing—review and editing, H.-T.K., J.R., C.Z., Z.Z., Á.R., H.M., and F.B.; supervision, K.-S.L. and F.B.; project administration, H.M.; funding acquisition, K.-S.L. All authors have read and agreed to the published version of the manuscript.

Funding: This research was funded by the Canadian Institutes of Health Research (grant number PJT-162243), Pancreas Centre BC, BC Cancer Foundation and VGH Foundation.

Acknowledgments: The authors would like to thank Nadine Colpo for her technical assistance.

Conflicts of Interest: The authors declare no conflict of interest.

References

1. Arano, Y. Renal brush border strategy: A developing procedure to reduce renal radioactivity levels of radiolabeled polypeptides. *Nucl. Med. Biol.* **2020**. [CrossRef] [PubMed]
2. Jackson, J.A.; Hungnes, I.N.; Ma, M.T.; Rivas, C. Bioconjugates of chelators with peptides and proteins in nuclear medicine: Historical importance, current innovations, and future challenges. *Bioconjugate. Chem.* **2020**, *31*, 483–491. [CrossRef] [PubMed]
3. Maleki, F.; Farahani, A.M.; Rezazedeh, F.; Sadeghzadeh, N. Structural modifications of amino acid sequences of radiolabeled peptides for targeted tumor imaging. *Bioorg. Chem.* **2020**, *99*, 103802. [CrossRef] [PubMed]
4. Tsai, W.T.; Wu, A.M. Aligning physics and physiology: Engineering antibodies for radionuclide delivery. *J. Label. Compd. Radiopharm.* **2018**, *61*, 693–714. [CrossRef]
5. Wissler, H.L.; Ehlerding, E.B.; Lyu, Z.; Zhao, Y.; Zhang, S.; Eshraghi, A.; Buuh, Z.Y.; McGuth, J.C.; Guan, Y.; Engle, J.W.; et al. Site-specific immuno-PET tracer to image PD-L1. *Mol. Pharm.* **2019**, *16*, 2028–2036. [CrossRef]
6. Dash, A.; Chakraborty, S.; Pillai, M.R.A.; Knapp, F.F.R. Peptide receptor radionuclide therapy: An overview. *Cancer Biother. Radiopharm.* **2015**, *30*, 47–71. [CrossRef]
7. Vegt, E.; Melis, M.; Eek, A.; De Visser, M.; Brom, M.; Oyen, W.J.G.; Gotthardt, M.; De Jong, M.; Boerman, O.C. Renal uptake of different radiolabelled peptides is mediated by megalin: SPECT and biodistribution studies in megalin-deficient mice. *Eur. J. Nucl. Med. Mol. Imaging* **2011**, *38*, 623–632. [CrossRef]
8. Maurer, A.H.; Elsinga, P.; Fanti, S.; Nguyen, B.; Oyen, W.J.G.; Weber, W.A. Imaging the folate receptor on cancer cells with 99mTc- etarfolatide: Properties, clinical use, and future potential of folate receptor imaging. *J. Nucl. Med.* **2014**, *55*, 701–704. [CrossRef]
9. Christensen, E.I.; Birn, H. Megalin and cubilin: Multifunctional endocytic receptors. *Nat. Rev. Mol. Cell Biol.* **2002**, *3*, 258–268. [CrossRef]
10. Fujioka, Y.; Satake, S.; Uehara, T.; Mukai, T.; Akizawa, H.; Ogawa, K.; Saji, H.; Endo, K.; Arano, Y. In vitro system to estimate renal brush border enzyme-mediated cleavage of peptide linkages for designing radiolabeled antibody fragments of low renal radioactivity levels. *Bioconjug. Chem.* **2005**, *16*, 1610–1616. [CrossRef]
11. Kanazawa, M.; Johnston, C.I. Distribution and Inhibition of Neutral Metalloendopeptidase (Nep) (Ec 3.4.24.11), the Major Degradative Enzyme for Atrial Natriuretic Peptide, in the Rat Kidney. *Clin. Exp. Pharmacol. Physiol.* **1991**, *18*, 449–453. [CrossRef] [PubMed]
12. Uehara, T.; Yokoyama, M.; Suzuki, H.; Hanaoka, H.; Arano, Y. A gallium-67/68–labeled antibody fragment for immuno-SPECT/PET shows low renal radioactivity without loss of tumor uptake. *Clin. Cancer Res.* **2018**, *24*, 3309–3316. [CrossRef] [PubMed]
13. Zhang, M.; Jacobson, O.; Kiesewetter, D.O.; Ma, Y.; Wang, Z.; Lang, L.; Tang, L.; Kang, F.; Deng, H.; Yang, W.; et al. Improving the Theranostic Potential of Exendin 4 by Reducing the Renal Radioactivity through Brush Border Membrane Enzyme-Mediated Degradation. *Bioconjug. Chem.* **2019**, *30*, 1745–1753. [CrossRef]
14. Kim, K.; Kim, S. Lu-177-Based Peptide Receptor Radionuclide Therapy for Advanced Neuroendocrine Tumors. *Nucl. Med. Mol. Imaging* **2018**, *52*, 208–215. [CrossRef]
15. Kuo, H.T.; Pan, J.; Zhang, Z.; Lau, J.; Merkens, H.; Zhang, C.; Colpo, N.; Lin, K.S.; Bénard, F. Effects of Linker Modification on Tumor-to-Kidney Contrast of 68GaLabeled PSMA-Targeted Imaging Probes. *Mol. Pharm.* **2018**, *15*, 3502–3511. [CrossRef]
16. Woods, M.; Leung, L.; Frantzen, K.; Garrick, J.G.; Zhang, Z.; Zhang, C.; English, W.; Wilson, D.; Benard, F.; Lin, K.S. Improveing the Stability of ^{11}C-labeled L-methionine with Ascorbate. *EJNMMI Radiopharm. Chem.* **2017**, *2*, 1–9. [CrossRef] [PubMed]
17. Shinoda, K.; Sohma, Y.; Kanai, M. Synthesis of chemically-tethered amyloid-β segment trimer possessing amyloidogenic properties. *Bioorganic Med. Chem. Lett.* **2015**, *25*, 2976–2979. [CrossRef] [PubMed]
18. Vaidyanathan, G.; Kang, C.M.; McDougald, D.; Minn, I.; Brummet, M.; Pomper, M.G.; Zalutsky, M.R. Brush Border Enzyme-cleavable Linkers: Evaluation for Reducing Renal Uptake of Radiolabeled Prostate-specific membrane Antigen Inhibitors. *Nucl. Med. Biol.* **2018**, *62*, 18–30. [CrossRef]
19. Barros, N.M.T.; Campos, M.; Bersanetti, P.A.; Oliveira, V.; Juliano, M.A.; Boileau, G.; Juliano, L.; Carmona, A.K. Neprilysin carboxydipeptidase specificity studies and improvement in its detection with fluorescence energy transfer peptides. *Biol. Chem.* **2007**, *388*, 447–455. [CrossRef]

20. Tsuchiya, Y.; Nakao, A.; Komatsu, T.; Yamamoto, M.; Shimokata, K. Relationship between Gallium 67 Citrate Scanning and Transferrin Receptor Expression in Lung Diseases. *Chest* **1992**, *102*, 530–534. [CrossRef]
21. Biber, J.; Stieger, B.; Stange, G.; Murer, H. Isolation of renal proximal tubular brush-border membranes. *Nat. Protoc.* **2007**, *2*, 1356–1359. [CrossRef] [PubMed]
22. Lin, K.; Pan, J.; Amouroux, G.; Turashvili, G.; Hundal-jabal, N.; Pourghiasian, M.; Lau, J.; Jenni, S.; Aparicio, S.; Benard, F. In vivo radioimaging of bradykinin receptor B1, a widely overexpressed molecule in human cancer. *Cancer Res.* **2015**, *75*, 387–393. [CrossRef] [PubMed]
23. Zhang, Z.; Amouroux, G.; Pan, J.; Jenni, S.; Zeisler, J.; Zhang, C.; Liu, Z.; Perrin, D.M.; Benard, F.; Lin, K. Radiolabeled B9958 derivatives for imaging bradykinin B1 receptor expression with positron emission tomography: Effect of the radiolabel-chelator complex on biodistribution and tumor uptake. *Mol. Pharm.* **2016**, *13*, 2823–2832. [CrossRef]
24. Benešová, M.; Bauder-Wüst, U.; Schäfer, M.; Klika, K.D.; Mier, W.; Haberkorn, U.; Kopka, K.; Eder, M. Linker Modification Strategies to Control the Prostate-Specific Membrane Antigen (PSMA)-Targeting and Pharmacokinetic Properties of DOTA-Conjugated PSMA Inhibitors. *J. Med. Chem.* **2016**, *59*, 1761–1775. [CrossRef] [PubMed]

Sample Availability: Samples of all the compounds are available from the authors.

Article

Radiochemical Synthesis and Evaluation of Novel Radioconjugates of Neurokinin 1 Receptor Antagonist Aprepitant Dedicated for NK1R-Positive Tumors

Paweł K. Halik [1,*], Piotr F. J. Lipiński [2], Joanna Matalińska [2], Przemysław Koźmiński [1], Aleksandra Misicka [2] and Ewa Gniazdowska [1]

1 Centre of Radiochemistry and Nuclear Chemistry, Institute of Nuclear Chemistry and Technology, 03-195 Warsaw, Poland; p.kozminski@ichtj.waw.pl (P.K.); e.gniazdowska@ichtj.waw.pl (E.G.)

2 Department of Neuropeptides, Mossakowski Medical Research Centre, Polish Academy of Sciences, 02-106 Warsaw, Poland; plipinski@imdik.pan.pl (P.F.J.L.); jmatalinska@imdik.pan.pl (J.M.); misicka@chem.uw.edu.pl (A.M.)

* Correspondence: p.halik@ichtj.waw.pl; Tel.: +48-22-504-1316

Academic Editor: Krishan Kumar

Received: 31 July 2020; Accepted: 15 August 2020; Published: 18 August 2020

Abstract: Aprepitant, a lipophilic and small molecular representative of neurokinin 1 receptor antagonists, is known for its anti-proliferative activity on numerous cancer cell lines that are sensitive to Substance P mitogen action. In the presented research, we developed two novel structural modifications of aprepitant to create aprepitant conjugates with different radionuclide chelators. All of them were radiolabeled with ^{68}Ga and ^{177}Lu radionuclides and evaluated in terms of their lipophilicity and stability in human serum. Furthermore, fully stable conjugates were examined in molecular modelling with a human neurokinin 1 receptor structure and in a competitive radioligand binding assay using rat brain homogenates in comparison to the aprepitant molecule. This initial research is in the conceptual stage to give potential theranostic-like radiopharmaceutical pairs for the imaging and therapy of neurokinin 1 receptor-overexpressing cancers.

Keywords: aprepitant; radiopharmaceuticals; neurokinin 1 receptor antagonist; radionuclide chelators

1. Introduction

The knowledge of a suitable molecular target and its specificity for a given pathology is a necessary condition in a targeted radionuclide therapy approach. Many malignant tumors possess an infiltrating character with no defined margins or spread out metastases around the whole body. Only the selective binding of a radiopharmaceutical to a molecular target allows for the reliable imaging or safe ablation of cancer lesions with minimal side effects.

Neurokinin 1 receptor (NK1R; tachykinin 1 receptor) is a well-known G protein-coupled receptor for neuropeptide Substance P (SP) and a promising system for an anticancer therapeutic molecular target [1,2]. The activation of the NK1R by its endogenous ligand creates significant proliferative impulses for tumor cells promoting growth and development, including angiogenesis and metastasis. At the same time, the frequent formation of SP-NK1R complexes stimulate the cellular up-regulation of NK1R on tumor cell surfaces [3], thus providing an even greater cell sensitivity for the mitogen action of SP. On the other hand, the blockage of SP action by using antagonists of NK1R on SP-sensitive tumor cells can selectively induce an anti-tumor effect through the mechanism of cell apoptosis [4,5].

Antagonists of NK1R are a very diverse and numerous group of compounds, though clinical applications have only been found for four compounds. They are applied to the prevention of nausea and vomiting induced by chemotherapy or surgical complications [2,6]. One of the best known and widely studied compounds in this group is aprepitant (APT; Figure 1)—a lipophilic and low molecular weight morpholine derivative with a high and selective affinity for NK1R. APT possesses anti-tumor activity, as has been determined in many cancer cell lines [5,7–12]. Moreover, the phenomenon of the synergism of the anti-tumor activity of NK1R antagonists with an inhibitory effect on the cancer cell growth of other agents has been confirmed [4]. It has been shown in vitro that the application of microtubule destabilizing agents in combination with antagonists of NK1R possess synergism in apoptotic effect in human glioblastoma, bladder, cervical and breast cancer cells [13]. More remarkable cytotoxic synergism has been proven in a combination of aprepitant and ritonavir (an antiretroviral agent) in the human glioblastoma GAMG cell line [14]. The application of these two drugs with temozolomide, an alkylating chemotherapeutic used clinically to treat glioblastoma, gives an even stronger synergistic effect.

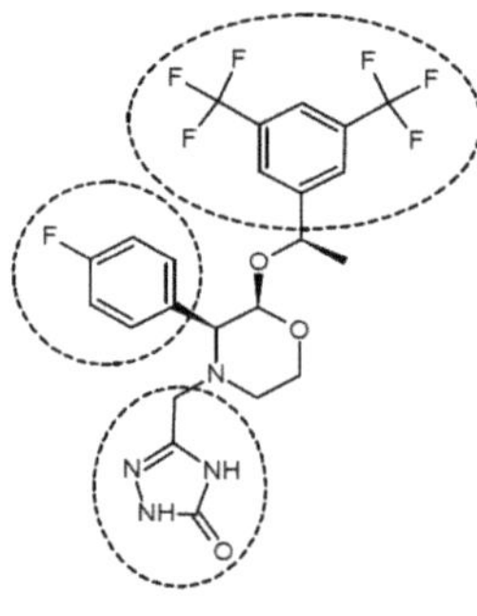

Figure 1. Structure of aprepitant with its key elements marked.

What is most relevant is that APT is a fairly safe drug with a known pharmacological profile, with tolerability similar to placebo- and dose- related action. This could be shown by the fact that aprepitant's half-maximal inhibitory concentration (IC_{50}) value determined for the human embryonic kidney (HEK) 293 cell line (a low expression NK1R control) is higher than the aprepitant IC_{100} values determined for numerous tumor cell lines overexpressing NK1R [15]. For the reasons described above, APT's structure is an interesting scaffold for creating conjugates for carrying radionuclides to NK1R-positive tumors.

By looking at aprepitant in terms of molecular structure, it can be seen that the compound (Figure 1) consists of a morpholine core decorated by three 'arms,' which are:

(i) *p*-fluorophenyl,
(ii) 3,5-bis-trifluoromethylphenyl suspended at an ether linker, and
(iii) a triazolinone moiety suspended at a methylene linker.

In the course of extensive structure-activity studies on NK1R antagonists [16–18], it has been established that the first two features (in particular: the distance and mutual positioning of two aromatic rings) are critical for high affinity and, therefore, for NK1R antagonism. On the other hand, the third element, triazolinone ring, can be, at least in some cases, safely modified without a significant loss of affinity [18]. This was exploited in attempts to improve the solubility of aprepitant derivatives, resulting in the derivative L-760,735.

That this site tolerates some modifications is now well-understood in terms of protein-ligand interactions. A recently reported X-ray structure of an NK1R-aprepitant complex [19] revealed that the triazolinone ring is located relatively close to the extracellular end of the receptor binding pocket, where it participates in hydrogen bonding to E193 and W184. However, E193A mutation has virtually no effect on aprepitant's affinity, thus suggesting that the interactions in this area are of less importance to high affinity binding. Therefore, it seemed the most rational that a convenient site for functionalizing the APT structure is at this very ring. Nevertheless, the performed functionalization of the APT molecule required confirmation that the obtained conjugate still had a sufficiently high affinity for the receptor.

Based on that knowledge, we focused our efforts on the syntheses and in vitro evaluation of newly designed radioconjugates of aprepitant with gallium-68 or lutetium-177 radionuclides. For this purpose, we have proposed two functionalization routes of the APT molecule, followed by conjugation of different macrocyclic chelators DOTA, Bn-DOTA, and Bn-DOTAGA, as well as acrylic chelator DTPA dedicated to ^{68}Ga and ^{177}Lu. For conjugates showing full stability in human serum, molecular modelling studies for human NK1R and preliminary in vitro examination were performed. These reported findings indicate new perspectives of aprepitant applications in the form of selective theranostic-like concept radiopharmaceuticals for NK1R-positive tumors.

2. Results and Discussion

2.1. Syntheses of Aprepitant-Based Radioconjugates

2.1.1. Syntheses of Aprepitant Derivatives

The first stage of synthesis concerned the modification of the APT structure in order to introduce a primary amine group. This was realized according to synthetic pathways presented below by using one of selected alkyl linkers (Scheme 1) or acetamide linkers (Scheme 2) so as to receive APT-alkylamine (**2A–C**) or APT-acetamide derivatives (**4D**,**E**).

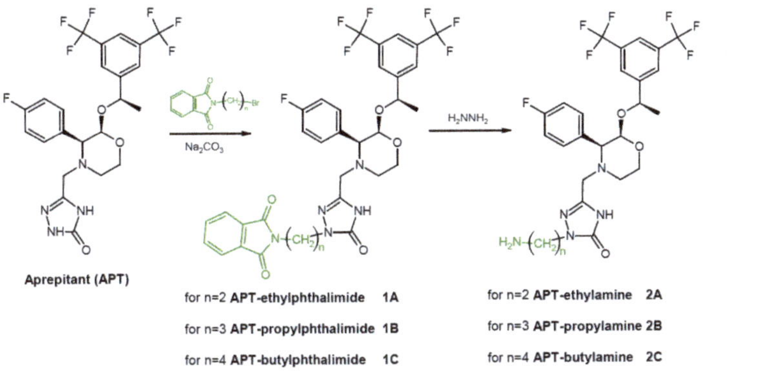

Aprepitant (APT)

Na_2CO_3

H_2NNH_2

for n=2 **APT-ethylphthalimide 1A**

for n=3 **APT-propylphthalimide 1B**

for n=4 **APT-butylphthalimide 1C**

for n=2 **APT-ethylamine 2A**

for n=3 **APT-propylamine 2B**

for n=4 **APT-butylamine 2C**

Scheme 1. Synthetic route of aprepitant derivatives with aminoalkyl linkers; where n = {2; 3; 4}.

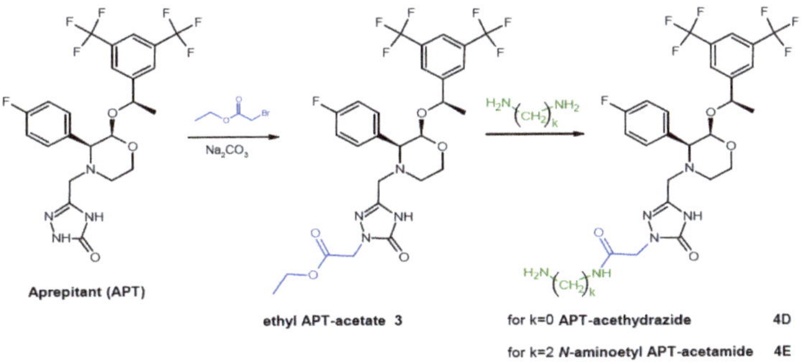

Aprepitant (APT)

Na_2CO_3

ethyl APT-acetate 3

for k=0 **APT-acethydrazide 4D**

for k=2 ***N*-aminoetyl APT-acetamide 4E**

Scheme 2. Synthetic route of aprepitant derivatives with acetamide linkers; where k = {0; 2}.

2.1.2. Syntheses of Aprepitant Conjugates

The coupling reactions of APT-ethylamine, **2A**, with different bifunctional chelating agents, were as follows: DOTA-NHS ester, *p*-SCN-Bn-DOTA, *p*-SCN-Bn-DOTAGA, or DTPA dianhydride, as presented in Scheme 3. The use of different chelators allowed for the evaluation of the effect of the chelating moiety on the physicochemical properties of later radioconjugates. Based on the stability results obtained for these radioconjugates (presented in a section below), all other obtained APT derivatives (**2B**, **2C**, **4D**, and **4E**) were only conjugated with selected macrocyclic chelator DOTA. The application of different linkers allowed for the evaluation of their influence on the physicochemical properties of later radioconjugates.

APT-Et-Bn-DOTAGA 7A

p-SCN-Bn-DOTAGA Et_3N

APT-Et-Bn-DOTA 6A

p-SCN-Bn-DOTA Et_3N

APT-ethylamine 2A

DOTA-NHS Et_3N

APT-Et-DOTA 5A

DTPA anhydride

APT-Et-DTPA 8A

+

APT-Et-DTPA-Et-APT 9A

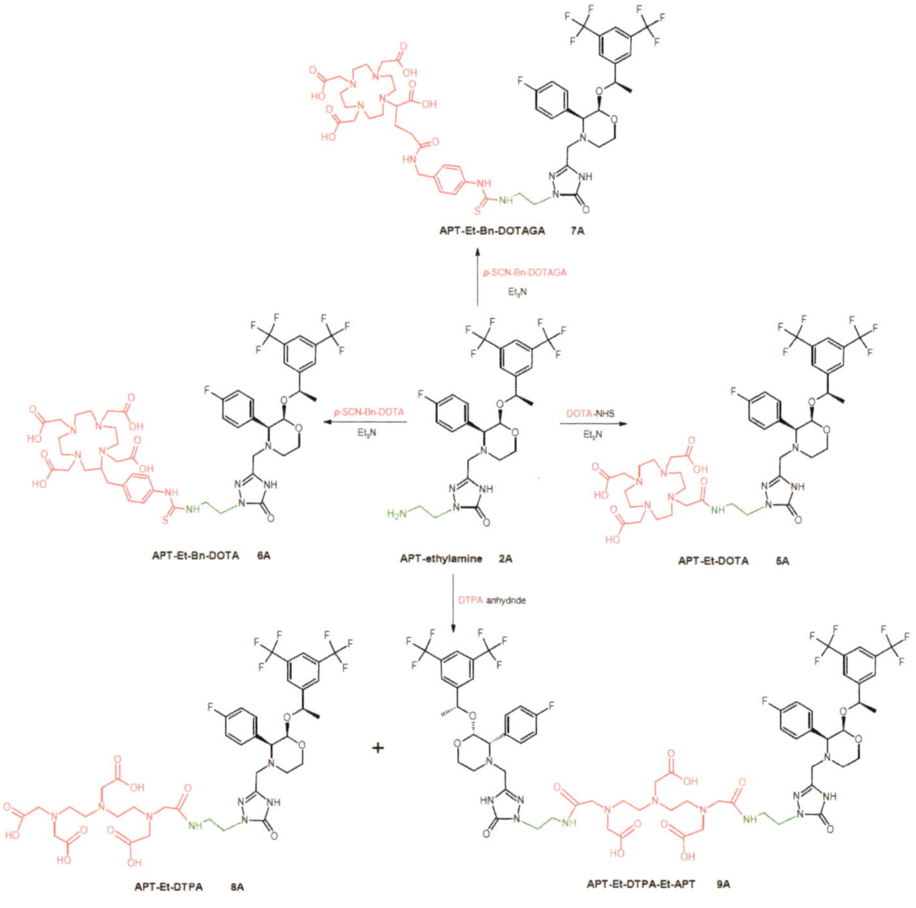

Scheme 3. Synthetic routes of conjugations of selected chelators to aprepitant-ethylamine **2A**.

2.1.3. Preparation of Radioconjugates

All APT conjugates with DOTA, Bn-DOTA, and Bn-DOTAGA were radiolabeled with ^{68}Ga and ^{177}Lu, while APT conjugates with DTPA were only radiolabeled with ^{68}Ga. Synthesized radioconjugates were purified using the solid phase extraction (SPE) method before HPLC identification (Figures 2 and 3) and further analyses.

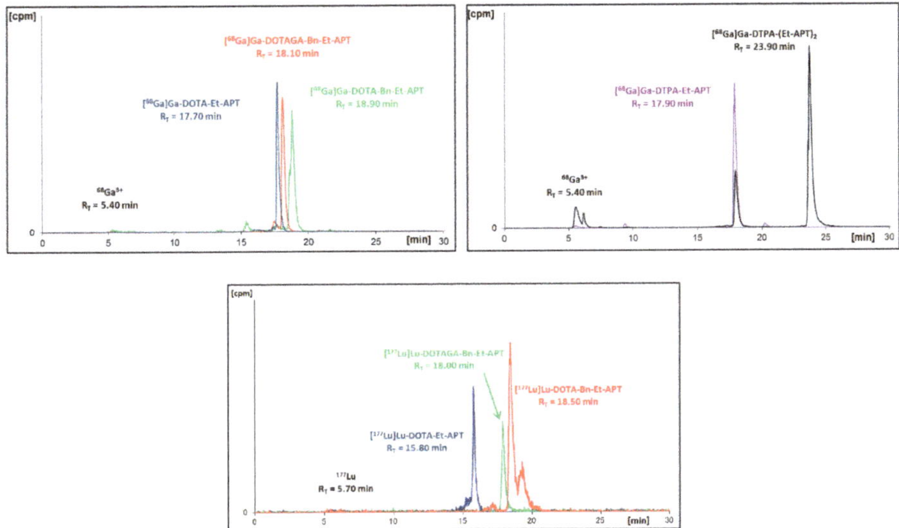

Figure 2. Radiochromatograms of aprepitant (APT)-ethylamine **2A** conjugates with DOTA, Bn-DOTA, Bn-DOTAGA or DTPA radiolabeled with gallium-68 (**upper** two) or with lutetium-177 (**bottom**).

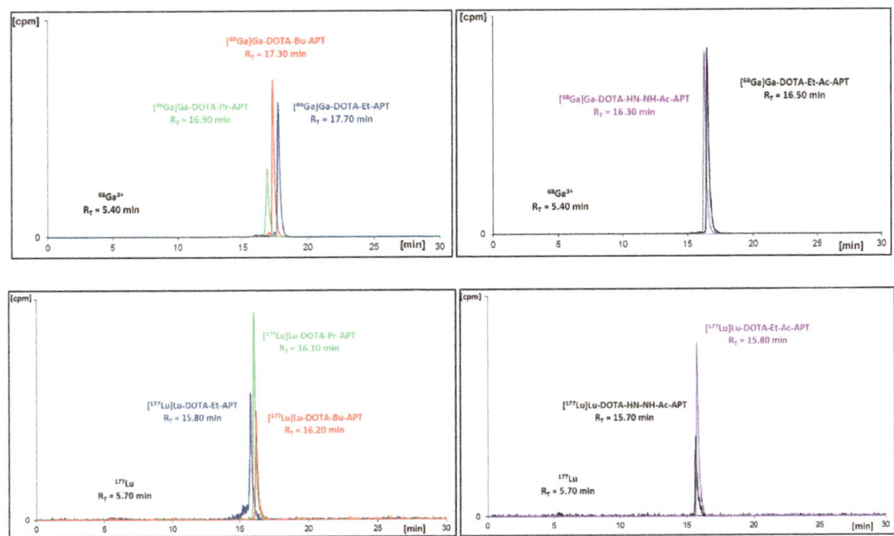

Figure 3. Radiochromatograms of DOTA conjugates with all APT derivatives radiolabeled with gallium-68 (**upper** two) or with lutetium-177 (**bottom** two).

As a result of the performed radiosyntheses, all radioconjugates were successfully obtained, except for [^{68}Ga]Ga-DTPA-(Et-APT)$_2$ (**[^{68}Ga]Ga-9A**), which proved to be immediately unstable. Moreover, in the radiochromatogram of [^{177}Lu]Lu-DOTA-Bn-Et-APT (**[^{177}Lu]Lu-6A**) one can see a small additional signal (about 19.3 min) that is recognized as an early by-product of an interaction with solvent (EtOH) from the purification process. To verify the identity of all synthesized [^{68}Ga]Ga-radioconjugates in a non-carrier added scale, the non-radioactive stable gallium reference compounds (**Ga-5A–Ga-9A** and **Ga-5A–Ga-5E**) were synthesized and characterized by mass spectrometry. The retention time values of the [^{68}Ga]Ga-radioconjugates and stable references presented below (Tables 1 and 2) overlapped, and the differences between them resulted from the serial connection of UV-Vis and gamma detectors only.

Table 1. Retention times (R_T) of stable gallium conjugates and [^{68}Ga]Ga-radioconjugates of aprepitant-ethylamine **2A**.

Stable Ga-Conjugate	R_T	[^{68}Ga]Ga-Radioconjugate	R_T
Ga-DOTA-Et-APT, Ga-5A	17.4 min	[^{68}Ga]Ga-DOTA-Et-APT, [^{68}Ga]Ga-5A	17.7 min
Ga-DOTA-Bn-Et-APT, Ga-6A	18.5 min	[^{68}Ga]Ga-DOTA-Bn-Et-APT, [^{68}Ga]Ga-6A	18.9 min
Ga-DOTAGA-Bn-Et-APT, Ga-7A	17.7 min	[^{68}Ga]Ga-DOTAGA-Bn-Et-APT, [^{68}Ga]Ga-7A	18.1 min
Ga-DTPA-Et-APT, Ga-8A	17.6 min	[^{68}Ga]Ga-DTPA-Et-APT, [^{68}Ga]Ga-8A	17.9 min
Ga-DTPA-(Et-APT)$_2$, Ga-9A	23.6 min	[^{68}Ga]Ga-DTPA-(Et-APT)$_2$, [^{68}Ga]Ga-9A	23.9 min

Table 2. R_T of stable gallium conjugates and [^{68}Ga]Ga-radioconjugates of all aprepitant derivatives.

Stable Ga-Conjugate	R_T	[^{68}Ga]Ga-Radioconjugate	R_T
Ga-DOTA-Et-APT, Ga-5A	17.4 min	[^{68}Ga]Ga-DOTA-Et-APT, [^{68}Ga]Ga-5A	17.7 min
Ga-DOTA-Pr-APT, Ga-5B	16.8 min	[^{68}Ga]Ga-DOTA-Pr-APT, [^{68}Ga]Ga-5B	17.0 min
Ga-DOTA-Bu-APT, Ga-5C	16.9 min	[^{68}Ga]Ga-DOTA-Bu-APT, [^{68}Ga]Ga-5C	17.3 min
Ga-DOTA-HN-NH-Ac-APT, Ga-5D	15.8 min	[^{68}Ga]Ga-DOTA-HN-NH-Ac-APT, [^{68}Ga]Ga-5D	16.3 min
Ga-DOTA-Et-Ac-APT, Ga-5E	16.1 min	[^{68}Ga]Ga-DOTA-Et-Ac-APT, [^{68}Ga]Ga-5E	16.5 min

2.2. *Physiochemical Evaluation of Radioconjugates*

2.2.1. Stability Study

The sine qua non condition of a radionuclide's application in vivo is its radiopharmaceutical stability in biological fluids like serum or cerebrospinal fluid. For this purpose, each isolated and solvent-free radioconjugate was incubated at 37 °C in human serum (HS). At specific time points, small samples of radioconjugate mixture were analyzed by the HPLC method for the assessment of the radioconjugate condition. The collected data presented on the charts below (Figure 4) point out that only the DOTA radioconjugates remained stable in the biological fluid; thus, these radioconjugates were selected for further analyses.

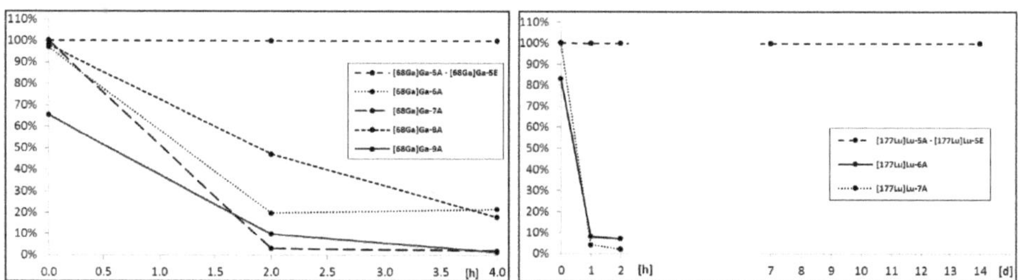

Figure 4. Percentage of intact [^{68}Ga]Ga-radioconjugates (**left**) and [^{177}Lu]Lu-radioconjugates (**right**) determined at specific time points during incubation in 37 °C human serum.

We concluded that for the demand of designed aprepitant radioconjugates, the acyclic chelator DTPA showed a poor radionuclide chelating ability during incubation in human serum. DOTA and its analogues presented a satisfactory radionuclide complex stability, however, for the overall stability of the radioconjugate results from the type of the formed chemical bond with the amine terminated aprepitant derivative and the presence of a negative charge on the chelator-metal complex moiety. The amide bond created by the DOTA-NHS ester and uncharged complex in the conjugates remained stable throughout the whole stability study, while the thiourea bonds and negatively charged complexes created by both *p*-SCN-Bn-DOTA and *p*-SCN-Bn-DOTAGA were found to gradually decompose in time. This phenomenon of instability in HS has been observed previously in various radiopharmaceuticals [20].

2.2.2. Lipophilicity Study

Drug distribution in vivo is highly related to both the lipophilicity and charge of a drug. The optimal radiotracer lipophilicity value for blood-brain barrier crossing lies within the range from 2.0 to 3.5 [21]. Non-peptide NK1R antagonists, like aprepitant, are characterized by a high lipophilicity (logD 4.8) [22], while the DOTA chelator is a highly hydrophilic moiety. In seeking to keep in lipophilicity of radioconjugates in a desired range, the choice of a proper linker (primary aprepitant modification) seems essential for distribution and pharmacokinetic aspects.

In the course of the lipophilicity study, each isolated DOTA radioconjugate (determined as fully stable in HS) was examined for distribution in the system of *n*-octanol and a phosphate-buffered saline (PBS) buffer (pH = 7.4) to estimate the lipophilicity of the radiocomplex. The lipophilicity of each radioconjugate (logD), defined as the logarithm of the distribution coefficient (D) is based on the ratio of the radioactivity of the organic phase to the radioactivity of the aqueous phase. The stability of the studied radioconjugate was verified simultaneously during the experiment through the HPLC analysis of the aqueous phase. LogD values of **[^{68}Ga]Ga-5A–[^{68}Ga]Ga-5E** and **[^{177}Lu]Lu-5A–[^{177}Lu]Lu-5E** are listed below in Table 3.

Table 3. LogD values of human serum stable radioconjugates determined in *n*-octanol/PBS buffer system.

Radioconjugate	logD	
	^{68}Ga-	^{177}Lu-
APT-Et-DOTA, 5A	0.141 ± 0.019	0.708 ± 0.021
APT-Pr-DOTA, 5B	0.058 ± 0.016	0.654 ± 0.022
APT-Bu-DOTA, 5C	0.290 ± 0.018	0.777 ± 0.021
APT-Ac-HN-NH-DOTA, 5D	−1.012 ± 0.017	−0.401 ± 0.015
APT-Ac-Et-DOTA, 5E	−0.231 ± 0.015	0.500 ± 0.017

The APT-alkylamine derivative-based radioconjugates showed similar lipophilicity values that were higher than those of the APT-acetamide derivative-based radioconjugates. The complexes with lutetium were more lipophilic by (on average) 0.6 logD units. However, the logD values for all radioconjugates significantly decreased in comparison to aprepitant, indicating possible divergences in the pharmacokinetic fate of the radioconjugates and the parent drug.

2.3. *Binding Affinity*

An important consideration in the search of conjugate vectors for radionuclides is whether the functionalization of a high affinity ligand would not reduce the binding strength for a desired receptor. For the preliminary addressing of this issue in the case of our conjugates, we measured the affinity of compounds **5A–E** (uncomplexed precursors) for the rat neurokinin-1 receptor. The human (hNK1R) and the rat (rNK1R) neurokinin 1 receptors differ in their sequences and pharmacology. It has been established that many (but not all) high affinity NK1R antagonists have a significantly lower affinity for the rat receptor than for that of human origin [23,24]. Still, the results presented below give some tentative insight into the affinity changes caused by the functionalization of the aprepitant structure at the triazolinone ring.

The results of the binding affinity determinations are given in Table 4. The parent compound, aprepitant, was found to exhibit IC_{50} = 128.4 nM. This value was roughly consistent with the reported potency of aprepitant in a functional assay. The compound was found to inhibit Substance P-evoked increases in intracellular Ca^{2+} mobilization in the cells expressing rNK1R with a pK_B reading 7.3 [25]. Note that in the assays with cells expressing hNK1R, aprepitant was significantly more potent (pK_B = 8.7), and the reported binding affinities for the human receptor were of the subnanomolar order (e.g., IC_{50} = 0.09 nM [17]).

Table 4. Binding affinity of aprepitant and compounds **5A–E** for the rat neurokinin 1 receptor.

Compound	IC_{50} ± SEM [a] [μM]	Ratio to APT
Aprepitant	0.13 ± 0.06	1.0
APT-Et-DOTA, 5A	6.2 ± 2.6	48.2
APT-Pr-DOTA, 5B	0.69 ± 0.07	5.3
APT-Bu-DOTA, 5C	1.8 ± 0.7	14.3
APT-Ac-HN-NH-DOTA, 5D	2.5 ± 0.7	19.1
APT-Ac-Et-DOTA, 5E	2.5 ± 0.5	19.7

[a] IC_{50} ± SEM: the half-maximal inhibitory concentration with the standard error of the mean of three independent experiments done in duplicate.

The aprepitant-based conjugates exhibited a diversified range of affinities. The strongest ligand in the set was the compound bearing a propylamine linker, **5B**. It was found to have an IC_{50} of 0.69 μM. This value was about five times worse than that of the parent compound. Interestingly, decreasing (**5A**) or increasing (**5C**) the linker length by one methylene unit was associated with much lower affinity of the micromolar order. The shorter **5A** exhibited the lowest binding in the set, with an IC_{50} of 6.2 μM. The analogue with the butylamine linker (**5C**) had an IC_{50} of 1.8 μM. Similar affinities (IC_{50}~2.5 μM) were found for the conjugates with the acylhydrazine (**5D**) or *N*-aminoethylacetamide (**5E**) linkers.

2.4. Molecular Modelling Study

In order to get insight into possible interactions between the aprepitant-DOTA conjugates reported herein and the NK1R, the complexes thereof were modelled by molecular docking. The applied procedure consisted in building the appropriate linker-DOTA fragments into the aprepitant structure crystallized with the receptor (Protein Data Bank (PDB) accession code: 6HLO [19]), followed by local search docking executed in AutoDock 4.2.6 [26].

According to this procedure, the presence of a linker-DOTA moiety in the aprepitant-based conjugates did not have a major impact on the interactions between the core of the molecule and the receptor. Only a slight repositioning of the morpholine core, 3,5-bis-trifluoromethylphenyl, or *p*-fluorophenyl moieties was observed compared to the 6HLO crystal structure (Figure 5A). Thus, the conjugates were predicted to bind with the 3,5-bis-trifluoromethylphenyl fragment located at the bottom of the ligand-binding pocket and the DOTA moiety closer to the extracellular side of the receptor (Figure 5A).

In the part that was common to all studied derivatives (and the parent aprepitant), the complexes were stabilized by (Figure 5B):

(i) hydrogen bonding between Q165 and the ether oxygen,
(ii) hydrophobic contacts of the morpholine ring and F268 and I182,
(iii) hydrophobic contacts with side chains of N109, P112, and I113,
(iv) hydrophobic contacts of the 3,5-bis-trifluoromethylphenyl with W261 and F264,
(v) hydrophobic contacts of the *p*-fluorophenyl ring and H197, V200, T201, I204, and H265.

These interactions were identical to those found for the parent aprepitant in 6HLO structure.

On the other hand, the presence of the linker-DOTA fragment was predicted to weaken the contacts that the triazolinone ring of the parent aprepitant had with the receptor in the crystal structure 6HLO [19]. In the optimized complexes for all the conjugates, this ring was displaced compared to the parent structure (Figure 6A,B), so hydrogen bonding to W184 was not possible. On the other hand, a better positioning of this ring for π–π stacking with H197 was predicted for the conjugates.

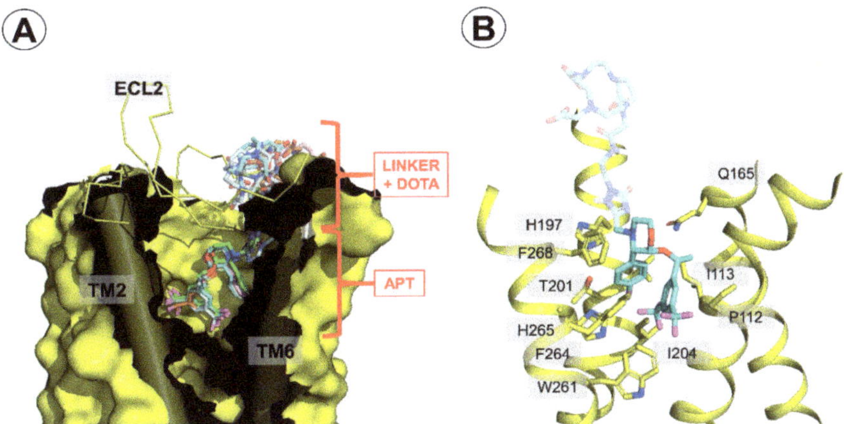

Figure 5. Binding mode of the reported conjugates in the neurokinin 1 receptor (NK1R) binding site. (**A**) A generalized view on the binding mode. The receptor is displayed as a yellow surface, with transmembrane helices (TMs) 2 and 6 shown as cylinders. The extracellular loop 2 (ECL2) is shown as a yellow ribbon. The conjugates are represented as colored sticks. (**B**) A view focused on the interactions of the part common to aprepitant and the conjugates. The conjugate shown is compound **5A** (pale blue sticks). Only several residues of the receptor are shown (yellow sticks).

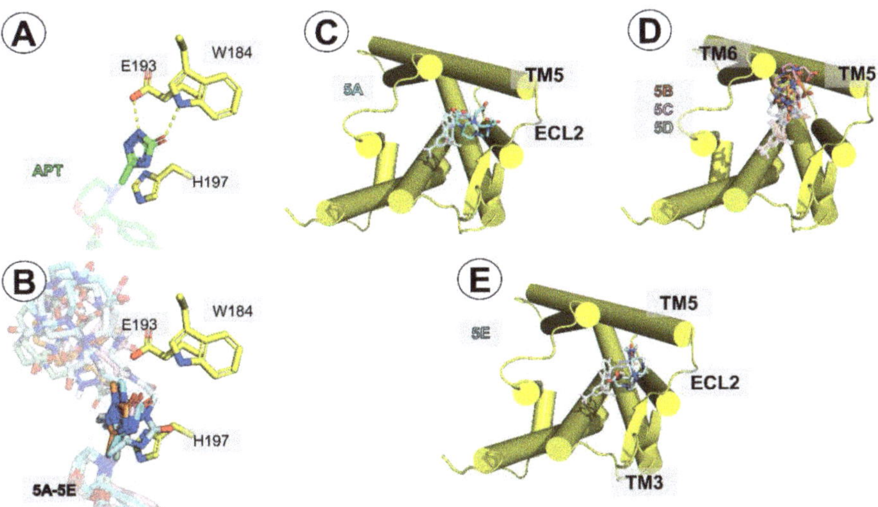

Figure 6. (**A**) Interactions of aprepitant's triazolinone ring with W184, E193, and H197 side-chains. (**B**) positioning of the triazolinone ring of the conjugates in the same projection as in (**A**). (**C–E**) Relative position of the DOTA moiety in conjugates **5A** (**C**), **5B–D** (**D**), and **5E** (**E**). The receptor helices are shown as yellow cylinders.

Regarding the positioning of the linker-DOTA part, in the case of **5A,** this fragment docked closely (Figure 6C) to the extracellular loop 2 (ECL2) and the extracellular terminus of the transmembrane helix 5 (TM5). One of the DOTA's carboxylate oxygens interacted with the side-chain of K190. For the analogues **5B–D**, the DOTA moiety was predicted to be located between the extracellular tips of TM5 and TM6 (Figure 6D). Its contacts included residues K194, K190, and P271. In the case of the longest derivative, **5E**, the docking placed the DOTA moiety close to TM5 and ECL2 (Figure 6E). Here, it could interact with K190 and M181.

Since in the crystal structures 6HLO, 6HLL, and 6HLP [19], several residues by the extracellular end of the receptor were found to adopt different rotamers upon the binding of different ligands, we wanted to see if the flexibility of these residues could affect the docking results. Therefore, the local docking procedure with the enabled flexibility of E193 and H197 was performed. It yielded similar results with only minor adjustments of the side chain rotamers. Its results (in terms of interactions and binding poses) are not discussed herein since they are almost perfectly accounted for by the description of the docking procedure with the rigid receptor.

Regarding the quantitative evaluation (Table 5), AutoDock scoring function predicted that aprepitant would bind with the free energy of −10.43 kcal/mol. For the conjugates, the estimated energy varied between −9.64 kcal/mol (**5D**) and −13.74 kcal/mol (**5E**). The predicted energies did not correlate with the experimental data. This was perhaps due to the problems with estimating the entropic contribution because the conjugates differed with respect to the number of the rotatable bonds.

Table 5. Scoring results from molecular docking. The values are the estimated free energy of binding (kcal/mol).

	Rigid Receptor		Enabled Flexibility of E193 and H197	
Compound	**Lowest [a]**	**Mean [b]**	**Lowest [a]**	**Mean [b]**
APT-Et-DOTA, 5A	−11.26	−11.26	−12.37	−10.24
APT-Pr-DOTA, 5B	−10.84	−9.86	−10.74	−9.27
APT-Bu-DOTA, 5C	−12.26	−11.90	−12.21	−11.83
APT-Ac-HN-NH-DOTA, 5D	−9.64	−9.07	−9.19	−8.68
APT-Ac-Et-DOTA, 5E	−13.74	−13.16	−14.31	−13.60
Aprepitant (APT)	−10.43	−10.31	−10.57	−10.40

[a] lowest energy in the best scored cluster; [b] mean energy in the best scored cluster.

Other sources of significant error may have been the way the DOTA moiety was modelled (aimed at mimicking the presence of the cation in a simplified manner) and the fact that the experimentally evaluated conjugates were uncomplexed.

3. Materials and Methods

Aprepitant (Santa Cruz Biotechnology Inc., Dallas, TX, USA), the DOTA-NHS ester (1,4,7,10-tetraazacyclododecane-1,4,7,10-tetraacetic acid mono-*N*-hydroxysuccinimide ester), *p*-SCN-Bn-DOTAGA (2,2′,2′′-(10-(1-carboxy-4-((4-isothiocyanatobenzyl)amino)-4-oxobutyl)-1,4,7,10-tetraaza-cyclododecane-1,4,7-triyl)triacetic acid) (CheMatech, Dijon, France), *p*-SCN-Bn-DOTA (*S*-2-(4-isothiocyanatobenzyl)-1,4,7,10-tetraazacyclododecane tetraacetic acid) (Macrocyclics, Plano, TX, USA), DTPA dianhydride (diethylenetriaminepentaacetic dianhydride), and other substances and solvents (Sigma Aldrich/Merck, Darmstadt, Germany) were commercially available, defined as reagent grade, and applied without further purification. $^{68}GaCl_3$ was eluted from the commercially available $^{68}Ge/^{68}Ga$ generator (Eckert & Ziegler, Berlin, Germany). The $^{177}LuCl_3$ solution in 0.04 M HCl was purchased at Radioisotope Centre POLATOM, National Centre for Nuclear Research, Otwock-Świerk, Poland. Sep-Pack® Classic Short C18 Cartridges were purchased from WATERS, Milford, MA, USA. Human serum was isolated and purified at the Centre of Radiobiology and Biological Dosimetry, INCT Warsaw, Poland.

The HPLC conditions and gradient were as follows: a semi-preparative Phenomenex Jupiter Proteo column, 4 μm, 90 Å, 250 × 10 mm, with UV/Vis (220 nm) or/and radio γ-detection at gradient elution: 0–20 min 20 to 80% solvent B; 20–30 min 80% solvent B; 2 mL/min; solvent A: 0.1% (*v*/*v*) trifluoroacetic acid (TFA) in water; and solvent B: 0.1% (*v*/*v*) TFA in acetonitrile.

Mass spectra were measured on a Bruker 3000 Esquire mass spectrometer equipped with electrospray ionization (ESI) (Bruker, Billerica, MA, USA).

3.1. Syntheses of Aprepitant Derivatives and Aprepitant-Based Conjugates

3.1.1. General Procedure of Syntheses of Aprepitant Derivatives with Alkyl Linker, **2A–C**

The slight molar excess of the selected *n*-(terminal-bromoalkyl) phthalimide was added into an equimolar mixture of APT and sodium carbonate in dimethylformamide (DMF). The reaction mixture was vigorously stirred in about 50 °C for 12–18 h. Then, the triple molar excess of hydrazine was added into the reaction mixture for an additional 3 h. The progress of the reaction was monitored by HPLC. The crude reaction mixture was evaporated, dissolved in the HPLC mobile phase, purified by the HPLC method, and lyophilized. The isolated main product was identified as a mono-substituted APT-alkylamine derivative (**2A–C**, ~75% reaction yield) by MS analysis confirmation.

MS: Calculated monoisotopic mass for **APT-Et-NH$_2$, 2A**, $C_{25}H_{26}F_7N_5O_3$: 577.19; found: 578.27 *m*/*z* [M + H$^+$]
MS: Calculated monoisotopic mass for **APT-Pr-NH$_2$, 2B**, $C_{26}H_{28}F_7N_5O_3$: 591.21; found: 592.12 *m*/*z* [M + H$^+$]
MS: Calculated monoisotopic mass for **APT-Bu-NH$_2$, 2C**, $C_{27}H_{30}F_7N_5O_3$: 605.22; found: 606.38 *m*/*z* [M + H$^+$]

3.1.2. General Procedure of Syntheses of Aprepitant Derivatives with Acetamide Linker, **4D** and **4E**

The slight molar excess of ethyl 2-bromoacetate was added into an equimolar mixture of APT and sodium carbonate in DMF. The reaction mixture was vigorously stirred in about 50 °C for 24 h. Then, the triple molar excess of hydrazine or ethylenediamine was added into the reaction mixture for an additional 3 h. The progress of the reaction was monitored by HPLC. The crude reaction mixture was evaporated, dissolved in the HPLC mobile phase, purified by the HPLC method, and lyophilized. The isolated main product was identified as a mono-substituted amino-terminated APT-acetamide derivative (**4D** and **4E**, 65–70% reaction yield) by MS analysis confirmation.

MS: Calculated monoisotopic mass for **APT-Ac-HN-NH$_2$, 4D**, $C_{25}H_{25}F_7N_6O_4$: 606.19; found: 608.07 *m*/*z* [M + H$^+$]
MS: Calculated monoisotopic mass for **APT-Ac-Et-NH$_2$, 4E**, $C_{27}H_{29}F_7N_6O_4$: 634.21; found: 635.31 *m*/*z* [M + H$^+$]

3.1.3. General Procedure of Syntheses of Aprepitant Conjugates with DOTA, **5A–E**

The obtained APT derivative (**2A–C**, **4D**, and **4E**) and the DOTA-NHS ester in similar molar ratios were dissolved in DMF purged from oxygen with technical nitrogen and supplemented with a triple molar excess of triethylamine. The reaction mixture was vigorously stirred in about 50 °C for 24 h. The progress of the reaction was monitored by HPLC. The crude reaction mixture was evaporated, dissolved in the HPLC mobile phase, purified by the HPLC method, and lyophilized. The isolated main product was identified as a DOTA conjugate with an APT derivative (**5A–E**, >90% reaction yield) by MS analysis confirmation.

MS: Calculated monoisotopic mass for **APT-Et-DOTA**, **5A**, $C_{41}H_{52}F_7N_9O_{10}$: 963.37; found: 964.27 *m*/*z* [M + H$^+$]
MS: Calculated monoisotopic mass for **APT-Pr-DOTA**, **5B**, $C_{42}H_{54}F_7N_9O_{10}$: 977.39; found: 978.42 *m*/*z* [M + H$^+$]
MS: Calculated monoisotopic mass for **APT-Bu-DOTA**, **5C**, $C_{43}H_{56}F_7N_9O_{10}$: 991.40; found: 992.41 *m*/*z* [M + H$^+$]
MS: Calculated monoisotopic mass for **APT-Ac-HN-NH-DOTA**, **5D**, $C_{41}H_{51}F_7N_{10}O_{11}$: 992.36; found: 993.17 *m*/*z* [M + H$^+$]
MS: Calculated monoisotopic mass for **APT-Ac-Et-DOTA**, **5E**, $C_{43}H_{55}F_7N_{10}O_{11}$: 1020.39; found: 1021.43 *m*/*z* [M + H$^+$]

3.1.4. Procedure of Syntheses of Aprepitant-Ethylamine Conjugates with *p*-SCN-Bn-DOTA and *p*-SCN-Bn-DOTAGA, **6A** and **7A**

The APT-ethylamine (**2A**) and bifunctional chelating agent in similar molar ratios were dissolved in DMF and supplemented with a 5-fold molar excess of triethylamine. The reaction mixture was vigorously stirred in about 50 °C for 24 h. The progress of the reaction was monitored by HPLC. The crude reaction mixture was evaporated, dissolved in the HPLC mobile phase, purified by the HPLC method, and lyophilized. The isolated main product was identified as a DOTA-Bn or DOTAGA-Bn conjugate with an APT derivative (**6A** and **7A**, > 90% reaction yield) by MS analysis confirmation.

MS: Calculated monoisotopic mass for **APT-Et-Bn-DOTA**, **6A**, $C_{49}H_{59}F_7N_{10}O_{11}S$: 1128.40; found: 1129.55 *m/z* [M + H$^+$]

MS: Calculated monoisotopic mass for **APT-Et-Bn-DOTAGA**, **7A**, $C_{52}H_{64}F_7N_{11}O_{12}S$: 1199.43; found: 1200.66 *m/z* [M + H$^+$]

3.1.5. Procedure of Syntheses of Aprepitant-Ethylamine Conjugates with DTPA Anhydride, **8A**, **9A**

The APT-ethylamine (**2A**) and DTPA anhydride in a 3:2 molar ratio were dissolved in DMF purged from oxygen with technical nitrogen. The reaction mixture was vigorously stirred in room temperature for 2 h. The progress of the reaction was monitored by HPLC. The crude reaction mixture was evaporated, dissolved in the HPLC mobile phase, purified by the HPLC method, and lyophilized. Two isolated main products were identified as DTPA conjugated with one or two molecules of the APT derivative (**8A** and **9A** with ~45% and ~40% reaction yields, respectively) by MS analysis confirmation.

MS: Calculated for monoisotopic mass **APT-Et-DTPA**, **8A**, $C_{39}H_{47}F_7N_8O_{12}$: 952.32; found: 953.40 *m/z* [M + H$^+$]

MS: Calculated for monoisotopic mass **APT-Et-DTPA-Et-APT**, **9A**, $C_{64}H_{71}F_{14}N_{13}O_{14}$: 1511.50; found: 1512.64 *m/z* [M + H$^+$]

3.2. *Preparation of Radioconjugates*

3.2.1. ^{68}Ga Radiolabeling

The ^{68}Ga radiolabeling of the DOTA, Bn-DOTA, and Bn-DOTAGA conjugates of APT was performed according to the following procedure: 145 µL of a concentrated solution of [^{68}Ga]GaCl$_3$ in 0.1 M HCl from the ^{68}Ge/^{68}Ga generator (4.9 ÷ 7.2 MBq) was added into the solution of 25 nmol of the selected conjugate in 200 µL of a 0.2 M acetate buffer (pH = 4.5) and heated for 5–10 min at 95 °C. After this time, each radioconjugate was purified using Sep-Pack® Classic Short C18 Cartridges according to producer recommendations, thereby obtaining an easily vaporized ethanolic solution of each radioconjugate. The effectiveness of the purification was monitored by HPLC. DTPA radioconjugates were obtained via an analogical procedure in room temperature.

3.2.2. ^{177}Lu Radiolabeling

The ^{177}Lu radiolabeling of the DOTA, Bn-DOTA, and Bn-DOTAGA conjugates of APT was performed according to the following procedure: 2.7 ÷ 5.3 µL of a [^{177}Lu]LuCl$_3$ n.c.a. solution in 0.04 M HCl (4.6 ÷ 5.2 MBq) was added into the solution of 2.5 nmol of the selected conjugate in 200 µL of a 0.02 M acetate buffer (pH 4.5) and heated for 10 min at 95 °C. After this time, each radioconjugate was purified using Sep-Pack® C18 Cartridges according to the producer recommendations, thereby obtaining an easily vaporized ethanolic solution of each radioconjugate. The effectiveness of the purification was monitored by HPLC.

3.2.3. Preparation of Non-Radioactive References

The non-radioactive Ga labelling of the DOTA, Bn-DOTA, and Bn-DOTAGA conjugates of APT was performed according to the following procedure: 145 µL of a concentrated solution of 20 mM

$GaCl_3$ in 0.1 M HCl was added into the solution of 50 nmol of the selected conjugate in 200 µL of a 0.2 M acetate buffer (pH = 4.5) and heated for 5–10 min at 95 °C. After this time, each reaction mixture was purified by the HPLC method, lyophilized, and characterized by mass spectrometry. DTPA conjugates were obtained via an analogical procedure in room temperature.

MS: Calculated for monoisotopic mass **APT-Et-DOTA-Ga, 5A-Ga**, $C_{41}H_{50}F_7N_9O_{10}Ga$: 1030.80 and 1032.28; found: 1030.40 and 1032.40 *m/z* [M^+]
MS: Calculated for monoisotopic mass **APT-Pr-DOTA-Ga, 5B-Ga**, $C_{42}H_{52}F_7N_9O_{10}Ga$: 1044.30 and 1046.30; found: 1044.38 and 1046.39 *m/z* [M^+]
MS: Calculated for monoisotopic mass **APT-Bu-DOTA-Ga, 5C-Ga**, $C_{43}H_{54}F_7N_9O_{10}Ga$: 1058.31 and 1060.31; found: 1058.37 and 1060.40 *m/z* [M^+]
MS: Calculated for monoisotopic mass **APT-Ac-HN-NH-DOTA-Ga, 5D-Ga**, $C_{41}H_{49}F_7N_{10}O_{11}Ga$: 1059.27 and 1061.27; found: 1059.31 and 1061.40 *m/z* [M^+]
MS: Calculated for monoisotopic mass **APT-Ac-Et-DOTA-Ga, 5E-Ga**, $C_{43}H_{53}F_7N_{10}O_{11}Ga$: 1087.30 and 1089.30; found: 1087.37 and 1089.44 *m/z* [M^+]
MS: Calculated for monoisotopic mass **APT-Et-Bn-DOTA-Ga, 6A-Ga**, $C_{49}H_{57}F_7N_{10}O_{11}SGa$: 1095.31 and 1097.31; found: 1195.54 and 1197.51 *m/z* [M^+]
MS: Calculated for monoisotopic mass **APT-Et-Bn-DOTAGA-Ga, 7A-Ga**, $C_{52}H_{62}F_7N_{11}O_{12}Sga$: 1266.34 and 1268.34; found: 1266.47 and 1268.47 *m/z* [M^+]
MS: Calculated for monoisotopic mass **APT-Et-DTPA-Ga, 8A-Ga**, $C_{39}H_{44}F_7N_8O_{12}Ga$: 1018.22 and 1020.22; found: 1019.35 and 1021.37 *m/z* [$M + H^+$]
MS: Calculated for monoisotopic mass **APT-Et-DTPA-(Ga)-Et-APT, 9A-Ga**, $C_{64}H_{68}F_{14}N_{13}O_{14}Ga$: 1577.40 and 1579.40; found: 1578.64 and 1580.66 *m/z* [$M + H^+$]

3.3. *Physiochemical Evaluation of Radioconjugates*

3.3.1. Stability Study

All obtained radioconjugates (isolated from the reaction mixtures using the SPE method and being solvent-free) were examined in terms of stability in human serum using HPLC analyses. A solution of each isolated selected radioconjugate in 100 µL of a 0.1M PBS buffer pH 7.40 was added to 900 µL of human serum and incubated at 37 °C for 4 h (^{68}Ga radioconjugates) or 14 days (^{177}Lu radioconjugates). At specific time points, 400 µL of the incubated mixture was added into 500 µL of ethanol, vigorously stirred to precipitate serum proteins, and centrifuged (13,500 rpm for 5 min) to separate the supernatant for HPLC analysis.

3.3.2. Lipophilicity Study

The lipophilicity values of the radioconjugates (logD), expressed as the logarithm of its D in the *n*-octanol/PBS (pH 7.40) system, mimicking the physiological conditions (Product Properties Test Guidelines of the Office of Prevention, Pesticides and Toxic Substances 830.7550, 1996), were determined right after the SPE method purification and ethanol evaporation processes. A solution of isolated selected radioconjugate in 500 µL of a 0.1 M PBS buffer at pH 7.40 and 500 µL of *n*-octanol was vigorously stirred and centrifuged (13,500 rpm for 5 min) to separate the immiscible phases. The radioactivities of the aqueous and organic layers were determined using a well-type NaI(Tl) detector. The distribution coefficient was calculated as the ratio of the radioactivity of the radioconjugate in the organic phase to that in the aqueous phase. Each measurement was performed in triplicate and averaged. Simultaneously, the aqueous phases were analyzed by HPLC to check whether the studied radioconjugate remained intact during the experiment.

3.4. Binding Affinity Determination

The binding affinity of aprepitant and compounds **5A–E** for rNK1R was determined in a competitive radioligand binding assay using rat brain homogenates, following a previously described method [27]. In brief, the membrane preparations obtained from rat brains were incubated at 25 °C for 60 min in the presence of a selective radioligand [^{3}H]-[Sar^9,Met(O_2)^{11}]-Substance P obtained from PerkinElmer, (Waltham, MA, USA) and the increasing concentrations of the tested compounds (each concentration in duplicate). Non-specific binding was measured in the presence of 10 μM cold Substance P. The assay buffer was composed of 50 mM Tris-HCl (pH 7.4), 5 mM $MnCl_2$, bovine serum albumin (BSA) (0.1 mg/mL), bacitracin (100 μg/mL), bestatin (30 μM), phenylmethylsulfonyl fluoride (30 μg/mL), and captopril (10 μM). The reaction total volume was 1 mL. With the incubation having been terminated, a rapid filtration through GF/B Whatman glass fiber strips was done with a M-24 Cell Harvester (Brandel, Gaithersburg, MD, USA). The filters were pre-soaked overnight with 0.5% polyethyleneimine so that the extent of non-specific binding could be minimized. After the filtration, the strips were dried, the filter discs were placed separately in 24-well plates, and a Betaplate Scint scintillation solution (PerkinElmer, Waltham, MA, USA) was added to each well. Radioactivity was measured with a MicroBeta LS scintillation counter, Trilux (PerkinElmer, Waltham, MA, USA). The data came from three independent experiments done in duplicate. The results are presented as IC_{50} with SEM.

3.5. Docking

In order to obtain the probable structures of the complexes of the neurokinin 1 receptor with the conjugates **5A–E**, the following modelling procedure was performed. The aprepitant structure (with neutral charge) in the complex with the receptor (PDB accession code: 6HLO [19]) was expanded by attaching to the triazolinone ring the appropriate linkers and the DOTA moiety. Such initial complexes were subjected to local search docking in AutoDock 4.2.6 [26].

The DOTA geometry was set based on the NOJYIU entry [28] of The Cambridge Structural Database [29]. This structure is a DOTA complex with Lu^{3+} (diaqua-lutetium(III)-sodium trihydrate). For the purposes of our modelling, DOTA carboxylate arms were protonated and frozen in the conformation found in the crystal structure of lutetium (III) chelate of DOTA (after removing the Lu^{3+} cation, Na^+ cations and waters). The rationale behind this gambit was the fact that the carboxylates would be primarily engaged in the interactions with a cation; therefore, they might have been expected to retain the conformation they had in the solid state structure. This approach could also give a rough approximation of the DOTA's steric influence on the binding of the conjugates despite a lack of properly scaled and validated parameters for modelling and scoring the complexes with the cations of interest.

The used receptor structure was a refined one (as provided by the GPCRdb service [30]) in order to have the mutated residues replaced with the native ones and to supply the side chains missing in the original PDB structure. The structure was pre-processed in AutoDock Tools [26]. The box was set around the experimental position of aprepitant in 6HLO and extended towards the extracellular part of the receptor so as to cover the expected length of the expanded conjugate. The grids were calculated with AutoGrid 4 [26]. We considered two variants of docking with respect to the flexibility of the receptor structure. In the first variant, all receptor residues were rigid. In the second variant, E193 and H197 side-chains were set to be flexible.

The docking procedure was the local search with the following parameters: 500 individuals in population, 500 iterations of the Solis-Wets local search, the *sw_rho* parameter of the local search space set to 20.0, and 1000 local search runs. The structures resulting from the local search were clustered, and the representative models of the lowest scored (on average) cluster were taken for further analysis. For the qualitative assessment of the binding energy, both the lowest and the mean energy of the clusters were collected. The molecular graphics were prepared in PyMol [31].

For comparative and validation purposes, the very same procedure of local docking (with and without the flexibility of the mentioned two residues) was performed for the parent aprepitant.

4. Conclusions

The presented paper describes the evaluation of aprepitant functionalization in order to provide an application of this NK1R antagonist in nuclear medicine.

Out of the corresponding ^{68}Ga/^{177}Lu radioconjugates of APT-ethylamine **2A** with DOTA, Bn-DOTA, Bn-DOTAGA, and DTPA, only the DOTA amide conjugates showed satisfactory stability in human serum throughout the whole incubation time. The evaluation of the linker effect on radioconjugate lipophilicity indicated APT-alkylamine derivatives as more promising biovectors with features closer to parent aprepitant. The physicochemical properties of obtained APT-alkylamine-DOTA derivatives labelled with ^{68}Ga (**[^{68}Ga]Ga-5A–[^{68}Ga]Ga-5C**) can be compared with those of [^{67}Ga]Ga-NOTA-NK1R radioligands based on another NK1R antagonist—L-733,060 [32]. The $^{67/68}$Ga-radioligands based on these two high affinity NK1R antagonists turned out to be very similar, as evidenced by the following parameters:

(i) they were labelled using macrocyclic chelators (DOTA and NOTA) incorporated in the same 'arm' of the antagonist molecule core,
(ii) the radioconjugates had similar molecular weights (about 1000),
(iii) they had comparable logD values (about 0.15 and 0.6 for the APT-radioligands and the L-733,060-radioligands, respectively),
(iv) all were fully stable in human serum examinations.

Regarding the affinity studies of the **5A–E** conjugates, on the assumption that the human NK1R affinities for aprepitant derivatives were generally much higher than the rat NK1R affinities and that structure-affinity trends were parallel in both species, all the synthesized compounds might be considered to retain reasonable NK1R affinity compared to their parent. In particular, the analogue **5B** (which only suffered a few-times decrease in affinity compared to APT) seems to be especially interesting for further development. Obtained results suggest that the functionalizing of the aprepitant structure via the triazolinone ring is the right strategy.

It is also worth mentioning that, in general, radiopharmaceuticals based on small non-peptide molecules (e.g., aprepitant and L-733,060) have many advantages over peptide-based radiopharmaceuticals [2]. They usually have lower molecular weights, higher lipophilicity values, and, hence, different pharmacokinetics; they are stable in vivo, but, more importantly, their radiosyntheses can be carried out at higher temperatures and in a wider pH range. Moreover, according to the literature, radiopharmaceuticals based on non-peptide antagonists interact with a receptor through more binding sites and accumulate better and for a longer time period in cancer cells [33,34]. Even though the further evaluation of aprepitant-based radiopharmaceuticals is still needed, the findings reported herein provide insight on the perspectives of their application in the theranostics paradigm.

5. Patent

In course of this study, the following national patent application was submitted: No. P430136 "The modified drug substance molecule, method of its production, diagnostic or therapeutic receptor radiopharmaceutical based on this molecule, method of its production and its application".

Author Contributions: Conceptualization, P.K.H. and P.F.J.L.; methodology, P.K.H., P.F.J.L., J.M., and P.K.; investigation, P.K.H. and J.M.; writing—original draft preparation, P.K.H. and P.F.J.L.; writing—review and editing, P.K. and E.G.; visualization, P.F.J.L.; supervision, A.M. and E.G.; project administration, A.M. and E.G.; funding acquisition, A.M. and E.G. All authors have read and agreed to the published version of the manuscript.

Funding: This research was funded by National Science Centre (Poland), grant number 2017/25/B/NZ7/01896.

Acknowledgments: The contribution of Paweł Krzysztof Halik has been done in the frame of the National Centre for Research and Development Project No POWR.03.02.00-00-I009/17 (Radiopharmaceuticals for molecularly targeted diagnosis and therapy, RadFarm. Operational Project Knowledge Education Development 2014–2020 co-financed by European Social Fund).

Conflicts of Interest: The authors declare no conflict of interest.

Abbreviations

6HLO, 6HLL, 6HLP	accession codes of co-ordinates and structure factors in PDB
APT	aprepitant
BBB	Blood-brain barrier
Bn	benzyl moiety
BSA	bovine serum albumin
DMF	dimethylformamide
DOTA	1,4,7,10-tetraazacyclododecane-1,4,7,10-tetraacetic acid
DOTAGA	1,4,7,10-tetraazacyclododecane, 1 glutaric acid - 4,7,10-acetic acid
DTPA	diethylenetriaminepentaacetic acid
ECL	extracellular loop
ESI	electrospray ionization
GAMG	human glioblastoma cell line
HEK 293	human embryonic kidney 293 cell line
hNK1R	human neurokinin 1 receptor
HS	human serum
IC_{100}	maximal inhibitory concentration
IC_{50}	half-maximal inhibitory concentration
logD	logarithm of distribution coefficient
MS	mass spectrometry
n.c.a.	non-carrier added
NHS	*N*-hydroxysuccinimide moiety
NK1R	neurokinin 1 receptor, tachykinin 1 receptor
PBS	phosphate-buffered saline
PDB	Protein Data Bank
pK_B	negative logarithm of Boltzmann constant
p-SCN	*para*-isocyanate group
rNK1R	rat neurokinin 1 receptor
R_T	retention time
SEM	standard error of the means
SP	Substance P
SPE	solid phase extraction
TFA	trifluoroacetic acid
TM	transmembrane helix

References

1. Muñoz, M.; Rosso, M.; Coveñas, R. The NK-1 receptor: A new target in cancer therapy. *Curr. Drug Targets* **2011**, *12*, 909–921. [CrossRef] [PubMed]
2. Majkowska-Pilip, A.; Halik, P.K.; Gniazdowska, E. The Significance of NK1 receptor ligands and their application in targeted radionuclide tumour therapy. *Pharmaceutics* **2019**, *11*, 443. [CrossRef] [PubMed]
3. Coveñas, R.; Muñoz, M. Cancer progression and substance P. *Histol. Histopathol.* **2014**, *29*, 881–890. [CrossRef] [PubMed]
4. Akazawa, T.; Kwatra, S.G.; Goldsmith, L.E.; Richardson, M.D.; Cox, E.A.; Sampson, J.H.; Kwatra, M.M. A constitutively active form of neurokinin 1 receptor and neurokinin 1 receptor-mediated apoptosis in glioblastomas. *J. Neurochem.* **2009**, *109*, 1079–1086. [CrossRef] [PubMed]
5. Berger, M.; Neth, O.; Ilmer, M.; Garnier, A.; Salinas-Martín, M.V.; de Agustín Asencio, J.C.; von Schweinitz, D.; Kappler, R. Hepatoblastoma cells express truncated neurokinin-1 receptor and can be inhibited by aprepitant in vitro and in vivo. *J. Hepatol.* **2014**, *60*, 985–994. [CrossRef] [PubMed]
6. Quartara, L.; Altamura, M.; Evangelista, S.; Maggi, C.A. Tachykinin receptor antagonists in clinical trials. *Expert Opin. Investig. Drugs* **2009**, *18*, 1843–1864. [CrossRef]
7. Muñoz, M.; Rosso, M.; Aguilar, F.J.; González-Moles, M.A.; Redondo, M.; Esteban, F. NK-1 receptor antagonists induce apoptosis and counteract substance P-related mitogenesis in human laryngeal cancer cell line HEp-2. *Investig. New Drugs* **2008**, *26*, 111–118. [CrossRef]

8. Muñoz, M.; Rosso, M.; Robles-Frías, M.J.; Salinas-Martín, M.V.; Coveñas, R. The NK-1 receptor is expressed in human melanoma and is involved in the antitumor action of the NK-1 receptor antagonist aprepitant on melanoma cell lines. *Lab Investig.* **2010**, *90*, 1259–1269. [CrossRef]
9. Muñoz, M.; González-Ortega, A.; Coveñas, R. The NK-1 receptor is expressed in human leukemia and is involved in the antitumor action of aprepitant and other NK-1 receptor antagonists on acute lymphoblastic leukemia cell lines. *Investig. New Drugs* **2012**, *30*, 529–540. [CrossRef]
10. Muñoz, M.; González-Ortega, A.; Salinas-Martín, M.V.; Carranza, A.; Garcia-Recio, S.; Almendro, V.; Coveñas, R. The neurokinin-1 receptor antagonist aprepitant is a promising candidate for the treatment of breast cancer. *Int. J. Oncol.* **2014**, *45*, 1658–1672. [CrossRef]
11. Muñoz, M.; Berger, M.; Rosso, M.; Gonzalez-Ortega, A.; Carranza, A.; Coveñas, R. Antitumor activity of neurokinin-1 receptor antagonists in MG-63 human osteosarcoma xenografts. *Int. J. Oncol.* **2014**, *44*, 137–146. [CrossRef] [PubMed]
12. Dikmen, M. Antiproliferative and apoptotic effects of aprepitant on human glioblastoma U87MG cells. *Marmara Pharm. J.* **2017**, *21*, 156–164. [CrossRef]
13. Kitchens, C.A.; McDonald, P.R.; Pollack, I.F.; Wipf, P.; Lazo, J.S. Synergy between microtubule destabilizing agents and neurokinin 1 receptor antagonists identified by an siRNA synthetic lethal screen. *FASEB J.* **2009**, *23*, 756–813.
14. Kast, R.E.; Ramiro, S.; Llado, S.; Toro, S.; Coveñas, R.; Muñoz, M. Antitumour action of temozolomide, ritonavir and aprepitant against human glioma cells. *J. Neurooncol.* **2016**, *126*, 425–431. [CrossRef] [PubMed]
15. Muñoz, M.; Rosso, M. The NK-1 receptor antagonist aprepitant as a broad-spectrum antitumor drug. *Investig. New Drugs* **2010**, *28*, 187–193. [CrossRef]
16. Hale, J.J.; Mills, S.G.; MacCoss, M.; Shah, S.K.; Qi, H.; Mathre, D.J.; Cascieri, M.A.; Sadowski, S.; Strader, C.D.; MacIntyre, D.E.; et al. 2(S)-((3,5-Bis(trifluoromethyl)benzyl)oxy)-3(S)-phenyl-4-((3-oxo-1,2,4-triazol-5-yl)methyl)morpholine (1): A potent, orally active, morpholine-based human neurokinin-1 receptor antagonist. *J. Med. Chem.* **1996**, *39*, 1760–1762. [CrossRef]
17. Hale, J.J.; Mills, S.G.; MacCoss, M.; Finke, P.E.; Cascieri, M.A.; Sadowski, S.; Ber, E.; Chicchi, G.G.; Kurtz, M.; Metzger, J.; et al. Structural optimization affording 2-(R)-(1-(R)-3,5-Bis(trifluoromethyl) phenylethoxy)-3-(S)-(4-fluoro)phenyl-4-(3-oxo-1,2,4-triazol-5-yl)methylmorpholine, a potent, orally active, long-acting Morpholine Acetal human NK-1 receptor antagonist. *J. Med. Chem.* **1998**, *41*, 4607–4614. [CrossRef]
18. Harrison, T.; Owens, A.P.; Williams, B.J.; Swain, C.J.; Williams, A.; Carlson, A.; Rycroft, W.; Tattersall, F.D.; Cascieri, M.A.; Chicchi, G.G.; et al. An orally active, water-soluble Neurokinin-1 receptor antagonist suitable for both intravenous and oral clinical administration. *J. Med. Chem.* **2001**, *44*, 4296–4299. [CrossRef]
19. Schöppe, J.; Ehrenmann, J.; Klenk, C.; Rucktooa, P.; Schütz, M.; Doré, A.S.; Plückthun, A. Crystal structures of the human neurokinin 1 receptor in complex with clinically used antagonists. *Nat. Commun.* **2019**, *10*, 17–27. [CrossRef]
20. Giannini, G.; Milazzo, F.M.; Battistuzzi, G.; Rosi, A.; Anastasi, A.M.; Petronzelli, F.; Albertoni, C.; Tei, L.; Leone, L.; Salvini, L.; et al. Synthesis and preliminary in vitro evaluation of DOTA-Tenatumomabconjugates for theranostic applications in tenascin expressing tumors. *Bioorg. Med. Chem.* **2019**, *27*, 3248–3253. [CrossRef]
21. Waterhouse, R.N. Determination of Lipophilicity and its use as a predictor of blood–brain barrier penetration of molecular imaging agents. *Mol. Imaging Biol.* **2003**, *5*, 376–389. [CrossRef] [PubMed]
22. Wu, Y.; Loper, A.; Landis, E.; Hettrick, L.; Novak, L.; Lynn, K.; Chen, C.; Thompson, K.; Higgins, R.; Batra, U.; et al. The role of biopharmaceutics in the development of a clinical nanoparticle formulation of MK-0869: A Beagle dog model predicts improved bioavailability and diminished food effect on absorption in human. *Int. J. Pharm.* **2004**, *285*, 135–146. [CrossRef] [PubMed]
23. Appell, K.C.; Fragale, B.J.; Loscig, J.; Singh, S.; Tomczuk, B.E. Antagonists that demonstrate species differences in neurokinin-1 receptors. *Mol. Pharmacol.* **1992**, *41*, 772–778. [PubMed]
24. Pradier, L.; Habert-Ortoli, E.; Emile, L.; Le Guern, J.; Loquet, I.; Bock, M.D.; Clot, J.; Mercken, L.; Fardin, V.; Garret, C. Molecular determinants of the species selectivity of neurokinin type 1 receptor antagonists. *Mol. Pharmacol.* **1995**, *47*, 314–321.
25. Leffer, A.; Ahlstedt, I.; Engberg, S.; Svensson, A.; Billger, M.; Oberg, L.; Bjursell, M.K.; Lindström, E.; von Mentzer, B. Characterization of species-related differences in the pharmacology of tachykinin NK receptors 1, 2 and 3. *Biochem. Pharmacol.* **2009**, *77*, 1522–1530. [CrossRef]

26. Morris, G.M.; Huey, R.; Lindstrom, W.; Sanner, M.F.; Belew, R.K.; Goodsell, D.S.; Olson, A.J. AutoDock4 and AutoDockTools4: Automated docking with selective receptor flexibility. *J. Comput. Chem.* **2009**, *30*, 2785–2791. [CrossRef]
27. Matalińska, J.; Lipiński, P.F.J.; Kotlarz, A.; Kosson, P.; Muchowska, A.; Dyniewicz, J. Evaluation of receptor affnity, analgesic activity and cytotoxicity of a hybrid peptide, AWL3020. *Int. J. Pept. Res. Ther.* **2020**. [CrossRef]
28. Aime, S.; Barge, A.; Botta, M.; Fasano, M.; Ayala, J.D.; Bombieri, G. Crystal structure and solution dynamics of the lutetium(III) chelate of DOTA. *Inorg. Chmica Acta* **1996**, *246*, 423–429. [CrossRef]
29. Allen, F.H. The cambridge structural database: A quarter of a million crystal structures and rising. *Acta Crystallogr. B* **2002**, *58*, 380–388. [CrossRef]
30. Pándy-Szekeres, G.; Munk, C.; Tsonkov, T.M.; Mordalski, S.; Harpsøe, K.; Hauser, A.S.; Bojarski, A.J.; Gloriam, D.E. GPCRdb in 2018: Adding GPCR structure models and ligands. *Nucleic Acids Res.* **2018**, *4*, D440–D446. [CrossRef]
31. Schrödinger LLC. The PyMOL Molecular Graphics System 2018. Available online: https://github.com/schrodinger/pymol-open-source (accessed on 31 July 2020).
32. Zhang, H.; Kanduluru, A.K.; Desai, P.; Ahad, A.; Carlin, S.; Tandon, N.; Weber, W.A.; Low, P.S. Synthesis and evaluation of a Novel ^{64}Cu- and ^{67}Ga-Labeled Neurokinin 1 receptor antagonist for in Vivo targeting of NK1R-Positive tumor Xenografts. *Bioconjug. Chem.* **2018**, *29*, 1319–1326. [CrossRef] [PubMed]
33. Ginj, M.; Zhang, H.; Waser, B.; Cescato, R.; Wild, D.; Wang, X.; Erchegyi, J.; Rivier, J.; Mäcke, H.R.; Reubi, J.C. Radiolabeled somatostatin receptor antagonists are preferable to agonists for in vivo peptide receptor targeting of tumors. *Proc. Natl. Acad. Sci. USA* **2006**, *103*, 16436–16441. [CrossRef] [PubMed]
34. Cescato, R.; Maina, T.; Nock, B.; Nikolopoulou, A.; Charalambidis, D.; Piccand, V.; Reubi, J.C. Bombesin receptor antagonists may be preferable to agonists for tumor targeting. *J. Nucl. Med.* **2008**, *49*, 318–326. [CrossRef] [PubMed]

Sample Availability: Samples of the compounds **2A–C**, **4D**, **E** are available from the authors.

Article

Involvement of Differentially Expressed microRNAs in the PEGylated Liposome Encapsulated ^{188}Rhenium-Mediated Suppression of Orthotopic Hypopharyngeal Tumor

Bing-Ze Lin [1,†], Shen-Ying Wan [1,2,†], Min-Ying Lin [1], Chih-Hsien Chang [1,3], Ting-Wen Chen [4,5,6], Muh-Hwa Yang [7,8,9] and Yi-Jang Lee [1,7,*]

1. Department of Biomedical Imaging and Radiological Sciences, National Yang-Ming University, No. 155, Sec. 2, Linong St. Beitou District, Taipei 11221, Taiwan; innovationmark1012@gmail.com (B.-Z.L.); sywang@mail.femh.org.tw (S.-Y.W.); milo841120@gmail.com (M.-Y.L.); chchang@iner.gov.tw (C.-H.C.)
2. Department of Nuclear Medicine, Far Eastern Memorial Hospital, New Taipei City 22000, Taiwan
3. Isotope Application Division, Institute of Nuclear Energy Research, Taoyuan 32546, Taiwan
4. Institute of Bioinformatics and Systems Biology, National Chiao Tung University, Hsinchu 30068, Taiwan; dodochen@nctu.edu.tw
5. Department of Biological Science and Technology, National Chiao Tung University, Hsinchu 30068, Taiwan
6. Center For Intelligent Drug Systems and Smart Bio-devices (IDS2B), National Chiao Tung University, Hsinchu 30068, Taiwan
7. Cancer Progression Research Center, National Yang-Ming University, Taipei 11221, Taiwan; mhyangymu@gmail.com
8. Institute of Clinical Medicine, National Yang-Ming University, Taipei 11221, Taiwan
9. Division of Medical Oncology, Department of Oncology, Taipei Veteran General Hospital, Taipei 11217, Taiwan

* Correspondence: yjlee2@ym.edu.tw
† These authors contributed equally to this work.

Academic Editor: Krishan Kumar
Received: 30 June 2020; Accepted: 6 August 2020; Published: 8 August 2020

Abstract: Hypopharyngeal cancer (HPC) accounts for the lowest survival rate among all types of head and neck cancers (HNSCC). However, the therapeutic approach for HPC still needs to be investigated. In this study, a theranostic ^{188}Re-liposome was prepared to treat orthotopic HPC tumors and analyze the deregulated microRNA expressive profiles. The therapeutic efficacy of ^{188}Re-liposome on HPC tumors was evaluated using bioluminescent imaging followed by next generation sequencing (NGS) analysis, in order to address the deregulated microRNAs and associated signaling pathways. The differentially expressed microRNAs were also confirmed using clinical HNSCC samples and clinical information from The Cancer Genome Atlas (TCGA) database. Repeated doses of ^{188}Re-liposome were administrated to tumor-bearing mice, and the tumor growth was apparently suppressed after treatment. For NGS analysis, 13 and 9 microRNAs were respectively up-regulated and down-regulated when the cutoffs of fold change were set to 5. Additionally, miR-206-3p and miR-142-5p represented the highest fold of up-regulation and down-regulation by ^{188}Re-liposome, respectively. According to Differentially Expressed MiRNAs in human Cancers (dbDEMC) analysis, most of ^{188}Re-liposome up-regulated microRNAs were categorized as tumor suppressors, while down-regulated microRNAs were oncogenic. The KEGG pathway analysis showed that cancer-related pathways and olfactory and taste transduction accounted for the top pathways affected by ^{188}Re-liposome. ^{188}Re-liposome down-regulated microRNAs, including miR-143, miR-6723, miR-944, and miR-136 were associated with lower survival rates at a high expressive level. ^{188}Re-liposome could suppress the HPC tumors in vivo, and the therapeutic efficacy was associated with the deregulation of microRNAs that could be considered as a prognostic factor.

Keywords: hypopharyngeal cancer; ^{188}Re-liposome; repeated therapy; NGS; microRNA

1. Introduction

Hypopharyngeal cancer (HPC) represents malignant growth in the hypopharynx region and accounts for about 5% of all head and neck cancers (HNSCC) [1]. As HPC is a rare cancer type with a late occurrence of symptoms and tumor spreading, it is not uncommon for it to be detected at advanced stages with a high mortality rate and poor prognosis [2]. HPC can be treated by conventional surgery, radiotherapy and chemotherapy, while radiotherapy alone is usually used at an early stage [3]. On the other hand, a combination of different therapeutic modalities can improve the five-year survival of this disease [4]. Several lines of evidence have claimed that a combination of radiotherapy and chemotherapy would provide better control of locoregional recurrence compared to surgical procedures [4,5]. A radiopharmaceutical named ^{99m}Tc-MIBI (methoxy-isobutyl-isonitrile) has been reported to detect HPC with up to a 95% sensitivity using single photon emission computed tomography (SPECT) [6]. However, nuclear medicine has not been reported to have assessed or monitored the efficacy of HPC therapy as mentioned above.

Radiopharmaceuticals are not only used for diagnostic purposes but also therapeutic purposes, so-called radiotheranostics [7]. Rhenium-188 (^{188}Re) belongs to this type of radionuclide as it emits 85% high-energy β-particles (2.12MeV) and 15% γ-rays (155keV) [8]. The average soft tissue penetration distance of β-particles is only around 3.8mm, suggesting that ^{188}Re is suitable for tumor ablation and will not have significant side effects on distant normal tissues [9,10]. ^{188}Re has been conjugated to hydroxyethylidene diphosphonate (HEDP) for bone pain palliation [11,12]. The antibody conjugated ^{188}Re has also been reported to treat different cancers [13]. Additionally, low immunogenic peptides such as somatostatin derivative conjugated ^{188}Re, have been investigated in clinics for patients with advanced pulmonary cancer [14]. ^{188}Re-labeled lipiodol and microspheres were used for the treatment of hepatocellular carcinoma [15–17]. ^{188}Re-labelled radiocolloids have also been developed to treat skin cancers within a brachytherapy device [18]. ^{188}Re-loaded lipid nanocapsules, also called ^{188}Re-liposome, have been demonstrated to be biocompatible and subjected to a phase 0 clinical study for patients with metastatic tumors [19]. ^{188}Re-liposome has shown a theranostic efficacy in various human cancers, including colorectal cancer, glioblastomas, lung cancer, ovarian cancer, esophageal cancer and head and neck cancer using xenograft tumor models [20–26]. The molecular mechanisms of rhenium-188 labelled radiopharmaceuticals are of interest for investigation aimed at interpreting the potent therapeutic efficacy.

Radiogenomics is defined as connecting radiomics and genetic profiles to apply the feature of medical imaging in radiation-mediated molecular responses [27]. The purpose of this discipline is to predict the association between gene expression and the radiotherapy-induced toxic effects of tumors [28]. Next-generation sequencing (NGS) analysis is the most cutting-edge technology that can decipher dramatic amounts of gene expressive alterations in tumors, with or without therapies [29,30]. NGS applications include RNA sequencing (RNA-Seq) that can analyze the expression of various RNA populations (mRNA, microRNA, long non-coding RNA, etc.) and their modification forms generated from alternative splicing, mutations or gene fusion [31]. Compared to the conventional cDNA expression microarray, RNA-Seq has a broader spectrum for finding novel and unidentified transcripts [32]. This method is important for providing more detailed and quantitative data for bioinformatics analysis of genetic profiling in novel drug development and predicting the signaling pathways of toxicity and therapeutic effects [33]. As a radioactive compound, ^{188}Re-liposome is believed to influence the genetic profile of tumors. However, related studies have rarely been reported.

In this study, we investigated the effects of ^{188}Re-liposome on the miRNA expressive profiles of HPC derived xenograft tumors using RNA-Seq technology. The changed miRNA profiles were analyzed using the Kyoto Encyclopedia of Genes and Genomes (KEGG) pathway database and Ingenuity Pathway

Analysis (IPA). The total expressive amount of miRNA from ^{188}Re-liposome-treated tumors was about 10% lower than that of untreated tumor. The changed miRNA profile led to 4498 differentially expressed genes (DEGs) that influenced 30 molecular pathways, including olfactory transduction, the cancer pathway, and taste transduction, which accounted for the top three pathways. We also found that ^{188}Re-liposome up-regulated several tumor suppressor microRNAs, such as miR-34-5p, miR-193a-5p, miR-125b-5p, miR-133a-5p, and miR-133b-5p. Concomitantly, several oncomirs, including miR-21-5p, miR-32-5p and miR-205-5p were down-regulated by ^{188}Re-liposome. An online Kaplan-Meier (K-M) plotter with The Cancer Genome Atlas (TCGA) database was also used to compare the expression of these miRNA and the survival of head and neck cancer patients. The results of ^{188}Re-liposome-induced miRNA dysregulation detected by RNA-Seq were discussed.

2. Results

2.1. Effects of ^{188}Re-Liposome on HPC Derived Orthotopic Tumors Using the Repeated Dose Regime

A flowchart of ^{188}Re-liposomal manufacturing and the chemical structures of the ^{188}Re-perrhenate precursor and BMEDA chelator are illustrated in Figure 1. The timeline was schemed for the establishment of HPC tumor-bearing mice using FaDu-3R cells (Supplementary Figure S2), the administration of ^{188}Re-liposome, the optical imaging of tumor responses, tumor resection for RNA extraction, and NGS analysis (Figure 2A). The growth of orthotopic tumors was significantly suppressed by ^{188}Re-liposome but not saline as detected using bioluminescent imaging (Figure 2B,C). The size of resected tumors also exhibited obvious differences between the saline control and ^{188}Re-liposome-treated mice after 30 days of initial implantation (Figure 2D). The body weights of tumor-bearing mice were not significantly affected by ^{188}Re-liposome (Figure 2E). Furthermore, we showed that DNA damage marker γ-H2AX was significantly up-regulated by ^{188}Re-liposome treatment (Figure 2F,G). These data suggest that the regime of ^{188}Re-liposome treatment with repeated doses exhibited therapeutic efficacy, with little systemic toxicity.

2.2. Use of NGS Analysis to Investigate the microRNA Expressive Profile of HPC Tumor Treated with ^{188}Re-Liposome

The resected FaDu HPC tumors treated with saline or repeated doses of ^{188}Re-liposome were subjected to RNA extraction to obtain a high quality of total RNA for RNA-seq analysis (Supplementary Figure S3). The quality of raw data obtained by this analysis was determined by the GC content, by which the raw data quality of the saline control and ^{188}Re-liposome-treated tumor was shown to be similar (Supplementary Figure S4). To confirm the expressive change in microRNA in tumors treated with or without ^{188}Re-liposome, the reads number of small RNA (15–55 mers) was extracted and counted using a Small RNA Analysis tool (see Materials and Methods). The average lengths of small RNA after extraction were 32.9 and 32 mers for the control and ^{188}Re-liposome treated tumor, respectively. Under this condition, the total reads number of the control and ^{188}Re-liposome treated tumor was 10,612,998 and 9,844,775, respectively. The reads were then subjected to the miRBase database (release 21) for the annotation of small RNA using the Annotate and Merge Count of Small RNA Analysis software. The annotated small RNAs of saline-treated tumors and ^{188}Re-liposome treated tumors were 639 and 572, respectively. The differentially expressed microRNAs were clustered in a heatmap for a comparison of saline-treated and ^{188}Re-liposome treated HPC tumor models (Supplementary Figure S5). Furthermore, we set the cutoff at five-fold change ($\log_2$) of microRNA and showed that 13 microRNA and 9 microRNA with mature forms were up-regulated and down-regulated by ^{188}Re-liposome normalized to the saline control, respectively (Table 1). The 13 up-regulated microRNAs ranked from highest to lowest fold change were miR-206-3p, mir668-3p, mit-485-3p, miR-382-5p, miR-1268b-5p, miR-193a-5p, miR-7-1-5p, miR-378a-5p, miR-1266-5p, miR-4510-5p, miR-370-3p, miR-34a-5p, and miR-342-5p. The nine down-regulated microRNAs ranked from highest to lowest fold change were miR-142-5p, miR-6723-5p, miR-944-3p, miR-142-3p, miR-136-3p, miR-151b-3p, miR-194-2-5p, miR-143-5p, and miR-3960-3p. According to the online analysis of dbDEMC, ^{188}Re-liposome-up-regulated microRNAs were mostly naturally down-regulated in HNSCC and were predicted as tumor suppressors (Table 2). MiR-193a, miR-7-1-5p and miR-342-5p were up-regulated in HNSCC, but could still be tumor suppressors in different types of cancer (see references in Table 2). For the nine ^{188}Re-liposome down-regulated microRNAs, only four of them (miR-136-3p, miR-142-3p, miR-944-3p, and miR-142-5p) were reported in the HNSCC of dbDEMC database. Interestingly, three out of these four microRNAs were found to be up-regulated in HNSCC and predicted as oncogenes (Table 3). However, most of the ^{188}Re-liposome-down-regulated microRNAs contain oncogenic properties (see references in Table 3). These results suggested that the therapeutic efficacy of ^{188}Re-liposome was associated with the deregulation of tumor-suppressive and/or oncogenic microRNAs.

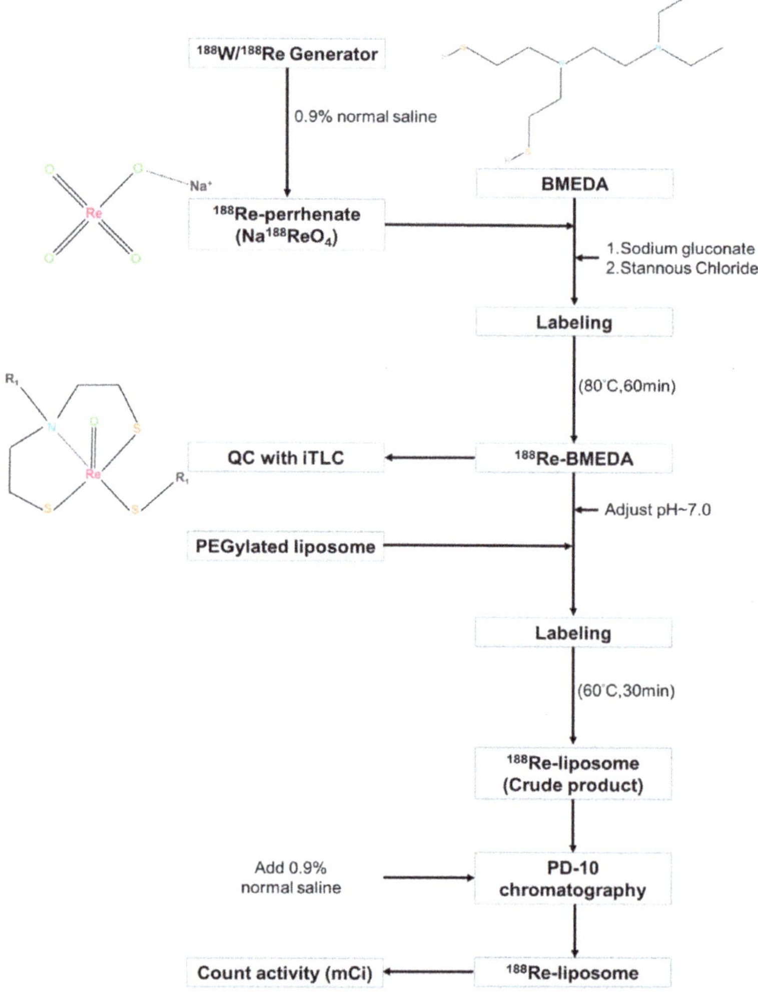

Figure 1. The flowchart of ^{188}Re-liposomal preparation. The chemical structures of $Na^{188}ReO_4$, BMEDA and formed ^{188}Re-BMEDA were also illustrated.

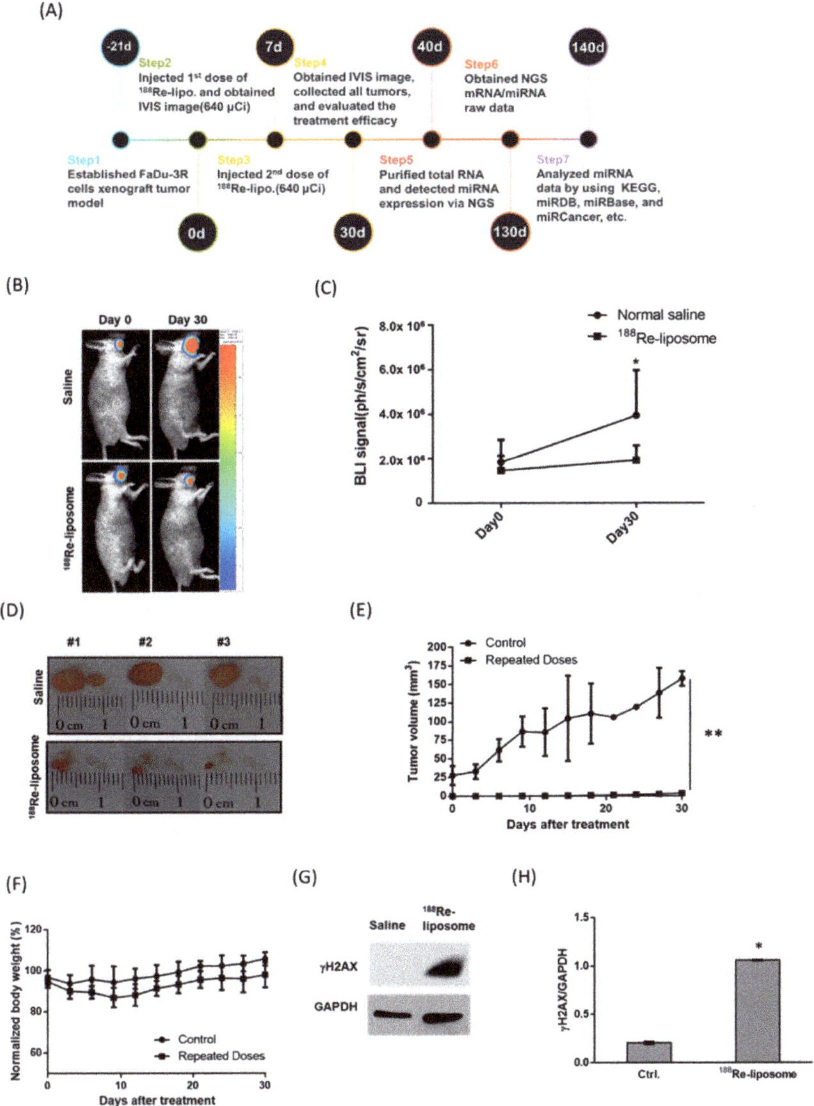

Figure 2. Comparison of PEGylated ^{188}Re-liposomal accumulation in orthotopic hypopharyngeal cancer (HPC) tumors after repeated injections. (**A**) The experimental scheme for ^{188}Re-liposome treatment and analysis. (**B**) Reporter gene imaging of tumor growth responding to repeated doses of ^{188}Re-liposome, and the saline-treated control. (**C**) Quantification of bioluminescent imaging (BLI) signals. *: $p < 0.05$. (**D**) Representative photos of excised orthotropic tumors with or without the treatment of ^{188}Re-liposomes. (**E**) Caliper measurement of tumor volumes. Data are represented as means ± S.D. **: $p < 0.01$. (**F**) Measurement of body weights of mice. (**G**) Comparison of the γ-H2AX protein expression in tumors with or without the treatment of ^{188}Re-liposomes. (**H**) Densitometric quantification of Western blots. *: $p < 0.05$.

Table 1. Expression of microRNAs in human hypopharyngeal tumor model treated with 188Re-liposome.

MicroRNA	Fold Change (188Re-lipo./Ctrl.) [a]	Ctrl. Norm. by TMM [b]	^{188}Re-lipo. Norm.by TMM	Ctrl. RPM [c]	^{188}Re-lipo. RPM
Hsa-miR-206-3p	40.19	76.05	3056.53	2.92	90.71
Hsa-miR-668-3p	11.16	2.45	27.38	0.09	0.81
Hsa-miR-485-3p	9.77	2.45	23.96	0.09	0.71
Hsa-miR-382-5p	9.15	22.08	201.94	0.85	5.99
Hsa-miR-1268b-5p	8.37	2.45	20.54	0.09	0.61
Hsa-miR-193a-5p	6.28	4.91	30.8	0.19	0.91
Hsa-miR-7-1-5p	6.28	9.81	61.61	0.38	1.83
Hsa-miR-378a-5p	5.78	51.52	297.78	1.98	8.84
Hsa-miR-1266-5p	5.58	2.45	13.69	0.09	0.41
Hsa-miR-4510-5p	5.58	2.45	13.69	0.09	0.41
Hsa-miR-370-3p	5.58	4.91	27.38	0.19	0.81
Hsa-miR-34a-5p	5.15	127.57	657.17	4.9	19.5
Hsa-miR-342-5p	5.12	7.36	37.65	0.28	1.12
Hsa-miR-3960-3p	−5.73	117.76	20.54	4.52	0.61
Hsa-miR-143-5p	−5.73	19.63	3.42	0.75	0.1
Hsa-miR-194-2-5p	−5.73	19.63	3.42	0.75	0.1
Hsa-miR-151b-3p	−6.45	22.08	3.42	0.85	0.1
Hsa-miR-136-3p	−7.17	24.53	3.42	0.94	0.1
Hsa-miR-142-3p	−7.88	26.99	3.42	1.04	0.1
Hsa-miR-944-3p	−9.32	31.89	3.42	1.22	0.1
Hsa-miR-6723-5p	−10.03	34.35	3.42	1.32	0.1
Hsa-miR-142-5p	−10.14	242.88	23.96	9.33	0.7

[a]—Fold change in microRNA over 5 or below −5 were selected. [b]—TMM: Trimmed mean of M values. [c]—RPM: Reads of exon model per million mapped reads. (ExonMappedReads × 10^6/TotalMapped Reads).

2.3. Validation of microRNA Identified in NGS Data Using qPCR

We next used qPCR to validate up-regulated and down-regulated microRNA in the FaDu HPC tumor model treated with ^{188}Re-liposome. We selected microRNAs that displayed over five-fold deregulation with the highest RPM, including miR-206-3p, miR-382-5p, miR-378a-5p, miR-3960-3p, and miR-142-5p to be validated. The results showed that the expressive patterns of these microRNAs obtained by qPCR were consistent with the observations of NGS analysis (Figure 3).

2.4. Investigation of Differentially Expressed microRNAs in Clinical Samples

As ^{188}Re-liposome could influence the expression of certain microRNAs that may correlate with the tumor-suppressive effect of the HPC model, we were interested in examining the status of these microRNAs in clinical HNSCC tumors. First, we obtained clinical HNSCC tissues from patients (n = 6) to investigate the differential expression of microRNAs in tumor tissues and adjacent normal tissues using qPCR analysis. We selected miR-206-3p, miR-378a-5p and miR-142-5p as they exhibited the highest differential deregulation by ^{188}Re-liposome (Figure 3). The results showed that the expression of miR-206-3p and miR-378a-5p was down-regulated (Figure 4A,B), while miR-142-5p was up-regulated in tumors compared to normal tissues (Figure 4C). Additionally, we employed the clinical information of HNSCC in the TCGA database to compare the differentially expressed microRNAs in HNSCC and normal tissues. Because the number of cases of hypopharynx cancer was too low to be analyzed, here we used clinical information on larynx cancer types instead. A heatmap was generated for the microRNAs displaying over five-fold change caused by ^{188}Re-liposome (Supplementary Figure S6). Accordingly, we found that miR-206 (equivalent to miR-206-3p) and miR-378a-5p were significantly down-regulated, while miR-143-5p, miR-142-3p, and miR-944 were significantly up-regulated in tumors (Figure 4D). MiR-142-5p also exhibited a trend of up-regulation, although the significance was marginal. This suggests that ^{188}Re-liposome can reverse the expression of the microRNAs that were originally deregulated in HNSCC.

Table 2. Functional prediction algorithm for microRNAs of human hypopharyngeal tumor model up-regulated by ^{188}Re-liposome.

MiRNAs up-Regulated in HPC	Original Change in HNSCC [a]	Role Prediction	Functional Importance	GEO ID	References [b]
Hsa-miR-206-3p	down-regulation	Tumor suppressor	Weaks cell proliferation, migration, invasion, promotes S phase cell arrest. Inhibits cell aggressiveness	TCGA_HNSC	[34,35]
Hsa-miR-668-3p	down-regulation	Tumor suppressor	Induces growth arrest and premature senescence	- [c]	[36,37]
Hsa-miR-485-3p	down-regulation	Tumor suppressor	Inhibits mitochondrial biogenesis or promotes cancer growth and migration	GSE75630	[38–40]
Hsa-miR-382-5p	down-regulation	Tumor Suppressor	Promotes lymph node metastasis and TNM stage; inhibition of proliferation and EMT in glioma cells	TCGA_HNSC	[41,42]
Hsa-miR-1268b-5p	-	Tumor suppressor	Increases chemosensitivity	-	[43]
Hsa-miR-193a-5p	up-regulation	Tumor Suppressor	Suppresses the growth and the metastasis of cancer cells	TCGA_HNSC	[44,45]
Hsa-miR-7-1-5p	up-regulation	Tumor suppressor	Inhibits proliferation, invasion and induces apoptosis in cancer cells	TCGA_HNSC	[46,47]
Hsa-miR-378a-5p	down-regulation	Tumor suppressor	Inhibits cellular proliferation and colony formation.	TCGA_HNSC	[48,49]
Hsa-miR-1266-5p	down-regulation	Tumor suppressor	Induces apoptosis and reduces proliferation	TCGA_HNSC	[50,51]
Hsa-miR-4510-5p	-	Tumor suppressor	Down-regulation in recurrent cancer and a potential cancer biomarker	-	[52,53]
Hsa-miR-370-3p	down-regulation	Tumor suppressor	Potential cancer biomarker	TCGA_HNSC	[54]
Hsa-miR-34a-5p	down-regulation	Tumor suppressor	Inhibits tumorigenesis and progression	TCGA_HNSC	[55,56]
Hsa-miR-342-5p	up-regulation	Tumor suppressor	Reduces cell cycle progression	TCGA_HNSC	[57,58]

[a]—The change of microRNA was determined by the dbDEMC online databases. [b]—The selected references may be not HPC or HNSCC related. [c]—The tumor suppressive function was concluded from clinical patients.

Table 3. Target prediction algorithm for microRNAs of human hypopharyngeal tumor model down-regulated by ^{188}Re-liposome.

MiRNAs Down-Regulated in HPC	Original Change in HNSCC	Role Prediction	Functional Importance	GEO ID	Reference
Hsa-miR-3960-3p	-	-	-	-	-
Hsa-miR-143-5p	-	Both	cell viability, colony formation	-	[59,60]
Hsa-miR-194-2-5p	-	Oncogene	cell proliferation, migration and invasion	-	[61,62]
Hsa-miR-151b-3p	-	Oncogene? [a]	Biomarker of sarcoma	-	[63]
Hsa-miR-136-3p	down-regulation	Oncogene	Biomarker of bladder cancer, promote cancer growth and migration	TCGA-HNSC	[64,65]
Hsa-miR-142-3p	up-regulation	Oncogene	Over-expression in OSCC, association with cancer growth and migration	TCGA-HNSC	[66,67]
Hsa-miR-944-3p	up-regulation	Oncogene	A biomarker of poor prognosis/Regulation of chemoresistance	TCGA-HNSC	[68,69]
Hsa-miR-6723-5p	-	-	-	-	-
Hsa-miR-142-5p	up-regulation	Oncogene	Deregulation of cell proliferation; SMAD3/TGF-β	GSE31277	[70,71]

[a] No direct evidence, but just an implication.

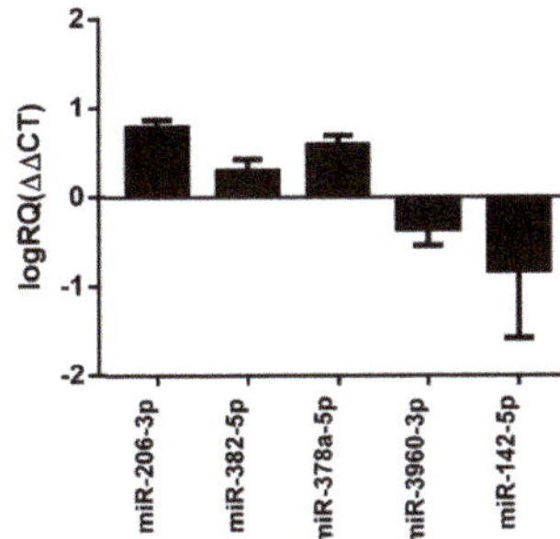

Figure 3. Validation of the next-generation sequencing (NGS) results of the HPC tumor model by qPCR analysis. The results were each microRNA of ^{188}Re-liposome-treated HPC normalized to that of untreated controls.

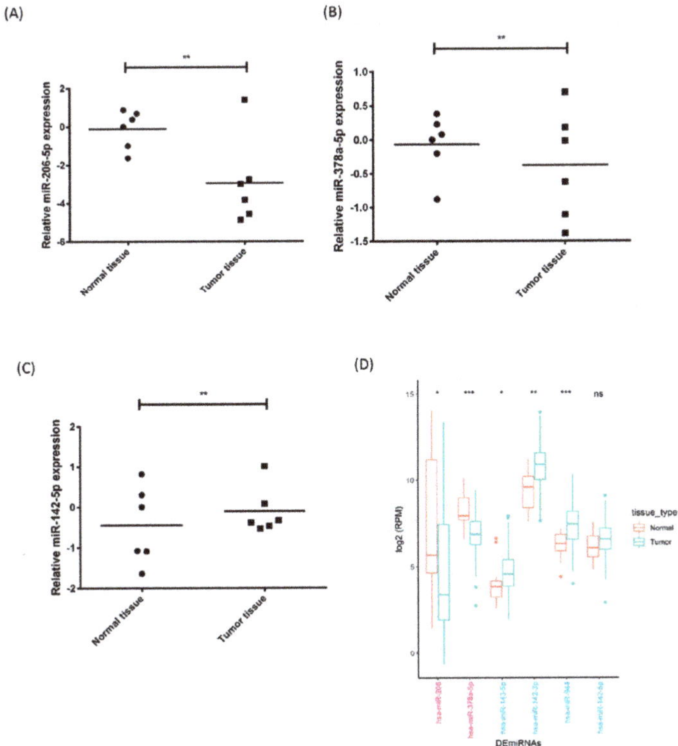

Figure 4. Comparison of the microRNA expression in clinical head and neck cancer (HNSCC) tissues and adjacent normal tissues by qPCR analysis. (**A**) MiR-206-3p. (**B**) MiR-378a-5p (**C**) MiR-142-5p. (**D**) Differentially expressed microRNA (DEmiRNA) in larynx tumor and normal tissues using the clinical information of The Cancer Genome Atlas (TCGA). *: $p < 0.05$, **: $p < 0.01$. ***: $p < 0.001$.

2.5. Prediction of Genes Targeted by ^{188}Re-Liposome-Deregulated microRNAs

We next examined the potent target genes that might be affected by ^{188}Re-liposome deregulated microRNAs. The miRDB online database was employed to rank the numbers of predicted genes targeted by microRNAs that were deregulated by ^{188}Re-liposome (see Materials and Methods). The ranking of target numbers affected by ^{188}Re-liposome up-regulated and down-regulated microRNAs was obtained, respectively (Figure 5A,B). Furthermore, miR-34a, miR-206-3p and miR-4510-5p were used to draw the Venn diagrams as their predicted target genes were ranked as the top three in ^{188}Re-liposome up-regulated microRNAs. Only one target, named *SLC44A2* encoded choline transporter-like protein 2 was co-regulated by these three microRNAs (Figure 5C). The same logic was used for ^{188}Re-liposome down-regulated microRNAs, and an only one target, named *U2SURP* encoded U2 snRNP-associated SURP motif-containing protein, was co-regulated by miR-142-5p, miR-194-5p and miR-944-3p (Figure 5D). Therefore, these bioinformatics analyses suggest that microRNAs and associated target genes of HPC tumors oppositely regulated by ^{188}Re-liposome are distinct.

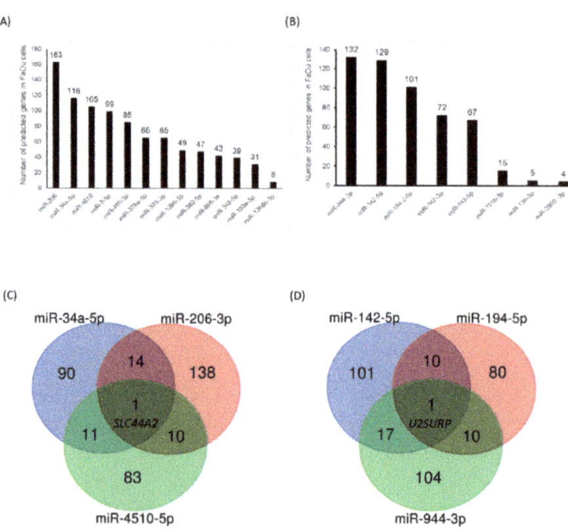

Figure 5. Analysis of miRNA-target interactions. (**A**) The individual target number of microRNAs up-regulated by ^{188}Re-liposome. (**B**) The individual target number of microRNAs down-regulated by ^{188}Re-liposome. (**C**,**D**) The Venn diagram calculated and drawn by the three microRNAs with the most targets.

2.6. Analysis of the Molecular Pathways Regulated by ^{188}Re-Liposome-Affected microRNA

We next used the pathview R package to investigate the genes influenced by microRNAs regulated by ^{188}Re-liposome. MicroRNA samples exhibiting over two-fold change were selected, and the affected target genes were subjected to the KEGG pathway database for determining the potent molecular pathways disturbed by ^{188}Re-liposome. We found that thirty pathways in resected HPC tumors were significantly affected by ^{188}Re-liposome ($p < 0.05$). The top three pathways with the lowest p values were genes involved in olfactory transduction, pathways in cancer, and taste transduction (Table 4). Additionally, ^{188}Re-liposome-influenced pathways could be categorized as cancer and carcinogenesis, cell adhesion and cytoskeletal organization, drug metabolism via cytochrome P450, tumor suppression and oncogenes. An integrated cancer pathway involved in the KEGG database was shown to demonstrate the related genes that could be regulated by ^{188}Re-liposome-induced or -suppressed microRNA expression (Figure 5). This revealed that genes associated with cell cycle progression, proliferation, and apoptosis were affected by ^{188}Re-liposome, including the down-regulation of *cyclin D*, *cyclin E*, cyclin-dependent kinase (*CDK*) *4/6*, *E2F* transcription factor, and *bcl-2* anti-apoptotic factor, and the up-regulation of *p15*, *p16*, and *p27* cell cycle inhibitors and the *Rb* tumor suppressor gene (Figure 6). This pathway analysis provides a potent profile of the molecular mechanism for ^{188}Re-liposome regulated microRNA expression and tumor suppression of the HPC tumor model.

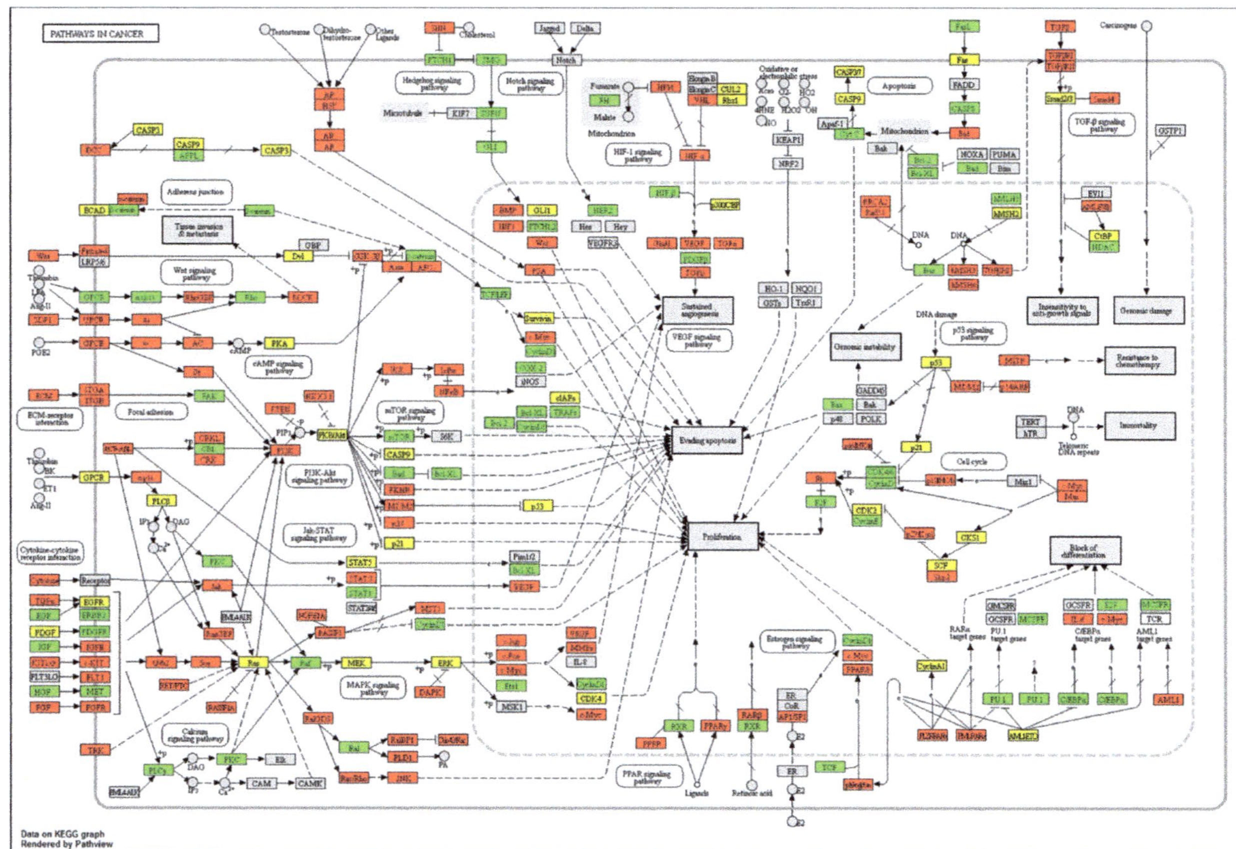

Figure 6. Use of the Kyoto Encyclopedia of Genes and Genomes (KEGG) database to display cancer pathway-associated genes affected by ^{188}Re-liposome mediated miRNA expression in HPC. Red: Up-regulated genes; green: down-regulated genes; yellow: unknown regulated genes; gray: unchanged genes.

Table 4. Thirty significant changed pathways after repeated doses of ^{188}Re-liposome treatment.

Pathways Description	No. of DEGs with Annotated Pathways (4498) [a]	Percentage of DEGs with Annotated Pathways (4498)	Down Regulated Gene	Up Regulated Gene	Unknown Regulated Gene	No. of All Genes with Annotated Pathways (6883)	Percentage of All Genes with Annotated Pathways (6883)	p-Value [b]
Olfactory transduction	51	1.13%	14	33	4	415	6.03%	6.37×10^{-45}
Pathways in cancer	329	7.31%	95	184	50	397	5.77%	0.00127
Taste transduction	16	0.36%	2	11	3	52	0.76%	0.00373
HTLV-I infection	217	4.82%	61	117	39	258	3.75%	0.00623
Proteoglycans in cancer	176	3.91%	60	93	23	203	2.95%	0.00739
Neuroactive ligand-receptor interaction	139	3.09%	40	84	15	277	4.02%	0.00779
MicroRNAs in cancer	151	3.36%	41	89	21	297	4.31%	0.00885
Viral carcinogenesis	175	3.89%	47	90	38	205	2.98%	0.0103
Chemical carcinogenesis	32	0.71%	4	21	7	81	1.18%	0.01124
Hippo signaling pathway	135	3.00%	34	82	19	154	2.24%	0.01459
Focal adhesion	174	3.87%	55	90	29	207	3.01%	0.01626
MAPK signaling pathway	209	4.65%	53	128	28	255	3.70%	0.01732
Drug metabolism - cytochrome P450	27	0.60%	5	17	5	68	0.99%	0.01947
Metabolism of xenobiotics by cytochrome P450	30	0.67%	4	20	6	74	1.08%	0.01954
ErbB signaling pathway	82	1.82%	30	40	12	87	1.26%	0.02121
Signaling pathways regulating pluripotency of stem cells	124	2.76%	29	80	15	142	2.06%	0.02202
Endocytosis	208	4.62%	65	116	27	258	3.75%	0.02587
FoxO signaling pathway	117	2.60%	32	75	10	134	1.95%	0.02612
Ras signaling pathway	185	4.11%	50	109	26	227	3.30%	0.0272
Neurotrophin signaling pathway	106	2.36%	28	64	14	120	1.74%	0.02764
Regulation of actin cytoskeleton	175	3.89%	49	98	28	214	3.11%	0.03031
Chronic myeloid leukemia	70	1.56%	22	40	8	73	1.06%	0.03053
Transcriptional misregulation in cancer	149	3.31%	43	80	26	179	2.60%	0.03365
Glioma	63	1.40%	23	30	10	65	0.94%	0.03554
Pancreatic cancer	64	1.42%	19	35	10	66	0.96%	0.03682
Colorectal cancer	60	1.33%	17	30	13	62	0.90%	0.03971
Acute myeloid leukemia	56	1.24%	17	28	11	57	0.83%	0.04123
Protein processing in endoplasmic reticulum	140	3.11%	40	78	22	169	2.46%	0.04447
Prostate cancer	80	1.78%	26	38	16	89	1.29%	0.04703
TGF-beta signaling pathway	76	1.69%	23	46	7	84	1.22%	0.04999

[a] DEG: Differentially expressed genes. [b] Only $p < 0.05$ was counted.

2.7. Association of ^{188}Re-Liposome-Regulated microRNA and Patients' Survival Rate

As ^{188}Re-liposome could up-regulate tumor suppressive microRNA or down-regulate oncogenic microRNA of the HPC tumor model, we considered whether these microRNAs could be considered as prognostic factors for patients' survival rates. The microRNAs deregulated by ^{188}Re-liposome that displayed over five-fold change were examined. Notably, the online dataset only includes precursor forms of microRNAs for an analysis of patients' survival. For ^{188}Re-liposome up-regulated microRNAs in the HPC model, only miR-342 and miR-378a potentially exhibited an increase in the survival rate in HNSCC patients, with a marginal significance (Figure 7A,B). Interestingly, a high expression of miR-342 was significantly associated with a better survival rate in female patients (HR = 0.61, 95% CI = 0.38–0.98, p = 0.038). On the other hand, ^{188}Re-liposome down-regulated microRNAs, including miR-143, miR-6723, miR-944, and miR-136 were associated with a reduced survival rate in HNSCC patients when they were highly expressed (Figure 7C–F). Additionally, miR-3960 was also associated with reduced survival, yet the expression was too low for meaningful analysis (data not shown). These data suggest that specific miRNAs affected by ^{188}Re-liposome may be important for perspective clinical evaluation.

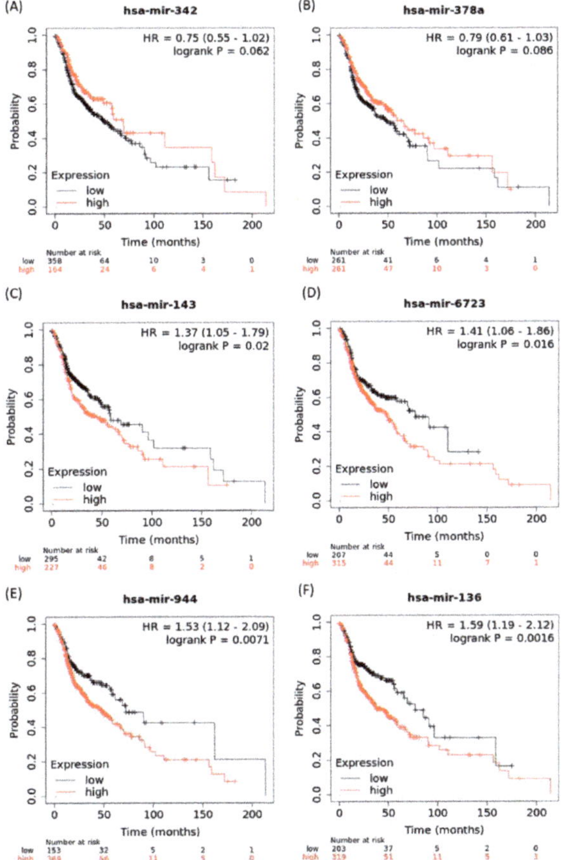

Figure 7. The association of ^{188}Re-liposome deregulated microRNAs with the survival rates of HNSCC patients. Kaplan–Meier (K-M) plot (**A**) hsa-miR-342; (**B**) hsa-miR-378a; (**C**) hsa-miR-143; (**D**) hsa-miR-6723; (**E**) hsa-miR-944; (**F**) hsa-miR-136.

3. Discussion

HPC mostly originates from mucosal squamous cells with a low incident rate. Because of unapparent early symptoms and a high metastatic ability, HPC accounts for the lowest survival rate of all head and neck cancers [72]. The FaDu cell line is a squamous cell carcinoma of the human hypopharynx commonly used for the study of molecular mechanisms of head and neck cancer in vitro and in vivo [73]. We have previously established an orthotopic tumor model using this cell line and found that *let-7* microRNA was associated with the therapeutic efficacy of ^{188}Re-liposome nanoparticles [24,74]. In this study, we used NGS analysis and found additional microRNAs that might be involved in the therapeutic efficacy of ^{188}Re-liposome after repeated administration. The expressions of the total RNA number and annotated small RNA were reduced in ^{188}Re-liposome treated tumors, suggesting that ^{188}Re-liposome would suppress gene transcription and expression.

As ^{188}Re is a high-energy β particles-emitter, it is expected to induce DNA damage. Indeed, the DNA damage marker γ H2AX was significantly induced by ^{188}Re-liposome compared to an untreated control. Interestingly, γ H2AX has been reported to be a tumor suppressor because of its role in the maintenance of genomic stability [75]. In our study, we found that ^{188}Re-liposome induced γ H2AX in HPC tumors. In the RNA-seq dataset, we also found that miR-138-2-5p, a potent inhibitor of H2AX [76], was down-regulated by over two-fold by ^{188}Re-liposome treatment. Together, these results are partially consistent with previous reports that might account for the therapeutic mechanisms of ^{188}Re-liposome from the viewpoint of γ-H2AX-mediated DNA damage responses.

According to the results of RNA-seq, ^{188}Re-liposome induced more than 200 microRNA to change their levels. To raise the selective criteria, we focused on the microRNA species exhibiting over five-fold up-regulation or down-regulation induced by ^{188}Re-liposome by comparing them to the untreated controls. The top three up-regulated microRNAs (miR-206-3p, miR-668-3p, and miR-485-3p) and down-regulated microRNAs (miR-142-5p, miR-944-3p, and miR-142-3p) displaying fold-change were categorized as tumor suppressors and oncogenes, respectively (Tables 2 and 3). Although miR-6723-5p was also highly suppressed by ^{188}Re-liposome, its role in cancer has not been interpreted in the dbDEMC database or microRNA Cancer association database (miRCancer) [77]. Most of the microRNAs deregulated by ^{188}Re-liposome were expressed oppositely in HNSCC, and they were reported to be potent tumor suppressors or oncogenes in different types of cancers. However, there were no available data for miR-4510-5p, miR-1268b-5p miR-6723-5p, miR-151b-3p, miR-143-5p, miR-194-2-5p and miR3960-3p in the database. As NGS is prominently used to find unknown genes that can be induced by treatment agents, the uncharacterized microRNAs shown to be significantly influenced by ^{188}Re-liposome treatment would be of interest for investigating their roles in the future.

The NGS analysis identified highly differentially expressed microRNAs that were also validated using FaDu tumors with or without ^{188}Re-liposome treatment, and clinical HNSCC samples using qPCR. The expression of analyzed microRNAs was consistent in these two different resources, although the case number of clinical samples was limited. The clinical information of the TCGA database could only be employed to analyze larynx cancer because the number of cases of HPC was too low to have normal tissues for analysis. Even so, we still found several microRNAs (e.g., miR-206-3p, miR-378-5p, and miR-142-5p) that were consistent with the results of NGS analysis and our clinical samples (Figure 4). In addition to ^{188}Re, ^{177}Lu is also a radionuclide that can emit 86% of β-particles and 14% of photons with a lower energy but longer half-life period. ^{177}Lu-octreotate has been reported to differentially regulate 57 specific microRNAs in mouse renal cortical tissue identified by the Mouse miRNA Oligo chip 4plex [78]. However, little microRNA was overlapped between their results and ours. Besides different types of radionuclides, the different animal models and microRNA mining methods may also account for the distinct observations in these two studies. Therefore, the differentially expressed microRNAs in the HPC model could be considered as specific prognostic factors for ^{188}Re-liposome treatment.

The predicted genes targeted by microRNAs were also analyzed by the miRDB public database. For the ^{188}Re-liposome up-regulated and down-regulated top three microRNAs, the *SLC44A2* gene and *U2SURP* gene were the targets recognized by these oppositely regulated microRNAs, respectively. Hence, it is expected that *SLC44A2* would exhibit an oncogenic property and *U2SURP* should be a tumor suppressor gene. *SLC44A2*, first discovered in the inner ear, is a member of the choline transporter-like protein family of membrane transporter proteins [79,80]. *U2SURP* is involved in RNA splicing as it is part of spliceosomes [81]. However, little is known about the association of these two genes with human cancers. It would be an interesting target gene to investigate for its role in mediating the efficacy of ^{188}Re-liposome.

Using the KEGG pathway database, we found that 30 pathways in orthotopic HPC tumors were significantly influenced by ^{188}Re-liposome treatment. Although the cancer-suppression-related pathways were expected to be regulated by ^{188}Re-liposome, most of the affected genes in the annotated pathway displayed olfactory transduction. Radiation therapy has been reported to cause olfactory loss in head and neck cancer patients [82]. However, the conclusion is that radiation can damage olfactory cells. The position of hypopharyngeal cancer was not in the olfactory tract, yet FaDu cells were orthotopically injected into the buccal position of the mouse. This operation was based on the fact that FaDu cells are also buccal carcinoma cells [83]. Whether the microenvironmental difference influences the gene regulatory pathway of tumors is unclear. We believed that the excised tumor should not be contaminated by olfactory cells because the human tumor was very small after ^{188}Re-liposome treatment, that is, the tumor size was not big enough to reach the olfactory tract.

An assessment of the correlation between the gene expression and survival rate of patients is important for evaluating the clinical relevance of preclinical study for novel genes and drugs. Here we used the Kaplan–Meier plotter online tool that includes the datasets of TCGA program, the Gene Expression Omnibus (GEO) and the European Genome-Phenome Archive (EGA) to find the ^{188}Re-liposome regulated primary microRNAs and their association with patients' survival rates [84]. Although several potent tumor suppressive or oncogenic microRNAs were influenced by ^{188}Re-liposome, only part of these microRNAs exhibited the expected association with patients' survival probability in HNSCC. For instance, ^{188}Re-liposome suppressed miR-142 exhibited higher survival probabilities in HNSCC when they were expressed at higher levels, although they were expected to be oncogenic microRNAs [85]. A potent limitation is the sample size (523 cases) of HNSCC patients, which may be too small to draw conclusions on the role of microRNAs in patients' survival rates. Besides, the online K-M plotter only analyzes the effects of precursor microRNA on patients' survival rate. The association of mature miRNAs with the survival rate remains unknown. Whether different forms of the same microRNAs will differentially influence the results of the survival rate remains to be addressed.

In summary, current data suggest that the de-regulation of microRNAs correlates with the therapeutic efficacy of ^{188}Re-liposome on human HPC tumors. Using NGS, we also found several microRNAs that have not been fully characterized for their roles in cancer development and therapy. Whether these microRNAs are important for mediating the efficacy of ^{188}Re-liposome would be interesting to further investigate. Additionally, the KEGG pathway analysis showed that not only cancer pathways but also olfactory and taste transduction were significantly changed in HPC tumors after they were treated by ^{188}Re-liposome. Although no clinical evidence showing that patients treated with ^{188}Re-liposome will lose olfactory and gustatory sensation, olfactory sensory dysfunction and gustatory impairment often occur after patients are treated with radiotherapy in the head and neck area [82,86–89]. To the best of our knowledge, this is the first study uncovering the therapeutic mechanisms of ^{188}Re-liposome by an investigation of the pan-expression of microRNA. As ^{188}Re-liposome has entered the clinical trial stage, these data may further extend the concept of precise medicine using this radiotheranostic agent and allow the affected microRNAs to be prognostic factors after cancer treatment.

4. Materials and Methods

4.1. Cell Lines and Plasmid

Human FaDu HPC cells (American Type Culture Collection, Manassas, VA, USA) were cultured in RPMI-1640 (Life Technologies Inc., Carlsbad, CA, USA) containing 10% fetal bovine serum (FBS) (Thermo Fisher Scientific Inc., Waltham, MA, USA), 1% penicillin (Sigma-Aldrich Co., St. Louis, MO, USA), and 1% L-glutamine (Sigma-Aldrich Co., St. Louis, MO, USA). FaDu-3R cells harboring multiple reporter genes were used and cultured as reported previously [24]. Cells were incubated at 37°C in a humidified incubator with 5% CO_2 and passaged every two days.

4.2. Preparation of ^{188}Re-Liposome

The procedure of ^{188}Re-liposomal preparation has been reported before [23]. In brief, ^{188}Re was milked from the ^{188}W/^{188}Re generator system (Institute National des Radioelements, Fleurus, Belgium) and conjugated with sodium perrhenate. Moreover, ^{188}Re was conjugated with *N,N*-bis(2-mercapatoethly)-*N′,N′*-diethylenediamine (BMEDA, ABX GmbH, Radeberg, Germany), and the quality of ^{188}Re-BMEDA was validated by using the instant thin-layer chromatography (iTLC) followed by a radioactive scanner (Bioscan AR2000; Bioscan, TriFoil Imaging Inc., Chatsworth, CA, USA). Furthermore, PEGylated liposome (NanoX; Taiwan Liposome Co. Ltd., Taipei, Taiwan) was used to encapsulate ^{188}Re-BMEDA and eluted using the PD-10 column (GE Health BioSciences, Pittsburgh, PA, USA) (Supplementary Figure S1). The average molecular weight of polyethylene glycol (PEG) was 2000. The particle size (84.6 ± 4.12 nm) and surface charge (1.1 ± 1.9 mV) were measured by the dynamic light scattering apparatus (Zetasizer Nano ZS90, Malvern Panalytical Ltd., Malvern, UK). The in vitro stabilities of ^{188}Re-liposome in normal saline and rat plasma were, respectively, over 92% and 82% in 72 h as reported before [20].

4.3. Establishment of HPC Orthotopic Tumor Model for Evaluating the Therapeutic Efficacy of ^{188}Re-Liposome

Six-week-old male BALB/c nude mice ($N = 5$ for each experimental group) were purchased from National Laboratory Animal Center, Taipei, Taiwan and used for the establishment of orthotopic HPC tumor model. FaDu-3R cells (1×10^6) were resuspended in 50 µL of OPTI-MEM (Sigma-Aldrich, St. Louis, MO, USA) and then injected into the buccal position of each mouse at right side using a 27 G insulin needle. For intravenous injection of ^{188}Re-liposome, 23.68 MBq (640 µCi) corresponding to 80% maximum tolerated dose (MTD), as we mentioned before [23]. To evaluate the therapeutic efficacy, the tumor viability and growth rate were measured using the luciferase reporter gene imaging and caliper measurement. The luminescent signals were acquired by the In Vivo Imaging System (*Optima*, Biospace Lab Inc., Paris, France). The tumor volume was calculated by the formula: $(\text{width}^2 \times \text{length})/2$ after caliper measurement every three days [90]. The animal experiments were approved by the Institutional Animal Care and Utilization Committee (IACUC) of National Yang-Ming University (No. 1061010).

4.4. Tumor Collection and Next-Generation Sequencing (NGS)

Tumors were harvested from tumor-bearing mice after four weeks of ^{188}Re-liposome treatment. Total RNA of both saline control and ^{188}Re-liposome treated group were extracted using the QIAGEN RNA mini kit (Thermo Fisher Scientific Inc., Waltham, MA, USA) according to manufacturer's instructions. Furthermore, the quality of RNA was detected using the Nanodrop spectrophotometer (Nanodrop Technologies LLC, Wilmington, DE, USA). The integrity and concentration of RNA samples were determined using the Agilent 2100 Bioanalyzer (Agilent Technologies, Santa Clara, CA, USA) with RNA 6000 nano kit (Agilent Technologies, Santa Clara, CA, USA). The TruSeq Small RNA Library kit (Illumina, Inc., San Diego, CA, USA) was then used to ligate RNA with adapters followed by the reverse transcription-PCR to generate cDNA library. The library was then sequenced by the HiSeq

4000 Sequencing System (2 × 150 bp paired-end Sequencing) and the results were processed with the Illumina software (Illumina, Inc., San Diego, CA, USA). To qualify the results of small RNA sequencing, the sequences were applied to the CLC Genomics Workbench v10 to obtain the qualified reads [91]. CLC Genomics Workbench counts different types of small RNAs in the data and compares them to databases of microRNAs or other small RNAs [92].

4.5. MicroRNA Expression Analysis

The 'miRBase' online source was used for the annotation of small RNA [93]. In the next step, all of the miRNAs were normalized by TMM (trimmed mean of M values) method using edgeR (R package: v.3.10.5) (Bioconductor, New York City, NY, USA) [94]. For a two-group experiment, we used the 'Fold Change' to tell how many times bigger the mean expression value in ^{188}Re-liposome group is relative to that of saline-only group. If the mean expression value in ^{188}Re-liposome group is small than that in saline-only group, the value will present negative sign, and vice versa. The criteria for microRNAs selection were fold change >5.

4.6. Western Blot Analysis

Tumors were collected from the tumor-bearing mice after 4 weeks of ^{188}Re-liposome treatment, and lysed in T-PER^{TM} Tissue Protein Extraction Reagent (Thermo Fisher Scientific, Waltham, MA, USA) containing 1% protease inhibitor cocktail (Sigma-Aldrich Co., St. Louis, MO, USA). Protein lysates (50 μg) were run on 8–12% SDS-PAGE, electro-transferred to nitrocellulose membrane, blocking and incubated with antibody as reported previously [95]. The primary antibodies were anti-γH2AX (GTX628789, GeneTex Inc., Irvine, CA, USA), and anti-glyceraldehyde3-phosphate dehydrogenase (GAPDH) (MA5-15738, Invitrogen Inc., Carlsbad, CA, USA).

4.7. Validation of microRNA Expression Using qPCR

To validate miRNA levels before and after HPC tumor treated with ^{188}Re-liposome and to compare normal tissue to HNSCC tissues using clinical samples, quantitative PCR (qPCR) of targeted miRNA was performed. Briefly, complementary DNA (cDNA) was generated from 2 μg total RNA using SuperScript II reverse transcriptase (Life-Technologies Co., Carlsbad, CA, USA). Then, the cDNA products were mixed with the Fast SYBR Green Master Mix (Life-Technologies Co., Carlsbad, CA, USA) and subjected to the StepOnePlus Real-time PCR System (Life-Technologies Co., Carlsbad, CA, USA) following the manufacturer's instructions. The sequences of stem loop primers, forward primers and universal primer used for miR-206-3p, miR-382-5p, miR-378a-5p, miR-3960-3p, and miR-142-5p amplification were summarized in Table 5. Use of human tissue samples was approved by the Institutional Review Board (No. 2019-01-010BC).

4.8. Heatmap Analysis of NGS Data

To gain the heatmap of total microRNAs' expression, 'TreeView' (v1.1.6r4) was used [96]. The heatmap analysis was based on the small RNA that has twice difference between individuals with or without the treatment of ^{188}Re-liposome, and then took Log2 value to draw out.

4.9. Analysis of microRNA Using the Cancer Genome Atlas (TCGA)

The expression levels of miRNAs and clinical information for TCGA Head-Neck Squamous Cell Carcinoma (HNSC) were downloaded from Broad GDAC Firehose [97]. The miRNA expression values for samples having anatomic subdivision labeled as normal larynx tissue (12 cases), larynx cancers (117 cases) or hypopharynx cancers (10 cases) were used. In house R scripts were used to parse and generate heatmap and boxplots [98]. The difference between normal and tumor samples were tested with Wilcoxon signed-rank test.

Table 5. The list of stem loop primers, forward primers, and reverse primer used for qPCR of microRNAs.

MicroRNA	Stem Loop Primer Sequence
miR-206-3p	5′-GTCGTATCCAGTGCAGGGTCCGAG GTATTCGCACTGGATACGACCCACAC-3′
miR-382-5p	5′-GTCGTATCCAGTGCAGGGTCCGA GGTATTCGCACTGGATACGACCGAATC-3′
miR-378a-5p	5′-GTCGTATCCAGTGCAGGGTCCGAG GTATTCGCACTGGATACGACACACAG-3′
miR-3960-3p	5′-GTCGTATCCAGTGCAGGGTCCGAG GTATTCGCACTGGATACGACCCCCCG-3′
miR-142-5p	5′-GTCGTATCCAGTGCAGGGTCCGAG GTATTCGCACTGGATACGACAGTAGT-3′
MicroRNA	**Forward Primer Sequence**
miR-206-3p	5′-CACGCATGGAATGTAAGGAAGT-3′
miR-382-5p	5′-CACGCAGAAGTTGTTCGTGGTG-3′
miR-378a-5p	5′-TGATTACTCCTGACTCCAGGTC-3′
miR-3960-3p	5′-TAATTATGGCGGCGGCGGAG-3′
miR-142-5p	5′-CACGCGCATAAAGTAGAAAGCA-3′
MicroRNA	**Reverse Primer Sequence**
Universal reverse primer	5′-CCAGTGCAGGGTCCGAGGT-3′

4.10. Characterization of miRNAs

The miRNA of interests were subjected to a Database of Differentially Expressed MiRNAs in human Cancer 2.0 (dbDEMC v2.0) that collects the data sets of Gene Expression Omnibus (GEO) and The Cancer Genome Atlas (TCGA) to exhibit differentially expressed miRNAs in human cancers detected by high throughput method [99]. The roles of miRNAs belonging to tumor suppressor genes or oncogenes were predicted accordingly. The miRDB online database for prediction of miRNA targets by a MirTarget bioinformatics tool was used to analyze putative downstream genes influenced by the miRNAs of interests [100,101]. Each microRNA was selected for target expression analysis for FaDu cells, and the targets expression level over 20 was counted as they were most relevant to FaDu cells. Target expression level was determined by RNA-Seq using the RPKM method (Reads Per Kilobase of transcript, per Million mapped reads). An online Venn diagrams drawing tool was exploited to calculate the intersections of list of miRNA targets [102].

4.11. The Pathway Analysis

The 'pathview' (R package: v1.4.2) software (Bioconductor, New York City, NY, USA) was used to draw pathways and find significant gene change (Fold change > 2) with the Kyoto Encyclopedia of Genes and Genomes (KEGG) [103,104]. Moreover, the pathways would be regarded significant if the p-value < 0.05.

4.12. Statistical Analysis

Statistical analysis was performed using GraphPad Prism 6.0 (GraphPad Software, San Diego, CA, USA). All data were represented as the means ± standard deviation (SD) with independent experiments. The Student's t-test was used for statistical analysis. Two-way Analysis of Variance (ANOVA) was used to compare the tumor growth curves. The Kaplan–Meier method with the log-rank test was used to analyze the association of miRNAs and patients' survival rates using the online K-M plotter with public datasets [84].

Supplementary Materials: The following are available online, Figure S1: Measurement of the combined efficiency between ^{188}Re and BMEDA chelator for embedding into liposome. Figure S2: Demonstration of reporter genes expression in FaDu-3R cells. Figure S3: Qualification of total RNA for NGS analysis. Figure S4: GC content analysis and sequencing qualification of RNA–Seq. Figure S5: A heatmap of differentially expressed microRNAs with or without ^{188}Re-liposomal treatment. Figure S6: A heatmap of selected differentially expressed microRNAs in normal tissues and larynx cancer tissues using the clinical information of TCGA.

Author Contributions: Conceptualization, Y.-J.L. and S.-Y.W.; methodology, B.-Z.L.; software, T.-W.C.; validation, B.-Z.L., M.-Y.L. and Y.-J.L.; formal analysis, B.-Z.L.; investigation, B.-Z.L.; resources, C.-H.C. and M.-H.Y.; data curation, M.-Y.L.; writing—original draft preparation, Y.-J.L.; writing—review and editing, Y.-J.L.; supervision, Y.-J.L. and S.-Y.W.; funding acquisition, Y.-J.L. and S.-Y.W. All authors have read and agreed to the published version of the manuscript.

Funding: This research was funded by National Yang-Ming University-Far Eastern Memorial Hospital Joint Research Program, grant number 108DN04; Ministry of Science and Technology grant number 105-2623-E-010-001-NU, 106-2623-E-010-002-NU, and 108-2314-B-010-016) and "The APC was funded by National Yang-Ming University-Far Eastern Memorial Hospital Joint Research Program".

Acknowledgments: We thanked Shen-Nan Lo, Ming-Hsuan Lin for helping the ^{188}Re preparation, production, and quality assurance. We thank Liang-Ting Lin for discussion of references. We also thank the Taiwan Animal Consortium (MOST 106-2319-B-001-004)—Taiwan Mouse Clinic which is funded by the Ministry of Science and Technology (MOST) of Taiwan for technical support in in vivo imaging experiments. We also thank the support from the Cancer Progression Research Center, National Yang-Ming University, from the Featured Areas Research Center Program within the framework of the Higher Education Sprout Project by the Ministry of Education (MOE) in Taiwan.

Conflicts of Interest: The authors declare no conflict of interest.

References

1. Patel, R.S.; Goldstein, D.P.; Brown, D.; Irish, J.; Gullane, P.J.; Gilbert, R.W. Circumferential pharyngeal reconstruction: History, critical analysis of techniques, and current therapeutic recommendations. *Head Neck* **2010**, *32*, 109–120. [CrossRef] [PubMed]
2. Chu, P.Y.; Chang, S.Y. Reconstruction of the hypopharynx after surgical treatment of squamous cell carcinoma. *J. Chin. Med. Assoc.* **2009**, *72*, 351–355. [CrossRef]
3. Iwamoto, H. Operative treatment of cancer of the hypopharynx. *Gan No Rinsho* **1968**, *14*, 660–662. [PubMed]
4. Kim, S.; Wu, H.G.; Heo, D.S.; Kim, K.H.; Sung, M.W.; Park, C.I. Advanced hypopharyngeal carcinoma treatment results according to treatment modalities. *Head Neck* **2001**, *23*, 713–717. [CrossRef] [PubMed]
5. Mura, F.; Bertino, G.; Occhini, A.; Benazzo, M. Surgical treatment of hypopharyngeal cancer: A review of the literature and proposal for a decisional flow-chart. *Acta Otorhinolaryngol. Ital.* **2013**, *33*, 299–306. [PubMed]
6. Medvedeva, A.; Chernov, V.; Zeltchan, R.; Sinilkin, I.; Bragina, O.; Chijevskaya, S.; Choynzonov, E.; Goldberg, A. Nuclear medicine imaging of locally advanced laryngeal and hypopharyngeal cancer. In *AIP Conference Proceedings*; AIP Publishing LLC: Melville, NY, USA, 2017; Volume 1882.
7. Jadvar, H.; Chen, X.; Cai, W.; Mahmood, U. Radiotheranostics in Cancer Diagnosis and Management. *Radiology* **2018**, *286*, 388–400. [CrossRef]
8. Argyrou, M.; Valassi, A.; Andreou, M.; Lyra, M. Rhenium-188 production in hospitals, by w-188/re-188 generator, for easy use in radionuclide therapy. *Int. J. Mol. Imaging* **2013**, *2013*, 290750. [CrossRef]
9. Liepe, K.; Hliscs, R.; Kropp, J.; Gruning, T.; Runge, R.; Koch, R.; Knapp, F.F., Jr.; Franke, W.G. Rhenium-188-HEDP in the palliative treatment of bone metastases. *Cancer Biother. Radiopharm.* **2000**, *15*, 261–265. [CrossRef]
10. Zhang, H.; Tian, M.; Li, S.; Liu, J.; Tanada, S.; Endo, K. Rhenium-188-HEDP therapy for the palliation of pain due to osseous metastases in lung cancer patients. *Cancer Biother. Radiopharm.* **2003**, *18*, 719–726. [CrossRef]
11. Guerra Liberal, F.D.C.; Tavares, A.A.S.; Tavares, J. Palliative treatment of metastatic bone pain with radiopharmaceuticals: A perspective beyond Strontium-89 and Samarium-153. *Appl. Radiat. Isot.* **2016**, *110*, 87–99. [CrossRef]
12. Maxon, H.R., III; Schroder, L.E.; Washburn, L.C.; Thomas, S.R.; Samaratunga, R.C.; Biniakiewicz, D.; Moulton, J.S.; Cummings, D.; Ehrhardt, G.J.; Morris, V. Rhenium-188(Sn)HEDP for treatment of osseous metastases. *J. Nucl. Med.* **1998**, *39*, 659–663. [PubMed]
13. Uccelli, L.; Martini, P.; Pasquali, M.; Boschi, A. Monoclonal Antibodies Radiolabeling with Rhenium-188 for Radioimmunotherapy. *BioMed Res. Int.* **2017**, *2017*, 5923609. [CrossRef] [PubMed]

14. Edelman, M.J.; Clamon, G.; Kahn, D.; Magram, M.; Lister-James, J.; Line, B.R. Targeted radiopharmaceutical therapy for advanced lung cancer: Phase I trial of rhenium Re188 P2045, a somatostatin analog. *J. Thorac. Oncol.* **2009**, *4*, 1550–1554. [CrossRef] [PubMed]
15. Kumar, A.; Bal, C.; Srivastava, D.N.; Thulkar, S.P.; Sharma, S.; Acharya, S.K.; Duttagupta, S. Management of multiple intrahepatic recurrences after radiofrequency ablation of hepatocellular carcinoma with rhenium-188-HDD-lipiodol. *Eur. J. Gastroenterol. Hepatol.* **2006**, *18*, 219–223. [CrossRef] [PubMed]
16. Lambert, B.; Bacher, K.; De Keukeleire, K.; Smeets, P.; Colle, I.; Jeong, J.M.; Thierens, H.; Troisi, R.; De Vos, F.; Van de Wiele, C. 188Re-HDD/lipiodol for treatment of hepatocellular carcinoma: A feasibility study in patients with advanced cirrhosis. *J. Nucl. Med.* **2005**, *46*, 1326–1332. [PubMed]
17. Nowicki, M.L.; Cwikla, J.B.; Sankowski, A.J.; Shcherbinin, S.; Grimmes, J.; Celler, A.; Buscombe, J.R.; Bator, A.; Pech, M.; Mikolajczak, R.; et al. Initial study of radiological and clinical efficacy radioembolization using 188Re-human serum albumin (HSA) microspheres in patients with progressive, unresectable primary or secondary liver cancers. *Med. Sci. Monit.* **2014**, *20*, 1353–1362. [CrossRef]
18. Jeong, J.M.; Lee, Y.J.; Kim, E.H.; Chang, Y.S.; Kim, Y.J.; Son, M.; Lee, D.S.; Chung, J.K.; Lee, M.C. Preparation of (188) Re-labeled paper for treating skin cancer. *Appl. Radiat. Isot.* **2003**, *58*, 551–555. [CrossRef]
19. Wang, S.J.; Huang, W.S.; Chuang, C.M.; Chang, C.H.; Lee, T.W.; Ting, G.; Chen, M.H.; Chang, P.M.; Chao, T.C.; Teng, H.W.; et al. A phase 0 study of the pharmacokinetics, biodistribution, and dosimetry of (188)Re-liposome in patients with metastatic tumors. *Ejnmmi. Res.* **2019**, *9*, 46. [CrossRef]
20. Chang, Y.J.; Chang, C.H.; Chang, T.J.; Yu, C.Y.; Chen, L.C.; Jan, M.L.; Luo, T.Y.; Lee, T.W.; Ting, G. Biodistribution, pharmacokinetics and microSPECT/CT imaging of 188Re-bMEDA-liposome in a C26 murine colon carcinoma solid tumor animal model. *Anticancer Res.* **2007**, *27*, 2217–2225.
21. Chang, C.H.; Liu, S.Y.; Chi, C.W.; Yu, H.L.; Chang, T.J.; Tsai, T.H.; Lee, T.W.; Chen, Y.J. External beam radiotherapy synergizes (1)(8)(8)Re-liposome against human esophageal cancer xenograft and modulates (1)(8)(8)Re-liposome pharmacokinetics. *Int. J. Nanomed.* **2015**, *10*, 3641–3649. [CrossRef]
22. Huang, F.Y.; Lee, T.W.; Chang, C.H.; Chen, L.C.; Hsu, W.H.; Chang, C.W.; Lo, J.M. Evaluation of 188Re-labeled PEGylated nanoliposome as a radionuclide therapeutic agent in an orthotopic glioma-bearing rat model. *Int. J. Nanomed.* **2015**, *10*, 463–473. [CrossRef]
23. Lin, L.T.; Chang, C.H.; Yu, H.L.; Liu, R.S.; Wang, H.E.; Chiu, S.J.; Chen, F.D.; Lee, T.W.; Lee, Y.J. Evaluation of the therapeutic and diagnostic effects of PEGylated liposome-embedded 188Re on human non-small cell lung cancer using an orthotopic small-animal model. *J. Nucl. Med.* **2014**, *55*, 1864–1870. [CrossRef]
24. Lin, L.T.; Chang, C.Y.; Chang, C.H.; Wang, H.E.; Chiou, S.H.; Liu, R.S.; Lee, T.W.; Lee, Y.J. Involvement of let-7 microRNA for the therapeutic effects of Rhenium-188-embedded liposomal nanoparticles on orthotopic human head and neck cancer model. *Oncotarget* **2016**, *7*, 65782–65796. [CrossRef]
25. Shen, Y.A.; Lan, K.L.; Chang, C.H.; Lin, L.T.; He, C.L.; Chen, P.H.; Lee, T.W.; Lee, Y.J.; Chuang, C.M. Intraperitoneal (188)Re-Liposome delivery switches ovarian cancer metabolism from glycolysis to oxidative phosphorylation and effectively controls ovarian tumour growth in mice. *Radiother. Oncol.* **2016**, *119*, 282–290. [CrossRef]
26. Chang, C.M.; Lan, K.L.; Huang, W.S.; Lee, Y.J.; Lee, T.W.; Chang, C.H.; Chuang, C.M. 188Re-Liposome Can Induce Mitochondrial Autophagy and Reverse Drug Resistance for Ovarian Cancer: From Bench Evidence to Preliminary Clinical Proof-of-Concept. *Int. J. Mol. Sci.* **2017**, *18*, 903. [CrossRef]
27. Rutman, A.M.; Kuo, M.D. Radiogenomics: Creating a link between molecular diagnostics and diagnostic imaging. *Eur. J. Radiol.* **2009**, *70*, 232–241. [CrossRef]
28. Kang, J.; Rancati, T.; Lee, S.; Oh, J.H.; Kerns, S.L.; Scott, J.G.; Schwartz, R.; Kim, S.; Rosenstein, B.S. Machine Learning and Radiogenomics: Lessons Learned and Future Directions. *Front. Oncol.* **2018**, *8*, 228. [CrossRef]
29. Lawrence, M.S.; Stojanov, P.; Mermel, C.H.; Robinson, J.T.; Garraway, L.A.; Golub, T.R.; Meyerson, M.; Gabriel, S.B.; Lander, E.S.; Getz, G. Discovery and saturation analysis of cancer genes across 21 tumour types. *Nature* **2014**, *505*, 495–501. [CrossRef]
30. Schulze, K.; Nault, J.C.; Villanueva, A. Genetic profiling of hepatocellular carcinoma using next-generation sequencing. *J. Hepatol.* **2016**, *65*, 1031–1042. [CrossRef]
31. Wang, Z.; Gerstein, M.; Snyder, M. RNA-Seq: A revolutionary tool for transcriptomics. *Nat. Rev. Genet.* **2009**, *10*, 57–63. [CrossRef]
32. Roh, S.W.; Abell, G.C.; Kim, K.H.; Nam, Y.D.; Bae, J.W. Comparing microarrays and next-generation sequencing technologies for microbial ecology research. *Trends Biotechnol.* **2010**, *28*, 291–299. [CrossRef]

33. Doostparast Torshizi, A.; Wang, K. Next-generation sequencing in drug development: Target identification and genetically stratified clinical trials. *Drug Discov. Today* **2018**, *23*, 1776–1783. [CrossRef]
34. Liu, F.; Zhao, X.; Qian, Y.; Zhang, J.; Zhang, Y.; Yin, R. MiR-206 inhibits Head and neck squamous cell carcinoma cell progression by targeting HDAC6 via PTEN/AKT/mTOR pathway. *Biomed. Pharm.* **2017**, *96*, 229–237. [CrossRef]
35. Koshizuka, K.; Hanazawa, T.; Fukumoto, I.; Kikkawa, N.; Matsushita, R.; Mataki, H.; Mizuno, K.; Okamoto, Y.; Seki, N. Dual-receptor (EGFR and c-MET) inhibition by tumor-suppressive miR-1 and miR-206 in head and neck squamous cell carcinoma. *J. Hum. Genet.* **2017**, *62*, 113–121. [CrossRef]
36. Momen-Heravi, F.; Trachtenberg, A.J.; Kuo, W.P.; Cheng, Y.S. Genomewide Study of Salivary MicroRNAs for Detection of Oral Cancer. *J. Dent. Res.* **2014**, *93*, 86S–93S. [CrossRef]
37. Shin, K.H.; Pucar, A.; Kim, R.H.; Bae, S.D.; Chen, W.; Kang, M.K.; Park, N.H. Identification of senescence-inducing microRNAs in normal human keratinocytes. *Int. J. Oncol.* **2011**, *39*, 1205–1211. [CrossRef]
38. Lou, C.; Xiao, M.; Cheng, S.; Lu, X.; Jia, S.; Ren, Y.; Li, Z. MiR-485-3p and miR-485-5p suppress breast cancer cell metastasis by inhibiting PGC-1alpha expression. *Cell Death Dis.* **2016**, *7*, e2159. [CrossRef]
39. Wang, Z.Q.; Zhang, M.Y.; Deng, M.L.; Weng, N.Q.; Wang, H.Y.; Wu, S.X. Low serum level of miR-485-3p predicts poor survival in patients with glioblastoma. *PLoS ONE* **2017**, *12*, e0184969. [CrossRef]
40. Zhang, Y.; Sui, R.; Chen, Y.; Liang, H.; Shi, J.; Piao, H. Downregulation of miR-485-3p promotes glioblastoma cell proliferation and migration via targeting RNF135. *Exp. Med.* **2019**, *18*, 475–482. [CrossRef]
41. Wang, J.; Chen, C.; Yan, X.; Wang, P. The role of miR-382-5p in glioma cell proliferation, migration and invasion. *Onco. Targets* **2019**, *12*, 4993–5002. [CrossRef]
42. Feng, J.; Qi, B.; Guo, L.; Chen, L.Y.; Wei, X.F.; Liu, Y.Z.; Zhao, B.S. miR-382 functions as a tumor suppressor against esophageal squamous cell carcinoma. *World J. Gastroenterol.* **2017**, *23*, 4243–4251. [CrossRef]
43. Zhu, W.J.; Chen, X.; Wang, Y.W.; Liu, H.T.; Ma, R.R.; Gao, P. MiR-1268b confers chemosensitivity in breast cancer by targeting ERBB2-mediated PI3K-AKT pathway. *Oncotarget* **2017**, *8*, 89631–89642. [CrossRef]
44. Pu, Y.; Zhao, F.; Cai, W.; Meng, X.; Li, Y.; Cai, S. MiR-193a-3p and miR-193a-5p suppress the metastasis of human osteosarcoma cells by down-regulating Rab27B and SRR, respectively. *Clin. Exp. Metastasis* **2016**, *33*, 359–372. [CrossRef]
45. Chou, N.H.; Lo, Y.H.; Wang, K.C.; Kang, C.H.; Tsai, C.Y.; Tsai, K.W. MiR-193a-5p and -3p Play a Distinct Role in Gastric Cancer: miR-193a-3p Suppresses Gastric Cancer Cell Growth by Targeting ETS1 and CCND1. *Anticancer Res.* **2018**, *38*, 3309–3318. [CrossRef]
46. Yin, C.Y.; Kong, W.; Jiang, J.; Xu, H.; Zhao, W. miR-7-5p inhibits cell migration and invasion in glioblastoma through targeting SATB1. *Oncol. Lett.* **2019**, *17*, 1819–1825. [CrossRef]
47. Shi, Y.; Luo, X.; Li, P.; Tan, J.; Wang, X.; Xiang, T.; Ren, G. miR-7-5p suppresses cell proliferation and induces apoptosis of breast cancer cells mainly by targeting REGgamma. *Cancer Lett.* **2015**, *358*, 27–36. [CrossRef]
48. Wang, Z.; Ma, B.; Ji, X.; Deng, Y.; Zhang, T.; Zhang, X.; Gao, H.; Sun, H.; Wu, H.; Chen, X.; et al. MicroRNA-378-5p suppresses cell proliferation and induces apoptosis in colorectal cancer cells by targeting BRAF. *Cancer Cell Int.* **2015**, *15*, 40. [CrossRef]
49. Li, H.; Dai, S.; Zhen, T.; Shi, H.; Zhang, F.; Yang, Y.; Kang, L.; Liang, Y.; Han, A. Clinical and biological significance of miR-378a-3p and miR-378a-5p in colorectal cancer. *Eur. J. Cancer.* **2014**, *50*, 1207–1221. [CrossRef]
50. Ostadrahimi, S.; Abedi Valugerdi, M.; Hassan, M.; Haddad, G.; Fayaz, S.; Parvizhamidi, M.; Mahdian, R.; Fard Esfahani, P. miR-1266-5p and miR-185-5p Promote Cell Apoptosis in Human Prostate Cancer Cell Lines. *Asian Pac. J. Cancer Prev.* **2018**, *19*, 2305–2311. [CrossRef]
51. Shen, X.P.; Ling, X.; Lu, H.; Zhou, C.X.; Zhang, J.K.; Yu, Q. Low expression of microRNA-1266 promotes colorectal cancer progression via targeting FTO. *Eur. Rev. Med. Pharm. Sci.* **2018**, *22*, 8220–8226. [CrossRef]
52. Hironaka-Mitsuhashi, A.; Matsuzaki, J.; Takahashi, R.U.; Yoshida, M.; Nezu, Y.; Yamamoto, Y.; Shiino, S.; Kinoshita, T.; Ushijima, T.; Hiraoka, N.; et al. A tissue microRNA signature that predicts the prognosis of breast cancer in young women. *PLoS ONE* **2017**, *12*, e0187638. [CrossRef] [PubMed]
53. Chen, L.; Yuan, L.; Wang, G.; Cao, R.; Peng, J.; Shu, B.; Qian, G.; Wang, X.; Xiao, Y. Identification and bioinformatics analysis of miRNAs associated with human muscle invasive bladder cancer. *Mol. Med. Rep.* **2017**, *16*, 8709–8720. [CrossRef] [PubMed]

54. Pedersen, N.J.; Jensen, D.H.; Lelkaitis, G.; Kiss, K.; Charabi, B.W.; Ullum, H.; Specht, L.; Schmidt, A.Y.; Nielsen, F.C.; von Buchwald, C. MicroRNA-based classifiers for diagnosis of oral cavity squamous cell carcinoma in tissue and plasma. *Oral. Oncol.* **2018**, *83*, 46–52. [CrossRef] [PubMed]
55. Wang, X.; Zhao, Y.; Lu, Q.; Fei, X.; Lu, C.; Li, C.; Chen, H. MiR-34a-5p Inhibits Proliferation, Migration, Invasion and Epithelial-mesenchymal Transition in Esophageal Squamous Cell Carcinoma by Targeting LEF1 and Inactivation of the Hippo-YAP1/TAZ Signaling Pathway. *J. Cancer* **2020**, *11*, 3072–3081. [CrossRef] [PubMed]
56. Gao, J.; Li, N.; Dong, Y.; Li, S.; Xu, L.; Li, X.; Li, Y.; Li, Z.; Ng, S.S.; Sung, J.J.; et al. miR-34a-5p suppresses colorectal cancer metastasis and predicts recurrence in patients with stage II/III colorectal cancer. *Oncogene* **2015**, *34*, 4142–4152. [CrossRef] [PubMed]
57. Yang, H.; Li, Q.; Niu, J.; Li, B.; Jiang, D.; Wan, Z.; Yang, Q.; Jiang, F.; Wei, P.; Bai, S. microRNA-342-5p and miR-608 inhibit colon cancer tumorigenesis by targeting NAA10. *Oncotarget* **2016**, *7*, 2709–2720. [CrossRef]
58. Soriano, A.; Masanas, M.; Boloix, A.; Masia, N.; Paris-Coderch, L.; Piskareva, O.; Jimenez, C.; Henrich, K.O.; Roma, J.; Westermann, F.; et al. Functional high-throughput screening reveals miR-323a-5p and miR-342-5p as new tumor-suppressive microRNA for neuroblastoma. *Cell Mol. Life Sci.* **2019**, *76*, 2231–2243. [CrossRef]
59. He, Z.; Yi, J.; Liu, X.; Chen, J.; Han, S.; Jin, L.; Chen, L.; Song, H. MiR-143-3p functions as a tumor suppressor by regulating cell proliferation, invasion and epithelial-mesenchymal transition by targeting QKI-5 in esophageal squamous cell carcinoma. *Mol. Cancer* **2016**, *15*, 51. [CrossRef]
60. Jin, X.; Chen, X.; Hu, Y.; Ying, F.; Zou, R.; Lin, F.; Shi, Z.; Zhu, X.; Yan, X.; Li, S.; et al. LncRNA-TCONS_00026907 is involved in the progression and prognosis of cervical cancer through inhibiting miR-143-5p. *Cancer Med.* **2017**, *6*, 1409–1423. [CrossRef]
61. Yang, F.; Xiao, Z.; Zhang, S. Knockdown of miR-194-5p inhibits cell proliferation, migration and invasion in breast cancer by regulating the Wnt/beta-catenin signaling pathway. *Int. J. Mol. Med.* **2018**, *42*, 3355–3363. [CrossRef]
62. Jiang, M.J.; Chen, Y.Y.; Dai, J.J.; Gu, D.N.; Mei, Z.; Liu, F.R.; Huang, Q.; Tian, L. Dying tumor cell-derived exosomal miR-194-5p potentiates survival and repopulation of tumor repopulating cells upon radiotherapy in pancreatic cancer. *Mol. Cancer* **2020**, *19*, 68. [CrossRef] [PubMed]
63. Kosela-Paterczyk, H.; Paziewska, A.; Kulecka, M.; Balabas, A.; Kluska, A.; Dabrowska, M.; Piatkowska, M.; Zeber-Lubecka, N.; Ambrozkiewicz, F.; Karczmarski, J.; et al. Signatures of circulating microRNA in four sarcoma subtypes. *J. Cancer* **2020**, *11*, 874–882. [CrossRef] [PubMed]
64. Gullu Amuran, G.; Tinay, I.; Filinte, D.; Ilgin, C.; Peker Eyuboglu, I.; Akkiprik, M. Urinary micro-RNA expressions and protein concentrations may differentiate bladder cancer patients from healthy controls. *Int. Urol. Nephrol.* **2020**, *52*, 461–468. [CrossRef] [PubMed]
65. Chen, X.; Huang, Z.; Chen, R. Microrna-136 promotes proliferation and invasion ingastric cancer cells through Pten/Akt/P-Akt signaling pathway. *Oncol. Lett.* **2018**, *15*, 4683–4689. [CrossRef] [PubMed]
66. Li, Y.; Chen, D.; Jin, L.U.; Liu, J.; Li, Y.; Su, Z.; Qi, Z.; Shi, M.; Jiang, Z.; Yang, S.; et al. Oncogenic microRNA-142-3p is associated with cellular migration, proliferation and apoptosis in renal cell carcinoma. *Oncol. Lett.* **2016**, *11*, 1235–1241. [CrossRef] [PubMed]
67. Zhang, H.; Li, T.; Zheng, L.; Huang, X. Biomarker MicroRNAs for Diagnosis of Oral Squamous Cell Carcinoma Identified Based on Gene Expression Data and MicroRNA-mRNA Network Analysis. *Comput. Math Methods Med.* **2017**, *2017*, 9803018. [CrossRef]
68. Park, S.; Kim, J.; Eom, K.; Oh, S.; Kim, S.; Kim, G.; Ahn, S.; Park, K.H.; Chung, D.; Lee, H. microRNA-944 overexpression is a biomarker for poor prognosis of advanced cervical cancer. *Bmc Cancer* **2019**, *19*, 419. [CrossRef]
69. He, H.; Tian, W.; Chen, H.; Jiang, K. MiR-944 functions as a novel oncogene and regulates the chemoresistance in breast cancer. *Tumour Biol.* **2016**, *37*, 1599–1607. [CrossRef]
70. Ma, Z.; Liu, T.; Huang, W.; Liu, H.; Zhang, H.M.; Li, Q.; Chen, Z.; Guo, A.Y. MicroRNA regulatory pathway analysis identifies miR-142-5p as a negative regulator of TGF-beta pathway via targeting SMAD3. *Oncotarget* **2016**, *7*, 71504–71513. [CrossRef]
71. Islam, F.; Gopalan, V.; Vider, J.; Lu, C.T.; Lam, A.K. MiR-142-5p act as an oncogenic microRNA in colorectal cancer: Clinicopathological and functional insights. *Exp. Mol. Pathol.* **2018**, *104*, 98–107. [CrossRef]

72. Lamperska, K.M.; Kolenda, T.; Teresiak, A.; Kowalik, A.; Kruszyna-Mochalska, M.; Jackowiak, W.; Blizniak, R.; Przybyla, W.; Kapalczynska, M.; Kozlowski, P. Different levels of let-7d expression modulate response of FaDu cells to irradiation and chemotherapeutics. *PLoS ONE* **2017**, *12*, e0180265. [CrossRef] [PubMed]
73. Rangan, S.R. A new human cell line (FaDu) from a hypopharyngeal carcinoma. *Cancer* **1972**, *29*, 117–121. [CrossRef]
74. Chang, C.Y.; Chen, C.C.; Lin, L.T.; Chang, C.H.; Chen, L.C.; Wang, H.E.; Lee, T.W.; Lee, Y.J. PEGylated liposome-encapsulated rhenium-188 radiopharmaceutical inhibits proliferation and epithelial-mesenchymal transition of human head and neck cancer cells in vivo with repeated therapy. *Cell Death Discov.* **2018**, *4*, 100. [CrossRef] [PubMed]
75. Bonner, W.M.; Redon, C.E.; Dickey, J.S.; Nakamura, A.J.; Sedelnikova, O.A.; Solier, S.; Pommier, Y. GammaH2AX and cancer. *Nat. Rev. Cancer* **2008**, *8*, 957–967. [CrossRef]
76. Wang, Y.; Huang, J.W.; Li, M.; Cavenee, W.K.; Mitchell, P.S.; Zhou, X.; Tewari, M.; Furnari, F.B.; Taniguchi, T. MicroRNA-138 modulates DNA damage response by repressing histone H2AX expression. *Mol. Cancer Res.* **2011**, *9*, 1100–1111. [CrossRef]
77. Xie, B.; Ding, Q.; Han, H.; Wu, D. miRCancer: A microRNA-cancer association database constructed by text mining on literature. *Bioinformatics* **2013**, *29*, 638–644. [CrossRef]
78. Schuler, E.; Parris, T.Z.; Helou, K.; Forssell-Aronsson, E. Distinct microRNA expression profiles in mouse renal cortical tissue after 177Lu-octreotate administration. *PLoS ONE* **2014**, *9*, e112645. [CrossRef]
79. Zajic, G.; Nair, T.S.; Ptok, M.; Van Waes, C.; Altschuler, R.A.; Schacht, J.; Carey, T.E. Monoclonal antibodies to inner ear antigens: I. Antigens expressed by supporting cells of the guinea pig cochlea. *Hear. Res.* **1991**, *52*, 59–71. [CrossRef]
80. He, L.; Vasiliou, K.; Nebert, D.W. Analysis and update of the human solute carrier (SLC) gene superfamily. *Hum. Genom.* **2009**, *3*, 195–206. [CrossRef]
81. De Maio, A.; Yalamanchili, H.K.; Adamski, C.J.; Gennarino, V.A.; Liu, Z.; Qin, J.; Jung, S.Y.; Richman, R.; Orr, H.; Zoghbi, H.Y. RBM17 Interacts with U2SURP and CHERP to Regulate Expression and Splicing of RNA-Processing Proteins. *Cell Rep.* **2018**, *25*, 726–736.e7. [CrossRef]
82. Bramerson, A.; Nyman, J.; Nordin, S.; Bende, M. Olfactory loss after head and neck cancer radiation therapy. *Rhinology* **2013**, *51*, 206–209. [CrossRef] [PubMed]
83. Reiss, M.; Stash, E.B.; Vellucci, V.F.; Zhou, Z.L. Activation of the autocrine transforming growth factor alpha pathway in human squamous carcinoma cells. *Cancer Res.* **1991**, *51*, 6254–6262. [PubMed]
84. Nagy, A.; Lanczky, A.; Menyhart, O.; Gyorffy, B. Author Correction: Validation of miRNA prognostic power in hepatocellular carcinoma using expression data of independent datasets. *Sci. Rep.* **2018**, *8*, 11515. [CrossRef] [PubMed]
85. Chang, S.S.; Jiang, W.W.; Smith, I.; Poeta, L.M.; Begum, S.; Glazer, C.; Shan, S.; Westra, W.; Sidransky, D.; Califano, J.A. MicroRNA alterations in head and neck squamous cell carcinoma. *Int. J. Cancer* **2008**, *123*, 2791–2797. [CrossRef] [PubMed]
86. Sagar, S.M.; Thomas, R.J.; Loverock, L.T.; Spittle, M.F. Olfactory sensations produced by high-energy photon irradiation of the olfactory receptor mucosa in humans. *Int. J. Radiat. Oncol. Biol. Phys.* **1991**, *20*, 771–776. [CrossRef]
87. Hua, M.S.; Chen, S.T.; Tang, L.M.; Leung, W.M. Olfactory function in patients with nasopharyngeal carcinoma following radiotherapy. *Brain Inj.* **1999**, *13*, 905–915. [CrossRef] [PubMed]
88. Ruo Redda, M.G.; Allis, S. Radiotherapy-induced taste impairment. *Cancer Treat. Rev.* **2006**, *32*, 541–547. [CrossRef]
89. Epstein, J.B.; Smutzer, G.; Doty, R.L. Understanding the impact of taste changes in oncology care. *Support. Care Cancer* **2016**, *24*, 1917–1931. [CrossRef]
90. Carlsson, G.; Gullberg, B.; Hafström, L. Estimation of liver tumor volume using different formulas—An experimental study in rats. *J. Cancer Res. Clin. Oncol.* **1983**, *105*, 20–23. [CrossRef]
91. Andrews, S.; Krueger, F.; Segonds-Pichon, A.; Biggins, L.; Krueger, C.; Wingett, S.; Montgomery, J. Babraham Bioinformatics-Fastqc a Quality Control Tool For High Throughput Sequence Data. Available online: http://www.bioinformatics.babraham.ac.uk/projects/fastqc/ (accessed on 22 January 2019).
92. Creighton, C.J.; Reid, J.G.; Gunaratne, P.H. Expression profiling of microRNAs by deep sequencing. *Brief. Bioinform.* **2009**, *10*, 490–497. [CrossRef]

93. Ambros, V.; Bartel, B.; Bartel, D.P.; Burge, C.B.; Carrington, J.C.; Chen, X.; Dreyfuss, G.; Eddy, S.R.; Griffiths-Jones, S.; Marshall, M.; et al. A uniform system for microRNA annotation. *RNA* **2003**, *9*, 277–279. [CrossRef] [PubMed]
94. Robinson, M.D.; McCarthy, D.J.; Smyth, G.K. edgeR: A Bioconductor package for differential expression analysis of digital gene expression data. *Bioinformatics* **2010**, *26*, 139–140. [CrossRef] [PubMed]
95. Tsai, C.H.; Lin, L.T.; Wang, C.Y.; Chiu, Y.W.; Chou, Y.T.; Chiu, S.J.; Wang, H.E.; Liu, R.S.; Wu, C.Y.; Chan, P.C.; et al. Over-expression of cofilin-1 suppressed growth and invasion of cancer cells is associated with up-regulation of let-7 microRNA. *Biochim. Biophys. Acta* **2015**, *1852*, 851–861. [CrossRef] [PubMed]
96. Saldanha, A.J. Java Treeview-extensible visualization of microarray data. *Bioinformatics* **2004**, *20*, 3246–3248. [CrossRef]
97. Deng, M.; Bragelmann, J.; Kryukov, I.; Saraiva-Agostinho, N.; Perner, S. FirebrowseR: An R client to the Broad Institute's Firehose Pipeline. *Database* **2017**, *2017*, baw160. [CrossRef]
98. Wickham, H. *ggplot2: Elegant Graphics for Data Analysis*; Springer: New York, NY, USA, 2016.
99. Yang, Z.; Wu, L.; Wang, A.; Tang, W.; Zhao, Y.; Zhao, H.; Teschendorff, A.E. dbDEMC 2.0: Updated database of differentially expressed miRNAs in human cancers. *Nucleic. Acids Res.* **2017**, *45*, D812–D818. [CrossRef]
100. Chen, Y.; Wang, X. miRDB: An online database for prediction of functional microRNA targets. *Nucleic Acids Res.* **2020**, *48*, D127–D131. [CrossRef]
101. Liu, W.; Wang, X. Prediction of functional microRNA targets by integrative modeling of microRNA binding and target expression data. *Genome Biol.* **2019**, *20*, 18. [CrossRef]
102. Lieven Sterck, V.d.P.Y. Draw Venn Diagram-Bioinformatics.psb.ugent.be. Available online: http://bioinformatics.psb.ugent.be/webtools/Venn/ (accessed on 25 June 2020).
103. Kanehisa, M.; Sato, Y.; Kawashima, M.; Furumichi, M.; Tanabe, M. KEGG as a reference resource for gene and protein annotation. *Nucleic Acids Res.* **2016**, *44*, D457–D462. [CrossRef]
104. Kanehisa, M.; Sato, Y.; Morishima, K. BlastKOALA and GhostKOALA: KEGG Tools for Functional Characterization of Genome and Metagenome Sequences. *J. Mol. Biol.* **2016**, *428*, 726–731. [CrossRef]

Sample Availability: Samples of the compounds BMEDA and liposomes are available from the authors.

Review

Pyrazoles as Key Scaffolds for the Development of Fluorine-18-Labeled Radiotracers for Positron Emission Tomography (PET)

Pedro M. O. Gomes, Artur M. S. Silva and Vera L. M. Silva *

LAQV-REQUIMTE, Department of Chemistry, University of Aveiro, 3810-193 Aveiro, Portugal; pm.gomes@ua.pt (P.M.O.G.); artur.silva@ua.pt (A.M.S.S.)

* Correspondence: verasilva@ua.pt; Tel.: +351-234-370714

Academic Editor: Krishan Kumar

Received: 15 March 2020; Accepted: 8 April 2020; Published: 9 April 2020

Abstract: The need for increasingly personalized medicine solutions (precision medicine) and quality medical treatments, has led to a growing demand and research for image-guided therapeutic solutions. Positron emission tomography (PET) is a powerful imaging technique that can be established using complementary imaging systems and selective imaging agents—chemical probes or radiotracers—which are drugs labeled with a radionuclide, also called radiopharmaceuticals. PET has two complementary purposes: selective imaging for diagnosis and monitoring of disease progression and response to treatment. The development of selective imaging agents is a growing research area, with a high number of diverse drugs, labeled with different radionuclides, being reported nowadays. This review article is focused on the use of pyrazoles as suitable scaffolds for the development of ^{18}F-labeled radiotracers for PET imaging. A brief introduction to PET and pyrazoles, as key scaffolds in medicinal chemistry, is presented, followed by a description of the most important [^{18}F]pyrazole-derived radiotracers (PET tracers) that have been developed in the last 20 years for selective PET imaging, grouped according to their specific targets.

Keywords: positron emission tomography (PET); pyrazoles; fluorine-18; radionuclides; PET probes; imaging pharmaceuticals

1. Introduction

Positron emission tomography (PET) is a nuclear medicine functional imaging technique that uses gamma rays (formed as a result of the annihilation of the emitted positrons) to provide three-dimensional images that give information about the functioning of a person's specific organs. PET is based on the detection of tiny amounts (picomolar) of a biological substance labeled with a short-lived positron-emitting radionuclide (PET tracer) without disturbing the biological system and has the advantage of being a noninvasive, functional and highly sensitive technique which provides a great wealth of information [1]. The PET tracers are drugs or biomolecules labeled with radionuclides such as ^{11}C, ^{13}N, ^{15}O, ^{18}F, ^{64}Cu, ^{68}Ga, ^{89}Zr and ^{124}I. Since radionuclides can replace the stable analogues, the PET probes have the same chemical structure as the drugs and biomolecules without altering their biological activity. The choice of the PET radionuclide must follow some criteria: (i) physical and chemical characteristics, (ii) availability and (iii) timescale of the biological process in the study. For different biological processes, some radionuclides are better than others. If the biological process in the study gives results hours or days after the injection of PET probe, ^{64}Cu, ^{89}Zr or ^{124}I can be used, because these three radionuclides have a long half-life time (12.8 h, 78.4 h and 4.17 days, respectively). On the contrary, radionuclides as ^{11}C or ^{18}F are used for labeling small organic compounds for faster biological processes [1]. To avoid unnecessary radiation, it is important to choose a radionuclide with a half-life that enable the reaction of the radionuclide with the carrier molecule and matches the timescale of the biological process in the study. ^{18}F is the ideal radionuclide for routine PET imaging, because it has a short half-life (109.8 min) but enough to allow all the process of synthesis, transport and imaging. Furthermore, it is a 97% positron-emitter, and the low positron energy of ^{18}F leads to a high resolution of PET imaging [1,2].

In the last three decades, the interest in PET has been growing, and nowadays, this technique is used in several areas of medicine, such as oncology [3], cardiology [4], neurology [5] and also in drug development and evaluation [6]. Besides the aforementioned advantages of PET, there are also some limitations; pregnant women should not undergo in PET imaging; the radioactivity of PET tracers limits the number of times one patient can undergo PET scans and is a very expensive treatment [7]. Moreover, the potential of PET strongly depends on the availability of suitable PET radiotracers. For instance, 2-[^{18}F]fluoro-2-deoxyglucose ([^{18}F]FDG) is one of the most used radiotracers for the clinical diagnosis and evaluation of cancer. However, it has inherent drawbacks, namely its high accumulation in inflamed and infected tissues giving false-positive results and its low uptake in tumors that grow slowly, which can lead to false-negative results [8]. Finally, the development of new PET-imaging probes is not trivial, and radiochemistry is a major limiting factor for this field. Hence, the research and development of new PET tracers is a challenging issue of major importance within the scientific community.

Following our interest in the synthesis and biological evaluation of pyrazole-type compounds envisaging their use in medicine [9–14] and considering their huge potential for application in PET field, herein we present an overview of pyrazole-derived PET radiotracers developed in the last 20 years. Pyrazoles labeled with ^{18}F radionuclide will be the focus of this review article, given the aforementioned advantages of this radionuclide for routine PET imaging.

2. Pyrazoles

This family of compounds is characterized by a simple heteroaromatic five-membered ring containing three carbon and two nitrogen atoms in adjacent positions [15]. The occurrence of the pyrazole core in nature is rare, probably due to the difficulty of living organisms to make a N-N bound [16]. Until now, there are only approximately 20 natural compounds with a pyrazole core on their structures [17]. Although scarce in nature, pyrazole (1*H*-pyrazole (**1**), Figure 1) and its reduced derivatives (pyrazolines **2–4**, Figure 1) are considered privileged structures in medicinal chemistry. Pyrazole scaffold is present in several synthetic drugs, being celecoxib (Celebrex®), sildenafil (Viagra®), rimonabant, fomepizole, penthiopyrad and sulfaphenazole the most notorious [15].

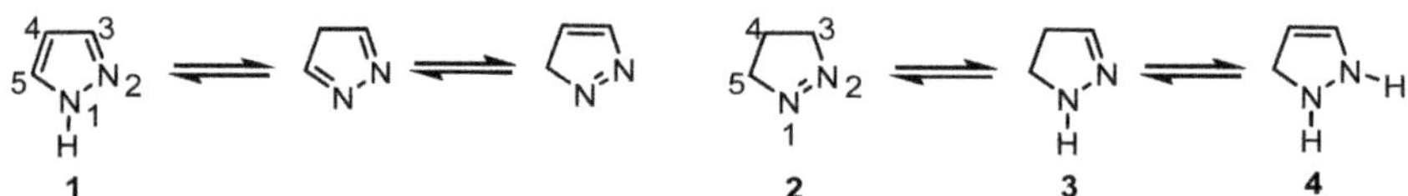

Figure 1. Chemical structures and numbering of pyrazole (**1**) and dihydropyrazole (pyrazoline) tautomers **2–4** [18].

Pyrazole derivatives act on diverse and relevant biological targets, which make them attractive for the development of PET tracers, and possess a wide range of pharmaceutical activities. Abdel-Maksoud et al. have demonstrated antitumoral activity of several compounds with a pyrazole and sulfonamide moieties [19]; antibacterial and antifungal activity of pyrazole was demonstrated by Chowdary et al. [20]. Pyrazoles also showed good activity as monoamine oxidase inhibitors, antidepressant and anticonvulsant agents [21], as BRAF inhibitors [22] and DNA gyrase inhibitors [23] and present antileishmanial, anti-inflammatory, analgesic, antidiabetic and cannabinoid activity; cyclin-dependent kinase and tissue-nonspecific alkaline phosphatase inhibitory activity and moderate antihepatotoxic activity, among others [9,11,24–29]. In 2013, Prabhu et. al. reported the antioxidant activity of pyrazole [30]. They demonstrated that pyrazole (1,2-diazole) can be used to prevent nephrotoxicity caused by cisplatin, a drug used to treat several cancers. Cisplatin provokes renal damage because of its toxicity to proximal tubule cells and can reduce glomerular filtration, resulting in renal failure. One of the reasons of nephrotoxicity induced by cisplatin is the decreasing concentration of glutathione (GSH). Pyrazole prevents nephrotoxicity induced by cisplatin by counteracting this effect, increasing the concentration of this enzyme [30]. Recently, Silva et al. reported the antioxidant activity of around three hundred pyrazoles [18].

2.1. *Pyrazoles as Probes for PET Imaging*

The labeling of pyrazole-type compounds with ^{18}F for PET imaging is a research issue that has been growing significantly, as evidenced by the number of papers and citations in the last years, according to our search on Web of Science using the keywords (pyrazole) and (positron emission tomography) and (^{18}F) (Figure 2). In this section are described the most important [^{18}F]-labeled pyrazoles developed for use as PET radiotracers. The compounds are presented according to their specific targets, which are indicated by alphabetical order and not by their relevance in the PET field. Whenever possible, the most promising applications of these compounds in this field are discussed.

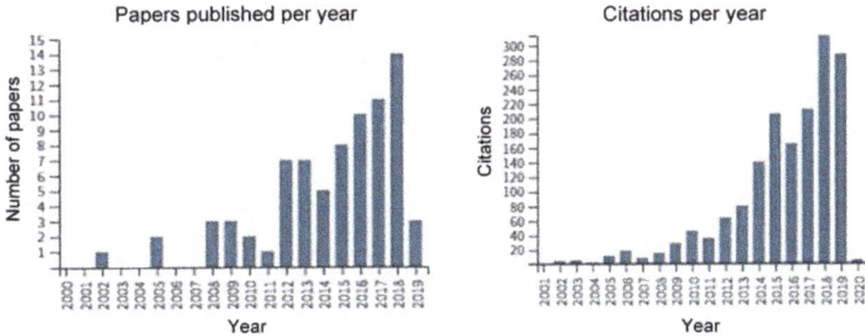

Figure 2. Number of papers published and citations per year found using the keywords "pyrazoles" and "positron emission tomography" and "^{18}F" in Web of Science in the period 2000–2020.

2.1.1. Adenosine Receptors Ligands

Adenosine, an endogenous-signaling substance, is a purine ribonucleoside composed of adenine (purine base) and ribose (sugar molecule), which is produced in response to metabolic stress and cell damage. It induces several physiopathological effects, regulating the central nervous, cardiovascular, peripheral, and immune systems due to the rapid generation of adenosine from cellular metabolism and the widespread distribution of its receptor subtypes in almost all organs and tissues [31]. There are four types of adenosine receptors: A_1, A_{2A}, A_{2B} and A_3. A_{2A}R are abundant in dopamine-rich regions of the brain, being the striatum the region with a higher concentration of A_{2A}R [31]. In preclinical studies, A_{2A}R antagonists showed potential benefits in the treatment of some neurodegenerative diseases like Alzheimer's disease (AD) and Parkinson's disease (PD), neuroinflammation, ischemia, spinal cord injury, drug addiction and other conditions [31]. Khanapur et al. developed a pyrazolo[4,3-*e*]-1,2,4-triazolo[1,5-*c*]pyrimidine labeled with carbon-11, [^{11}C]SCH442416 (**5**) and its [^{18}F]fluoroethyl and [^{18}F]fluoropropyl derivatives ([^{18}F]FESCH (**6**) and [^{18}F]FPSCH (**7**), respectively) (Figure 3), and both radioligands showed a distribution in the rat brain corresponding to the regional A_{2A}R densities, as evidenced by in vitro autoradiography (ARG) experiments and binding assays [32]. These two tracers showed a similar ratio specific/nonspecific binding, using the striatum as the specific binding and cerebellum as the nonspecific binding (4.6 at 25 and 37 min after injection, respectively) and reversible binding in rat brains, although their kinetics were slightly different [32]. Recently, the same authors studied the full kinetics of radioligands **6** and **7** in rat brains. On the basis of the Akaike information criterion, they have found that 1TCM was the most appropriate model for describing [^{18}F]FPSCH kinetics, whereas 2TCM was the most suitable model for [^{18}F]FESCH kinetics. Using dynamic PET imaging, under baseline and full blocking conditions, they proved that **6** is the most suitable PET radioligand for quantifying $A_{2A}R$ receptor expression in the rat brain. However, before starting the clinical use of **6**, it will be necessary to reevaluate the brain uptake in humans due to possible interspecies differences in tracer kinetics and metabolisms [33].

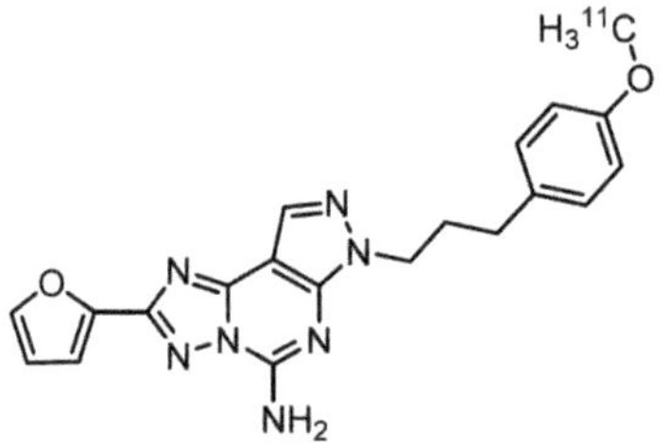

[^{11}C]SCH442416
5

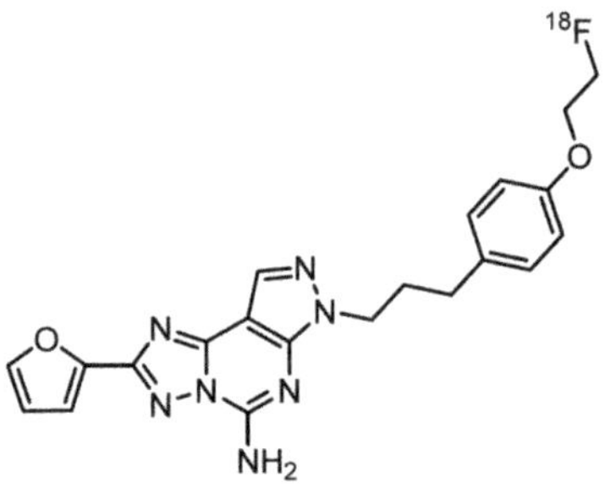

[^{18}F]FESCH
6

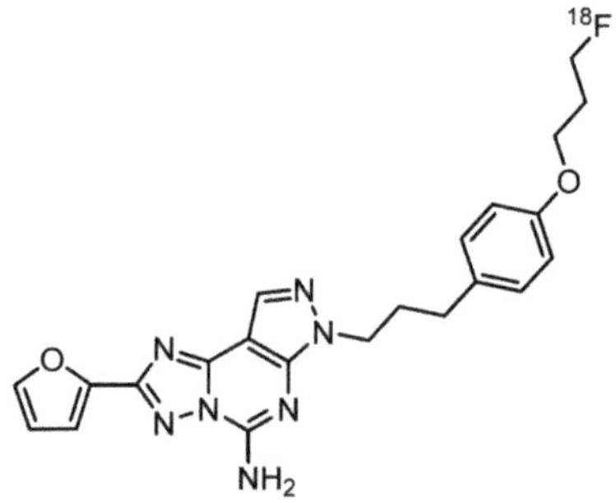

[^{18}F]FPSCH
7

Figure 3. Adenosine receptor ($A_{2A}R$) radioantagonists **5–7** for PET imaging [32,33].

2.1.2. Cannabinoid Receptors Ligands

Products derived from *Cannabis sativa* are some of the oldest and widely used drugs in the world. These products are known as natural cannabinoids, but several synthetic cannabinoids have been developed as well. Cannabinoids have been used as analgesics for more than 100 years [34]. Additionally, they have been used as antiemetic agents to prevent chemotherapy-induced nausea and vomiting, because they can bind to opiate receptors in the forebrain, blocking the vomiting center in the medulla [35]. In 1998, Williams and Kirkham have demonstrated that anandamide, an endogenous cannabinoid, provokes hyperphagia in satiated rats [36]. Cannabinoid-type compounds bind to cannabinoid receptors, which can be divided in two groups—cannabinoid receptor type 1 (CB_1), predominantly found in the brain, and the peripheral cannabinoid receptor type 2 (CB_2), mainly expressed in immune tissues, and both are G-protein-coupled membrane receptors [37].

The imaging of the CB_1 receptor is of great importance for studying its role in neuropsychiatric disorders, including depression, and in obesity, drug or alcohol addiction, and is an active target for in vivo imaging development [38]. The first selective CB_1 receptor antagonist was

[*N*-(piperidin-1-yl)-5-(4-chlorophenyl)-1-(2,4-dichlorophenyl)-4-methyl-1*H*-pyrazole-3-carboxamide] SR141716 (**8**), known as rimonabant (Figure 4) [39]. This compound was approved in Europe, in 2006, to treat obesity by reducing the patient's appetite. Two years later, the European Medicines Agency (EMA) withdrew rimonabant from sale due to its evident secondary effects. Some studies demonstrated that SR141716 induced anxiety, depression, agitation, eating disorders, irritability, aggression and insomnia. Rimonabant was not approved in the USA by the Food and Drug Administration (FDA) [40]. However, some analogs of SR141716 (**8**) have been labeled with radionuclides for PET imaging. [^{18}F]SR144385 (**9**) and [^{18}F]SR147963 (**10**) [37,41], [^{18}F]NIDA-42033 (**11**) and its related ester derivative (**12**) [42], [^{18}F]O-1302 (**13**) [43] and [^{18}F]AM5144 (**14**) [44] are some examples of PET tracers labeled with ^{18}F (Figure 4), while [^{123}I]AM251 (**15**) and [^{123}I]AM281 (**16**) [44] are two examples of PET tracers labeled with ^{123}I (Figure 5).

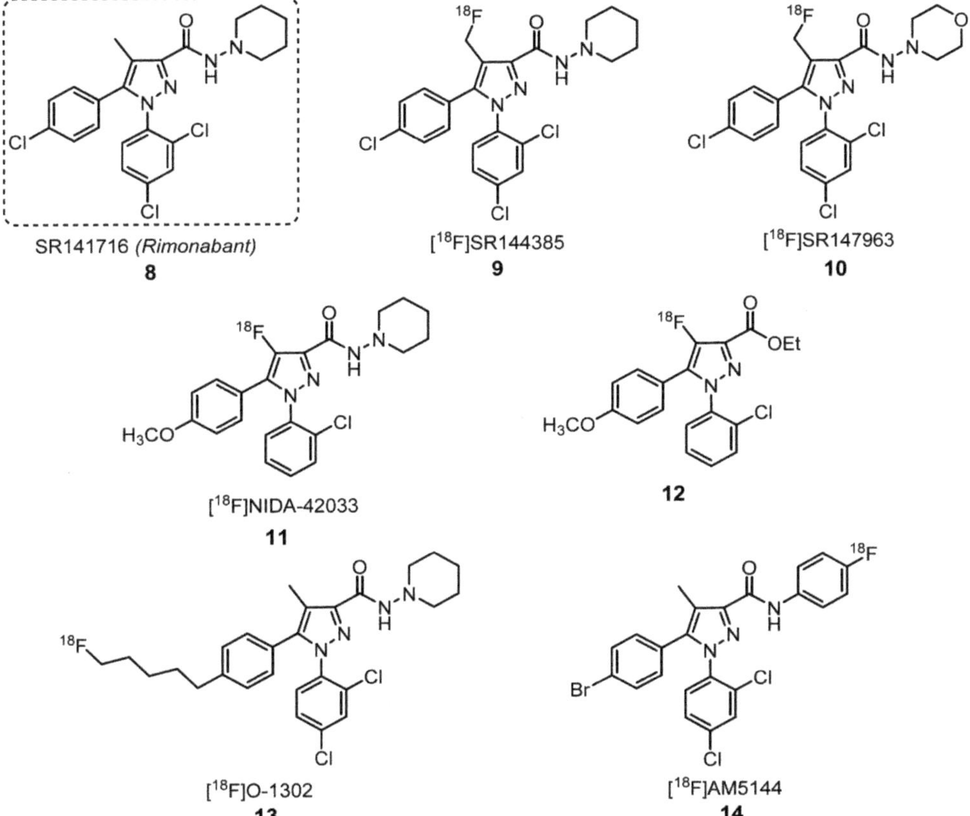

Figure 4. Structure of rimonabant (**8**) and structurally related [^{18}F]pyrazole PET tracers **9–14** [37,39,41–44].

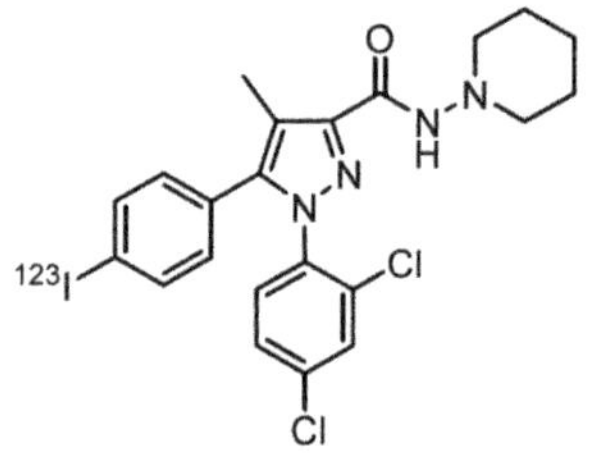

Figure 5. Pyrazole-derived PET tracers **15** and **16** labeled with ^{123}I [44].

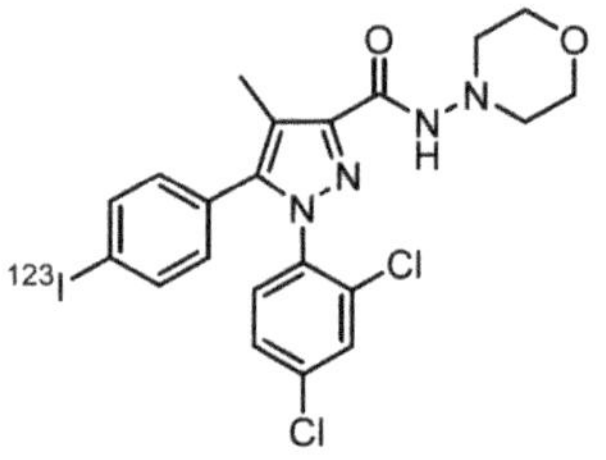

Studies were performed to evaluate the specificity of cannabinoids [^{18}F]SR144385 (**9**) and [^{18}F]SR147963 (**10**) (Figure 4) for CB_1 receptors [37]. After 15 min of injection in male CD-1 mice (25–30 g), compound **10** showed a higher brain uptake compared to compound **9** (5.70% ID/g and 3.06% ID/g, respectively). The hippocampus was the brain region that exhibited the highest uptake of both tracers, followed by the striatum, cerebellum, frontal cortex, cortex, olfactory tubercles and hypothalamus. Both the brain stem and thalamus showed low uptakes of the tracers, and the thalamus showed the fastest decrease of %ID/g. These results are in accordance with the knowledge of CB_1 receptors' brain density in rats, which was found to be lower in the thalamus region. For evaluation of the target to nontarget ratio, the thalamus was used as indicator of nonspecific binding. Compounds **9** and **10** showed differences in the post-injection time they reached the maximum target to nontarget ratio (90 min for compound **9** with a ratio of 2.50 and 60 min for compound **10** with a ratio of 1.69) [37]. In vivo selectivity and specificity studies of [^{18}F]SR144385 (**9**) showed that significant blocking of this tracer uptake was achieved when a 1-mg/kg dose of the structurally similar blocking agent SR141716 (**8**) was given 30 min prior to the radiotracer. Likewise, the uptake of [^{18}F]SR147963 (**10**) at either 30 or 60 min post-injection was also blocked by a 1-mg/kg dose of SR141716 (**8**) given 30 min before the radiotracer.

In 2005, Li et al. synthetized the radioactive pyrazole [^{18}F]AM5144 (**14**) (Figure 4) as a PET radioligand candidate for cannabinoid CB_1 receptors. Using baboons, they demonstrated that the highest radioactivity concentration of compound **14** was found on the cerebellum, and the lowest concentration was found in the thalamus. They concluded that there was a poor brain uptake of **14** and of other related [^{18}F]pyrazoles because of their high lipophilicity, although this property is not the key factor in brain uptake. This study proved that there is no close relationship between the clogP value and brain uptake, indicating that there are other uptake key factors [44]. Despite the high in vitro-binding affinity and moderate lipophilicity of [^{18}F]O-1302 (**13**) (Figure 4), it is not suitable for imaging the CB_1 receptor in the brain due to poor brain entry and high levels of nonspecific binding [43]. Although radiotracer uptake in the brain is often considered a function of the log P that peaks between a log P of 2 and 3, the investigation of the factors that affect brain uptake beyond lipophilicity is crucial to a better comprehension of this process. Some studies highlighted that there may be large species differences in brain penetration for a given PET radiotracer; for example, the brain uptake of [^{123}I]AM251 (**15**) (Figure 5) is very different comparing mice and monkeys, being higher for mice [44,45].

Chang et al. observed that [^{18}F]DBPR211 (**17**) (Figure 6) was distributed in the brain, liver, heart, thigh muscle and kidney when intravenously injected in mice. In the brain, it was found in a very low percentage (less than 1%) over 90-min scans. This result is evidence of its action as a peripherally restricted CB_1 antagonist [46]. Taking into account these observations, this radiotracer can be seen as a promising tool to study metabolic processes associated with peripheral CB_1 receptors.

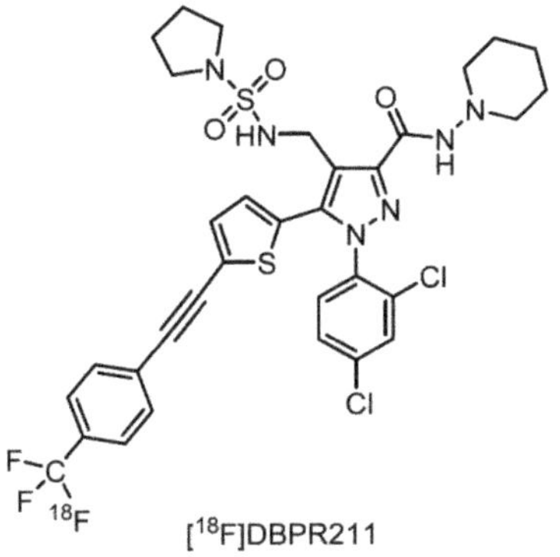

[18F]DBPR211
17

Figure 6. [^{18}F]pyrazole-derived peripheral CB_1 antagonist **17** [46].

2.1.3. Cyclooxygenase Inhibitors

Cyclooxygenase-2 (COX-2) is an enzyme with high-level expressions at sites of inflammation and cancer and is also a promising target for neuroinflammation imaging [47,48]. The development of agents capable of monitoring COX-2 levels is highly desirable for cancer prevention and therapy. To achieve this goal, Uddin et al. synthetized several indomethacin and celecoxib derivatives and evaluated their IC_{50} values for the inhibition of COX-2 in vivo and in intact cells [49]. Indomethacin derivatives were effective COX-2 inhibitors in intact cells (0.09–0.26 μmol/L), but the synthesis of the respective radiotracer has some drawbacks, because the *p*-chlorobenzoyl group is not stable in the conditions of radiochemical synthesis. Among celecoxib derivatives, the more effective COX-2 inhibitor was compound **18** with an IC_{50} value of 0.16 μmol/L. After the synthesis of ^{18}F-**18** (Figure 7), in vivo biodistribution was studied in the inflamed and contralateral footpad of male Sprague Dawley rats. The major advantage of this inflammation model is the possibility to image the inflamed footpad in comparison with the noninflamed contralateral footpad. Compound **18** had a higher accumulation in the inflamed footpad in comparison with the noninflamed. The selective binding between **18** and COX-2 was proved with an assay using celecoxib to block COX-2 before the injection of **18**. To confirm the COX-2 specificity of **18**, COX-2-null mice were injected with carrageenan to promote inflammation. COX-2-null feet demonstrated a 1.08 ± 0.09 ratio of inflamed/noninflamed feet. On other hand, the wild-type mice group, used as the control, had an uptake ratio of 1.48 ± 0.04. The ability of **18** to target COX-2 in human tumor xenografts was also demonstrated using 1483 HNSCC cells and HCT116 cells injected in the left and right hip of female nude mice. In the COX-2-null HCT116 tumor cells, the uptake of **18** was minimal, while, in the 1483 HNSCC cells, a significative uptake of the radiotracer was observed. Once again, to prove that the difference in the uptake of **18** was related with the expression of COX-2, mice bearing 1483 xenografts were pretreated with celecoxib to block the active site of COX-2. There was a significative lower uptake of **18** for mice pretreated with celecoxib when compared with the control group [49].

Lebedev et al. developed a new, high-affinity ^{18}F-COX-2 inhibitor **19** (Figure 7) that is radiolabeled directly on a heteroaromatic ring with the purpose to increase biodistribution and metabolic stability [50]. In vitro studies demonstrated a clear correlation between COX-2 expression and uptake of this radiotracer. Moreover, pharmacokinetic studies in healthy mice revealed no bone retention or defluorination within 2 h of injection, significant blood clearance, since the molecule is excreted from blood within an hour mainly through the hepatobiliary excretion pathway, crossing of blood-brain barrier (BBB) and no significant metabolites in major organs. Although these properties make this probe ideal for PET imaging, some aspects related to radiochemical synthesis severely limit the application of this compound as a probe. In fact, the use of $Et_4NF^{+}4HF$ has limited the specific activity to 3 Ci/mmol, and the current decay-corrected radiochemical yield of 2%, although enough for preclinical studies and the production of a single patient dose, need further improvement to achieve the successful use of this compound as a PET tracer.

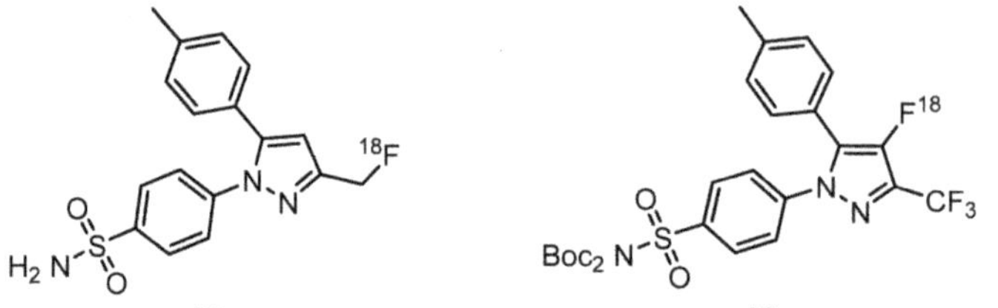

Figure 7. [^{18}F]-labeled COX-2 inhibitors **18** and **19** [49,50].

Aiming to develop selective probes for the two COX enzyme subtypes, COX-1 and COX-2, McCharty et al. have prepared [^{18}F]SC63217 (**20**) and [^{18}F]SC58125 (**21**) (Figure 8), starting from the corresponding nonradioactive COX-1 and COX-2 selective inhibitors, SC63217 and SC58125, respectively [51]. Both compounds are structurally similar, presenting only a different substituent on only one aromatic ring. In vitro binding studies of both compounds, using J774 macrophages, revealed that compound **20** is a potential probe for COX-2, while **21** was not a good marker of COX-1 due to high nonspecific binding. In vivo studies showed that, for each tracer, rat biodistribution was well-matched with the known distribution of these enzymes [51].

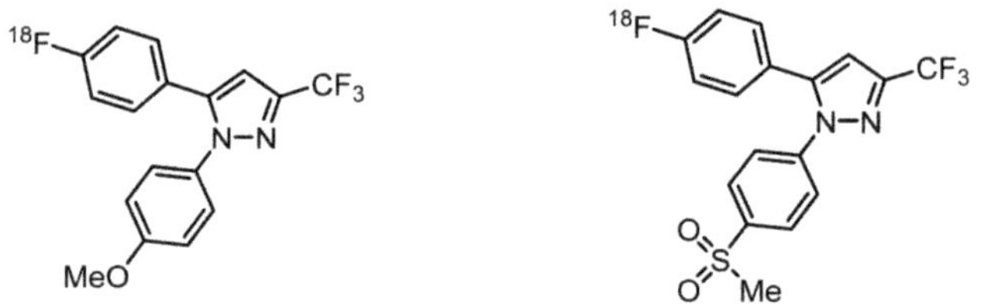

Figure 8. [^{18}F]-labeled pyrazoles as probes for COX-1 **20** and COX-2 **21** [51].

2.1.4. Dopamine Receptors Ligands

The subfamily of D_2-like dopamine receptors includes the D_2-, D_3- and D_4-receptor subtypes and mediates the action of dopamine in the brain by the inhibition of adenylate cyclase activity. While the distribution of D_2- and D_3- receptors is well-known, there are still some uncertainties regarding D_4-receptor (D_4R) expression. If D_4R selective radioligands are available, PET can be suited to gain deeper knowledge into the distribution and pathophysiological role of D_4R in humans.

In 2008, the pyrazolo[1,5-*a*]pyridine derivatives, FAUC 113 (**22**) and FAUC 213 (**23**) (Figure 9), bearing a (4-chlorophenyl)piperazinylmethyl moiety in the 3 and 2 positions of the pyrazolo[1,5-*a*]pyridine core, were synthetized. After an evaluation of their binding affinities for D_4-like dopamine receptors, **22** and **23** were chosen as lead compounds for the development of [^{18}F]-labeled PET tracers. A novel compound with a fluoroethoxyphenyl substituent in the *para*-position of the phenylpiperazinyl moiety showed the highest specificity to the D_4 receptor. Due to these results, the next step was radiochemical synthesis to prepare the same labeled derivative, [^{18}F]FAUC F41 (**24**) (Figure 9). Using coronal rat brains, in vivo AGR studies were performed to evaluate the specific binding of **24** to D_4R. An increased uptake of this compound was detected in the hippocampus, hypothalamus, cortex, medial habenular nucleus and central medial thalamic nucleus. The observed binding pattern was mainly consistent with the known D_4R distribution in the rat brains. The log P value for this compound was found to be 2.9, which may indicate adequate BBB penetration. Moreover, this ligand revealed high stability in human serum, even after long-term incubation for up to 90 min [52]. These results demonstrate that [^{18}F]FAUC F41 (**24**) represents a potential radioligand for studying the D_4R in vivo by PET imaging.

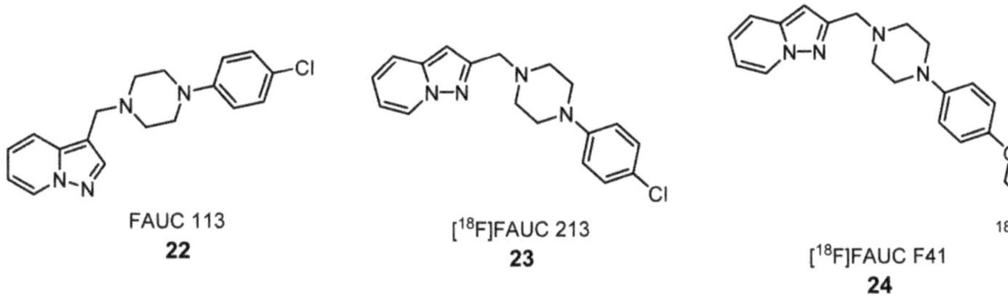

FAUC 113
22

[^{18}F]FAUC 213
23

[^{18}F]FAUC F41
24

Figure 9. [^{18}F]-labeled high-affinity dopamine receptor (D_4R) tracers **22–24** [52].

With the aim to develop a radiotracer with high selectivity and favorable lipophilicity for imaging of the D_3-receptor in the brain, Stößel et al. have synthetized compound **26** as the radioactive [^{18}F]-labeled analogue of the D_3 ligand FAUC 329 (**25**) (Figure 10) [53]. In vitro AGR studies showed that **26** successfully visualized D_3-rich brain regions, including the islands of Calleja. However, in vivo PET imaging revealed that it does not significantly accumulate in the CNS structures of rat brains, probably due to a low BBB penetration. Instead, a significant uptake occurred in the brain ventricular system, due to a significant penetration of this compound in the blood-liquor barrier and, more noticeable, in the pituitary gland, outside the BBB [53]. The results of PET studies also suggest that the main mode of action of FAUC 329, in vivo, could be due to binding to the dopamine receptors in the pituitary gland.

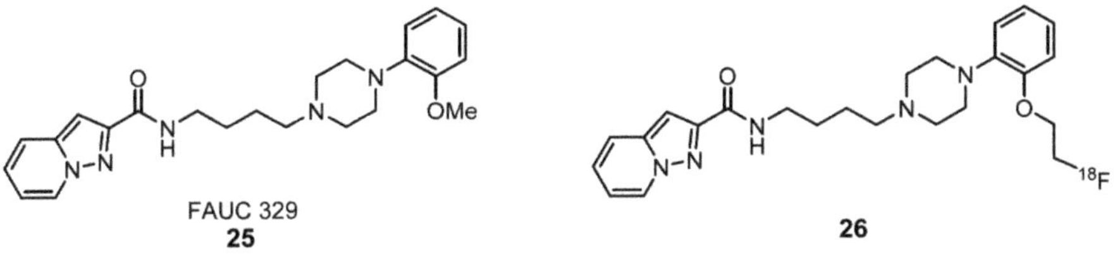

FAUC 329
25

26

Figure 10. Radiotracer **26** derived from the dopamine D_3 receptor ligand FAUC 329 (**25**) [53].

2.1.5. Glucocorticoid Receptor Ligands

In 1998, Hoyte et al. reported a series of aryl-pyrazolo steroids similar to the potent glucocorticoid cortivazol **27** and evaluated the affinity of these compounds for glucocorticoid receptors (GRs). Among them, the fluoro analog **28,** which showed good binding and was a very potent glucocorticoid, was labeled with ^{18}F to be used as a glucocorticoid receptor-mediated imaging agent for PET (Figure 11) [54].

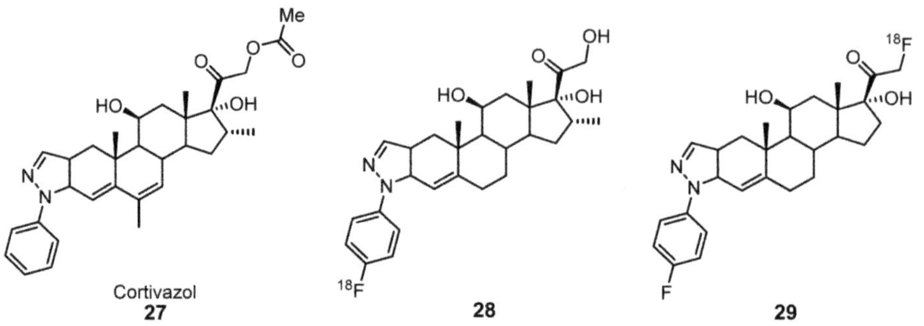

Cortivazol
27

28

29

Figure 11. Cortivazol (**27**) and related [^{18}F]arylpyrazolo steroids **28–29** as potential glucocorticoid receptor ligands for PET imaging [54,55].

Later, Würst et al. used the 2′-(4-fluorophenyl)-21-[^{18}F]fluoro-20-oxo-11β,17α-dihydroxypregn-4-eno[3,2-*c*]pyrazole (**29**) (Figure 11) as a ligand for studying brain GRs. Biodistribution and radiopharmacological studies with male Wistar rats revealed promising brain uptakes and low in vivo radiodefluorinations in comparison with other PET radioligands for brain GRs [55].

2.1.6. Insulin-Like Growth Factor-1 Receptor Ligands

Insulin-like growth factor-1 receptor (IGF-1R) is a potential therapeutic target, because it is overexpressed in many cancers, AD, traumatic brain injury, amyotrophic lateral sclerosis (ALS), Friedreich ataxia and aging. In 2013, Majo et al. synthetized and evaluated in vitro [^{18}F]BMS-754807 (**30**) (Figure 12), a potent and reversible small molecule inhibitor of the IGF-1R/IR kinases' family, currently in phase II clinical trials. ARG studies using surgically removed and pathologically identified grade IV glioblastoma, breast cancer and pancreatic tumors demonstrate that **30** binds to IGF-1R. The selectivity over other kinases, the presence of metabolically stable fluorine in the 2-substituted pyridine ring, which is amenable for radiolabeling by nucleophilic displacement with [^{18}F]fluoride and a calculated lipophilicity (clogP) of 3.5, make this ligand a potential PET-imaging agent for in vivo monitoring of IGF-1R [56].

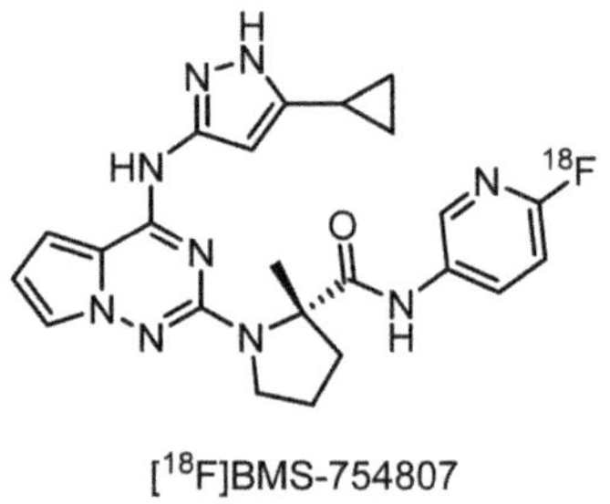

[^{18}F]BMS-754807
30

Figure 12. Potential PET radiotracer **30** for IGF-IR imaging [56].

2.1.7. Phosphodiester-10A Enzyme Inhibitors

Phosphodiester-10A (PDE10A) is an enzyme that hydrolyzes adenosine and/or guanosine 3′,5′-cyclic monophosphates (cAMP and cGMP, respectively). In the medium spiny neurons of the striatum, PDE10A messenger RNA and the corresponding protein are highly abundant. PDE10A inhibitors are a potential target for the diagnosis of schizophrenia, Huntington's disease, PD, obsessive-compulsive disorder and addiction.

In 2010, Tu et al. made a first attempt to visualize PDE10A using a ^{11}C-radiolabeled PDE10A inhibitor named papaverine (**31**) (Figure 13). In vitro ^{11}C-papaverine showed selective PDE10A binding, but in vivo failed because of rapid washout of the tracer [57]. To overcome this problem, Celen et al. synthetized a specific and selective radioligand for PDE10A, the ^{18}F-quinoline-labeled [^{18}F]JNJ-41510417 (**32**) (Figure 13). They used rats and PDE10A knockout mice to show that **32** binds specifically and reversibly to PDE10A in the striatum, presenting high accumulation therein. This brain region was the only to show an increase of tracer concentration after the injection (1.6 SUV after 2 min vs. 2.6 SUV after 60 min). Other brain regions, the hippocampus, cortex and cerebellum showed a washout of the radiotracer. These results are in accordance with the distribution of PDE10A. Despite the [^{18}F]JNJ-41510417 (**32**) good target specificity and signal-to-noise characteristics, slow brain kinetics due to its high potency is a limitation [58].

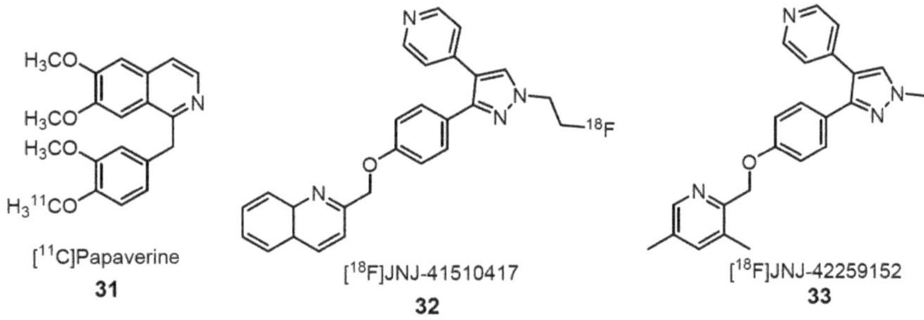

Figure 13. Radiotracers **31–33** for phosphodiester-10A (PDE10A) visualization in the brain by PET imaging [57–62].

Then, Laere et al., in collaboration with Janssen Pharmaceuticals, synthetized and studied the human biodistribution of another radioligand [^{18}F]JNJ-42259152 (**33**) [2-(4-(1-(2[^{18}F]fluoroethyl)-4-(4-pyridinyl)-1*H*-pyrazol-3-yl)phenoxy)methyl)-3,5-dimethylpyridine] (Figure 13), which belongs to a series of structurally related analogues of MP10 in which the 2-quinolinyl heterocycle was replaced with substituted 2-pyridinyl moieties, resulting in a slightly lower potency compared to [^{18}F]JNJ-41510417 (**32**) [59,60]. Six healthy male Caucasians (23–69 years; 3 younger than 40 years and 3 older than 60 years) have been volunteers in the study managed by Laere. Compound **33** showed a rapid uptake in the striatum, a few minutes after the injection, with a high clearance rate consistent with specific binding in this target region. The results of this pilot study are in accordance with the distribution of PDE10A in the human brain and show a promising kinetics and biodistribution of [^{18}F]JNJ-42259152 (**33**) [60].

Laere et al. have demonstrated that PDE10A activity in the brain can be reliably quantified using [^{18}F]JNJ-42259152 (**33**) [61]. A relatively fast kinetics for the striatal region was observed, followed by a subsequent moderately fast washout. The regional in vivo distribution of this radiotracer was in agreement with the known distribution of PDE10A, being found predominantly in the putamen followed by the caudate nucleus, ventral striatum and substantia nigra. Compared with the activity in the striatum, the cortical and cerebellar activity were more than 10-fold lower. Later, Ooms et al. have investigated the effect of alterations in cAMP levels on [^{18}F]JNJ-42259152 binding to PDE10A in the striatum homogenates. Increased affinity of this radiotracer for PDE10A was observed in the presence of cAMP, which seems to have an important role in the allosteric regulation of PDE10A [62].

Stepanov et al. described the synthesis of two [^{18}F]-labeled PET radioligands to target PDE10A, the [^{18}F]FM-T-773-d_2 (**34**) and [^{18}F]FE-T-773-d_4 (**35**) (Figure 14), and their in vivo evaluation in nonhuman primates [63]. High brain uptake was measured for both radioligands and a fast washout. Specific binding reached the maximum after 30 min for [^{18}F]FM-T-773-d_2 (**34**) and after 45 min for

[^{18}F]FE-T-773-d_4 (**35**). On account of brain uptake specific binding and kinetics, [^{18}F]FM-T-773-d_2 (**34**) was considered the more promising PET radioligand for further clinical evaluation.

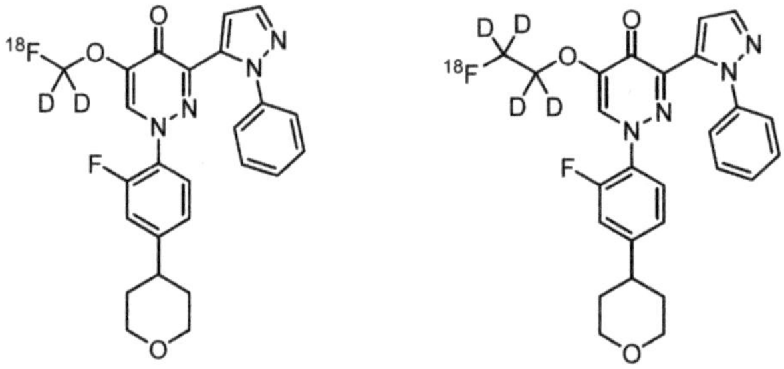

[^{18}F]FM-T-773-d_2
34

[^{18}F]FE-T-773-d_4
35

Figure 14. PET radiotracers **34** and **35** for in vivo phosphodiester-10A (PDE10A) evaluation in nonhuman primates' brains [63].

2.1.8. Translocator Protein Receptor Ligands

The translocator protein (TSPO) receptor is an 18 kDa protein organized around five large transmembrane α-helices and located on the mitochondrial outer membrane [2]. This protein has a key role in the regulation of several cellular processes: steroid biosynthesis, cholesterol metabolism, apoptosis and cellular metabolism [64]. TSPO is highly expressed in organs involved in steroid synthesis as adrenal glands, testis, ovaries and pituitary glands [65]. In the central nervous system (CNS) and liver, TSPO expression is modest. However, in the case of acute or neurodegenerative pathologies associated with microglia or astrocytes, levels of TSPO in the brain are upregulated. The upregulation of this protein is directly correlated with the degree of neuronal damage. For these reasons, TSPO is considered a very promising target for the early imaging of neuroinflammation [66] and a possible indirect marker of neuronal loss progression, multiple sclerosis and AD [66–68] and has high relevance in neuroscience. The expression of this protein is also elevated in several cancers: colon, breast, glioma, prostate, colorectal, liver and ovary cancer, relating TSPO with disease progression and survival [2,64,65,69]. These evidences increased the interest in TSPO and led to the development of several radiolabeled ligands for the evaluation of the expression of this protein and detection of some of the aforementioned diseases by PET imaging.

Since the discovery of the first nonbenzodiazepine ligand for TSPO, the isoquinoline carboxamide PK11195 (**36**) [70–72], several families of compounds were evaluated as TSPO ligands—among them, [^{11}C]DPA-713 (**37**), [^{18}F]DPA-714 (**38**), [^{18}F]DPA (**39**), [^{18}F]VUIIS1008 (**40**) and [^{18}F]DPA-716 (or [^{18}F]PBR146) (**41**), which are pyrazolo[1,5-*a*]pyrimidine acetamides (Figure 15) [72,73].

Chauveau et al. compared compound **38** with **36** and **37** using a rat model of acute neuroinflammation. (*R,S*)-α-amino-3-hydroxy-5-methyl-4-isoxazolopropionic acid (AMPA) was used to provoke neuroinflammation. Compounds **37** and **38** were specifically localized in the neuroinflammatory site with a similar signal-to-noise ratio in vitro. However, the fluorine-labeled tracer **38** achieved a higher bioavailability in the brain, higher uptake and higher binding potential than the other radiotracers. In the reference zone (contralateral area), a lower uptake of **37** and **38** were found when compared with **36**. These results showed that **37** and **38** have lower nonspecific bindings than **36** [74].

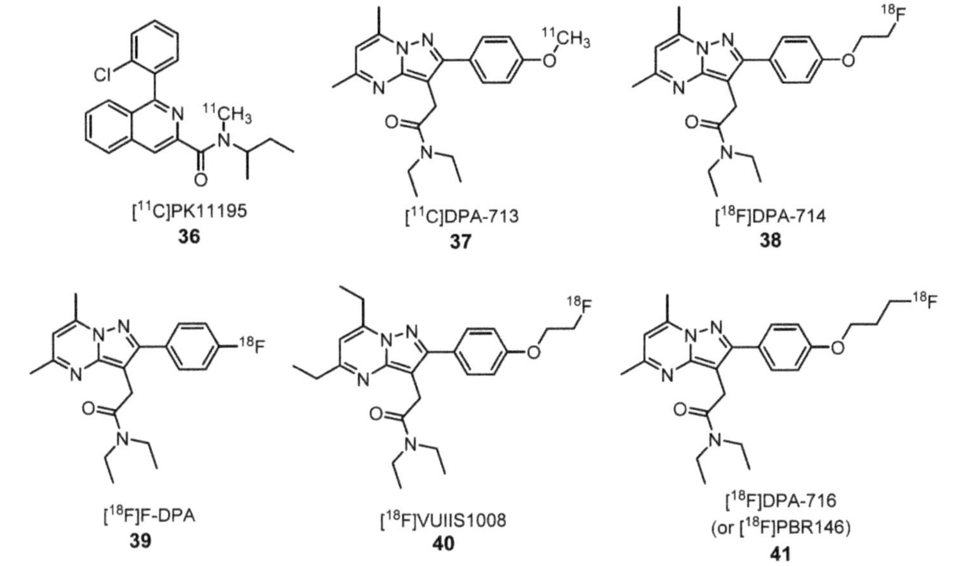

Figure 15. [^{11}C]PK11195 (**36**) and pyrazolopyrimidine-derived radiotracers **37–41** for translocator protein (TSPO) PET imaging [72–82].

Compound **38** was also used to monitor the TSPO levels after the injection of some antibiotic like minocycline. This drug inhibits the activation of microglial cells [75]. Furthermore, the radiotracer **38** was used on studies of rodent models of excitotoxicity, herpes encephalitis [76], astrocytic activation, excitotoxically lesioned nonhuman primate brains, abdominal aortic aneurysm [77], rheumatoid arthritis [78,79] and neuroinflammatory changes in the brain after morphine exposure [65,67]. It demonstrated a good uptake in the primate brain and an eight-fold higher uptake in the lesioned striatum of a quinolinic acid-lesioned rat model of activated microglia. Both **37** and **38** showed better imaging properties than **36** in the striatum of lesioned rats. A study using a rat model of herpes simplex encephalitis (HSE) suggested that **38**, as an agonist of TSPO, is potentially suitable for visualizing mild neuroinflammation [76].

Kuszpit et al. demonstrated that [^{18}F]DPA-714 (**38**) is a sensitive tool for the detection of neuroinflammation, induced by Zika virus (ZIKV) infection, using a mice model of ZIKV neurological disease. Moreover, the evaluation of therapeutics being developed for the treatment of the disease is also possible by [^{18}F]DPA-714 (**38**) imaging [80].

In 2017, Keller et al. compared the activity of **38** with its analogue [^{18}F]F-DPA (**39**) (Figure 15), which presents the radionuclide directly linked to the phenyl ring. This study showed that compound **39** is metabolically more stable than its analogue **38** in rats' brains, being regarded as a promising TSPO radiotracer, because it shows a higher ratio between specific and nonspecific binding [81]. Later, in 2018, the same author described the potential of [^{18}F]F-DPA **39** to visualize activated microglia at an early stage of AD pathology. The in vivo PET imaging and ex vivo brain AGR data indicated and increased uptake of this radiotracer **39** with age in the brains of transgenic animals (APP/PS1-21 mouse models of AD) in comparison with wild-type animals [68].

In 2012, Tang et al. used glioma-bearing rats to study the feasibility of using [^{18}F]DPA-714 (**38**) for the visualization of TSPO expressing in brain tumors. This PET tracer showed to be highly specific to TSPO in glioma cell line homogenates. In vivo studies showed a higher uptake of **38** in tumor tissues than in other brain regions, suggesting that this compound could be used as a novel predictive cancer imaging tool [82]. In 2013, the same author carried out the synthesis and structure-activity relationship study of pyrazolopyrimidine-derived radioligands, presenting different substituents, namely at the 5, 6 and 7 positions, in order to find the more robust PET ligand. The best result was achieved with compound [^{18}F]VUIIS1008 (**40**) (Figure 15), which showed a negligible binding in normal brains but a much higher binding in tumor tissues [64]. This result demonstrated that this radiotracer is a promising PET ligand for TSPO in tumors.

In 2014, Médran-Navarrete et al. evaluated a new ^{18}F-labeled analogue of [^{18}F]DPA-714 (**38**), the [^{18}F]DPA-C5yne (**42**) (Figure 16), as a TSPO radioligand. In vitro studies revealed that **42** was stable in plasma at 37 °C for at least 90 min. AGR studies, using slices of AMPA-lesioned rat brains, showed a high specificity of binding and selectivity for TSPO, highlighting the potential of **42** as a radiotracer for TSPO [83]. One year later, Damont et al. synthetized a series of novel pyrazolo[1,5-*a*]pyrimidines and evaluated, in vitro, their affinity, lipophilicity and metabolism. Based on the results obtained, two of the synthetized compounds were chosen for ^{18}F-radiolabeling affording ligands [^{18}F]DPA-C5yne (**42**) and (**43**) an analogue of [^{18}F]DPA-714 (**38**), where the oxygen atom was replaced by a methylene group (Figure 16). Neuroinflammation PET images, using anesthetized Wistar rats seven days after AMPA-induced brain lesions in the right striatum, showed that both radiotracers **42** and **43** have a high in vivo-specific binding for TSPO. Compound **43** presented an ipsilateral-to-contralateral ratio of 3.57 ± 0.48 comparable to **38** (3.71 ± 0.39), and **42** showed an ipsilateral-to-contralateral ratio of 4.62 ± 0.44. These results suggest that both **42** and **43** are appropriate for neuroinflammation imaging [72].

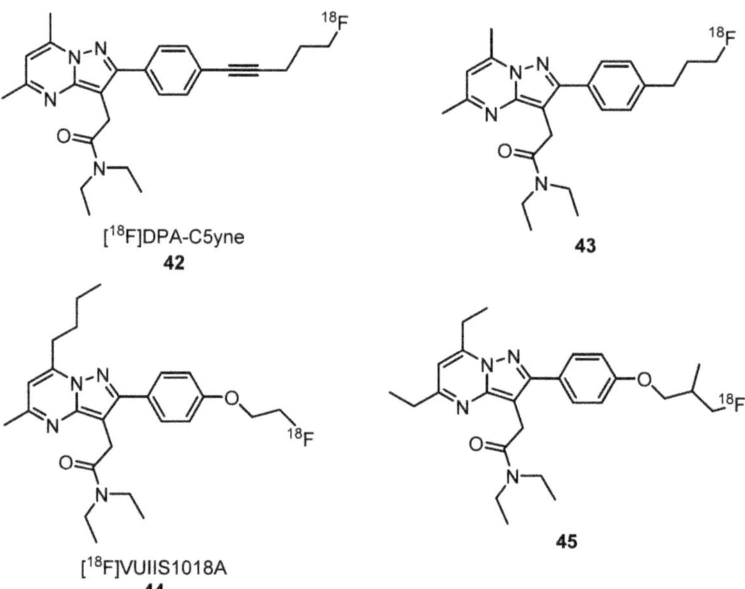

Figure 16. Pyrazolopyrimidine-derived radiotracers **42**–**45** for TSPO PET imaging [83–87].

Recently, Tang et al. synthetized a new TSPO PET tracer [^{18}F]VUIIS1018A (**44**), an analogue of [^{18}F]DPA-714 (**38**) where the 7-methyl group of the pyrazolopyrimidine ring was replaced by a *n*-butyl group (Figure 16), and evaluated its behavior in a preclinical model of neuroinflammation. These authors concluded that this structural modification increased the lipophilicity of **44** compared with **38** (3.7 vs. 2.4, respectively). After 60 min of injection of **44**, more than 85% of this probe was intact, indicating that it is very stable in the brain. It is noteworthy that, for **38**, just 50% of the probe was intact after 60 min of injection. In vivo blocking experiments and in vitro AGR assays confirmed a high binding specificity of **44** for TSPO, showing that it can be a promising TSPO PET tracer [84]. Tang et al.

developed a preclinical evaluation to image glioma. In this work, **44** exhibited a lower accumulation in healthy brains, what was regarded as an advantage to distinguish lower-grade gliomas. Compared with [^{18}F]DPA-714 (**38**) and [^{18}F]VUIIS1008 (**40**), **44** had an improved tumor-to-background ratio, a higher specific-to-nonspecific binding ratio and a higher tumor-binding potential. These results showed that **44** is a promising candidate to detect tumors with modest TSPO expression [85].

A series of novel 2-phenylpyrazolo[1,5-α]pyrimidin-3-ylacetamides were synthesized, and their in vitro binding affinities for TSPO and lipophilicity (log $P_{7.5}$) were evaluated using [^{18}F]DPA-714 (**38**) as the control. Based on the results of these assays, the tosylated precursor was selected for radiofluorination, affording **45** (Figure 16). Using LPS induced neuroinflammation rat models, a dynamic micro-PET study was performed and demonstrated a higher uptake of **45** in the ipsilateral region and a higher ratio of target-to-background than **38**. These results suggest that **45** can be a promising PET probe candidate for TSPO imaging [86].

In 2018, Verweij et al. measured the cellular response in patients after an acute coronary syndrome by PET imaging using [^{18}F]DPA-714 (**38**) as a probe. TSPO receptor is highly expressed in myeloid cells. Using a PET tracer with a high affinity for the TSPO receptor as **38** was possible to determine the hematopoietic activity. In the acute phase, the treatment with **38** revealed a higher uptake in the bone marrow and the spleen. Three months after, the bone marrow uptake decreased to levels comparable to the healthy control. On the other hand, the spleen uptake remained elevated [87].

3. Conclusions

Considering that ^{18}F is the most suitable radionuclide for routine PET imaging and that pyrazoles are a key motif in medicinal chemistry and drug design, we made a compilation of [^{18}F]pyrazole-derived imaging probes that have been developed and evaluated in the last 20 years. Regarding the examples presented herein, it is evident that pyrazoles are important scaffolds for the development of radiotracers for the diagnosis of several pathologies. In fact, pyrazoles interact with remarkable targets, namely with adenosine receptors, cannabinoid receptors, cyclooxygenase enzymes, dopamine receptors, glucocorticoid receptor, insulin-like growth factor-1 receptor, phosphodiester-10A enzyme and the translocator protein receptor. Among the probes presented herein, most are cannabinoid ligands and dopamine ligands, with a huge potential for brain imaging and TSPO ligands, namely pyrazolopyrimidine-derived radiotracers, that have shown important applications for the imaging of neuroinflammation and cancer.

The successful use of the described pyrazoles-imaging probes in clinics may help to increase the understanding of several diseases such as AD, PD, Huntington's disease, atherosclerosis, neuropsychiatric disorders, neuroinflammation, cardiovascular diseases and cancer, among others, and to identify ways to improve therapy. In this sense, the pyrazoles described herein can be regarded as potential theragnostic agents and as important templates for the development of novel radiotracers with improved properties for PET imaging.

Author Contributions: V.L.M.S. and P.M.O.G. idealized the manuscript. P.M.O.G. did the bibliographic research and prepared the original draft. V.L.M.S. and A.M.S.S. supervised the manuscript preparation, revised it and did the necessary corrections. All authors have read and agreed to the published version of the manuscript.

Funding: The authors would like to thank the University of Aveiro and FCT/MEC for the financial support to the QOPNA (FCT UID/QUI/00062/2019) and LAQV-REQUIMTE (UIDB/50006/2020) research projects, financed by national funds and, when appropriate, co-financed by FEDER under the PT2020 Partnership Agreement to the Portuguese NMR network and, also, to the assistant professor position of Vera L. M. Silva (within CEE-CINST/2018; since 01/09/2019). Vera L. M. Silva also thanks the funding from national funds through the FCT-I.P. in the framework of the execution of the program contract provided in paragraphs 4, 5 and 6 of art. 23 of Law no. 57/2016 of 29 August, as amended by Law no. 57/2017 of 19 July and the Integrated Programme of SR&TD "pAGE—Protein Aggregation Across the Lifespan" (reference CENTRO-01-0145FEDER-000003), co-funded by the Centro 2020 program, Portugal 2020 and European Union through the European Regional Development Fund.

Conflicts of Interest: The authors declare no conflicts of interest.

References

1. Li, Z.; Conti, P.S. Radiopharmaceutical chemistry for positron emission tomography. *Adv. Drug Deliv. Rev.* **2010**, *62*, 1031–1051. [CrossRef] [PubMed]
2. Damont, A.; Roeda, D.; Dollé, F. The potential of carbon-11 and fluorine-18 chemistry: Illustration through the development of positron emission tomography radioligands targeting the translocator protein 18 kDa. *J. Label. Compd. Radiopharm.* **2013**, *56*, 96–104. [CrossRef] [PubMed]
3. Coleman, R.E. Clinical PET in oncology. *Clin. Positron Imaging* **1998**, *1*, 15–30. [CrossRef]
4. Anderson, C.J.; Bulte, J.W.M.; Chen, K.; Chen, X.; Khaw, B.-A.; Shokeen, M.; Wooley, K.L.; VanBrocklin, H.F. Design of targeted cardiovascular molecular imaging probes. *J. Nucl. Med.* **2010**, *51*, 3S–17S. [CrossRef]
5. Xiong, K.L.; Yang, Q.W.; Gong, S.G.; Zhang, W.G. The role of positron emission tomography imaging of β-amyloid in patients with Alzheimer's disease. *Nucl. Med. Commun.* **2010**, *31*, 4–11. [CrossRef]
6. Murphy, P.S.; McCarthy, T.J.; Dzik-Jurasz, A.S.K. The role of clinical imaging in oncological drug development. *Br. J. Radiol.* **2008**, *81*, 685–692. [CrossRef]
7. Cazzato, R.L.; Garnon, J.; Shaygi, B.; Koch, G.; Tsoumakidou, G.; Caudrelier, J.; Addeo, P.; Bachellier, P.; Namer, I.J.; Gangi, A. PET/CT-guided interventions: Indications, advantages, disadvantages and the state of the art. *Minim. Invasive Ther. Allied Technol.* **2018**, *27*, 27–32. [CrossRef]
8. Wood, K.A.; Hoskin, P.J.; Saunders, M.I. Positron emission tomography in oncology: A review. *Clin. Oncol.* **2007**, *19*, 237–255. [CrossRef]
9. Silva, V.L.M.; Silva, A.M.S.; Pinto, D.C.G.A.; Jagerovic, N.; Callado, L.F.; Cavaleiro, J.A.S.; Elguero, J. Synthesis and pharmacological evaluation of chlorinated *N*-alkyl-3- and -5-(2-hydroxyphenyl)pyrazoles as CB_1 cannabinoid ligands. *Monatsh. Chem.* **2007**, *138*, 797–811. [CrossRef]
10. Silva, V.L.M.; Silva, A.M.S.; Pinto, D.C.G.A.; Cavaleiro, J.A.S.; Elguero, J. Synthesis of (*E*)- and (*Z*)-3(5)-(2-hydroxyphenyl)-4-styrylpyrazoles. *Monatsh. Chem.* **2009**, *140*, 87–95. [CrossRef]
11. Silva, V.L.M.; Silva, A.M.S.; Pinto, D.C.G.A.; Rodríguez, P.; Gómez, M.; Jagerovic, N.; Callado, L.F.; Cavaleiro, J.A.S.; Elguero, J.; Fernández-Ruiz, J. Synthesis and pharmacological evaluation of new (*E*)- and (*Z*)-3-aryl-4-styryl-1*H*-pyrazoles as potential cannabinoid ligands. *Arkivoc* **2010**, 226–247. [CrossRef]
12. Marques, J.; Silva, V.L.M.; Silva, A.M.S.; Marques, M.P.M.; Braga, S.S. Ru(II) trithiacyclononane 3-(2-hydroxyphenyl)-5-(4-methoxystyryl)pyrazole, a complex with facile synthesis and high cytotoxicity against PC-3 and MDA-MB-231 cells. *Complex Metals* **2014**, *1*, 7–12. [CrossRef]
13. Soengas, R.G.; Silva, V.L.M.; Ide, D.; Kato, A.; Cardoso, S.M.; Almeida Paz, F.A.; Silva, A.M.S. Synthesis of 3-(2-nitrovinyl)-4*H*-chromones: Useful scaffolds for the construction of biologically relevant 3-(pyrazol-5-yl)chromones. *Tetrahedron* **2016**, *72*, 3198–3203. [CrossRef]
14. Carreira, A.R.F.; Pereira, D.M.; Andrade, P.B.; Valentão, P.; Silva, A.M.S.; Braga, S.S.; Silva, V.L.M. Novel styrylpyrazole-glucosides and their unexpected dioxolo-bridged doppelgangers: Synthesis and cytotoxicity. *New J. Chem.* **2019**, *43*, 8299–8310. [CrossRef]
15. Abrigach, F.; Touzani, R. Pyrazole derivatives with NCN junction and their biological activity: A review. *Med. Chem.* **2016**, *6*, 292–298. [CrossRef]
16. Suri, M.; Jousseaume, T.; Neumann, J.J.; Glorius, F. An efficient copper-catalyzed formation of highly substituted pyrazoles using molecular oxygen as the oxidant. *Green Chem.* **2012**, *14*, 2193–2196. [CrossRef]
17. Kumar, V.; Kaur, K.; Gupta, G.K.; Sharma, A.K. Pyrazole containing natural products: Synthetic preview and biological significance. *Eur. J. Med. Chem.* **2013**, *69*, 735–753. [CrossRef]
18. Silva, V.L.M.; Elguero, J.; Silva, A.M.S. Current progress on antioxidants incorporating the pyrazole core. *Eur. J. Med. Chem.* **2018**, *15*, 394–429. [CrossRef]
19. Abdel-Maksoud, M.S.; El-Gamal, M.I.; Gamal El-Din, M.M.; Oh, C.H. Design, synthesis, in vitro anticancer evaluation, kinase inhibitory effects, and pharmacokinetic profile of new 1,3,4-triarylpyrazole derivatives possessing terminal sulfonamide moiety. *J. Enz. Inhib. Med. Chem.* **2019**, *34*, 97–109. [CrossRef]
20. Chowdary, B.N.; Umashankara, M.; Dinesh, B.; Girish, K.; Baba, A.R. Development of 5-(aryl)-3-phenyl-1*H*-pyrazole derivatives as potent antimicrobial compounds. *Asian J. Chem.* **2019**, *31*, 45–50. [CrossRef]
21. Secci, D.; Bolasco, A.; Chimenti, P.; Carradori, S. The state of the art of pyrazole derivatives as monoamine oxidase inhibitors and antidepressant/anticonvulsant agents. *Curr. Med. Chem.* **2011**, *18*, 5114–5144. [CrossRef] [PubMed]

22. Liu, J.J.; Zhang, H.; Sun, J.; Wang, Z.C.; Yang, Y.S.; Li, D.D.; Zhang, F.; Gong, H.B.; Zhu, H.L. Synthesis, biological evaluation of novel 4,5-dihydro-2*H*-pyrazole 2-hydroxyphenyl derivatives as BRAF inhibitors. *Bioorg. Med. Chem.* **2012**, *20*, 6089–6096. [CrossRef] [PubMed]
23. Tanitame, A.; Oyamada, Y.; Ofuji, K.; Fujimoto, M.; Iwai, N.; Hiyama, Y.; Suzuki, K.; Ito, H.; Terauchi, H.; Kawasaki, M.; et al. Synthesis and antibacterial activity of a novel series of potent DNA gyrase inhibitors. Pyrazole derivatives. *J. Med. Chem.* **2004**, *47*, 3693–3696. [CrossRef] [PubMed]
24. Ansari, A.; Ali, A.; Asif, M.; Shamsuzzaman, S. Review: Biologically active pyrazole derivatives. *New J. Chem.* **2016**, *41*, 16–41. [CrossRef]
25. Liu, J.-J.; Zhao, M.-Y.; Zhang, X.; Zhao, X.; Zhu, H.-L. Pyrazole derivatives as antitumor, anti-inflammatory and antibacterial agents. *Mini Rev. Med. Chem.* **2013**, *13*, 1957–1966. [CrossRef]
26. Küçükgüzel, S.G.; Senkardes, S. Recent advances in bioactive pyrazoles. *Eur. J. Med. Chem.* **2015**, *97*, 786–815. [CrossRef]
27. Santos, C.M.M.; Silva, V.L.M.; Silva, A.M.S. Synthesis of chromone-related pyrazole compounds. *Molecules* **2017**, *22*, 1665. [CrossRef]
28. Faria, J.V.; Vegi, P.F.; Miguita, A.G.C.; dos Santos, M.S.; Boechat, N.; Bernardino, A.M.R. Recently reported biological activities of pyrazole compounds. *Bioorg. Med. Chem.* **2017**, *25*, 5891–5903. [CrossRef]
29. Perez-Fernandez, R.; Goya, P.; Elguero, J. A review of recent progress (2002–2012) on the biological activities of pyrazoles. *Arkivoc* **2014**, 233–293. [CrossRef]
30. Prabhu, V.V.; Kannan, N.; Guruvayoorappan, C. 1,2-Diazole prevents cisplatin-induced nephrotoxicity in experimental rats. *Pharmacol. Reports* **2013**, *65*, 980–990. [CrossRef]
31. Borea, P.A.; Gessi, S.; Merighi, S.; Vincenzi, F.; Varani, K. Pharmacology of adenosine receptors: The state of the art. *Physiol. Rev.* **2018**, *98*, 1591–1625. [CrossRef] [PubMed]
32. Khanapur, S.; Paul, S.; Shah, A.; Vatakuti, S.; Koole, M.J.B.; Zijlma, R.; Dierckx, R.A.J.O.; Luurtsema, G.; Garg, P.; van Waarde, A.; et al. Development of [^{18}F]-labelled pyrazolo[4,3-*e*]-1,2,4-triazolo[1,5-*c*]pyrimidine (SCH442416) analogs for the imaging of cerebral adenosine A_2A receptors with positron emission tomography. *J. Med. Chem.* **2014**, *57*, 6765–6780. [CrossRef] [PubMed]
33. Khanapur, S.; van Waarde, A.; Dierckx, R.A.J.O.; Elsinga, P.H.; Koole, M.J.B. Preclinical evaluation and quantification of ^{18}F-fluoroethyl and ^{18}F-fluoropropyl analogs of SCH442416 as radioligands for PET imaging of the adenosine A_2A receptor in rat brain. *J. Nucl. Med.* **2017**, *58*, 466–472. [CrossRef] [PubMed]
34. Li, J.; Daughters, R.S.; Bullis, C.; Bengiamin, R.; Stucky, M.W.; Brennan, J.; Simone, D.A. The cannabinoid receptor agonist WIN 55,212-2 mesylate blocks the development of hyperalgesia produced by capsaicin in rats. *Pain* **1999**, *81*, 25–33. [CrossRef]
35. Gonzalez-Rosales, F.; Walsh, D. Intractable nausea and vomiting due to gastrointestinal mucosal metastases relieved by tetrahydrocannabinol (Dronabinol). *J. Pain Symptom Manag.* **1997**, *14*, 311–314. [CrossRef]
36. Williams, C.M.; Kirkham, T.C. Anandamide induces overeating: Mediation by central cannabinoid (CB_1) receptors. *Psychopharmacology* **1999**, *143*, 315–317. [CrossRef]
37. Mathews, W.B.; Scheffel, U.; Finley, P.; Ravert, H.T.; Frank, R.A.; Rinaldi-Carmona, M.; Barth, F.; Dannals, R.F. Biodistribution of [^{18}F]SR144385 and [^{18}F]SR147963: Selective radioligands for positron emission tomographic studies of brain cannabinoid receptors. *Nucl. Med. Biol.* **2000**, *27*, 757–762. [CrossRef]
38. Horti, A.G.; Fan, H.; Kuwabara, H.; Hilton, J.; Ravert, H.T.; Holt, D.P.; Alexander, M.; Kumar, A.; Rahmim, A.; Scheffel, U.; et al. ^{11}C-JHU75528: A radiotracer for PET imaging of CB_1 cannabinoid receptors. *J. Nucl. Med.* **2006**, *47*, 1689–1696.
39. Rinaldi-Carmona, M.; Barth, F.; Héaulme, M.; Shire, D.; Calandra, B.; Congy, C.; Martinez, S.; Maruani, J.; Néliat, G.; Caput, D.; et al. SR141716A, a potent and selective antagonist of the brain cannabinoid receptor. *FEBS Lett.* **1994**, *350*, 240–244. [CrossRef]
40. Moreira, F.A.; Crippa, J.A.S. The psychiatric side-effects of rimonabant. *Braz. J. Psychiatry* **2009**, *31*, 145–153. [CrossRef]
41. Mathews, W.B.; Ravert, H.T.; Musachio, J.L.; Frank, R.A.; Rinaldi-Carmona, M.; Barth, F.; Dannals, R.F. Synthesis of [^{18}F]SR144385: A selective radioligand for positron emission tomography studies of brain cannabinoid receptors. *J. Label. Compd. Radiopharm.* **1999**, *42*, 589–596. [CrossRef]
42. Katoch-Rouse, R.; Horti, A.G. Synthesis of *N*-(piperidin-1-yl)-5-(4-methoxyphenyl)-1-(2-chlorophenyl)-4-[^{18}F]fluoro-1*H*-pyrazole-3-carboxamide by nucleophilic [^{18}F] fluorination: A PET radiotracer for studying CB_1 cannabinoid receptors. *J. Label. Compd. Radiopharm.* **2003**, *46*, 93–98. [CrossRef]

43. Nojiri, Y.; Ishiwata, K.; Qinggeletu; Tobiishi, S.; Sasada, T.; Yamamoto, F.; Mukai, T.; Maeda, M. Radiosynthesis and biodistribution in mice of a ^{18}F-labelled analog of O-1302 for use in cerebral CB_1 cannabinoid receptor imaging. *Biol. Pharm. Bull.* **2008**, *31*, 1274–1278. [CrossRef] [PubMed]
44. Li, Z.; Gifford, A.; Liu, Q.; Thotapally, R.; Ding, Y.-S.; Makriyannis, A.; Gatley, S.J. Candidate PET radioligands for cannabinoid CB_1 receptors: [^{18}F]AM5144 and related pyrazole compounds. *Nucl. Med. Biol.* **2005**, *32*, 361–366. [CrossRef]
45. Gatley, S.J.; Lan, R.; Volkow, N.D.; Pappas, N.; King, P.; Wong, C.T.; Gifford, A.N.; Pyatt, B.; Dewey, S.L.; Makriyannis, A. Imaging the brain marijuana receptor: Development of a radioligand that binds to cannabinoid CB1 receptors in vivo. *J. Neurochem.* **1998**, *70*, 417–423. [CrossRef]
46. Chang, C.-P.; Huang, H.-L.; Huang, J.-K.; Hung, M.-S.; Wu, C.-H.; Song, J.-S.; Lee, C.-J.; Yub, C.-S.; Shia, K.-S. Fluorine-18 isotope labeling for positron emission tomography imaging. Direct evidence for DBPR211 as a peripherally restricted CB_1 inverse agonist. *Bioorg. Med. Chem.* **2019**, *27*, 216–223. [CrossRef]
47. Taketo, M.M. Cyclooxygenase-2 inhibitors in tumorigenesis (Part II). *J. Natl. Can. Inst.* **1998**, *90*, 1529–1536. [CrossRef]
48. Minghetti, L. Cyclooxygenase-2 (COX-2) in inflammatory and degenerative brain diseases. *J. Neuropath. Exp. Neur.* **2004**, *63*, 901–910. [CrossRef]
49. Uddin, M.J.; Crews, B.C.; Ghebreselasie, K.; Huda, I.; Kingsley, P.J.; Ansari, M.S.; Tantawy, M.N.; Reese, J.J.; Marnett, L.J. Fluorinated COX-2 inhibitors as agents in PET imaging of inflammation and cancer. *Cancer Prev. Res.* **2011**, *4*, 1536–1545. [CrossRef]
50. Lebedev, A.; Jiao, J.; Lee, J.; Yang, F.; Allison, N.; Herschman, H.; Sadeghi, S. Radiochemistry on electrodes: Synthesis of an ^{18}F-labelled and in vivo stable COX-2 inhibitor. *PLoS ONE* **2017**, *12*, e0176606. [CrossRef]
51. McCarthy, T.J.; Sheriff, A.U.; Graneto, M.J.; Talley, J.J.; Welch, M.J. Radiosynthesis, in vitro validation, and in vivo evaluation of ^{18}F-labeled COX-1 and COX-2 inhibitors. *J. Nucl. Med.* **2002**, *43*, 117–124. [PubMed]
52. Prante, O.; Tietze, R.; Hocke, C.; Löber, S.; Hübner, H.; Kuwert, T.; Gmeiner, P. Synthesis, radiofluorination, and in vitro evaluation of pyrazolo[1,5-*a*]pyridine-based dopamine D_4-receptor ligands: Discovery of an inverse agonist radioligand for PET. *J. Med. Chem.* **2008**, *51*, 1800–1810. [CrossRef] [PubMed]
53. Stößel, A.; Box, R.; Purkayastha, N.; Hübner, H.; Hocke, C.; Prante, O.; Gmeiner, P. Development of molecular tools based on the dopamine D_3 receptor ligand FAUC 329 showing inhibiting effects on drug and food maintained behavior. *Bioorg. Med. Chem.* **2017**, *25*, 3491–3499. [CrossRef]
54. Hoyte, R.M.; Labaree, D.C.; Fede, J.-M.; Harris, C.; Hochberg, R.B. Iodinated and fluorinated steroid 2′-aryl-[3,2-*c*] pyrazoles as potential glucocorticoid receptor imaging agents. *Steroids* **1998**, *63*, 595–602. [CrossRef]
55. Würst, F.; Kniess, T.; Kretzschmar, M.; Bergmann, R. Synthesis and radiopharmacological evaluation of 2′-(4-fluorophenyl)-21-[^{18}F]fluoro-20-oxo-11β,17α-dihydroxypregn-4-eno[3,2-c]pyrazole as potential glucocorticoid receptor ligand for positron emission tomography (PET). *Bioorg. Med. Chem. Lett.* **2005**, *15*, 1303–1306. [CrossRef] [PubMed]
56. Majo, V.J.; Arango, V.; Simpson, N.R.; Prabhakaran, J.; Kassir, S.A.; Underwood, M.D.; Bakalian, M.; Canoll, P.; Mann, J.J.; Dileep Kumar, J.S. Synthesis and in vitro evaluation of [^{18}F]BMS-754807: A potential PET ligand for IGF-1R. *Bioorg. Med. Chem. Lett.* **2013**, *23*, 4191–4194. [CrossRef]
57. Tu, Z.; Xu, J.; Jones, L.A.; Li, S.; Mach, R.H. Carbon-11 labelled papaverine as a PET tracer for imaging PDE10A: Radiosynthesis, in vitro and in vivo evaluation. *Nucl. Med. Biol.* **2010**, *37*, 509–516. [CrossRef]
58. Celen, S.; Koole, M.; De Angelis, M.; Sannen, I.; Chitneni, S.K.; Alcazar, J.; Dedeurwaerdere, S.; Moechars, D.; Schmidt, M.; Verbruggen, A.; et al. Preclinical evaluation of ^{18}F-JNJ41510417 as a radioligand for PET imaging of phosphodiesterase-10A in the brain. *J. Nucl. Med.* **2010**, *51*, 1584–1591. [CrossRef]
59. Andres, J.-I.; De Angelis, M.; Alcazar, J.; Iturrino, L.; Langlois, X.; Dedeurwaerdere, S.; Lenaerts, I.; Vanhoof, G.; Celen, S.; Bormans, G. Synthesis, in vivo occupancy, and radiolabeling of potent phosphodiesterase subtype-10 inhibitors as candidates for positron emission tomography imaging. *J. Med. Chem.* **2011**, *54*, 5820–5835. [CrossRef]
60. Laere, K.V.; Ahmad, R.U.; Hudyana, H.; Celen, S.; Dubois, K.; Schmidt, M.E.; Bormans, G.; Koole, M. Human biodistribution and dosimetry of ^{18}F-JNJ42259152, a radioligand for phosphodiesterase 10A imaging. *Eur. J. Nucl. Med. Mol. Imaging* **2013**, *40*, 254–261. [CrossRef]

61. Laere, K.V.; Ahmad, R.U.; Hudyana, H.; Dubois, K.; Schmidt, M.E.; Celen, S.; Bormans, G.; Koole, M. Quantification of ^{18}F-JNJ-42259152, a novel phosphodiesterase 10A PET tracer: Kinetic modelling and test-retest study in human brain. *J. Nucl. Med.* **2013**, *54*, 1285–1293. [CrossRef] [PubMed]
62. Ooms, M.; Altili, B.; Celen, S.; Koole, M.; Verbruggen, A.; Laere, K.V.; Bormans, G. [^{18}F]JNJ-42259152 binding to phosphodiesterase 10A, a key regulator of medium spiny neuron excitability, is altered in the presence of cyclic AMP. *J. Neurochem.* **2016**, *139*, 897–906. [CrossRef] [PubMed]
63. Stepanov, V.; Takano, A.; Nakao, R.; Amini, N.; Miura, S.; Kimura, H.; Taniguchi, T.; Halldin, C. Development of two fluorine-18 labeled PET radioligands targeting PDE10A and in vivo PET evaluation in nonhuman primates. *Nucl. Med Biol.* **2018**, *57*, 12–19. [CrossRef] [PubMed]
64. Tang, D.; McKinley, E.T.; Hight, M.R.; Uddin, M.I.; Harp, J.M.; Fu, A.; Nickels, M.L.; Buck, J.R.; Manning, H.C. Synthesis and structure-activity relationships of 5,6,7-substituted pyrazolopyrimidines: Discovery of a novel TSPO PET ligand for cancer imaging. *J. Med. Chem.* **2013**, *56*, 3429–3433. [CrossRef]
65. Zheng, J.; Winkeler, A.; Peyronneau, M.A.; Dollé, F.; Boisgard, R. Evaluation of PET imaging performance of the TSPO radioligand [^{18}F]DPA-714 in mouse and rat models of cancer and inflammation. *Mol. Imaging Biol.* **2016**, *18*, 127–134. [CrossRef]
66. Hagens, M.H.J.; Golla, S.V.; Wijburg, M.T.; Yaqub, M.; Heijtel, D.; Steenwijk, M.D.; Schober, P.; Brevé, J.J.P.; Schuit, R.C.; Reekie, T.A.; et al. In vivo assessment of neuroinflammation in progressive multiple sclerosis: A proof of concept study with [^{18}F]DPA714 PET. *J. Neuroinflammation* **2018**, *15*, 314–324. [CrossRef]
67. Lavisse, S.; Inoue, K.; Jan, C.; Peyronneau, M.A.; Petit, F.; Goutal, S.; Dauguet, J.; Guillermier, M.; Dollé, F.; Rbah-Vidal, L.; et al. [^{18}F]DPA-714 PET imaging of translocator protein TSPO (18 kDa) in the normal and excitotoxically-lesioned nonhuman primate brain. *Eur. J. Nucl. Med. Mol. Imaging* **2015**, *42*, 478–494. [CrossRef]
68. Keller, T.; López-Picón, F.R.; Krzyczmonik, A.; Forsback, S.; Kirjavainen, A.K.; Takkinen, J.S.; Alzghool, O.; Rajander, J.; Teperi, S.; Cacheux, F.; et al. [^{18}F]F-DPA for the detection of activated microglia in a mouse model of Alzheimer's disease. *Nucl. Med. Biol.* **2018**, *67*, 1–9. [CrossRef]
69. Gungor, H.; Saleem, A.; Babar, S.; Dina, R.; El-Bahrawy, M.A.; Curry, E.; Rama, N.; Chen, M.; Pickford, E.; Agarwal, R.; et al. Dose-finding quantitative ^{18}F-FDG PET imaging study with the oral Pan-AKT inhibitor GSK2141795 in patients with gynecologic malignancies. *J. Nucl. Med.* **2015**, *56*, 1828–1835. [CrossRef]
70. Inoue, O.; Yamasaki, T.; Hashimoto, K.; Kojima, M. Evaluation of ^{3}H-PK11195 as a radioligand for the in vivo study of peripheral benzodiazepine receptor. *Kaku Igaku.* **1985**, *22*, 1385–1389.
71. Camsonne, R.; Moulin, M.A.; Crouzel, C.; Syrota, A.; Maziere, M.; Comar, D. Carbon-11 labelling of PK11195 and visualization of benzodiazepine peripheral receptors using positron emission tomography. *J. Pharmacol.* **1986**, *17*, 383.
72. Damont, A.; Médran-Navarrete, V.; Cacheux, F.; Kuhnast, B.; Pottier, G.; Bernards, N.; Marquet, F.; Puech, F.; Boisgard, R.; Dollé, F. Novel pyrazolo[1,5-*a*]pyrimidines as translocator protein 18 kDa (TSPO) ligands: Synthesis, in vitro biological evaluation, [^{18}F]-labeling, and in vivo neuroinflammation PET images. *J. Med. Chem.* **2015**, *58*, 7449–7464. [CrossRef] [PubMed]
73. Fookes, C.J.R.; Pham, T.Q.; Mattner, F.; Greguric, I.; Loc'h, C.; Liu, X.; Berghofer, P.; Shepherd, R.; Gregoire, M.-C.; Katsifis, A. Synthesis and biological evaluation of substituted [^{18}F]imidazo[1,2-*a*]pyridines and [^{18}F]pyrazolo[1,5-*a*]pyrimidines for the study of the peripheral benzodiazepine receptor using positron emission tomography. *J. Med. Chem.* **2008**, *51*, 3700–3712. [CrossRef] [PubMed]
74. Chauveau, F.; Camp, N.V.; Dollé, F.; Kuhnast, B.; Hinnen, F.; Damont, A.; Boutin, H.; James, M.; Kassiou, M.; Tavitian, B. Comparative evaluation of the translocator protein radioligands ^{11}C-DPA-713, ^{18}F-DPA-714, and ^{11}C-PK11195 in a rat model of acute neuroinflammation. *J. Nucl. Med.* **2009**, *50*, 468–476. [CrossRef] [PubMed]
75. Martín, A.; Boisgard, R.; Kassiou, M.; Dollé, F.; Tavitian, B. Reduced PBR/TSPO expression after minocycline treatment in a rat model of focal cerebral ischemia: A PET study using [^{18}F]DPA-714. *Mol. Imaging Biol.* **2011**, *13*, 10–15. [CrossRef]
76. Doorduin, J.; Klein, H.C.; Dierckx, R.A.; James, M.; Kassiou, M.; de Vries, E.F.J. [^{11}C]-DPA-713 and [^{18}F]-DPA-714 as new PET tracers for TSPO: A comparison with [^{11}C]-(R)-PK11195 in a rat model of herpes encephalitis. *Mol. Imaging Biol.* **2009**, *11*, 386–398. [CrossRef]

77. Sarda-Mantel, L.; Alsac, J.M.; Boisgard, R.; Hervatin, F.; Montravers, F.; Tavitian, B.; Michel, J.B.; Le Guludec, D. Comparison of ^{18}F-fluoro-deoxy-glucose, ^{18}F-fluoro-methyl-choline, and ^{18}F-DPA714 for positron-emission tomography imaging of leukocyte accumulation in the aortic wall of experimental abdominal aneurysms. *J. Vasc. Surg.* **2012**, *56*, 765–773. [CrossRef]
78. Pottier, G.; Bernards, N.; Dollé, F.; Boisgard, R. [^{18}F]DPA-714 as a biomarker for positron emission tomography imaging of rheumatoid arthritis in an animal model. *Arthritis Res. Ther.* **2014**, *16*, R69. [CrossRef]
79. Gent, Y.Y.J.; Weijers, K.; Molthoff, C.F.M.; Windhorst, A.D.; Huisman, M.C.; Kassiou, M.; Jansen, G.; Lammertsma, A.A.; van der Laken, C.J. Promising potential of new generation translocator protein tracers providing enhanced contrast of arthritis. *Arthritis Res. Ther.* **2014**, *16*, R70. [CrossRef]
80. Kuszpit, K.; Hollidge, B.S.; Zeng, X.; Stafford, R.G.; Daye, S.; Zhang, X.; Basuli, F.; Golden, J.W.; Swenson, R.E.; Smith, D.R.; et al. [^{18}F]DPA-714 PET imaging reveals global neuroinflammation in Zika Virus-infected mice. *Mol. Imaging Biol.* **2018**, *20*, 275–283. [CrossRef]
81. Keller, T.; Krzyczmonik, A.; Forsback, S.; Picón, F.R.L.; Kirjavainen, A.K.; Takkinen, J.; Rajander, J.; Cacheux, F.; Damont, A.; Dollé, F.; et al. Radiosynthesis and preclinical evaluation of [^{18}F]F-DPA, a novel pyrazolo[1,5-*a*]pyrimidine acetamide TSPO radioligand, in healthy Sprague Dawley rats. *Mol. Imaging Biol.* **2017**, *19*, 736–745. [CrossRef] [PubMed]
82. Tang, D.; Hight, M.R.; McKinley, E.T.; Fu, A.; Buck, J.R.; Smith, R.A.; Tantawy, M.N.; Peterson, T.E.; Colvin, D.C.; Ansari, M.S.; et al. Quantitative preclinical imaging of TSPO expression in glioma using *N*,*N*-diethyl-2-(2-(4-(2-^{18}F-fluoroethoxy)phenyl)-5,7-dimethylpyrazolo[1,5-*a*]pyrimidin-3-yl)acetamide. *J. Nucl. Med.* **2012**, *53*, 287–294. [CrossRef] [PubMed]
83. Médran-Navarrete, V.; Bernards, N.; Kuhnast, B.; Damont, A.; Pottier, G.; Peyronneau, M.A.; Kassiou, M.; Marguet, F.; Puech, F.; Boisgard, R.; et al. [^{18}F]DPA-C5yne, a novel fluorine-18-labelled analogue of DPA-714: Radiosynthesis and preliminary evaluation as a radiotracer for imaging neuroinflammation with PET. *J. Label Compd. Radiopharm.* **2014**, *57*, 410–418. [CrossRef] [PubMed]
84. Tang, D.; Fujinaga, M.; Hatori, A.; Zhang, Y.; Yamasaki, T.; Xie, L.; Mori, W.; Kumata, K.; Liu, J.; Manning, H.C.; et al. Evaluation of the novel TSPO radiotracer 2-(7-butyl-2-(4-(2-([^{18}F]fluoroethoxy)phenyl)-5-methylpyrazolo[1,5-*a*]pyrimidin-3-yl)-*N*,*N*-diethylacetamide in a preclinical model of neuroinflammation. *Eur. J. Med. Chem.* **2018**, *150*, 1–8. [CrossRef] [PubMed]
85. Tang, D.; Li, J.; Nickels, M.L.; Huang, G.; Cohen, A.S.; Manning, H.C. Preclinical evaluation of a novel TSPO PET ligand 2-(7-butyl-2-(4-(2-[^{18}F]fluoroethoxy)phenyl)-5-methylpyrazolo[1,5-*a*]pyrimidin-3-yl)-*N*,*N*-diethylacetamide (^{18}F-VUIIS1018A) to image glioma. *Mol. Imaging Biol.* **2019**, *21*, 113–121. [CrossRef] [PubMed]
86. Kwon, Y.D.; Kang, S.; Park, H.; Cheong, I.K.; Chang, K.A.; Lee, S.Y.; Jung, J.H.; Lee, B.C.; Lim, S.T.; Kim, H.K. Novel potential pyrazolopyrimidine based translocator protein ligands for the evaluation of neuroinflammation with PET. *Eur. J. Med. Chem.* **2018**, *159*, 292–306. [CrossRef]
87. Verweij, S.L.; Stiekema, L.C.A.; Delewi, R.; Zheng, K.H.; Moens, S.J.B.; Kroon, J.; Stroes, C.I.; Versloot, M.; Piek, J.J.; Verberne, H.J.; et al. Prolonged hematopoietic and myeloid cellular response in patients after an acute coronary syndrome measured with ^{18}F-DPA-714 PET/CT. *Eur. J. Nucl. Med. Mol. Imaging* **2018**, *45*, 1956–1963. [CrossRef]

MDPI
St. Alban-Anlage 66
4052 Basel
Switzerland
Tel. +41 61 683 77 34
Fax +41 61 302 89 18
www.mdpi.com

Molecules Editorial Office
E-mail: molecules@mdpi.com
www.mdpi.com/journal/molecules

www.ingramcontent.com/pod-product-compliance
Lightning Source LLC
LaVergne TN
LVHW070134170726
843515LV00007B/1792